The Correspondence of John Tyndall

VOLUME 10

THE CORRESPONDENCE OF JOHN TYNDALL

VOLUME 10

The Correspondence, January 1867–December 1868

Edited by
Michael D. Barton, Ken Corbett, and Roland Jackson

General Editors
Roland Jackson, Bernard Lightman, and Michael S. Reidy

Associate Editors
Michael D. Barton, Erin Grosjean, and Michael Laurentius

UNIVERSITY OF PITTSBURGH PRESS

Published by the University of Pittsburgh Press, Pittsburgh, Pa., 15260

Manufactured in the United States of America
Printed on acid-free paper
10 9 8 7 6 5 4 3 2 1

Cataloging-in-Publication data is available from the Library of Congress

ISBN 13: 978-0-8229-4658-8
ISBN 10: 0-8229-4658-0

CONTENTS

ACKNOWLEDGMENTS

As ever, this volume of *The Correspondence of John Tyndall* is built on the work of many colleagues whose willingness to share their time and expertise is deeply appreciated. We especially acknowledge the work of many graduate students and other colleagues in providing the original transcriptions and many initial translations of foreign-language letters. We greatly valued the work of James Braund in checking the transcriptions and translations of German letters and of Caley McCarthy and Eleanor Coulter for French letters, and the additional assistance of Elizabeth Neswald, Efram Sera-Shriar, and Irena Brezowski. We sometimes required new scans of letters, and for this we warmly thank Jane Harrison, formerly of the Royal Institution of Great Britain. We appreciate the continuing support of Charlotte New at the Royal Institution and Frank James, formerly of the Royal Institution. For help in creating high-quality image files of the drawings within letters in this volume, in addition to her tireless work of moving letters between the volume editors, we thank Erin Grosjean, and we thank Michael Laurentius for his work in collating the letters and manuscript. For invaluable help in answering a variety of questions about letters in this volume, and their reviews of material, we would like to acknowledge the general editors of *The Correspondence of John Tyndall*, James Elwick, Bernard Lightman, and Michael Reidy. Many other individuals have given help on information for specific letters and we thank, in no particular order, James Ungureanu, Norman McMillan, Martin Nevin, Cyril Smyth, Richard Carter, Jonathan A. Smyth, Aileen Christianson (editor of the Carlyle Letters), Gordon Campbell (Maynooth University), Gowan Dawson, Robert M. Seiler (Pater scholar), Johannes Orphal (for help identifying Clausius's children), Virginia Mills (Archivist at the Royal Society of London), Anne Cameron (Archives Assistant, Archives and Special Collections, University of Strathclyde), Jennie De Protani (archivist at the Athenaeum), Lauranne Lefrançois, Sarah Mitchell of the Royal Archives (for bringing our attention to an 1867 letter in their collection), Tara Craig (Head of Public Services, Rare Book & Manuscript Library, Butler Library, Columbia University), and Harry Bysshe Vedge for suggesting that the word abbreviations in two letters from James Wilson that eluded us—and we assumed were in a style of shorthand—in fact contained thorn letters (*ye* for *the* and *yt* for *that*).

LIST OF ABBREVIATIONS

ADB	*Allgemeine Deutsche Biographie*
Alum. Cantab.	J. A. Venn, *Alumni Cantabrigienses, 1752–1900*, 6 vols. (Cambridge: Cambridge University Press, 1940–54)
Alum. Oxon.	Joseph Foster, *Alumni Oxonienses: The Members of the University of Oxford, 1715–1886*, 4 vols. (Oxford: Parker and Co., 1888–92)
ANB	*American National Biography*
Archives Nationales	Archives Nationales (France), Base Léonore (Légion d'honneur), http://www2.culture.gouv.fr/
Ascent of John Tyndall	Roland Jackson, *The Ascent of John Tyndall: Victorian Scientist, Mountaineer, and Public Intellectual* (Oxford: University of Oxford Press, 2018)
Athenaeum Rules (1862)	*Athenaeum. Rules and Regulations, and List of Members* (London: 1862)
Athenaeum Rules (1874)	*Athenaeum. Rules & Regulations & List of Members* (London: 1874)
BAAS	British Association for the Advancement of Science
BBAW	Berlin-Brandenburg Academy of Sciences
BL	British Library
Brit. Assoc. Rep.	*Reports of the British Association for the Advancement of Science*
Brock & MacLeod, *Hirst Journals*	William H. Brock and Roy M. MacLeod, eds., *Natural Knowledge in Social Context: The Journals of Thomas Archer Hirst FRS* (London: Mansell microfiche, 1980)
CDSB	*Complete Dictionary of Scientific Biography*, 27 vols. (Detroit: Charles Scribner's Sons, 2008)
CLO	*The Carlyle Letters Online*, ed. Brent E. Kinser (Durham, NC: Duke University Press, 2007–16), www.carlyleletters.org
Darwin Correspondence, vol. 16	Frederick Burkhardt et al., eds., *The Correspondence of Charles Darwin* (New York: Cambridge University Press, 2008), vol. 16, parts 1 and 2
DBF	*Dictionnaire de biographie française*, 21 vols. (Paris: Letouzey et Ané, 1932–2016)
DBI	*Dizionario Biografico Degli Italiani*, 94 vols. (Rome: Istituto della Enciclopedia Italiana, 1960–2019)

Faraday Correspondence	Frank A. J. L. James, ed., *The Correspondence of Michael Faraday*, vol. 5, 1855–1860 (London: Institution of Engineering and Technology, 2008); Frank A. J. L. James, ed., *The Correspondence of Michael Faraday*, vol. 6, 1861–1867 (London: Institution of Engineering and Technology, 2011)
Faraday as a Discoverer	John Tyndall, *Faraday as a Discoverer* (London: Longmans, Green, and Co., 1868)
Fragments of Science	John Tyndall, *Fragments of Science for Unscientific People: A Series of Detached Essays, Addresses, and Reviews*, 6th ed., 2 vols (London: Longmans, Green, and Co., 1879)
FRS	Fellow of the Royal Society
Glaciers of the Alps	John Tyndall, *The Glaciers of the Alps* (London: John Murray, 1860)
Heat (1863)	John Tyndall, *Heat Considered as a Mode of Motion: Being a Course of Twelve Lectures Delivered at the Royal Institution of Great Britain in the Season of 1862* (London: Longman, Green, Longman, Roberts, & Green, 1863)
Heat (1865)	John Tyndall, *Heat as a Mode of Motion*, 2nd ed. (London: Longman, Green, Longman, Roberts, & Green, 1865)
Heat (1868)	John Tyndall, *Heat as a Mode of Motion*, 3rd ed. (London: Longmans, Green, and Co., 1868)
Helmholtz	David Cahan, *Helmholtz: A Life in Science* (Chicago: University of Chicago Press, 2018)
HLS	*Historisches Lexikon der Schweiz*
Hours of Exercise in the Alps	John Tyndall, *Hours of Exercise in the Alps*, 1st ed. (London: Longmans, Green, and Co., 1871) [The second edition of this book was also published in 1871]
IC HP	Imperial College of Science, Technology and Medicine, College Archives, Huxley Papers
Irish Genealogy	Department of Culture, Heritage and the Gaeltacht, Irish Genealogy, https://www.irishgenealogy.ie/
Journal	John Tyndall's journal, 1855–72, typescript (RI MS JT/2/13c)
LT	Louisa Tyndall, John Tyndall's widow
Mountaineering in 1861	John Tyndall, *Mountaineering in 1861. A Vacation Tour* (London: Longman, Green, Longman, and Roberts, 1862)
NDB	*Neue Deutsche Biographie*
NHM WP	Natural History Museum (London), Wallace Papers
NLS	National Library of Scotland
ODNB	*Oxford Dictionary of National Biography*
OED	*Oxford English Dictionary*
Oxford	Oxford University Archives
Phil. Mag.	*Philosophical Magazine*
Phil. Trans.	*Philosophical Transactions of the Royal Society of London*

PMG	*Pall Mall Gazette*
Poggendorff's Annal.	*Annalen der Physik und Chemie*
RI	Royal Institution of Great Britain
RI MS JT	Royal Institution, London, John Tyndall Papers
Roy. Inst. Proc.	*Proceedings of the Royal Institution of Great Britain*
Roy. Soc. Proc.	*Proceedings of the Royal Society of London*
RS	Royal Society of London
RS HS	Royal Society, Herschel Papers
Sound (1867)	John Tyndall, *Sound: A Course of Eight Lectures Delivered at the Royal Institution of Great Britain* (London: Longmans, Green, and Co., 1867)
TRC Letters	Tennyson Research Center
Tyndall Correspondence, vol. 2	Melinda Baldwin and Janet Browne, eds., *The Correspondence of John Tyndall, volume 2: The Correspondence, September 1843–December 1849* (Pittsburgh: University of Pittsburgh Press, 2016)
Tyndall Correspondence, vol. 4	Ian Hesketh and Efram Sera-Shriar, eds., *The Correspondence of John Tyndall, volume 4: The Correspondence, January 1853–December 1854* (Pittsburgh: University of Pittsburgh Press, 2018)
Tyndall Correspondence, vol. 5	William H. Brock and Geoffrey Cantor, eds., *The Correspondence of John Tyndall, volume 5: The Correspondence, January 1855–October 1856* (Pittsburgh: University of Pittsburgh Press, 2018)
Tyndall Correspondence, vol. 6	Michael D. Barton, Janet Browne, Ken Corbett, and Norman McMillan, eds., *The Correspondence of John Tyndall, Volume 6: The Correspondence, November 1856–February 1859* (Pittsburgh: University of Pittsburgh Press, 2019)
Tyndall Correspondence, vol. 7	Roland Jackson, Nanna Katrine Lüders Kaalund, and Diarmid A. Finnegan, eds., *The Correspondence of John Tyndall, Volume 7: The Correspondence, March 1859–May 1862* (Pittsburgh: University of Pittsburgh Press, 2019)
Tyndall Correspondence, vol. 9	Michael D. Barton, Iwan Rhys Morus, and James Ungureanu, eds., *The Correspondence of John Tyndall, volume 9: The Correspondence, February 1865–December 1866* (Pittsburgh: University of Pittsburgh Press, 2022)
E. Wadge, 'Spiritualism'	Elisabeth Wadge, "A Fair Trial for Spiritualism?: Fighting Dirty in the *Pall Mall Gazette*," in *Repositioning Victorian Sciences: Shifting Centres in Nineteenth-Century Scientific Thinking*, ed. David Clifford, Elisabeth Wadge, Alex Warwick, and Martin Willis (London: Anthem, 2006)
Wellcome	Wellcome Library
X Club	Ruth Barton, *The X Club: Power and Authority in Victorian Science* (Chicago: University of Chicago Press, 2018)

Frontispiece: John Tyndall, photographed by John Robert Mather Sawyer (1828–88) on 2 September 1868, while Tyndall was in Norwich for the annual meeting of the BAAS. Tyndall mentioned this photograph in a letter to his fellow X Club member John Lubbock (letter 2864). Credit: Royal Institution of Great Britain, Science Photo Library.

INTRODUCTION TO VOLUME 10

The tenth volume of *The Correspondence of John Tyndall* contains 402 letters covering a two-year period from January 1867 to December 1868. The period centers on the death of Michael Faraday in August 1867. This was a great personal loss for Tyndall, and it led to substantial changes in his professional and personal circumstances, as he succeeded Faraday as superintendent of the house and director of the laboratory at the Royal Institution of Great Britain (RI) and moved into accommodation in the building. He remained there until his resignation in 1887.

The volume begins with Tyndall publicly and visibly enmeshed in the continuation of the controversial "Eyre Affair," in defense of the mode of suppression of a rebellion in Jamaica by Governor Edward Eyre, revealing Tyndall's views on race, politics, and justifiable violence. Tyndall found himself in opposition to close friends such as Thomas Henry Huxley and Herbert Spencer. Further pressure was exerted on him personally by the death of his mother, Sarah, and the numerous calls on him to support relatives in Ireland. Faraday's death added to these demands, as Tyndall took over his responsibilities at the RI as well as with the Board of Trade and Trinity House as scientific adviser. Throughout all this, Tyndall had to balance the commitments of his lecturing, writing, and research. In this period, Tyndall wrote his second major popular book, *Sound*, as well as *Faraday as a Discoverer*, his memorial to Faraday. Many letters detail his negotiations over translation and publication of his books in French and German, and the hidden role of women in that process. Eventually the pressure of lecturing at the Royal School of Mines weighed heavily on Tyndall, and he resigned to concentrate on his research into the interaction of light with matter, leading to significant discoveries about the cause of the polarization of light and the blue color of the sky, and on the nature of matter suspended in the air, what he referred to as the "floating matter of the air." He retained his mountaineering interests, returning to the Swiss Alps in the summers of 1867 and 1868 to make a notable ascent of the Eiger and the first traverse of the Matterhorn.

This volume also sees Tyndall making substantive pronouncements about the nature of science and about materialism, spiritualism, and religion.

He had, in 1862, already published under his own name the chapter titled "Reflections" in his book *Mountaineering in 1861*, and published related articles anonymously, although his authorship may well have been known to many. His set-piece lectures to the British Association for the Advancement of Science (BAAS) meetings in 1867 and 1868 in particular, in Dundee and Norwich respectively, made these views more explicit and public. The fact that these both took place after Faraday's death is noteworthy, and perhaps beyond coincidental. It has been argued that the respect Tyndall held for Faraday, and the unwritten rule that controversial matters relating to religion should not be discussed in RI lectures, resulted in Tyndall being particularly careful not to offend.[1]

Tyndall's correspondents in the extant letters of this period, exchanged with fifty different individuals, encompass scientific relationships, social acquaintances, and personal friends. The largest number of letters with a single correspondent (twenty-eight) is with Jane Barnard, Michael Faraday's niece, reflecting Tyndall's care and concern to ensure that his lectures and book in memory of Faraday caused not one iota of offense. Scientific correspondents feature widely, both continental and British. The most extensive letters on scientific matters are with Rudolf Clausius, Gustav Magnus, and Hermann Helmholtz on the Continent, and with John Herschel, George Stokes, and George Airy in Britain. Many letters with continental correspondents, including François-Napoléon-Marie Moigno, Jules Jamin, Henry Gauthier-Villars, Heinrich Vieweg, Gustav Wiedemann, and Anna Helmholtz, relate to the translation of Tyndall's books into German and French. Tyndall's close relationships with several society women can be followed in this volume. They include Mary Egerton, Juliet Pollock, Emily Peel, Louisa Baring (Lady Ashburton), and Mary Adair, the latter of whom would turn down Tyndall's offer of marriage in 1869. There are many letters to close friends, especially to members of the X Club such as Thomas Hirst, Thomas Huxley, and Joseph Hooker. Further letters deal with family matters in Ireland, the death of his mother and the need to look after his sister, and respond to the requests of many Irish relatives. In a sign of the future, six letters to the editor of the *Pall Mall Gazette* show Tyndall on the cusp of becoming one of the most visible participants in public debate on the significance and meaning of scientific materialism and its relationship to theological and religious matters.

Personal and Scientific Relationships

The death of Michael Faraday on 25 August 1867 was a turning point in Tyndall's life.[2] His mentor and friend, the man he looked upon as a father, was gone. More letters in this volume relate to Faraday's death and its aftermath than to any other subject, and a good number of these were exchanged with Barnard. Other letters are Tyndall's notes to close friends,[3] discussing the preparations and strain of delivering two Friday Evening Discourses in honor of Faraday at the RI on 17 and 24 January 1868,[4] a further series of nine lectures titled "The Discoveries of Faraday" in January and February 1868,[5] and the writing of his book *Faraday as a Discoverer*,[6] published on 5 March 1868. Several letters relating to *Faraday as a Discoverer* show Tyndall trying to be fair to William Thomson in particular, following previous altercations over priority.[7] In addition, there are exchanges over a suitable memorial for Faraday.[8]

Faraday's death followed the death of Tyndall's own mother on 30 April 1867,[9] and he went to Ireland for the funeral in early May, remembering "the love, the fortitude, and the natural dignity of my dear mother."[10] Given his commitments in London he had rarely been able to visit his mother and sister, and now had to ensure the care of his sister Emma, whose health required medical attention. Many letters, particularly with Elizabeth Dawson Steuart in Leighlinbridge, refer to arrangements he made for his sister,[11] and the extensive financial support he gave to members of his extended family in Ireland.[12] A further responsibility was owed to Margaret Ginty and her children, following the death of Tyndall's old surveying friend William Ginty, who had moved to Brazil. Tyndall was the executor of William Ginty's will, and several letters attest to Tyndall's and Margaret's difficulties and frustration in these matters.[13]

Over Easter 1868 Tyndall made a trip to Italy with John Lubbock and two others, including visits to Naples and Vesuvius, which he also described in his journal.[14] Relationships with other X Club members had not broken down over the divisive Eyre Affair, which had started in 1866, when Tyndall found himself on the same side as Joseph Dalton Hooker and Thomas Archer Hirst, but in opposition to Huxley, Lubbock, and Spencer. Letters in this volume starkly lay out common views, shared by Tyndall, of the inferiority of "negroes."[15] The position Tyndall took was strongly influenced by his hero and friend Thomas Carlyle, who led the public defense of Eyre.[16]

Professional Matters

The death of Faraday led immediately to Tyndall assuming his role as superintendent of the house and director of the laboratory, becoming the leading figure at the RI. In consequence, he left the house at 14 Waverley Place, which he had shared with Hirst since June 1861,[17] and moved in January 1868 into rooms in the RI where he remained, even after his marriage in 1876, until his retirement in 1887. This was an anticipated transition. Yet it might not have happened quite as envisaged. Several letters imply that Tyndall seriously considered taking up an invitation from the government in India to investigate its meteorology for one to two years, though he came to the view that he should not go.[18] In addition to assuming Faraday's responsibilities at the RI, Tyndall effectively inherited his position as scientific adviser to the Board of Trade and Trinity House on lighthouses around the British Isles.[19] Over time, this role would lead the typically outspoken Tyndall into controversy and arguments with his political masters, which can be followed in subsequent volumes.

Tyndall found his lecturing responsibilities a constant demand,[20] and Huxley had the same challenges.[21] In the two years covered by this volume, Tyndall gave twelve lectures on "vibratory motion with special reference to sound" in January and February 1867, the Christmas Lectures for 1867–78 on sound, nine lectures on the discoveries of Faraday in January and February 1868, and four Friday Evening Discourses,[22] in addition to his commitments at the Royal School of Mines. By the autumn of 1868, with his government responsibilities also increasing, and a new research program using the radiation from light ahead of him, Tyndall resigned his professorship at the School of Mines,[23] despite regretting the loss of income (though this was balanced out by the salary for his new post as scientific adviser). He tried to have Alexander Herschel, Sir John's son, appointed in his place, but was unsuccessful.[24] These lectures at the School of Mines had been notable, given the almost exclusively male audiences, for the presence of a few of Tyndall's female friends, including Lady Tankerville (Olivia Bennet), Lady Emily Peel, and Mary Egerton.[25]

Tyndall's broader educational expertise was called upon in several inquiries. At the BAAS meeting in 1866 he had been appointed to the Committee on the Teaching of Natural Science in Schools. The committee reported to the Dundee meeting in 1867. Various letters reveal elements of the discussions and differences of opinion, and information that Tyndall sought from others.[26] He also made his views clear to the Parliamentary Schools Enquiry Commission, on the superiority of continental scientific education.[27]

We see few direct examples of the machinations of the X Club members to advance the interests of their members, as they would have been in regular personal contact, but one letter reveals Tyndall's involvement in approaching Hooker to be president of the BAAS in Norwich in 1868,[28] an approach that was eventually successful.

Publications

As a part of his educational mission, extending the impact of his popular lectures, Tyndall had embarked on writing *Sound*, publishing it in 1867, following the publication in 1863 of *Heat*. This book, based also on lectures at the RI, was dedicated to his early educational mentor Richard Dawes, rector of King's Somborne when Tyndall was teaching at Queenwood College in the 1840s. Dawes, who became dean of Hereford in 1850, had just died in 1867.

As with his previous book, Tyndall sought advice from friends such as Wilhelm Weber and Rudolf Clausius.[29]

Tyndall's reputation as a communicator and author of popular science books generated wide interest on the Continent.[30] His book *Heat* was translated into German by Clara Wiedemann and Anna Helmholtz, although this information was concealed even from the publisher and their husbands were publicly credited,[31] and published by Vieweg in 1867. *Sound* soon followed with the same arrangement.[32] Meanwhile *Sound* was translated into French by François Moigno and eventually published by Gauthier-Villars in 1869.[33] Others vied for the rights.[34] Relationships with Moigno were not the smoothest,[35] nor were they with Gauthier-Villars, who had wanted to sell copies of *Heat* (1863) in Germany as well.[36] It was not only foreign editors and publishers with whom Tyndall had differences, however. John Murray had published Tyndall's first book, *Glaciers of the Alps*, in 1860, but Tyndall took forceful issue with his manner of conducting business and explained why he had published his subsequent books in Britain, including *Sound*, with William Longman.[37]

Translations were soon extended to *Faraday as a Discoverer*. The German edition, translated by Anna Helmholtz, was arranged through Hermann Helmholtz[38] and Vieweg.[39] The French edition was brokered by Jules Jamin.[40] Tyndall found writing these popular books particularly demanding, as he wrote to Rudolf Clausius: "I do not intend to publish anything of the kind for a long time to come—perhaps never again, because such work takes my thoughts so much away from original investigation. Still I think those two books on Heat & Sound have done great good, by giving numbers of people an interest in scientific matters who had previously no interest in them."[41] Tyndall did not entirely keep to that resolution, publishing *The Forms of Water in Clouds and Rivers, Ice and Glaciers* in 1872 and *Six Lectures on Light. Delivered in America in 1872–1873* in 1873.[42]

Mountaineering

Tyndall made his usual summer visit to the Alps in both years covered by this volume, as he had every year since the start of his glacier investigations in 1856. By now, the attraction of climbing mountains in their own right had relegated scientific exploration to an afterthought. In 1867 he made an early ascent of the famed Eiger, sending a detailed description to Juliet Pollock. It was a challenging climb, as he described on facing the exposed slopes above the north face while descending: "[F]or the first time in my life I turned my face to the ice slopes and went backwards down them."[43] He considered an attempt on the Matterhorn, but bad weather drove him away.[44] The following year he planned a further attack on the Matterhorn.[45] This was a bold undertaking, and in 1868 he made the first traverse of the mountain from the Italian to the Swiss side,[46] described also by Felice Giordano, who repeated the feat a little later.[47]

Blue Color of the Sky

Soon after his return from the Matterhorn in the autumn of 1868, Tyndall embarked on a series of experiments that would lead to one of the discoveries for which he is best known: his explanation of the blue color of the sky. On 9 October, Tyndall noticed that solar light seemed to cause the decomposition of the vapor of amyl nitrite. He soon noticed also that the clouds formed in his tubes appear initially to be blue. He submitted his first paper "On a New Series of Chemical Reactions Produced by Light" on 24 October,[48] and two more papers by the end of January 1869.[49] Letters of the period attest to how he sought help for his experiments[50] and established awareness and support for his ideas from scientific friends, including Clausius,[51] John Herschel,[52] and others.[53] In the exchanges with Herschel, Tyndall effectively shared his discoveries and their possible implications in real time with a man whose expertise and influence he much respected.[54] It is in a letter of 2 December 1868 that Herschel suggested a link between his discoveries and the formation of the tail of a comet, which have may have stimulated Tyndall's later explanation for the phenomenon.[55] Letters exchanged with Stokes, as he reviewed Tyndall's work for possible publication by the Royal Society of London, show how important those exchanges were for Tyndall,[56] as it had been in his groundbreaking work on the absorption of radiant heat.[57] That work on the absorption of radiant heat followed Tyndall even into this period, as he sought to counter Magnus's continued objections to his findings on the absorption of heat by water vapor.[58]

Materialism and Spiritualism

This volume includes the first two of Tyndall's most significant public pronouncements on materialism, both to meetings of the BAAS. In Dundee, on 5 September 1867, he gave the inaugural lecture to working men, "On Matter and Force."[59] Here he laid out his claim that the operation of living systems can be understood in purely physical terms, through the medium of scientific experiment and explanation. The talk contained a remarkable statement: "Depend upon it, if a chemist, by bringing the proper materials together in a retort or crucible, could make a baby, he would do it. There is no law, moral or physical, forbidding him to do it—his enquires in this direction are limited solely by his own capacity and the laws of matter and force."[60] This controversial public statement does not seem to have elicited substantial comment, either privately or in the newspapers. There is, however, a glimpse of its reception in a letter to Lady Colville, in which Tyndall wrote to her that he would be reprising the lecture for William Spottiswoode, who has "forbidden all heresy which might have added spice to the occasion."[61] The second lecture, given in Norwich on 19 August 1868 in his role as president of the Mathematical and Physical Section, extended and deepened the arguments advanced in Dundee. It was published in 1871 as "Scope and Limit of Scientific Materialism."[62]

These two lectures followed a related article that Tyndall had published in the *Fortnightly Review* for June 1867, titled "Miracles and Special Providences,"[63] a dismissal of the reality of miracles on the basis of the observed constancy of natural laws. This picked up a theme Tyndall had raised in an earlier article of December 1865, titled "The Constitution of the Universe."[64] Tyndall was a visible opponent of the Victorian craze for "spirit-rapping," as was Faraday.[65] An exchange of letters in the *Pall Mall Gazette*,[66] as well as letters with Alfred Russel Wallace and others,[67] attest to the public interest and to Tyndall's willingness to engage in public debate.

Conclusion

This volume straddles a pivotal moment in Tyndall's life, the death of Michael Faraday. Taking over from Faraday, both at the RI and at the Board of Trade and Trinity House as scientific adviser, put Tyndall into a more visible, prestigious, and authoritative position. While he took great care to pay homage to Faraday, and sought to smooth over his feuds with Peter Guthrie Tait and William Thomson, he nevertheless made statements about materialism amounting to heresy in some ears in his public address to the large audience of working men of Dundee just a few weeks after Faraday's death. In the literary periodicals, and in public addresses in Norwich in 1868, Liverpool in 1870, and most notably in Belfast in 1874 those views would soon be developed further, as Tyndall became associated with Thomas Huxley in the public mind as a leading figure speaking out about the place of science in society, education, and culture.

That public visibility had been highlighted when Tyndall spoke out so strongly in defense of Governor Eyre, in support of Thomas Carlyle and of the manner in which Eyre had violently suppressed a rebellion in Jamaica. That issue split both society in general and the X Club in particular. Tyndall found himself at odds with his close friend Huxley, although they managed to avoid a rupture of their personal relationship. Thomas Hirst, also an admirer of Carlyle's, supported Tyndall's views, but the need for Tyndall to move from the house they had shared for six and a half years into the RI inevitably led to less close contact between them.

Tyndall was invariably under pressure of work, made more challenging by his attraction to social gatherings, which he was generally not able to resist in the manner that Faraday had. The demands of taking over at the RI and of inheriting Faraday's scientific advisory roles added to these pressures. They left him physically drained, verging on severe illness. Given those demands and the prospect of a substantial research program ahead of him on the interaction of light with matter, Tyndall resigned his position at the Royal School of Mines.

The Alps, as ever, provided some respite. Tyndall continued to travel there each summer for an extended period to recharge and to settle his mind in the landscape which he loved. His mountaineering achievements were extended with an early ascent of the Eiger, by the only route up the mountain at that time, and by the first traverse of the Matterhorn. This traverse, from the Italian side to the Swiss, and made at a time when the only previous descent on the Swiss side had resulted in four deaths, was a bold undertaking and places him among the most notable mountaineers of his generation.

By the end of this volume, Tyndall had taken his place as Faraday's heir at the RI. He had embarked on a major program of research into the interaction of light with matter, and become visible as a public intellectual in the controversial domain of scientific materialism, and in debates about the relationship between science, theology, and religion.

Notes

1. Bernard Lightman, "Fashioning the Victorian Man of Science: Tyndall's Shifting Strategies," *Journal of Dialectics of Nature* 38 (2015): 25–38. But see also Ian Hesketh, "The Making of John Tyndall's Darwinian Revolution," *Annals of Science* 77, no. 4 (2020): 524–48.
2. Letters 2638 and 2639.
3. Letters 2640 and 2641.
4. "On Faraday as a Discoverer," *Roy. Inst. Proc.* 5 (1868): 199–272. See letters 2665, 2671, 2673, 2676, 2684, 2703, 2717, 2733, 2743, and 2745.
5. Letters 2767 and 2770.
6. Letters 2654, 2692, 2694, 2710, 2711, 2712, 2713, 2754, 2757, 2759, 2760, 2762, 2771, 2780, 2782, 2788, 2790, 2792, 2793, 2794, 2795, 2796, 2797, 2798, 2823, and 2892.
7. Letters 2715, 2726, 2731, 2735, 2736, and 2786. See also letter 2744.
8. Letters 2774, 2776, 2778, 2781, 2784, and 2785. See also *The Ascent of John Tyndall*, 209.
9. Letters 2587, 2589, and 2612.
10. Letter 2627.
11. Letters 2734, 2768.
12. Letters 2644, 2675, 2681, 2718, 2730, 2777, 2810, 2827, 2828, 2831, 2845, 2857, 2875, 2877, and 2889.
13. Letters 2539, 2548, 2555, 2557, and 2570.
14. Letters 2566, 2801, 2802, 2803, and 2810. See also Tyndall Journal, RI MS JT/2/13c/1341–6.
15. Letters 2567, 2571, 2572, 2576, and 2586. See also *The Ascent of John Tyndall*, 199–202.
16. *The Ascent of John Tyndall*, 199–202.
17. Letter 2770.

18. Letters 2656, 2658, 2659, 2661, 2663, 2665, and 2667.
19. Letters 2590, 2614, 2624, 2789, 2791, and 2829.
20. Letters 2707 and 2907.
21. Letters 2592 and 2593.
22. "On Sounding and Sensitive Flames," *Roy. Inst. Proc.* 5 (1868): 6–12; "On Some Experiments of Faraday, Biot, and Savart," *Roy. Inst. Proc.* 5 (1868): 188–90; "On Faraday as a Discoverer," *Roy. Inst. Proc.* 5 (1868): 199–272.
23. Letters 2833, 2834, and 2835.
24. Letters 2833, 2834, 2836, 2852, and 2911
25. Letters 2799, 2807, 2808, 2809, 2810, 2830, and 2831.
26. Letters 2577, 2579, 2584, 2636, and 2822. See also *The Ascent of John Tyndall*, 288–89.
27. Letter 2601.
28. Letter 2554.
29. Letters 2551, 2562, 2574, 2581, 2585, 2594, and 2735.
30. Letter 2613.
31. Letters 2607, 2677, and 2709.
32. Letters 2684, 2687, and 2691.
33. Letters 2617, 2792, 2795, and 2818
34. Letters 2660, 2680, 2697.
35. Letters 2632, 2664, 2706, 2838, and 2848.
36. Letters 2690 and 2700.
37. Letters 2866, 2868, and 2872.
38. Letter 2686.
39. Letter 2687.
40. Letters 2722, 2750, 2762, 2779, 2795, and 2858.
41. Letter 2840.
42. *The Forms of Water in Clouds & Rivers, Ice & Glaciers* (London: H. S. King & Co., 1872); *Six Lectures on Light delivered in America in 1872–1873* (London: Longmans, Green, and Co., 1873).
43. Letter 2622.
44. Letters 2623 and 2651.
45. Letter 2855.
46. Letters 2856, 2869, and 2905.
47. Letter 2935.
48. "On a New Series of Chemical Reactions Produced by Light," *Roy. Soc. Proc.* 17 (1868–9): 92–102. In his experimental notebook (RI MS JT 3/48/99), Tyndall described the result, which he called Frankland into his laboratory to see.
49. "On the Blue Colour of the Sky, the Polarization of Skylight, and on the Polarization of Light by Cloudy Matter Generally," *Roy. Soc. Proc.* 17 (1868–9): 223–33 (received 16 December 1868); and "Note on the Formation and Phenomena of Clouds," *Roy. Soc. Proc.* 17 (1868–9): 317–19 (received 25 January 1869).
50. Letters 2901 and 2902.
51. Letters 2905, 2925, and 2931.
52. Letter 2910.
53. Letters 2915 and 2934.
54. Letters 2910, 2911, 2912, 2914, 2916, 2918, 2927, and 2929.
55. Letter 2916. "On Cometary Theory," *Phil. Mag.* 37 (1869): 241–45.
56. Letters 2894, 2917, 2920, 2921, 2922, 2924, 2926, and 2928.
57. See *Tyndall Correspondence*, vol. 7, xv–xvi.
58. Letters 2578, 2580, 2598, and 2905. See also *Tyndall Correspondence*, vol. 7, xix.
59. Letters 2623, 2641, 2646, and 2647.

60. "On Matter and Force," in *Fragments of Science: A Series of Detached Essays, Lectures, and Reviews* (London: Longmans, Green, and Co., 1871), 69–94, on 92.
61. Letter 2692.
62. "Scope and Limit of Scientific Materialism," in *Fragments of Science: A Series of Detached Essays, Lectures, and Reviews* (London: Longmans, Green, and Co., 1871), 107–23. For several responses, see letters 2862, 2864, 2865, 2876, and 2882.
63. "Miracles and Special Providences," *Fortnightly Review* 7 (1867): 645–60. See also letters 2537, 2597, 2620, 2627, and 2882.
64. "The Constitution of the Universe," *Fortnightly Review* 3 (1865): 129–44.
65. See letter 2537, n. 5 for extensive references.
66. Letters 2811, 2812, 2813, 2814, 2815, 2820, 2824, and 2825.
67. Letters 2817, 2819, and 2822, and 2906.

EDITORIAL PRINCIPLES

Our aim is to include all letters to and from John Tyndall that are currently extant. Some of the letters are from published sources, such as letters written by Tyndall to newspapers or journals, but the majority are either the originals or typescripts produced by Louisa Tyndall from originals. One central editorial principle has been to transcribe, wherever possible, from original letters rather than Louisa's typescripts. This is because of Louisa's conscious and unconscious editorial interventions, a point discussed in more detail below. We have also aimed to reproduce, as accurately as possible, the text of the letters as they were written. Spelling mistakes that have appeared in the handwritten letters have therefore been preserved; mistakes appearing where only the typescript letter has survived are silently corrected, as in our judgment they are almost always typographical errors. But, in general, nonstandard spellings (e.g., the use of dialect) has been retained even when in a typescript-only letter (though there is the possibility that the spelling is Louisa's, not her husband's). Where the transcription is based on a typescript transcription by Louisa (or one of her assistants), this will be indicated by "LT Transcript Only."

We have included illustrations that are part of the letter. In cases where it was possible to reprint the original handwritten image, we have done so. In other cases, volume editors have reconstructed drawings because it was not possible to reprint an image from the handwritten letter (due, for example, to deterioration of the original document).

The letters are presented in chronological order of writing. Where more than one letter has the same date, letters from Tyndall (in alphabetical order of addressee) come first, followed by letters to Tyndall (in alphabetical order of writer), with the exception that if the order of writing can be determined, that order takes precedence. Where letters cannot be dated precisely they are placed at the latest likely date of writing. (In some cases if it is possible, but unlikely, that they were written later, notes indicate such possibilities.) To indicate the beginning of a new year, the date of that year has been inserted in a large font before the first letter from that year. Readers should note that although volumes are arranged in chronological order, letters that are discovered from years already covered in published volumes will appear in the final volume of the correspondence.

For non-English-language letters, both the transcription of the original and the English translation are presented. The transcription of the non-English letter appears first, followed by the English translation in a smaller font. Editorial notes are inserted in the English translation. Some foreign letters are in an original hand, while others exist only as transcriptions by Louisa or others who often did not have a full command of the language. In the latter case these letters contain frequent spelling, grammatical, and interpretation errors that do not seem to have been committed by the original author. In these typescript-only cases, and only in these cases, spelling and grammatical errors have been corrected, as have poor interpretations where a better word choice was obvious from the context.

There is a standard format for each letter. The first line lists the author, if it is to Tyndall, or the recipient, if it is from Tyndall; the date; and the letter number. The second line lists any information supplied by the writer in their salutation, such as the place, date, or time of day when the letter was written. (In some cases, though, where the writer put the date at the foot of the letter we have left it there in order to reproduce the text accurately.) The opening salutation, the text of the letter, and closing then follow. Then come the source and editorial notes, the latter indicated in the text by arabic numerals.

Editorial notes are intended to provide the contextualization necessary to understand the contents of each letter. There are basically six types of editorial notes. Many of them are informative notes on persons mentioned in the letters. If a person referred to in a letter is not described in a note, they have been mentioned more than twice in the volume and they will have an entry in the biographical register at the back of the volume. For the rest—those mentioned once or twice—biographical information will appear in the editorial note appended to the letter in which he or she is first mentioned. When there is a second reference to that person a note will refer back to the letter in which he or she was first mentioned. Biographical notes on well-known figures such as Faraday or Huxley stress what that person was doing at the time when the letter was written rather than providing a full overview of the person's life.

A second type of editorial note identifies allusions and quotations. In the case of quotations, where possible, the first edition of the work is cited unless a different edition is mentioned in the letter. If a quote is not in English a translation is given. Glosses on obscure places, abstruse words, and words used in an unfamiliar sense constitute the third type of editorial note. The fourth type provides information about the context of the letter, drawing on such sources as contemporary publications, scholarly sources, Tyndall's personal journals and field notes, and other letters in the *Tyndall Correspondence*. The fifth type of editorial note is cross-references to other letters in the correspondence, referring to the letter number. The last type is information on significant textual changes that Tyndall made to the original that is thought to be relevant to the meaning of the letter.

There are a series of conventions for indicating when editors have given additional information about the text of a letter:

When Louisa Tyndall has made an annotation or marginal insertion that the editors have deemed important enough to retain, we have included her annotation in an editorial footnote.

When editors other than she have made insertions, they are indicated by unitalicized square brackets; for example, [word].

When certain words are ambiguous and we have had to conjecture their meaning, they are indicated by italicized square brackets; for example, *[*words*]*; when we have considered such words illegible the number of illegible words is indicated in full italicization and square brackets; for example, *[3 words illeg]*.

If a word or series of words are not present or have been destroyed—by inkblots, holes, or mold, for example—they are indicated as missing, italicized and in angular brackets; for example, <*3 words missing*>.

In extremely rare cases where it is clear that Louisa herself has intentionally destroyed a few words, they are indicated as excised, italicized and in angular brackets; for example, <*3 words excised*>.

Finally, a brief word about Louisa. She was responsible for collecting together many of Tyndall's letters in order to write a biography of her late husband. This biography was not completed. It is important to recognize that Louisa was not always reliable as a transcriber. She often threw out the manuscript letter when she had made her transcription, but some original manuscripts remain, which indicate her editing processes. Like any transcriber, she was sometimes inaccurate, but corrections show that she proofread her transcriptions. However, she also edited Tyndall's writing to make his letters seem more consistent or grammatically correct. For example, she made punctuation more formal, corrected grammar, and, occasionally, corrected quotations. She consistently corrected Tyndall's usual (deliberate) lowercase for "god" to "God." Her typewriter was unable to insert the superscripts that Tyndall used in abbreviations, or the "&c" symbols that stood for an address at the foot of the letter. (Modern typefaces are unable to deal with all the variant scrawls that the "etc." symbol took and they are all formalized here to "&c.") Louisa also left out details that she considered too private or too salacious for public consumption. She equated postmark and date of writing and overlooked the time taken to write some letters. When she inferred dates from internal evidence, her reasoning was sometimes faulty. Similarly, her identifications of people are sometimes in error. Independent evidence has always been sought for her dates and identifications. Lastly, the patterns of punctuation differ between LT sources and manuscript sources. Tyndall himself used dashes more often than commas, semicolons, and stops, and he used dashes of many different lengths, which are all standardized here to em-dashes. It is often difficult to distinguish between very short dashes and commas or very short dashes and stops. Similarly, capitalization is often ambiguous; distinguishing *S*/*s*, for "science," for example, is a particular problem. Therefore, scholars for whom punctuation and capitalization are significant will need to consult the manuscripts. Users should also be aware that throughout this project the transcribers and editors have generally worked from scans and photographs rather than the original manuscripts.

NOTE ON MONEY

In the nineteenth century the currency used in England was the pound sterling, usually abbreviated to £ (or, occasionally, to *L*). A pound consisted of 20 shillings (abbreviated *s*) and each shilling contained 12 pennies (*d*). A guinea was 21 shillings; that is, £1 1s. A crown was 5 shillings (5s), and half a crown equaled two shillings and sixpence (2s 6d). The smallest coin was a farthing, one quarter of a penny (¼d).

Several forms of representation were in contemporary use. Sometimes the numbers for pounds, shillings, and pence were separated by dots, forward slashes, or spaces; thus, for example, 3 pounds, 4 shillings, and 5 pennies could be written as £3.4.5, £3/4/5, or £3 4 5. A zero could be represented by a dash; for example, 7 shillings was often written as 7/- or just 7/. A half crown was often written as 2/6.

TIMELINE OF JOHN TYNDALL'S LIFE

Year	Event
c. 1822	John Tyndall born at Leighlin Bridge, County Carlow, Ireland
c. 1836	Attends John Conwill's National School in Ballinabranna
1839	Begins employment as civil assistant with Ordnance Survey of Ireland in Carlow
1840	Becomes civil assistant in the Ordnance Survey Office at Youghal, Cork County
1841	First appearance in print (poem in *Carlow Sentinel* under pseudonym "W[alter] S[nooks]")
1842	Joins English Ordnance Survey in Preston
1843	Leads written criticism of Ordnance Survey in *Liverpool Mercury*; dismissed November
1844	Home in Ireland, mostly unemployed, until obtaining a position with the firm of Nevins and Lawton, Surveyors, of Manchester
1845–47	Works for Richard Carter, Surveyor, of Halifax
1845	Meets Thomas Archer Hirst in Halifax
1847	Begins teaching mathematics and surveying at Queenwood College, Hampshire
1848–50	At University of Marburg, working for PhD with Robert Bunsen and Friedrich Stegmann
1850	Returns to England; meets Michael Faraday and William Francis and attends his first British Association meeting, before returning to Marburg
1851	Moves from Marburg to Berlin (April–June); then returns to Queenwood
1852	Elected fellow of the Royal Society
1852	Friendship developed with Thomas Henry Huxley
1853	First lecture at Royal Institution (RI)
1853	Meets Herbert Spencer
1853	Appointed professor of natural philosophy at RI
1853	Awarded Royal Medal for magnetic work; declines when controversy arises over award
1856	Meets Thomas Carlyle
1857	Climbs Mont Blanc for first time with Hirst
1859	Demonstrates existence of greenhouse gases and climatic implications
1859	Joins Government School of Mines as professor of natural philosophy
1860	*The Glaciers of the Alps* published
1860	Meets John Lubbock and his wife, Ellen Lubbock
1861	First ascent of the Weisshorn

Year	Event
1861	Delivers his first Christmas Lectures at RI
1862	*Mountaineering in 1861* published
1862	Begins to champion Mayer's priority in discovery of conservation of energy
1863	*Heat Considered as a Mode of Motion* published
1864	Awarded Royal Society's Rumford Medal
1864	First meeting of X Club
1865	*On Radiation* published
1865	Engages in public controversy over the efficacy of prayer
1866	Succeeds Michael Faraday as scientific adviser to Trinity House
1867	Faraday dies; Tyndall becomes superintendent of the RI
1867	*Sound: A Course of Eight Lectures* published
1868	Successfully climbs Matterhorn
1868	Discovers the cause of light scattering, to be known as "Tyndall Effect"
1868	*Faraday as a Discoverer* published
1868	"Scientific Materialism" lecture at the British Association
1868	Resigns from Royal School of Mines
1869	Joins Metaphysical Society
1870	"On the Scientific Use of the Imagination" lecture at British Association
1870	*Three Scientific Addresses* published
1870	*Researches on Diamagnetism and Magne-Crystallic Action* published
1871	Meets Louis Pasteur for first time while in Paris
1871	*Hours of Exercise in the Alps* published
1871	*Light and Electricity* published
1871	*Fragments of Science* published
1872	Debates over how to measure the efficacy of prayer, aka the "Prayer-Gauge Debate"
1872	*Contributions to Molecular Physics in the Domain of Radiant Heat* published
1872	*The Forms of Water in Clouds and Rivers, Ice and Glaciers* published
1872–73	USA Lecture Tour
1873	*Six Lectures on Light* published
1874	As president of the BAAS, delivers the "Belfast Address"
1876	Marries Louisa Charlotte Hamilton
1877	Develops "Tyndallization" (discontinuous heating and cooling process to destroy heat-resistant spores)
1877	Tyndall and Louisa build summer cottage, Alp Lusgen, at Belalp, northern side of Valais, above Brig
1877	*Fermentation and Its Bearings on Phenomena of Disease* published
1879	*Fragments of Science*, which had gradually expanded, first published in two volumes
1881	*Essays on the Floating Matter of the Air* published
1885	Tyndall and Louisa build Hindhead retreat, Surrey Downs
1887	Resigns from RI
1890	Fight with Gladstone in the *Times* over Irish Home Rule
1892	Hirst dies
1892	*New Fragments* published
1893	Dies at Haslemere, Surrey, accidentally poisoned

TIMELINE OF EVENTS IN JOHN TYNDALL'S LIFE SPECIFIC TO VOLUME 10

Month and year	Biographical details	Notable social, political, and cultural events
December 1866	Travels to Mentone, France, with Thomas Carlyle to visit Louisa Baring (leaves London 21st, night in Paris 22nd, in Mentone 23rd, back in London 29th)	Edward Frankland begins Christmas Lectures (Adapted for a Juvenile Auditory) at the RI, "On the Chemistry of Gases" (27 December 1866–8 January 1867)
1867	German translation of *Heat* (1865) is published as *Die Wärme, betrachtet als eine Art der Bewegung* Tyndall's "Matter and Force" lecture in Dundee and an 1862 lecture on force are published in French as *La matière et la force. La Force: deux conferences* Edward Youmans publishes *Modern Culture: Its True Aims and Requirements; A Series of Addresses and Arguments on the Claims of Scientific Education*, which includes Tyndall's 1854 lecture at the RI, "On the Importance of the Study of Physics as a Branch of Education"	Henry Enfield Roscoe isolates the element vanadium Joseph Dalton Hooker publishes second volume of *Handbook of the New Zealand Flora* (1864–67) Frederick Farrar publishes *Essays on a Liberal Education* RS publishes first volume of *Catalogue of Scientific Papers (1800–1863)* George Campbell (Duke of Argyll) publishes *The Reign of Law*
January 1867	Attends X Club meeting (3rd) Dines with Debus, Wright, and Hirst (9th) Gives a Friday Evening Discourse at the RI, "On Sounding and Sensitive Flames" (18th) Dines at Huxley's with Farrar (Harrow), Wilson (Rugby), Griffith, and Hirst (who wrote that they all "form a sub-committee of Brit. Ass. to draw up a report on the introduction of Science into schools") (20th)	William Odling gives Friday Evening Discourse at the RI, "On Mr. Graham's Recent Discoveries on the Diffusion of Gases" (25th) French-American explorer Paul Belloni Du Chaillu publishes *A Journey to Ashango-Land: and Further Penetration into Equatorial Africa*

Month and year	Biographical details	Notable social, political, and cultural events
January 1867	Begins twelve lectures "On Vibratory Motion, with Special Reference to Sound" at the RI (22 January–28 February)	
February 1867	Walks with Hirst to Kew to encourage Joseph Dalton Hooker to accept the presidency of the BAAS at Norwich in 1868 (3rd); Hooker at first declines but later accepts Attends X Club meeting (7th) Speaks at a meeting of the Eyre Defence Committee, held at Willis's Rooms (18th) Publishes RI lecture "On Sounding and Sensitive Flames" in the *Phil. Mag.*	Frederick Farrar gives Friday Evening Discourse at the RI, "On Some Defects in Public School Education" (1st) American physicist and superintendent of the US Coast Survey Alexander Dallas Bache dies (17th) First ship passes through the Suez Canal (17th); canal opens in 1869
March 1867	Attends X Club meeting (7th)	George Henslow begins "Six Lectures on the Practical Study of Botany" at the RI (5–9 April) British educationalist and dean of Hereford Richard Dawes, a friend of Tyndall, dies (10th) Joseph Lister publishes article outlining the discovery of antiseptic surgery in the *Lancet* (16th) Edward Frankland gives Friday Evening Discourse at the RI, "The Water Supply of the Metropolis" (29th) First prosecution of Governor Edward Eyre in Jamaica affair results in his discharge (29th) Constitution Act passes, creating the Dominion of Canada (29th)
April 1867	Attends X Club meeting (4th) Visits Paris for the International Exposition (19th–26th); serves as a juror for Apparatus and Processes for Heating and Lighting Takes long walk with Hirst in the Forest of Saint-Germain outside of Paris (21st)	International Exposition (or Exposition Universelle) begins in Paris (1 April–3 November) Balfour Stewart gives Friday Evening Discourse at the RI, "On the Sun as a Variable Star" (12th)

Month and year	Biographical details	Notable social, political, and cultural events
April 1867	Introduced by Hirst to French mathematician Michel Chasles at the International Exposition (23rd) Tyndall's mother, Sarah Tyndall, dies (30th)	
May 1867	Travels to Ireland for his mother's funeral (2nd–5th) Publishes "On the Action of Sonorous Vibrations on Gaseous and Liquid Jets" (his sixth lecture on sound) in the *Phil. Mag.*	Thomas Henry Huxley begins twelve lectures "On Ethnology" at the RI (2–8 June) Alfred Nobel patents dynamite London Conference results in Treaty of London (11th) William Allen Miller begins four lectures "On Spectrum Analysis, including its application to Astronomy" at the RI (14 May–4 June) William Odling gives Friday Evening Discourse at the RI, "On the Occlusion of Gases by Metals" (17th) John Stuart Mill's law for women's suffrage rejected (20th) Alexander Herschel gives Friday Evening Discourse at the RI, "On the Shooting-stars of the Years 1866–7" (24th) T. Sterry Hunt gives Friday Evening Discourse at the RI, "On the Chemistry of the Primeval Earth" (31st)
June 1867	Publishes "Miracles and Special Providences" in the *Fortnightly Review* (1st) Attends X Club meeting (6th) Gives Friday Evening Discourse at the RI, "On Some Experiments of Faraday, Biot and Savart" (21st) Visits Maidenhead (west of London) with the X Club (29th–30th) Publishes "Note on Magnus's Paper. The Influence of the Adhesion of Vapour in Experiments on the Absorption of Heat" in the *Phil. Mag.*	

Month and year	Biographical details	Notable social, political, and cultural events
July 1867	Publishes *Sound: A Course of Eight Lectures* (5th) Departs London for Switzerland (5th) Is officially offered position of scientific adviser to the Board of Trade and the three lighthouse boards it oversees (he had already done work as Faraday's health declined) (10th) Summits the Eiger in the Alps (11th)	Otto von Bismarck becomes chancellor of the North German Confederation (1st) William Thomson and Peter Guthrie Tait publish *Treatise on Natural Philosophy*
August 1867	Accepts the position of scientific adviser to the Board of Trade (4th) Returns to London from Switzerland (8th)	New Reform Act passes (15th) British physicist Michael Faraday, Tyndall's friend and mentor, dies at Hampton Court (25th); is buried at Highgate Cemetery (30th)
September 1867	Arrives in Dundee for BAAS meeting (4th) BAAS Committee on Scientific Education, of which Tyndall was a member, submits report at 1867 meeting in Dundee Gives lecture titled "Force and Matter" to working-class men at Kinnaird Hall, Dundee during BAAS meeting (5th) Visits Thomson and Tait at St Andrews (7th–8th) Explores in the Scottish Highlands with Scottish lawyer and folklorist John Stuart Stuart-Glennie (10th–21st) Visits the Armstrongs at Cragside Morpeth, Scotland (21st–25th) Returns to London from Scotland (25th) Succeeds Michael Faraday as superintendent of the house and director of the laboratory at the RI (~25th)	BAAS holds its thirty-seventh annual meeting in Dundee (4th–11th) Karl Marx publishes first volume of *Das Kapital* (14th)
October 1867	Attends X Club meeting (10th) Declines offer of a position in British India to undertake meteorological research (21st)	English astronomer and president of the RS (1854–58) John Wrottesley dies (27th)

Month and year	Biographical details	Notable social, political, and cultural events
November 1867	Attends X Club meeting (7th) Sees walrus at the Zoological Garden in London (17th) Sees the play *A Widow's Hunt* (with American actor John Sleeper Clarke) at St James Theatre with Hirst (27th); four days later this theatre will be completely destroyed by fire Begins intensive work on Faraday lectures for the RI in January, declines most social invitations	Herbert Spencer publishes 2nd edition of *First Principles* Polish physicist Maria Skłodowska, later Marie Curie, born (7th) RS awards Copley Medal to Karl Ernst von Baer "for his discoveries in embryology and comparative anatomy, and for his contributions to the philosophy of zoology" (30th) RS awards Royal Medals to John Bennet Lawes and Joseph Henry Gilbert "for their researches in agricultural chemistry," and William Edmond Logan "for his geological researches in Canada, and the construction of a geological map of that colony" (30th)
December 1867	Sees the opera *Linda di Chamouni* (with American singer Clara Louise Kellogg) at Her Majesty's Theatre with Hirst (2nd) Lectures at the Whittington Club to the printers employed by Spottiswoode (4th) Attends X Club meeting (5th) Begins Christmas Lectures (adapted for a Juvenile Auditory) at the RI, "Heat and Cold" (26 December–7 January 1868)	British military forces begin rescue mission and retaliation against the Ethiopian Empire (4 December 1867–13 May 1868) Henry Roscoe delivers Bakerian Lecture at the RS, "Researches on Vanadium" (19th) French photographer and inventor Antoine François Jean Claudet dies (27th) Augustes de la Rive publishes "Michael Faraday, his Life and Works" in the *Phil. Mag.* (original in French published in October)
1868	Publishes *On Experimental Physics*, an 1861 lecture to teachers at the South Kensington Museum Publishes 1867–68 Christmas Lectures at the RI, on "Heat and Cold," in French as *Chaleur et froid: six leçons faites devant un jeune auditoire pendant les vacances de Noël 1867* *Faraday as a Discoverer* is published in French as *Faraday Inventeur*	Cecilia Anne Barlow (née Law), wife of RI secretary John Barlow and a friend of Tyndall, dies RS publishes second volume of *Catalogue of Scientific Papers (1800–1863)* Thomas Henry Huxley is appointed president of the Geological Society Louisa May Alcott publishes first volume of *Little Women*

Month and year	Biographical details	Notable social, political, and cultural events
1868		Ernst Haeckel publishes *Natürliche Schöpfungsgeschichte* (*The History of Creation*)
January 1868	Attends X Club meeting (9th) Sees the play *How She Loves Him* at the Prince of Wales's Theatre with Hirst (16th) Gives Friday Evening Discourse at the RI, "Faraday as a Discoverer" (first part) (17th) Visits Michael Faraday's grave at Highgate Cemetery in London (24th) Gives Friday Evening Discourse at the RI, "Faraday as a Discoverer" (second part) (24th) Moves out of 14 Waverley Place to occupy Faraday's old rooms at the RI (25th) Begins nine lectures "On the Discoveries of Faraday" at the RI (30 January–27 February)	Thomas Henry Huxley gives lecture "A Liberal Education; and Where to Find It" at the Working Men's College (4th) Cornish chemist John Davy, younger brother of Humphry, dies (24th) Henry Roscoe begins eleven lectures "On the Chemistry of Non-Metallic Elements" at the RI (25 January–4 April) Darwin publishes two-volume *The Variation of Animals and Plants under Domestication* (30th) Frederick William Farrar gives Friday Evening Discourse at the RI, "On Public School Education" (31st)
February 1868	Attends X Club meeting (6th) Sees the play *Chimney Corner* at the St James Theatre with Hirst (25th)	Thomas Henry Huxley gives Friday Evening Discourse at the RI, "On the Animals Which Are Most Nearly Intermediate between Birds and Reptiles" (7th) Scottish physicist and inventor David Brewster dies (10th) French physicist Léon Foucault dies (11th) Henry Roscoe gives Friday Evening Discourse at the RI, "On Vanadium, One of the Trivalent Group of Elements" (14th) Herbert Spencer elected to the Athenaeum Club (14th)
March 1868	Visits and dines with painter Henry Phillips and Hirst at Crystal Palace at Sydenham (1st) Visits the Tennysons on the Isle of Wight (4th–5th) Publishes *Faraday as a Discoverer* (5th)	US House of Representatives impeaches President Andrew Johnson (2nd–3rd); in May, Senate fails to convict French geologist Louis Lartet discovers the first identified skeletons of Cro-Magnon, the first anatomically modern humans, in Les Eyzies, Dordogne, France

Month and year	Biographical details	Notable social, political, and cultural events
March 1868	Misses X Club meeting when Charles Darwin was a guest (5th)	
	Attends a soiree for the Chemical Society of London at Willis's Rooms (11th)	
	Dines at the dean of Westminster's with Thomas Carlyle and Prince and Princess Christian (17th)	
	Begins effort with Hooker and Huxley to honor Michael Faraday with a memorial statue	
April 1868	Visits Birmingham to inspect lighthouse lanterns (4th)	Edward Frankland gives Friday Evening Discourse at the RI, "On the Proposed Water Supply for the Metropolis" (3rd)
	Leaves London for Naples, Pompeii, and Vesuvius (9th)	William Odling begins four lectures "On Chemical Combustion" at the RI (23 April–2 May)
	Joins John Lubbock in Amiens, France, for trip to Naples and Vesuvius; together visit the "pits where the famous flint implements which have caused so much sensation are found in some abundance"; then to Paris and further through the night to Marseille (10th)	
	Continues journey to Italy by boat from Marseille (12th)	
	Arrives at Livorno, Italy; then to Pisa; ascends the Leaning Tower; back to Livorno; to Naples overnight by boat (15th)	
	Arrives at Naples, Italy; visits the extinct volcanic crater Solfaterra (16th)	
	Arrives at Pompeii and Vesuvius (17th)	
	Climbs Vesuvius (17th or 18th)	
	Climbs Vesuvius again (19th)	
	Leaves Naples for London (21st); returns no later than 27 April	
	Begins his last lecture course at the Royal School of Mines, thirty-two lectures on "magnetism, electricity, sound, light, and heat" (27th)	
	Gives an account of his trip to Mount Vesuvius at a meeting of the Philosophical Club of the RS (27th)	

Month and year	Biographical details	Notable social, political, and cultural events
May 1868	Speaks at the RI on his and Lubbock's trip to Mount Vesuvius (after Francis Turner Palgrave's Friday Evening Discourse, "How to Form a Good Taste in Art") (1st) Writes to *Pall Mall Gazette* about the spiritualist Daniel Home, a dispute that continues through the month (4th) Attends X Club meeting (7th) Is interviewed by the Select Committee on Scientific Instruction about his lecture-based course in general physics at the Royal School of Mines (11th)	Jean Baptiste Dumas gives a eulogy for Faraday at the Académie des Sciences in Paris (11th) Robert Grant begins four lectures "On Astronomy, Viewed in Connexion with the Establishment of the Theory of Gravitation" at the RI (21st–30th) German mathematician and physicist Julius Plücker dies (22nd) William Odling gives Friday Evening Discourse at the RI, "On Some Effects of the Heat of the Oxy-hydrogen Flame" (22nd) Second prosecution of Governor Edward Eyre in Jamaica affair results in dismissal of the case
June 1868	Attends X Club meeting (5th) Writes to Huxley that he had written to Roderick Murchison resigning his position at the Royal School of Mines (10th) Visits Oxford with the X Club (20th–21st) With Hirst, sits for FRS portraits (29th)	John Lubbock begins four lectures, "On Savages," at the RI (4th–13th) Edward Frankland gives Friday Evening Discourse at the RI, "On the Source of Light in Luminous Flames" (12th) French physicist Claude Pouillet dies (14th) Italian physicist and neurophysiologist Carlo Matteucci dies (25th)
July 1868	Arrives in Paris and joins Hirst (5th) With Hirst, leaves Paris for Switzerland (6th) Arrives in Switzerland (7th) Climbs the Diablerets (18th) Traverses the Matterhorn from the Italian to Swiss side, the first party to do so (26th–28th)	William Odling is appointed Fullerian Professor of Chemistry at the RI (6th) United States adopts fourteenth amendment to Constitution, entitling all persons born or naturalized in the United States to citizenship and equal protection under the law (9th) Irish and Scotch Reform Acts pass (13th) Public Schools Act enacted (31st)
August 1868	Returns to London from Switzerland (15th) Arrives in Norwich for BAAS meeting (18th)	French astronomer Jules Janssen detects helium in the spectrum of the Sun's chromosphere during a total eclipse in Guntur, India (18th)

Month and year	Biographical details	Notable social, political, and cultural events
August 1868	Gives lecture on scientific materialism as president of the Mathematical and Physical Section at the BAAS meeting in Norwich (20th)	BAAS holds its thirty-eighth annual meeting in Norwich (19th–26th) Joseph Dalton Hooker gives Presidential Address championing Darwinian theories at the BAAS meeting in Norwich (19th) Thomas Henry Huxley gives lecture "On a Piece of Chalk" at the BAAS meeting in Norwich (26th) German-Swiss chemist Christian Schönbein dies (29th)
September 1868	Begins experimental researches on the effect of light on vapors and their connection to the blue color of the sky, the polarization of light, and the formation of clouds (11th)	
October 1868	Attends X Club meeting (1st) Discovers that "floating matter" in the air can be burned by a flame and is thus organic (5th) Dines at Hooker's with Hirst and American botanist Asa Gray and his wife, Jane Loring Gray (7th) Discovers that solar light can decompose the vapor of amyl nitrite (10th) Submits paper "On a New Series of Chemical Reactions Produced by Light" to the RS (24th) Stays at Charles Darwin's home in Downe with the Hookers and Grays (24th) Dines at Spencer's with Hirst and American science writer and editor Edward Youmans (31st)	English astronomer Norman Lockyer detects helium in the solar spectrum (20th) American inventor Thomas Edison applies for his first patent (28th)
November 1868	Attends X Club meeting (3rd) Visits the Tennysons on the Isle of Wight (3rd–8th) Sees the play *Monte Cristo* at the Adelphi Theatre with Hirst (13th)	Ulysses S. Grant wins US presidential election (3rd) Thomas Henry Huxley lectures "On the Physical Basis of Life" at Edinburgh (8th) Rudolf Clausius elected Foreign FRS (30th)

Month and year	Biographical details	Notable social, political, and cultural events
November 1868	Sees the play *King of Scots* at the Drury Lane Theatre with Hirst (15th) Sees the play *Leah, the Forsaken* at the Haymarket Theatre with Hirst (18th) Publishes third edition of *Heat*	RS awards Copley Medal to Charles Wheatstone "for his researches in acoustics, optics, electricity and magnetism" (30th) RS awards Royal Medal to Alfred Russel Wallace "for his labours in practical and theoretical zoology" and George Salmon "for his researches in analytical geometry and the theory of surfaces" (30th) RS awards Rumford Medal to Balfour Stewart "for his researches on the qualitative as well as quantitative relation between the emissive and absorptive powers of bodies for heat and light" (30th)
December 1868	Submits paper "On the Blue Colour of the Sky, the Polarization of Skylight, and on the Polarization of Light by Cloudy Matter Generally" to the RS (16th) Visits Stratford on Avon with Hirst (21st) Visits Shakespeare's house, then to Great Malvern (21st) Walks over the Malvern Hills to Eastnor with Hirst (22nd)	Third Derby-Disraeli ministry ends (1st) First Gladstone ministry begins (3rd) World's first traffic signal installed outside the Houses of Parliament (9th) William Odling begins Christmas Lectures (adapted for a Juvenile Auditory) at the RI, "On the Chemical Changes of Carbon" (29 December 1868–9 January 1869) Scottish physicist and glaciologist James David Forbes, a rival of Tyndall's, dies (31st)
January 1869	Reads paper "On the Blue Colour of the Sky, the Polarization of Skylight, and on the Polarization of Light by Cloudy Matter Generally" at the RS (14th) Gives Friday Evening Discourse at the RI, "On Chemical Rays and the Light of the Sky" (15th) Submits paper "Note on the Formation and Phenomena of Clouds" to the RS (25th)	

The Correspondence of John Tyndall

1867

To Ellen Busk — 2 January [1867][1] — 2535

Royal Institution of Great Britain | 2nd. January

My dear Mrs. Busk

I fear to say yes[2]—in fact I hardly ever dine out on the day of my lecture[3]—Though it would give me delight to meet you all I hardly think I ought to risk it. I might be compelled to back out of it even if I promised to go.

We[4] had a flourishing passage to Mentone.[5] We went day & night passing in a few hours from dismal fog-crammed winter to splendid cloudless summer. Mentone is charming & Lady Ashburton[6] is charming & her little daughter[7] is charming & her brandy was charming & the mountains & sea are charming and the orange trees loaded with fruit are charming. On Xmas day Carlyle astride a donkey reached a height of 2000 feet[8] accompanied by Lady A & her daughter. We had luncheon there. They returned & I went on to the top of the highest Aiguille whence I had a most glorious view of Alps and sea. They drove me to Monaco[9] on Wednesday—There I took steamer for Nice[10]—Thence I had 27 hours continuous travelling to Paris. But that was nothing to the horrible night passage I had from Calais to Dover.[11]

A happy new year—But I err. There is the ring of steel in what you say about happiness. Old Carlyle would grunt approval if he heard the utterance.

Yours ever | John Tyndall

NLS, Mrs. George Busk, Ms. 13836, ff. 175–6
RI MS JT/1/TYP/1/186

---

1. *[1867]:* the year is given by reference to the trip to Mentone.
2. *I fear to say yes:* if this is in response to a previous letter from Busk, the letter is missing.
3. *my lecture:* on 18 January 1867, Tyndall gave a Friday Evening Discourse at the RI, 'On Sounding and Sensitive Flames' (*Roy. Inst. Proc.,* 5 (1866–9), pp. 6–12). Thomas Hirst wrote in his journal for 18 January 1867: 'At Tyndall's Friday evening lecture on "Sounding and Sensitive Flames." Sat between my two best lady-friends Lady Lubbock and Mrs.

Busk. I need not say the hour was a pleasant one. Tyndall's experiments were new and most interesting. Every body was highly entertained if not instructed' (Brock & MacLeod, *Hirst Journals,* p. 1797).

4. *We:* Thomas Carlyle joined Tyndall on the trip to visit Louisa Baring in Mentone, where she had a home (see letter 2546, n. 11). They left London on 21 December, spent the evening of 22 December in Paris, and arrived in Mentone very late on 23 December. Tyndall returned to London on 29 December 1866, as he noted in a letter to Eliza Spottiswoode, the wife of William Spottiswoode (RI MS JT/1/TYP/3/1243). See also *The Ascent of John Tyndall,* p. 202.
5. *Mentone:* a Mediterranean coastal town in southeastern France, at the border with Italy; *Menton* to the French and *Mentone* to Italians.
6. *Lady Ashburton:* Louisa Baring.
7. *her little daughter:* Mary Florence Baring (1860–1902), daughter of Louisa Baring, Lady Ashburton, and Bingham Baring, 2nd Baron Ashburton. She married William Compton, 5th Marquess of Northampton in 1884, and held the title Marchioness of Northampton from 1897 ('Death of the Marchioness of Northampton', *Northampton Mercury,* 6 June 1902, p. 5).
8. *reached a height of 2000 feet:* Carlyle wrote in his journal for 20 January 1867: 'Xmas day, hot and bright, the gracious Lady took us all, on asses by the rugged cliffs & *sierras* to a village and peak called Saint-Agnore (strangest village in the world, with a strange old castle perched on the very point of the cliff and of it), where we *lunch'd,* in sight of the whole populat*n,* gathered round us like a theatre with pit, gallery and boxes,—very civil, tho' so eagerly curious;—and in the ev*g* we all dined with a Lady Marion Alford, not known to me before, but elegant, gifted, and blandly high in her way' (J. A. Froude (ed), *Thomas Carlyle: a History of His Life in London,* 2 vols (New York: Charles Scribner's Sons, 1910), vol. 2 (1834–1881), p. 284). The editors of the Carlyle correspondence give the following information: '"Saint-Agnore" is presumably the village and castle of Sainte-Agnes, about 6 mi. (10 km.) NNW of Menton; the village is 2,700 ft. (800 m.) above sea level' (*CLO*).
9. *Monaco:* an independent city-state on France's Mediterranean coast, situated between Mentone and Nice.
10. *Nice:* a Mediterranean coastal city in southeastern France.
11. *Calais to Dover:* passage of the English Channel by ferry from the French coastal town of Calais to Dover in England.

To Auguste de la Rive [3 January 1867][1] 2536

Thursday

My dear Friend,

The moment I received your papers[2] I forwarded them without delay to D^r^. Francis,[3] the editor of the Philosophical Magazine. He now informs me that even before I wrote to him he had resolved on publishing translations of the memoirs. So that you see their intrinsic merits were discovered without any interference of mine.

I hope soon to be able to send you something in return in the shape of a little book on Sound,[4] which I trust will not invoke your censure.

And now I wish you a happy new year, and with kindest regards to Madame DelaRive[5] and your excellent children[6]

Believe me | Yours most sinc[er]ely | John Tyndall

Lady Ashburton[7] whom I saw a few days ago at Mentone[8] spoke to me of a very pleasant visit which she had to[9] paid to Presinges.[10]

RI MS JT/T/382
RI MS JT/1/TYP/1/365

---

1. *[3 January 1867]:* the date is given by the mention of seeing Lady Ashburton 'a few days ago at Mentone'. Tyndall travelled from Mentone to Monaco on Wednesday, 26 December and thus would still be travelling the following day, making this 'Thursday letter' fall in the next week, on 3 January.
2. *your papers:* the following papers of de la Rive's appeared in the *Phil. Mag.* in 1867: 'On the Propagation of Electricity in highly Rarefied Elastic Fluids, and in particular on the Stratifications of the Electric Light in very rare media', *Phil. Mag.*, 33 (1867), pp. 241–61; 'On the Action of Magnetism upon the Electric Discharge in highly Rarefied Gaseous Media', *Phil. Mag.*, 33 (1867), pp. 512–30; 'Note on the Electrical Condition of the Terrestrial Globe', *Phil. Mag.*, 34 (1867), pp. 322–5; 'On a Photometer for Measuring the Transparence of the Air', *Phil. Mag.*, 34 (1867), pp. 241–4; 'Michael Faraday, his Life and Works', *Phil. Mag.*, 34 (1867), pp. 409–37.
3. *Dr Francis:* William Francis.
4. *little book on Sound: Sound* (1867).
5. *Madame DelaRive:* Louise-Victoire-Marie de la Rive (née Fatio, 1808–74), the second wife of Auguste de la Rive. They married in 1855.
6. *excellent children:* the de la Rives had six children: Eugène de la Rive (*c.* 1827), Jeanne Adèle de la Rive (1829–95), William de la Rive (*c.* 1831), Lucien Charles de la Rive (1834–1924), Augusta de la Rive (b. *c.* 1836), and Alice de la Rive (b. *c.* 1838).
7. *Lady Ashburton:* Louisa Baring.
8. *Mentone:* see letter 2535, n. 4 and n. 5.
9. *to:* a writing mistake by Tyndall.
10. *Presinges:* a village in the Swiss canton of Geneva where de la Rive lived.

To the Editor of *PMG*[1] 3 January 1867 2537

CORRESPONDENCE. | THE LAWS OF NATURE. |
To the EDITOR *of the* PALL MALL GAZETTE.

SIR,—The review[2] of the "Reign of Law"[3] which has recently appeared in your excellent journal induces me to send you the following brief extract from an unfinished essay on a kindred theme:[4]—

"If you ask me who is to limit the outgoings of divine power, my answer is, 'Not I.' Should you urge that if the Maker of this universe chose to put an end to cholera, cattle plague, and bad harvests, there is nothing to prevent Him from doing so, I am not prepared to contradict you. I neither agree with you nor differ from you, because this is a subject of which I know nothing. But I would invite you to consider whether these questions of yours regarding almighty power really relate to that power as it is actually displayed in the universe, and not to the capabilities of your own imagination. Your question, I imagine, is not, Has the Omnipotent done so and so? or, Is it at all likely that under the circumstances the Omnipotent should do so and so? but, Is my imagination competent to picture a being capable of doing so and so? I am not prepared to deny your competence. To the human mind belong the faculty of enlarging and diminishing, of distorting and combining indefinitely the revelations of the senses or of its own consciousness. It can imagine a mouse as large as an elephant, an elephant as large as a mountain, and a mountain as high as the stars. It can separate congruities and unite incongruities. We see a fish and we see a woman; we can mentally drop one half of each and unite the other two halves to make a mermaid. We see a horse and we see a man; we are able to drop one half of each and unite the two remaining halves to make a centaur. Thus also the pictorial representations of the Deity; of the bodies and wings of angels; of the hoof, horns, and tail of the Evil One; of the joys of the blessed and the tortures of the damned, have been all elaborated—and for the most part unworthily elaborated—by the mythologic imagination from materials furnished by the senses. And it behoves you and me to take care that our conceptions regarding the power which rules the universe are not mere ignorant or fanciful enlargements of human power. The capabilities of what you call your reason are not to be denied. If the mythologic imagination be the measure of what is possible in nature, all things are possible—the sorceries of the middle ages, as well as the spirit-rapping of to-day.[5] In fact, this conceded, no law can withstand you, no science can trammel you. With perfect consistency you may call science 'a weapon of air.' The physical student, however, while accepting the substance of your figure, will probably reverse its parts; thinking that it is not science which is here the thing of air, but the unsubstantial figments of the fancy to which its solidity is opposed.

"Did the hold upon the human mind of this belief in the changeability of nature depend solely on the material benefits seen to be derived from it, it could not, in my opinion, last half a century. As a purely objective fact it would soon become evident that the distribution of natural phenomena is unaffected by the moral merits or demerits of man; that the law of gravitation crushes the simple worshippers of

Ottery St. Mary[6] while singing their hymns just as surely as if they had been engaged in a midnight brawl. The hold of this belief upon the mind derives its strength from the inner warmth, force, and elevation with which it is so often associated.* It is plain, however, that these feelings may exist under the most various forms. They are not limited to Protestantism; they are not even limited to Christianity. They are the common property of worshippers in every age and clime, pervading in his noblest moments the thoughts of even the scientific man. They constitute, in fact, the primal powers of man's religious nature. They find utterance in the battle-cry of the Mahometan.[7] They take form and colour in the hunting-grounds of the Red Indian, lifting both of them upon a tide of victory above the terror of the grave."

JOHN TYNDALL. | *Athenaeum Club,*[8] *Jan. 3,* 1867.

* And which are so conspicuous in the writer[9] of "Reign of Law."

PMG, 5 January 1867, p. 5

---

1. *Editor of PMG:* Frederick Greenwood.
2. *review:* 'The Duke of Argyll on the Reign of Law', *PMG,* 31 December 1866, p. 10, and 1 January 1867, p. 12.
3. *"Reign of Law":* G. Campbell, *The Reign of Law* (London: A. Strahan, 1867).
4. *brief extract from an unfinished essay on a kindred theme:* various parts of this extract are similar to parts of an article Tyndall published the following June: J. Tyndall, 'Miracles and Special Providences', *Fortnightly Review,* 7 (1867), pp. 645–60.
5. *spirit-rapping:* the communication from or with spirits by raps or knockings (*OED*). Tyndall held a special interest in denouncing spiritualism, having published an account of his experience of a séance he attended in 1864 ([J. Tyndall], 'Science and the Spirits', *Reader,* 4 (1864), pp. 725–6). On Tyndall and spiritualism, see S. Delorme, 'Physiology or Psychic Powers? William Carpenter and the Debate over Spiritualism in Victorian Britain', *Studies in History and Philosophy of Science Part C: Studies in History and Philosophy of Biological and Biomedical Sciences,* 48 (2014), pp. 57–66; R. Noakes, 'The "Bridge Which Is between Physical and Psychical Research": William Fletcher Barrett, Sensitive Flames, and Spiritualism', *History of Science,* 42 (2004), pp. 419–64; R. Noakes, 'The Sciences of Spiritualism in Victorian Britain: Possibilities and Problems', in T. Kontou and S. Wilburn (eds), *The Ashgate Research Companion to Nineteenth-Century Spiritualism and the Occult* (London: Routledge, 2012), pp. 25–54, Tyndall on pp. 42–3; and E. Wadge, 'A Fair Trial for Spiritualism? Fighting Dirty in the *Pall Mall Gazette*', in D. Clifford, E. Wadge, A. Warwick, and M. Willis (eds), *Repositioning Victorian Science: Shifting Centers in Nineteenth-Century Scientific Thinking* (London: Anthem Press, 2006), pp. 95–106. On Victorian science and spiritualism in general, see R. Noakes, *Physics and Psychics: The Occult and the Sciences in Modern Britain* (Cambridge: Cambridge University Press, 2019). Other letters in this volume that address spiritualism are: 2558 and 2560, 2811, 2812, 2813, 2814, 2815, 2817, 2819, 2820, 2822, 2823, 2824, 2825, 2876, and 2906.

6. *Ottery St. Mary:* a town in Devon, England, on the River Otter. On 2 September 1866, some months after a great fire on 25 May 1866 destroyed a substantial portion of the town, a crowd was listening to a Plymouth Brethren preacher in the ruins of a shop. The chimney collapsed, killing eight of the crowd (see R. Packer, *Long Ago in Ottery* (Exmouth: E.R.D. Publications Ltd, 1980)).
7. *Mahometan:* an adherent of the religion of the prophet Muhammad; a follower of Islam, a Muslim (*OED*).
8. *Athenaeum Club:* a private members' club located at 107 Pall Mall in London, particularly for those with intellectual interests, and accomplishments in art, science, or literature. The club was founded in 1824, at the instigation of John Wilson Croker. Humphry Davy became the first chairman and Michael Faraday the first secretary. Faraday became a member after a few months when he was able to resign the secretaryship. Tyndall became a member in 1860, Thomas Huxley in 1858, Joseph Hooker in 1851, and Thomas Hirst in 1866. On the club, see M. Wheeler, *The Athenaeum: More Than Just Another London Club* (New Haven, CT: Yale University Press, 2020).
9. *the writer:* George Campbell.

From Henry Bence Jones 4 January [1867][1] 2538

Royal Institution of Great Britain | 4 Jan[uar]y | 11.30 PM

My dear friend

I must have expressed myself very badly to day and I shall not make it any better by writing. As far as I can possibly do it I & everyone else will let you & your assistant[2] alone, but changes must of necessity occur & poor Andersons departure[3] must change arrangements that were under him.

This & nothing else has led to this subject[4] being mentioned now & it was necessary something should be done at once. Had I thought it would have troubled you I would have kept it until Saturday evening. Pray think no more about it but consider that I am sure to try to do the best I can for you & every one in the house. It is much better I think that your letter[5] did not reach Miss Barnard.[6]

I shall try & see you to morrow but please do not let a thought that I or anyone else had any desire to interfere with you enter your mind.

If I had known that you had any thought about the managers committee[7] I should have spoken about it when I became secretary.

Pray do not let this subject trouble you tomorrow or any day.

I am | Y[ou]rs ever | H Bence Jones

RI MS JT/1/J/107
RI MS JT/1/TYP/2/727

1. *[1867]:* the year is given based on the year that Charles Anderson died (1866).
2. *your assistant:* probably John Cottrell.
3. *Andersons departure:* Charles Anderson (1790–1866), a laboratory assistant at the RI. He was first employed at the RI in 1827 to assist Faraday in the project (on behalf of the Board of Longitude) of improving optical glass. With the termination of that project in 1830 Anderson became Faraday's laboratory assistant and also assisted him in his RI lectures (*Faraday Correspondence,* vol. 5, p. xlv).
4. *this subject:* not identified.
5. *your letter:* letter missing.
6. *Miss Barnard:* Jane Barnard.
7. *managers committee:* Tyndall sometimes attended Managers' meetings at the RI. The following were elected Managers of the RI on 1 May 1866: Warren De la Rue (1815–89), George Everest (1790–1866), John Hall Gladstone (1827–1902), William Robert Grove (1811–96), Caesar H. Hawkins (1798–1884), John Lubbock (1834–1913), John Carrick Moore (1805–98), William Pole (1814–1900), William Frederick Pollock (1815–88), Lewis Powell (*c.* 1797–1867), Edward Sabine (1788–1883), Philip Henry Stanhope (1805–75), Joseph William Thrupp (1799–1873), Charles Wheatstone (1802–75), and Philip James Yorke (1799–1874). Henry Holland (1788–1873) was elected President, William Spottiswoode (1823–83) was elected Treasurer, and Henry Bence Jones (1813–73) was elected Secretary ('Annual Meeting', *Roy. Inst. Proc.,* 4 (1862–6), p. 615; *ODNB*).

From Margaret Ginty 7 January 1867 2539

Rio de Janeiro | Jan. 7th 67.

Dear Dr. Tyndall,

I was very much disappointed that you did not write to Mr James[1] or the Baron[2] by this Mail, I am sure if you knew of what importance it is to me that you would have written before this time. Perhaps you are not aware of it, but the money which ought to be in your hands now is losing in interest as Mr James is leaving Rio (perhaps for good) very soon I will lose the chance of having it well invested. Our property is not yet sold so I have only this money to depend on for the education of Georgie and Fred.[3] The Baron and Mr James say they can do nothing until they hear from you. If you have not written, will you do so at once? If our property remains much longer in the hands of the Brazilian Authorities it will give me endless trouble. I will not enter into all the details for I am in hopes that your letter may be on the way. I do hope it is for there will be a chance of my getting a little settled. I am so anxious that I count the days and hours until the Mail comes in. There is a great deal that I ought to tell you but I will wait until your letter arrives. Your note written on the 12th of November[4] was very short and very vague. I see after all that no one knows any thing and even you—you are not wiser than I am. It is a dreadful mystery to me that the one so full of life and energy should be taken away when he was so much wanted.[5] To where and for what I do not know—I only know—oh so well that I am alone with this fearful dull dark void always before me. If he had only gone on a journey somewhere and I could write and tell my darling of all my troubles how different it would be. It is harder to bear every day, Oh I wish I could forget or even hope for something—but it is no use. I have lost him and I do miss him more and more. You might have given me some comfort but I see there is none to be found. I am tired with all my trouble, but it will end some time, for I feel changed and ill. For the sake of my children I must arrange my affairs as soon as possible—will you write? Until I hear from you I will say no more but remain

dear Dr. Tyndall | Yours sincerely | M. Ginty

RI MS JT/1/TYP/11/3653
LT Typescript Only

1. *Mr James:* Bartlett James.
2. *Baron:* Irineu Evangelista de Sousa (1813–89), a Brazilian industrialist who received the title of Baron (1854) and later Viscount (1874) of Mauá. William Ginty, Margaret's husband, worked for Baron Mauá after the family relocated to Rio de Janeiro in 1854. See A. Marchant, *Viscount Mauá and the empire of Brazil: a biography of Irineu Evangelista de Sousa, 1813–1889* (Berkeley: University of California Press, 1965).
3. *Georgie and Fred:* sons of William and Margaret Ginty. They also had another son, William Robert, and two daughters, Kathleen and Fanny (these three were listed on the 1851 English census).
4. *Your note written on the 12th of November:* the letter was actually from 13 November 1866. See letter 2504, *Tyndall Correspondence,* vol. 9.
5. *the one so full of life and energy should be taken away when he was so much wanted:* William Gilbert Ginty (*c.* 1820–66), Margaret's husband, and a close friend of Tyndall's, had died suddenly on 13 July 1866.

From Mary Adair 9 January 1867 2540

Heatherton Park.[1] | Jan[ua]ry 9th /67

My dear Mr. Tyndall,

Thank you very much for so kindly sending me the Lectures,[2] which seem most interesting, and will no doubt enlighten my ignorance on some of the knotty points connected with these curious asteroids. I fear, though, Mr. Herschel does not enter into the 'Meteoric Theory',[3] which is especially puzzling to me . . . but <u>as far as they go</u>, the Lectures seem most clear and interesting.

Our friends the Hamiltons[4] have all gone to Torquay[5] for a few weeks—I hope you enjoyed your visit to Mentone[6] very much. It must be such a lovely place!

Thank you too, many times for your good wishes for the New Year. We wish you the same most sincerely.

My Mother[7] joins me in kind regards, and believe me

Yours very truly | H. Mary Adair

RI MS JT/1/A/3

---

1. *Heatherton Park:* the Adair family estate in Somerset, England. Mary's brother Alexander William Adair (1829–89) inherited the estate when their father Alexander Adair (1791–1863) died.
2. *the Lectures:* J. Herschel, *Familiar Lectures on Scientific Subjects* (London: Alexander Strahan, 1866). Adair's mention of 'curious asteroids' is probably a reference to Herschel's lecture on 'The Sun' (pp. 47–90).
3. *'Meteoric Theory':* the theory that the Sun's heat and light were the result of asteroids falling into it, proposed by the German physicist Julius Robert Mayer (1814–78) (J. R. Mayer, 'On Celestial Dynamics', *Phil. Mag.,* 25 (1863), pp. 241–8, 387–409, 417–28; originally published in German in 1848). Tyndall had discussed the meteoric theory in *Heat* (1863), in lecture XII (pp. 436–9). Herschel did mention the theory in his lecture on the Sun, when writing about the source of the Sun's energy, but indeed as Adair wrote, 'does not enter into' it: 'The theory at present current of it is founded on what may not unfairly be considered a further development of this idea, the friction being supposed to arise from meteoric matter circulating round the sun, and gradually subsiding into it, and either tearing up its surface, or ploughing into its atmosphere. But on this we cannot dilate, as nothing has been hitherto said about the appearance of the sun in telescopes, and the strange phaenomena its surface, so examined, exhibits' (J. Herschel, *Familiar Lectures on Scientific Subjects,* p. 74). For more on the meteoric theory, see F. A. J. L. James, 'Thermodynamics and Sources of Solar Heat, 1846–1862', *British Journal for the History of Science,* 15 (1982), pp. 155–81; and H. Kragh, 'The Source of Solar Energy, ca. 1840–1910: From Meteoric Hypothesis to Radioactive Speculations', *European Physical Journal H,* 41 (2016), pp. 365–94.
4. *Hamiltons:* not identified. In his journal entry for 28 November 1869, Tyndall wrote that 'Years and years ago I remember her friend Miss Hamilton sending me down to Ventnor a draft of Miss Adair's notes of my lecture' (RI MS JT/2/13c/1357), so it is possible the Hamiltons in question had a daughter who attended Tyndall's lectures.
5. *Torquay:* a seaside resort town in southeast England, on the English Channel.
6. *your visit to Mentone:* see letter 2535, n. 4 and n. 5.
7. *My Mother:* Harriet Adair.

From Charles Lyell[1] 11 January 1867 2541

73, Harley St.[2] | London. W. | Jan[uar]y 11 | 1867

My dear Dr Tyndal

Our friend Hooker has been reading Ch. 13 of my new (or 10th) Ed[itio]n. of my Principles of Geology[3] & is afraid that you might think that what I have said at p. 288[4] of the Glacial Period having been a hot one, or of the error of those who have positively held that doctrine, was intended as an objection to you. I therefore write to say that not only I did not allude to you whether by name or in thought but that I always felt that your proposition[5] was a sound one, namely that the greater the quantity of snow the greater must have been the quantity of heat required to produce so much vapour.

The writer of whom I was thinking was Dr Frankland[6] & as his paper[7] was a strictly geological one severely handled by Prof[esso]r Ramsay in the (Saturday?) review[8] my friends understand what I meant. You may remember how Dr F. warmed up the sea round Norway to a subtropical heat at the very period when we know paleontologically that the waters were quite as cold as the sea is at present on the north side of Spitzbergen.[9] Our evidence being derived from fossil shells of living species found as far south as the south of Sweden it was a great oversight of Franklands.

Others have since fallen into the same mistake & therefore I found it necessary to allude to the theory, but I did not think myself called upon to name Dr Frankland. So far as I remember what you have said you have not applied your doctrine of a great snow-fall implying great evaporation in such a manner as any geologist can object to. But had I imagined that any one could have supposed that I alluded to you I might easily have stated that I did not & could not controvert your proposition by which I & others have profited greatly. I shall take an early opportunity of showing that I agree with you without alluding to Dr Frankland's very rash paper.

I do not believe that you would have objected to what I have said at p. 289[10] if you had seen it in proof.

believe me | ever most t[ru]ly y[ou]rs | Cha Lyell

Prof. Tyndal[11]

RI MS JT/1/L/53
RI MS JT/1/TYP/3/852–3

---

1. *Charles Lyell:* Charles Lyell (1797–1875), a Scottish geologist and lawyer, whose *Principles of Geology,* 3 vols (London: John Murray, 1830–3) advocated a 'uniformitarian' view of the processes of geological change. Educated at Exeter College, Oxford, he studied law at Lincoln's Inn, London. He was briefly professor of geology at King's College, London

(1831–3), after which he lived on private means. He married Mary Elizabeth Horner in July 1832; she assisted her husband in his scientific work (*ODNB*).

2. *73, Harley S^t.:* Lyell's residence from 1854–75.
3. *my new (or 10th) Ed[itio]^n. of my Principles of Geology:* C. Lyell, *Principles of Geology: or the Modern Changes of the Earth and its Inhabitants,* 10th edn, 2 vols (London: John Murray, 1867).
4. *what I have said at p. 288:* 'Several writers have fallen into the strange error of supposing that the Glacial Period must have been one of a higher mean temperature than usual, because an excess of snow implies an excess of evaporation, and consequently of heat. This fallacy has arisen from omitting the element of time from the calculation. No doubt the formation of so much snow requires that a corresponding quantity of water should be converted by heat into vapours; but we must take into account the number of years over which the process of distillation was spread. If the summer's warmth cannot get rid of all the winter's snow, even by a few feet in a century, there will, in the course of a thousand years, be as large a store of ice formed as geologists may require' (C. Lyell, *Principles of Geology,* p. 288).
5. *your proposition:* in lecture VI of *Heat* (1863), Tyndall wrote of the 'solar action involved in the production of ancient glaciers' (pp. 205–7). He noted, 'Cold will not produce glaciers. You could have the bitterest north-east winds here in London throughout the winter without a single flake of snow. Cold must have the fitting object to operate upon, and this object—the aqueous vapour of air—is the direct product of heat' (p. 206).
6. *D^r. Frankland:* Edward Frankland.
7. *his paper:* E. Frankland, 'On the Physical Cause of the Glacial Epoch', *Phil. Mag.*, 27 (1864), pp. 321–41. Read at the RI on 29 January 1864. Frankland's paper was critiqued in J. Croll, 'On the Physical Cause of the Change of Climate during Geological Epochs', *Phil. Mag.*, 28 (1864): pp. 121–37.
8. *severely handled by Prof[essor]^r Ramsay in the (Saturday?) review:* see the unattributed essay written by Andrew Crombie Ramsay, 'Changes in Climate', *Saturday Review,* 17 (1864), pp. 591–2, 719–21 (A. Geikie, *Memoir of Sir Andrew Crombie Ramsay* (London: MacMillan and Co., 1895), p. 370; *ODNB*).
9. *Spitzbergen:* the largest island of the Svalbard archipelago in northern Norway.
10. *289:* Lyell mistakenly wrote '289' instead of '288'.
11. *Tyndal:* misspelled by Lyell.

From Mary Adair [14 January 1867][1] 2542

Heatherton Park.[2] Monday.

Dear M^r. Tyndall,

I should fear to presume upon your kindness were I not sure that my difficulty will be none at all to you.

The Meteoric Theory[3] is very grand and beautiful but very puzzling. I cannot reconcile it with what I have been told is <u>the</u> theory of Aerolites.[4] Lately at Worthing[5] M^r^. Read the Astronomer[6] told us that Aerolites are the "fragments of a shattered planet, held together in groups by a feeble attraction & pursuing paths parallel to their original orbit". Now either (as I believe) M^r^. Read's theory was erroneous, or else the Meteoric theory would be broken down for want of sufficient supply.

(1). Is it not unphilosophical to imagine that at least <u>one</u> planet has been destroyed to give fuel to the Sun which is the Mainspring of life to all the fellow planets? Especially as in all likelihood the Suns of all the other systems are supported by similar means.

(2). But again, you have proved to us that the destruction of all the planets & moons in one system would only cover the Sun's radiation for less than 50,000 years. It is therefore clear that the débris of a single planet would only suffice it for a much shorter period.

50000 years however is but a fraction of the ages reckoned by geologists to have elapsed since the foundations of the round world were laid.

<u>Many times</u> the <u>mass</u> of all the planets & moons must of necessity have been consumed...

Suppose the aerolites to have had an origin similar to the planets is it not singular that such an immense amount of small planets should have been generated in proportion to the very few large ones?

I think these are the chief points that puzzle me—smaller difficulties would be easily solved if I could grasp the main ones.

Hoping you will not think me very troublesome, and with our kind regards,[7] believe me

Sincerely your's | H. Mary Adair

RI MS JT/1/A/9

---

1. *[14 January 1867]:* the date is given based on a note on the letter from LT: 'prob[ab]ly Jan 14, 1867', as well as relation to letters 2543 (15 January) and 2544 (20 January).
2. *Heatherton Park:* see letter 2540, n. 1.
3. *Meteoric Theory:* see letter 2540, n. 3.
4. *theory of Aerolites:* as Tyndall would write in a later reply, 'The notion of the asteroids being the fragments of a larger planet' (see letter 2544).
5. *Worthing:* a seaside town in West Sussex, England.
6. *M^r^. Read the Astronomer:* William Read (*c.* 1799–1884), a Manchester tobacco merchant who later became a clergyman. He was a Fellow of the Royal Astronomical Society and Royal Microscopical Society and served as chaplain of the Chapel of Ease, Worthing, Sussex from 1851–82 (*Alum. Cantab.*).
7. *our kind regards:* in other letters, Adair would close with regards from her mother.

To Mary Adair [15 January 1867][1] 2543

Tuesday

My dear Miss Adair.

Give me grace until after Friday evening next, when I preach a sermon which I am now preparing,[2] and then I will go into your difficulties.[3] I do not promise to solve them to your satisfaction, but what you want really is a frank statement of the limits of human knowledge regarding the meteors. If that knowledge should appear incomplete, as it assuredly will do, you must only bow your proud head and accept it as it is. Taking comfort however from the hope that a brighter light will dawn by and by.

Give my very kind regards to your mother[4] and believe me always

Yours most since[rel]y | John Tyndall

RI MS JT/1/T/91
RI MS JT/1/TYP/1/7

---

1. *[15 January 1867]:* the date is given based on a note on the typescript letter from LT: '(Post mark on the envelope) "Jan.15.67"'.]
2. *when I preach a sermon which I am now preparing:* on 18 January 1867, Tyndall gave a Friday Evening Discourse at the RI, 'On Sounding and Sensitive Flames' (*Roy. Soc. Proc.,* 5 (1866–9), pp. 6–12).
3. *I will go into your difficulties:* see letter 2542.
4. *your mother:* Harriet Adair.

To Mary Adair 20 January [1867][1] 2544

Sunday morning 20th Jan.

My dear Miss Adair.

The notion of the asteroids being the fragments of a larger planet arose thus.

After Kepler's[2] time many attempts were made to find some law regulating the distances of the planets from the sun

A Professor named Titius[3] who lived at Wittenberg[4] found the following curious relation to exist between the distances of the planets. The relation was afterwards published[5] by an astronomer named Bode,[6] and after him was called the law of Bode[7]

Excuse this blot, and take the numbers

0, 3, 6, 12, 24, 48, 96, 192, 384

Each of these numbers you will observe, except the second, is double of that which precedes it.

Add 4 to each of these numbers, we obtain the following series:—

4, 7, 10, 16, 28, 52, 100, 196, 388

These numbers nearly represent the distances of the planets from the sun.

The following table will show you this:—

Names of planets	True relative distance from sun	Distance according to Bode's law.
Mercury	3.87	4.0
Venus	7.23	7.0
Earth	10.0	10.0
Mars	15.23	16.0
		28.0
Jupiter	52.03	52.0
Saturn	95.39	100.0
Uranus	191.82	196.0

You see between Mars and Jupiter exists a great gap: we have a distance according to the law of Bode but we have as yet no planet answering to this distance.

It is a most extraordinary fact that examination of the space between Mars & Jupiter subsequently revealed, not the existence of a large planet, but of that group of asteroids so many of which have been discovered by Mr. Hind.[8] Now to fill this gap a planet was expected—and inasmuch as instead of a planet this group of asteroids was discovered, the supposition at once started into the mind that they were the fragments of the missing planet.

When however the problem is closely considered there are the strongest mechanical grounds for inferring that they are not the fragments of a single planet.

All this is apart from the meteorites—but it shows you how this notion of an exploded planet has arisen. As regards the meteorites themselves there is no ground for assigning them to such an origin.

How matter was first introduced into space transcends the power of man to determine. Here it is, and we can but go back to a more remote state of things without ever reaching the most remote. The nebular hypothesis[9] carries us back to a bewildering extent—the hypothesis namely that the sun & planets are due to the condensation of nebulous matter. I looked a few evenings ago at the nebula of Orion—There is something inexpressibly solemn in the aspect of those veils of matter, and in the thoughts which they excite. But without going so far back as the nebular hypothesis, I should be inclined to reverse the views of the astronomer you mention and to regard planets as formed by the collision of meteoric masses rather than that meteors are formed by the explosion of planets. Given the dead dark matter in space with no form of force except that of gravitation, it is demonstrable that out of that dead dark matter suns & planets could be formed. The velocity imparted by the pull of gravity would produce this effect. Now whether the solid masses of the planets were ever thus scattered through space as little pocket planets or not is a matter of pure hypothesis—It is incapable of demonstration—but we can demonstrate that if they were thus scattered the attraction of gravity w[oul][d]. be competent to produce the present state of things. The sun may have been kindled by this collision—the earth may have been once molten through a similar cause. We have to choose between this hypothesis and other hypotheses, or between this hypothesis and no hypothesis at all. Supposing all the meteorites that have fallen upon the earth during the last thousand million years to be carted away into space I should like to know how the size of the earth would be affected. It might then cease to be a large planet. The "generation" of large planets may in fact be due to the agglomeration of small ones. And in this light your difficulty as regards the enormous proportion in point of numbers of the latter to the former would disappear.

To sum up. With regard to the condition of the Universe which preceded its present condition we can only make guesses more or less wise. But we are able to assign a state of things which if it existed would be demonstrably followed by the present state of things. As knowledge grows & light brightens we shall probably arrive at greater certainty in these matters than we now possess. All that we now know is that from the cold dark matter of the Universe, and the action of the law of gravitation the pomp of the starry heavens might be derived. Good bye, and remember if you think I can further help you write to me again.[10]

Yours ever | John Tyndall.

RI MS JT/1/T/90
RI MS JT/1/TYP/1/8–11

1. *[1867]:* the year is given based on relation to letters 2542, 2543, and 2545.
2. *Kepler's:* Johannes Kepler (1571–1630), an astronomer and mathematician best known for his three laws of planetary motion and for his support of Copernican heliocentrism (*CDSB*).
3. *Titius:* Johann Daniel Titius (1729–96), a German astronomer and professor at the University of Wittenberg, best known for proposing the Titius–Bode law (see n. 7) (*CDSB*).
4. *Wittenberg:* a town in eastern Germany.
5. *published:* J. E. Bode, *Anleitung zur Kenntniss des Gestirnten Himmels* (Hamburg: Harmsen, 1768).
6. *Bode:* Johann Elert Bode (1747–1826), a German astronomer and director of the Berlin

Observatory (1786–1825). Known for his elaboration of the Titius–Bode law (see n. 7), and for his role in William Herschel's discovery of Uranus in 1781 (*CDSB*).

7. *law of Bode:* the Titius–Bode law (sometimes, simply Bode's law) describes the geometrical progression in the distance between planets, suggested by Johann Titius (1766) and popularized by Johann Bode (1772). The law successfully predicted the location of the orbits of both Uranus (1781) and Ceres (1801).
8. *M^r. Hind:* John Russell Hind (1823–95), an English astronomer known for his discovery of ten asteroids and his superintendence of the *Nautical Almanac* from 1853–91 (*ODNB*).
9. *nebular hypothesis:* a model for the formation and evolution of the Solar System, now widely accepted. First proposed by Immanuel Kant in the mid-eighteenth century and independently suggested by the French mathematician Pierre-Simon Laplace in 1796, the model explains the formation of stars and planets as matter condensing through gravitational force from a cloud, or 'nebula', of gas and dust.
10. *write to me again:* Adair's reply is letter 2545.

From Mary Adair [22 January 1867][1] 2545

Heatherton.[2] Tuesday

My dear M^r. Tyndall,

I am extremely obliged to you for all the trouble you have taken[3] to explain my difficulties[4] ∴[5] and, thanks to your kindness, I begin to perceive my way much more clearly towards their solution. It is a great help to know where the knowledge of Facts ends, & Hypothesis commences—and now I shall hope with patience gradually to unravel the entanglements of the subject.

Your idea about the formation of the large planets was quite new to me. And I feel very stupid not to have realized it before. Of course that quite accounts for my puzzle about the enormous quantity of the meteors. My mistake arose from my having imagined that according to the Nebular hypothesis[6] the fire mist condensed at a few foci, and that each focus was the centre of one of the larger planets.

I shall much enjoy reading the Paper[7] you have been good enough to send me—and wish there had been any chance of our being in Town while the Lectures on Vibration are delivered.[8]

Again thanking you, believe me Very truly your's | H. Mary Adair.

RI MS JT/1/A/22

1. *[22 January 1867]:* the date is given based on a note on the letter from LT that states this letter is a reply to Tyndall's letter of 20 January 1867, a Sunday, making this 'Tuesday' letter from 22 January.

2. *Heatherton:* see letter 2540, n. 1.
3. *the trouble you have taken:* letter 2544. Adair had asked Tyndall to explain the Meteoric theory (letter 2542). He replied (letter 2543) asking for some time to answer her questions as he was preparing a lecture, which he did so in letter 2544.
4. *to explain my difficulties:* see letter 2542.
5. ∴ : symbol for *therefore.*
6. *Nebular hypothesis:* see letter 2544, n. 9.
7. *Paper:* not identified.
8. *the Lectures on Vibration are delivered:* Tyndall delivered twelve lectures 'On Vibratory Motion, with Special Reference to Sound' at the RI in 1867, on Tuesday and Thursday afternoons from 22 January to 28 February (see RI Guard Book, vol. 3, p. 59; and *Roy. Inst. Proc.,* 5 (1866–9), p. 4).

To Louisa Baring [20 or 27 January 1867][1] 2546

London Sunday morning

Dear Lady Ashburton,

Your very brief but very precious little letter[2] is now lying before me. I have never forgotten it, and should have answered it long ago had I not been overwhelmed with work.

I had a prosperous voyage to Nice,[3] and thence a drowsy journey of five and twenty hours to Paris. I passed from the glorious sunshine of Mentone[4] to the *[1 word illeg]* fog of Lyon. It was an exact reversal of our journey to Mentone. At Macon[5] the end of the carriage facing my coupé was sheeted with ice.

In Paris I stayed a day—and by a night boat passed over to Dover.[6] We were awfully knocked about, I hating the crowded cabin stayed upon deck and there endured all the tortures of vibratory motion, with an occasional splash of brine to keep me awake.

We have had extraordinary weather in London since. You have heard all about it in the newspapers. You have also heard of the terrible accident in Regent's Park.[7] It makes one smile rather grimly to hear the *[1 word illeg]* liberal *[1 word illeg]* of self government calling out for supervision in the Parks. Had the Police or Park Keepers kept the people off the ice they would probably have been snubbed by these same papers.

You surely must have had your share of this weather at Mentone—perhaps not. I can believe anything of Mentone. Morally and physically it was so beautiful to me that I can fancy a cloud never dims the heavens there.

I was so pleased to hear that the philosopher[8] was well. He <u>ought</u> to be—the greeting he received on the night of his arrival would "wake the dead". I have been thinking how he is to be got back. My lectures[9] forbid me from quitting London. Otherwise I would volunteer to take charge of you all. It would be very pleasant to me. But I have been arranging with another of the General's Lieutenants,[10] who is free from the 19th to the 29th of April. And who will join him in that interval—if he will—and will carry him through the row of Paris. No man will *[1 word illeg]* him more tenderly—no man will administer brandy & water & tobacco more judiciously. In short he could not be in better hands. My fellow Lieutenant is a far better nurse than I am, and will never be in the least fractious—The General knows him a little—not much—His name is Hirst. Hirst offers to take charge of him, and engages to carry him triumphantly from the Villa Madonna to 5 Cheyne Row.[11]

I hope the pipes[12] arrived all safe–if a single one of them was broken I shall be awfully humiliated.

I hope little May[13] is well. When quitting the carriage at Monaco,[14] I ventured timidly to kiss her hand, but on the landing place, where you could not see us, I stole a real kiss. I wish you would repeat it for me.

Goodbye—vive May! vive le Philosopher—vive L[ad]y Imperatrice! I still hold on to my theory

Yours ever | John Tyndall

I have opened the envelope to express my sympathy for the hand I hope it is well.

NLS, Ashburton Papers, Acc. 11388 N0.101

1. *[20 or 27 January 1867]:* the date is given based on both the mention of the accident in Regent's Park on 15 January 1867 and Tyndall's recent trip to Mentone in December 1866. Sundays following 16 January were 20 or 27 January.
2. *Your very brief but very precious little letter:* letter missing.
3. *Nice:* see letter 2535, n. 10.
4. *Mentone:* see letter 2535, n. 4 and n. 5.
5. *Macon:* a city in the Burgundy region of France.
6. *Dover:* see letter 2535, n. 11.
7. *terrible accident in Regent's Park:* on 15 January 1867, 40 people died when the ice cover of the lake in London's Regent's Park collapsed ('The Catastrophe in the Regent's Park', *Times,* 22 January 1867, p.12).
8. *the philosopher:* Thomas Carlyle.
9. *My lectures:* see letter 2545, n. 8.
10. *another of the General's Lieutenants:* Thomas Hirst.
11. *from the Villa Madonna to 5 Cheyne Row:* Villa Madonna was the house in Mentone that Baring rented (V. Surtees, *The Ludovisi Goddess: The Life of Louisa Lady Ashburton* (Michael Russell, 1984), p. 126), and 5 Cheyne Row was Carlyle's home in Chelsea, central London since 1834.

12. *pipes:* see letters 2549, 2550, and 2553.
13. *May:* Mary Baring.
14. *Monaco:* see letter 2535, n. 9.

From Anne Guthrie[1] 21 January [1867][2] 2547

The Deanery, Hereford. | January 21st

Dear Mr. Tyndall

The dear, everybody calls him the dear, Dean[3] was very much affected by the warmth of your letter[4] and he has read it several times, and he wishes me to tell you that there is no man living for whom he has a more affectionate regard, and that no day passes without his thinking of you, and Frankland,[5] and all the pleasant genial hours you had passed together.

I had written so far when the Dean sent for me again to tell me to be sure "to say everything that was kindest to Tyndall and not to forget to add that he could not read his letter without tears". I can do no better than give his own words.

He is better today, bright and cheerful, but alas there is no hope of recovery. He has for some weeks been gradually losing the use of his left arm from some kind of paralysis, caused by the want of vital power, and it is slowly extending down the same side. These words are easily written, but they have a terrible meaning! Still we hope the danger is not immediate, his mind is quite clear, he has little or no pain and is only too resigned, and tries to teach his own beautiful resignation to all around him. With the warmest regards of the Dean and my sister,[6] believe me, dear Mr. Tyndall,

Very truly yours | Anne L. Guthrie.

RI MS JT/1/TYP/1/327B
LT Typescript Only

---

1. *Anne Guthrie:* Anne Leonora Guthrie (*c.* 1807–93), sister of Mary Helen Dawes (see n. 6), and daughter of the surgeon George James Guthrie (1785–1856) ('Deaths', *Morning Post,* 1 April 1893, p. 1).
2. *[1867]:* the year is given based on LT noting the year on her typescript, but also that Richard Dawes was soon to die.
3. *Dean:* Richard Dawes (1793–1867), Dean of Hereford, an educationalist concerned with improving secondary schools in England (*ODNB*). Dawes was Rector and Headmaster of King's Somborne near Queenwood, where he originally met Tyndall, before moving to Hereford in 1850. Dawes died soon after this letter, on 10 March 1867.
4. *your letter:* letter missing.

5. *Frankland:* Edward Frankland.
6. *my sister:* Mary Helen Dawes (née Gordon), daughter of Alexander Gordon, Esq. of Logie, Aberdeenshire and step-daughter of the surgeon George James Guthrie (1785–1856). She married Richard Dawes in 1836 (Westminster Marriages Record; the *ODNB* entry for Richard Dawes incorrectly states 1838).

From Margaret Ginty 22 January 1867 2548

Rio de Janeiro | January 22nd 1867

Dear Mr. Tyndall,

Another Mail has come in and still no letter from you.[1] Perhaps you do not know that there are but two Mails from England every month. One from Southampton[2] on the 8th and another viâ Bordeaux[3] on the 16th.

Mr. James[4] is going to Europe in a couple of months and I want to use his influence whilst I can—he was a great friend of my poor husband's[5] and I think he is ready to do any thing in his power for me. If I can get our property out of the hands of the judges here as quickly as possible it will save me a great deal of trouble. Mr. James can do nothing until you answer his letter which he wrote last July.[6] This country is in such a bad state now that I cannot sell our place without making a sacrifice, so they advise me to let it if I can get a safe tenant for it. Before I can do so I must I believe pay all the debts. This I cannot do without drawing some of the money in England. I do not quite understand how it is to be done, but Mr. James will manage it when he hears from you. So far I am quite puzzled to know what to do. I am not quite certain am I justified in keeping my two boys[7] at school—our house is not let and I am getting no income. As I am living at present depending on others is almost intolerable to me. I am very sorry to give you so much trouble but I cannot help it. The longer my affairs are left in the hands of the Brazillian Authorities the much more difficult it will be for me, when I am more settled I can I know get the Baron de Mauá[8] to use his interest for me and place Georgie with some Mechanical Engineer. I find it hard to have to think of all these things now after having had one to anticipate my wishes. If I could only blot out the last few months and live my happy life over again. Ah how different I would feel—but no the sad reality is too plain I can feel nothing but the greatest rebellion when I see useless people left and one so good so true and loving taken away. Please do write to Mr. James for he can do a great deal for me.

Pray do not consider me too troublesome
and believe me | dear Dr. Tyndall | Yours sincerely | M. Ginty.

RI MS JT/1/TYP/11/3654
LT Typescript Only

1. *still no letter from you:* see letter 2539.
2. *Southampton:* a city on the south coast of England in the county of Hampshire.
3. *Bordeaux:* a port city in southwestern France.
4. *Mr. James:* Bartlett James.
5. *my poor husband's:* William Ginty (see letter 2539, n. 5).
6. *his letter which he wrote last July:* letter missing (presumably a letter from Bartlett James to Tyndall from July 1866, but there are none in *Tyndall Correspondence,* vol. 9).
7. *my two boys:* George and Fred Ginty. The Gintys also had another son and two daughters (see letter 2539, n. 3).
8. *Baron de Mauá:* Irineu Evangelista de Sousa (see letter 2539, n. 2). In the typescript letter, LT had mistranscribed 'Mauá' as 'Manâ'.

From Louisa Baring [before 26 January 1867][1] 2549

Mentone[2] | Monday—

Dear Professor Tyndall,

I enclose you the address where Mr Carlyle's[3] pipes &c. are to be sent.

Rev[eren]d. R. Wingfield Digby[4] | Mentone. | care of Lady Smith[5] | 30. Berkeley Square. | London.

Mentone is very dull without you. The beloved old philosopher[6] is well, & accommodating himself to his circumstances. We are anxiously looking for your news[7]—which please send us soon. I have hurt my hand, & cannot say more today, than that I am ever

Yours most truly | L. Ashburton

Elton Hall, Ashburton Collection
RI MS JT/TYP/1/131

1. *[before 26 January 1867]:* the date is given based on the reference to sending pipes to Carlyle. In letter 2550 dated 26 January 1867, Carlyle wrote to Tyndall that 'Weeks ago I received from you one of the briefest and kindest of letters; announcing that your man had packed me 24 unexceptionable pipes, with other bountiful equipments of foreign existence, and sent them punctually to their appointed address'. Baring wrote again to Tyndall concerning the pipes in letter 2553.
2. *Mentone:* see letter 2535, n. 5.
3. *Mr Carlyle's:* Thomas Carlyle.
4. *Rev[eren]d. R. Wingfield Digby:* probably Richard Henry Wingfield Digby (1837–76), rector of Thornford (B. Burke (ed.), *A Genealogical and Heraldic History of the Landed*

Gentry of Great Britain & Ireland, 7th edn, 2 vols (London: Harrison and Sons, 1886), vol. 1, p. 515).

5. *Lady Smith:* Frances Smith (née Pinney, 1803–95), Lady Smith, daughter of John Frederick Pinney, married John James Smith (1800–62), 3rd baronet in 1828 ('Deaths', *Times*, 3 June 1895, p. 1).
6. *beloved old philosopher:* Thomas Carlyle.
7. *your news:* possibly a reference to Tyndall's involvement with the Eyre Defence Committee (see letters 2561, 2563, 2565, 2566, 2567, 2571, 2572, and 2576).

From Thomas Carlyle 26 January 1867 2550

MENTONE,[1] 26 January 1867.

Dear Tyndall,

Weeks ago I received from you one of the briefest and kindest of letters;[2] announcing that your man[3] had packed me 24 unexceptionable pipes,[4] with other bountiful equipments of foreign existence, and sent them punctually to their appointed address. All is right, I dare say, and the Package or Packages will arrive: but in the meanwhile not a whisper of their Bearer or of them is heard ever since; and I decide to wait no longer for the perfect development of that, as my intention was, but to write at once. Not to "thank" you only, which you might justly count <u>superfluous</u>;—but to say (what is really a bit of practical history, and ends record) that your black lambskin "<u>breeches with one leg</u>", your Glacier journey-breeches, were noticed by me here, laid by among <u>my</u> portmanteaus,[5] a day or two after your departure;—noticed, I may add with a pang, and with a strict determination to restore them to the owner, if I lived to meet him again. This very night I will inquire of Bailey,[6] and see them again with eyes;—for, sinner that I am I doubt if I said anything on first discovery of the article; and now it has been laid away somewhere, and I ought to know that it is in some safer place!

Seriously the real meaning of my writing at first is, to assure you, what I know will be perceptibly delightful and ought to be so, that your heroic journey out hitherward in charge of me has not been a labour in vain, but promises, by all the symptoms hitherto, to be a real profit to me. Our climate fairly passes expectation,—today is radiant and beautiful as leafy June; the airy pinnacles and olive-coated heights a thing to fall in love with:—in the black tumult of the elements elsewhere, we have not had a touch of "bad weather", now and then a plunge of warm <u>rain</u>, calm, perpendicular, grandly copious, constant for many hours, I don't call this <u>bad</u>, but good too, and it makes everything else twice as good on the morrow! In about a week after you landed me I had, as you predicted, got all my bits of adjustments made, and ever since I have been going on what you would call well,—the continual breathing and bathing in such an atmosphere gradually making itself sensible even on such a broken-down subject as me. It is certain I have not had as much sleep in the time, for a long while back, and sometimes I almost fancy there may be before I have done here, an actual fraction of improvement in general health achieved; which would be an undeserved blessing to me; and to the friendly heart, I know well, exceeding great reward. This is all I had to say at present. Lady A.,[7] and your little May[8] are well, often speaking of a certain distinguished Professor, and wishing to see more and more of him!

Yours always, | T. Carlyle.

RI MS JT/1/TYP/1/200–2
LT Typescript Only

---

1. *MENTONE:* see letter 2535, n. 4 and n. 5.
2. *briefest and kindest of letters:* letter missing. Carlyle had gone with Tyndall to Mentone in December 1866 (see letter 2535, n. 4) and remained there after Tyndall returned to London. In a December 1866 letter to Roderick Murchison, Tyndall requested to secure for Carlyle the ability to have in France specific pipes, tobacco, and brandy (see letter 2528, *Tyndall Correspondence,* vol. 9). See also this volume, letters 2549 and 2553.
3. *your man:* not identified.
4. *24 unexceptionable pipes:* see also letters 2549 and 2553. Carlyle wrote to his wife Jane on 24 February 1868, referring back to 10 February: 'The Pipes came, that wet sunday, as I perhaps told you: 24 of them, *charge* 12 francs, ab*t* 4 *d* each—after waiting some 6 weeks on a Lady Smith that was to bring them *gratis:* but put off, put off,—just *till* the very day the Pipes had set out on their own footing' (*CLO*).
5. *portmanteaus:* a case or bag for carrying clothing and other belongings when travelling (*OED*).
6. *Dailey:* James Dailey, Louisa Baring's butler (*CLO*).
7. *Lady A.:* Louisa Baring.
8. *your little May:* Mary Baring.

From Wilhelm Weber 26 January 1867 2551

Göttingen 26 Jan. 1867.

Hochverehrter Freund!

Beiliegend übersende ich Ihnen ein Portrait von Chladni, welches in der Ausführung zwar Manches zu wünschen übrig lässt unter den mir bekannten Portraits aber ihm am ähnlichsten ist. Es ist in demselben Verlage (Breitkopf u Härtel in Leipzig) wie seine deutsche Akustik erschienen und ist viel besser als das in diesem Werke ursprünglich enthaltene Bild. Doch ist es möglich, dass in Ihrem Exemplare der Akustik dieses neuere Bild schon enthalten ist, wenn dasselbe nämlich erst nach dessen Erscheinen von der Buchhandlung verkauft worden wäre.—Jedenfalls kann ich versichern, dass kein besseres Bild von Chladni hier existiert.

Ausser diesem Bilde und dem älteren in der Akustik sind mir nämlich noch zwei französische Portraits bekannt, ein kleineres, von Chretien in Paris im Jahre 1809, und ein grösseres lithographirtes und in Braun gedrucktes, von Louis de Montmorillon nach der Natur gezeichnet. Letzteres ist besser als ersteres, aber weniger ähnlich als das hier beigelegte.

Ich füge noch bei, dass ich Chladni persönlich wohl gekannt und in seinen späteren Jahren oft gesehen habe. Noch länger und näher mit ihm befreundet war mein älterer Bruder, Prof. E. H. Weber in Leipzig, von dem ich dieses Bild erhalten habe und welcher dem obigen Urtheile vollkommen beistimmt.—Auch Prof. Wöhler hat Chladni gut gekannt und erkannte ihn augenblicklich, als ich eben dieses Bild ihm zeigte, und erklärte es für das ähnlichste was er gesehen.

An dem von Ihnen unternommenen Werke nehme ich das grösste Interesse und freue mich herzlich über die wohlverdiente Anerkennung, welche Sie Chladnis scharfsinnigen und genialen Forschungen zu Theil werden lassen,—um so mehr als Chladni mir von Jugend auf persönlich Führer und Vorbild auf dem Wege der Naturforschung gewesen und noch in seinen letzten Lebensjahren durch freundliche Anregung und Aufmunterung auch meine geringen wissenschaftlichen Bestrebungen, wie ich es nicht dankbar genug erkennen kann, gefördert hat.

Empfangen Sie freundlichst den Ausdruck der aufrichtigsten Hochachtung

Ihres | treu ergebenen | Wilhelm Weber.

Göttingen[1] 26 Jan. 1867.

Esteemed Friend!

I am sending you enclosed a portrait of Chladni,[2] which in its execution admittedly leaves something to be desired, but among the portraits I know of is most like him. It appeared in the same publishing house (Breitkopf and Härtel in Leipzig[3]) as his German Akustik[4] and is much better than the picture originally contained in that work. However, it is possible that this new picture is already contained in your copy of the Akustik,[5] namely if the latter had not been sold by the bookshop until after the former's appearance.—In any case, I can affirm that no better picture of Chladni exists here.

Apart from this picture and the older one in the Akustik, I know of two more

French portraits, a smaller one, by Chretien in Paris in 1809,[6] and a larger lithographed one, printed in brown, by Louis de Montmorillon and drawn from nature.[7] The latter is better than the former, but of lesser likeness than the one enclosed here.

I may further add that I knew Chladni well personally and often saw him in his later years. My older brother, Prof. E. H. Weber in Leipzig,[8] from whom I obtained this picture, and who completely agrees with the above view, was an even older and closer friend of his.—Prof. Wöhler[9] also knew Chladni well and recognised him immediately when I showed him this picture just now, and declared it the best likeness that he had seen.

I take the greatest interest in the work you have undertaken and am thoroughly pleased at the well-deserved recognition which you have given to Chladni's astute and ingenious research,[10]—all the more as, from my youth, Chladni was my personal guide and model on my path to the natural sciences, and even in the final years of his life also promoted, through kind suggestion and encouragement, my humble scientific endeavours in a manner I cannot recognise gratefully enough.

Accept most kindly the expression of sincerest respect from

Your | faithfully devoted | Wilhelm Weber.

RI MS JT/1/W/15

1. *Göttingen:* a city in central Germany, where Weber was professor of physics at the University of Göttingen.
2. *Chladni:* Ernst Chladni (1756–1827), a German physicist and musician, best known for his work in acoustics (*CDSB*).
3. *Breitkopf and Härtel in Leipzig:* a music publisher founded in Leipzig in 1719 by Bernhard Christoph Breitkopf (1695–1777), which became Breitkopf and Härtel in 1796 when the firm was purchased by Gottfried Christoph Härtel (1763–1827) (*ADB*).
4. *German Akustik:* E. F. F. Chladni, *Die Akustik* (Leipzig: Breitkopf and Härtel, 1802).
5. *your copy of the Akustik:* probably E. F. F. Chladni, *Die Akustik,* 2nd edn (Leipzig: Breitkopf and Härtel, 1830).
6. *a smaller one, by Chretien in Paris in 1809:* Gilles-Louis Chrétien (1754–1811), a French musician and engraver who invented the physionotrace, an instrument for making profile portraits. See *Bryan's Dictionary of Painters and Engravers,* rev. G. Williamson, 4th edn, 5 vols (New York: Macmillan, 1903–5), vol. 1, p. 292. A copy of the portrait produced by Chrétien in 1809 is held at the Bibliothèque nationale de France, and can be viewed at https://gallica.bnf.fr/ark:/12148/btv1b8416444r.
7. *larger lithographed one, printed in brown, by Louis de Montmorillon and drawn from nature:* Ludwig Albert von Montmorillon (1794–1854), a German-French lithographer born in Erlangen, Bavaria to parents who had fled the French Revolution.

8. *Prof. E. H. Weber in Leipzig:* Ernst Heinrich Weber (1795–1878), a German physician considered to be one of the founders of experimental psychology. He was the eldest brother of Wilhelm Weber (*CDSB*).
9. *Prof. Wöhler:* Friedrich Wöhler (1800–82), a German chemist and professor of chemistry at the University of Göttingen (*CDSB*).
10. *well-deserved recognition which you have given to Chladni's astute and ingenious research:* Tyndall shared much about Chladni's acoustic experiments in his lecture course at the RI, 'On Sound', given as the annual Christmas Lectures for children in 1865–6 (*Roy. Inst. Proc.*, 4 (1862–6), p. 485).

To Moncure Conway[1] [January 1867][2] 2552

Tuesday ev[enin]g

Dear Mr Conway.

A large map <u>can</u> be hung up with ease.

I would strongly advise you to limit yourself to one hour.[3] That is our time—people come wound up for that time & get fidgety afterwards.

Had I the leisure I should certainly feel the armour of my friends of the Pall-Mall Gazette.[4]

Yours very sincerely | John Tyndall | over

The general[5] bore his journey[6] admirably. He slept, smoked, eat & drank prosperously. I left him in the care of a devoted woman[7] with a glorious sky overhead, noble mountains & a blue sea near him. I put him astride a donkey on Xmas day and took him to a height of more than 2000 feet among the mountains[8]

John Tyndall to Moncure Conway; [January 1867]; Columbia University Manuscripts; Moncure Daniel Conway Papers, Box 21; University Archives, Rare Book & Manuscript Library, Columbia University Libraries, Rare Book & Manuscript Library, Columbia University in the City of New York

---

1. *Moncure Conway:* Moncure Daniel Conway (1832–1907), an American minister, social reformer, and author. While his father was a Virginia slave owner, Conway was an ardent abolitionist. In 1863 he left the United States for England where he spent twenty-two years (*ODNB*; *ANB*).
2. *[January 1867]:* the month and year are given based on Conway's Friday Evening Discourse at the RI (see n. 3), as well as mention of Tyndall's trip to Mentone, France with Thomas Carlyle (see n. 6).
3. *limit yourself to one hour:* Conway delivered a Friday Evening Discourse at the RI, 'On New England', on 22 February 1867 (M. D. Conway, 'On New England', *Roy. Inst. Proc.*, 5 (1866–9), pp. 59–62).

4. *feel the armour of my friends of the Pall-Mall Gazette:* probably a reference to his letter to the *PMG* of 3 January 1867 (see letter 2537).
5. *general:* Thomas Carlyle.
6. *his journey:* Tyndall and Carlyle traveled to Mentone, France to visit Louisa Baring at the end of December 1866 (see letter 2535, n. 4 and n. 5).
7. *devoted woman:* Louisa Baring.
8. *height of more than 2000 feet among the mountains:* see letter 2535, n. 8.

From Louisa Baring 2 February [1867][1] 2553

Mentone[2]—Feb. 2nd

Dear Professor Tyndall—

I am so sorry again to trouble you about Lady Smith[3] and the packages—but as the stock of Pipes is exhausted, will you kindly oblige Mr Carlyle,[4] by sending out that Box[5] (addressed to me,) by the (Grande Vitesse—) Globe Express Company, as soon as you can—& the other Box, please send to my name, to care of Mrs. Mildmay[6]—31. Great Cumberland Place—who in due time will bring it out.

Mr. Carlyle has not been so well, the last few days, but I hope, it is only a temporary derangement. I am not quite sure that Mentone has proved to him on the longer acquaintance, what it seemed to you, on a slight one.[7] But it is a great blessing to me, having him. I wish it were you, & not "Hirst", that were coming south. With all kindest regards

I am ever | Yours truly | L. Ashburton

Elton Hall, Ashburton Collection
RI MS JT/TYP/1/130

---

1. *[1867]:* the year is given by reference to the December 1867 trip to Mentone (see letter 2535, n. 4).
2. *Mentone:* see letter 2535, n. 4 and n. 5.
3. *Lady Smith:* Frances Smith (see letter 2549, n. 5).
4. *Mr Carlyle:* Thomas Carlyle.
5. *Box:* see letters 2549 and 2550.
6. *Mrs. Mildmay:* Lord Ashburton's sister, Anne Eugenia, who died in 1839, had married Humpfrey St John Mildmay (1794–1853) in 1823. 'Mrs Mildmay' may refer to the wife of one of their sons, but is unidentified (Aileen Christianson of *CLO,* personal communication).
7. *what it seemed to you, on a slight one:* while Tyndall went to Mentone with Carlyle to join Baring, he returned to London on his own (see letter 2535, n. 4).

To Joseph Dalton Hooker [3 February 1867][1] 2554

Royal Gardens Kew.[2] Sunday evening.

My dear Hooker

Hirst and myself came out to talk to you for five minutes about an affair of importance to the world generally, though not to you. Without preface the affair is this—The British Association[3] is to meet next year in Norwich,[4] and we consider it exceedingly desirable that you should be president on that occasion.[5] This is a duty that the gods have laid upon you and you can by no means shirk it. So make up your mind to say that you will accept the duty, and thus delight the hearts of Hirst and your old friend

Tyndalides[6]

RI MS JT/1/TYP/8/2595
LT Typescript Only

---

1. *[3 February 1867]:* the date is given by Hirst's journal entry of 3 February 1867: 'Walked to Kew with Tyndall, we tried but failed to induce Hooker to accept the Presidentship of the British Association at Norwich in 1868' (Brock & MacLeod, *Hirst Journals,* p. 1798). While Tyndall wrote that they had gone to Kew and missed Hooker, Hirst wrote that they 'failed to induce' him, indicating the possibility of their having gone to Kew twice on or around 3 February 1867.
2. *Royal Gardens Kew:* the Royal Botanic Gardens at Kew in southwest London, where Hooker served as the garden's Assistant Director from 1855 and followed his father William Jackson Hooker (1785–1865) as Director from 1865–85.
3. *British Association:* the BAAS.
4. *meet next year in Norwich:* the BAAS held its thirty-eighth annual meeting at Norwich in August 1868.
5. *you should be president on that occasion:* despite his initial reluctance, Hooker indeed became President of the BAAS for 1868, and at the annual meeting at Norwich delivered the Presidential Address on 19 August 1868 in which he discussed the role of museums and natural history collections, as well as championed Darwin's theory of evolution by natural selection (*Brit. Assoc. Rep. 1868,* pp. lviii–lxxv). On Hooker as President of the BAAS for 1868, see *X Club,* pp. 252–6.
6. *Tyndalides:* a nickname for Tyndall; Hooker used it in at least four other letters to Tyndall.

From Bartlett James 6 February 1867 2555

Rio de Janiero | 6th February 1867
Mr. John Tyndall Esquire, F.R.S. &c. &c.

Dear Sir,

Your note of the 17th December[1] reached me on the 2nd Inst and I am happy to learn that the amount insured by our late esteemed friend W. Ginty[2] will be received on the 14th March next.

I have made enquires of Mrs. Ginty[3] regarding any sums that may be unpaid in England and received this morning the reply which I now enclose[4] and I think the amount of these accounts should be paid as soon as you receive the money from the Insurance Company. The Balance should be invested in some <u>very safe</u> stock or shares and I know of none that I could more confidently recommend than the shares of the Rio de Janeiro Gas Company, Limited,[5] of which our late friend was Engineer in Chief.

I learn from Mrs. Ginty that her son George[6] now at school in England, has a decided inclination for scientific pursuits. He is now more than 14 years of age, and it would therefore be every way desirable that some plan be fixed on for his future career, without any great loss of time. Will you therefore kindly favour us with your invaluable advice on this important topic? The eldest son Willie,[7] is well placed in a Merchant's office in this city and is now gaining his livelihood—I presume the third son, Fred,[8] will also be placed in mercantile life as soon as he can leave school.

I hope to be in England in June next when I shall be most happy to confer with you regarding the welfare of our late friend's family in which I need not say I take a deep interest.

With great esteem and respect,

I remain | dear Sir | yours very faithfully | B. James

Gas shares can be got <u>here</u> at £21 to 22 each if dearer[9] in London, the purchase had better be effected in Rio.

RI MS JT/1/TYP/11/3655
LT Typescript Only

1. *Your note of the 17th December:* letter missing.
2. *W. Ginty:* William Ginty.
3. *Mrs. Ginty:* Margaret Ginty.
4. *the reply which I now enclose:* enclosed letter from Margaret Ginty missing.
5. *Rio de Janeiro Gas Company, Limited:* William Ginty was appointed engineer for the Imperial Gas Works in Rio de Janeiro in 1854.

6. *son George:* George Ginty.
7. *eldest son Willie:* William Roberts Ginty.
8. *third son, Fred:* Fred Ginty. The Gintys also had two daughters (see letter 2539, n. 3).
9. *dearer:* at a high price; at great cost (*OED*).

To Connop Thirlwall[1] 7 February 1867 2556

7th February 1867

My dear Lord

A letter lies before me[2] dated a long time back thanking you for the copy of your excellent address[3] which you were good enough to send me.

The case of Science could hardly be stated with greater fairness, and you have said nothing regarding the limits of scientific research to which any true philosopher could take objection.

With regard to a somewhat amusing passage which you quote from my earlier writings[4] I may say that a dozen years have made me a sadder & a wiser man. In a late reprint of that lecture by Macmillan,[5] the passage to which you have referred is omitted.[6]

Pray accept this tardy acknowledgement of your kindness and excuse the accident[7] which rendered it tardy.

Yours very faithfully | John Tyndall

Mss.Ms.C011.200. American Philosophical Society

---

1. *Connop Thirlwall:* while the American Philosophical Society had this letter as addressed to a 'Lord Scott', we've determined that the recipient is Connop Thirlwall (1797–1875), an English bishop and historian (*ODNB*). The full letter was published in J. H. S. Perowne, (ed), *Remains Literary and Theological of Connop Thirlwall: Essays, Speeches, Sermons, etc.*, 3 vols (London: Daldy, Isbister & Co., 1877–8), vol. 3 (1878), p. 311.
2. *A letter lies before me:* letter missing.
3. *your excellent address:* C. Thirlwall, *The Present State of Relations between Science and Literature: An Address, delivered at the Royal Institution of South Wales, December 6, 1866* (London: Rivingtons, 1867).
4. *amusing passage which you quote from my earlier writings:* on pp. 9–10 of his address, Thirlwall quoted from Tyndall's 'On the Importance of the Study of Physics as a Branch of Education for All Classes', in *Lectures on Education, delivered at the Royal Institution of Great Britain* (London: John W. Parker and Son, 1854), pp. 171–211: 'When I saw your Lecturer [the Oxford Professor of Chemistry] reduced to the necessity of pleading for Science, and meekly claiming for it, from the Institution which we are accustomed to regard as the highest in this land [the University of Oxford], recognition equal to that accorded

to Philology, I confess that the effect on me was to excite a certain revolutionary tendency in a mind which is usually tranquil almost to apathy in these matters. Science behind Philology! The knowledge of the laws by which God's universe is sustained, and the perpetual advancement of humanity secured, inferior to that of the manner in which ancient and savage tribes put their syllables together, and express the varieties of mood, tense, and case' (pp. 208–9).

5. *late reprint of that lecture by Macmillan:* J. Tyndall, 'On the Importance of the Study of Physics', in E. Youmans (ed), *Modern Culture: Its True Aims and Requirements; a Series of Addresses and Arguments on the Claims of Scientific Education* (London: Macmillan and Co., 1867), pp. 1–29.
6. *passage to which you refer is omitted:* in the 1867 reprint, the following words were omitted: 'Science behind Philology! The knowledge of the laws by which God's universe is sustained, and the perpetual advancement of humanity secured, inferior to that of the manner in which ancient and savage tribes put their syllables together, and express the varieties of mood, tense, and case' (see p. 28).
7. *accident:* not identified.

To Margaret Ginty 8 February 1867 2557

8th. Feb. 1867.

My dear Mrs. Ginty,

On the receipt of your note[1] I immediately sent it to Mr. Worthington[2] in whose hands I have placed your affairs, asking him how I should reply to it. He has provided me with the answer marked A.[3]

I beg to state that I have neglected nothing in connexion with your affairs. I am utterly ignorant of legal formalities and on that account I resorted to my friend the eldest son of the late Lord Chief Baron[4] for advice. His advice was distinct—namely that I should hand over the technical management of your affairs to some respectable solicitor—This I have done.

You complain of my not having written to my co-executor[5]—pray what am I to write about? The moment I received intelligence from Mr. Worthington[6] that the Insurance company was satisfied with the proofs presented to them I communicated that result to my co-executor[7]—Whatever am I to write about? If you are desirous that I should take any distinct step; or do any thing special in connexion with your affairs then I would beg for a distinct intimation to that effect from my co-executor.

I trust that he will bear in mind that I am entirely inexperienced in legal matters, and that if he possessing greater experience wishes any thing to be done he will have the goodness to give me clear information with regard to his proposals or wishes.

A bill was sent to me some time ago from Faulkners of Manchester[8]—I wish to know whether it is to be paid?

After writing to me the letter marked A. Mr. Worthington, on the same date favoured me with the letter marked B.[9] I wish you to hand it over to my Co-executor.

The letter with which you lately favoured me[10] was on the point of being lost. I receive papers and letters unpaid, and they are sometimes returned by the person who receives them at the Royal Institution. This was very near being the fate of yours, as it presented no appearance which would lead me to expect that it was from you. It will be much safer to prepay the letters in Rio.[11]

Again I would urge the necessity of giving me very distinct information as to what my Co-executor propose regarding the management of your affairs. You simply bewilder me when you complain of my not having written to my Co-executor; he never having given me the slightest intimation of what his proposals and wishes are.

I am | Your sincere friend | John Tyndall.

RI MS JT/1/TYP/11/3657–7a
LT Typescript Only

---

1. *your note:* see letters 2539 and 2548.
2. *Mr. Worthington:* a solicitor, of Sale, Worthington and Shipman (or sometimes Sale, Worthington, Shipman and Seddon) in Manchester, England.
3. *the answer marked A:* letter missing.
4. *my friend the eldest son of the late Lord Chief Baron:* Tyndall's close friend William Frederick Pollock, son of Jonathan Frederick Pollock.
5. *co-executor:* Bartlett James.
6. *received intelligence from Mr. Worthington:* letter missing.
7. *I communicated that result to my co-executor:* letter missing.
8. *Fallkeners of Manchester:* possibly Kendal, Milne & Faulkner, a department store in Manchester, England, first opened in 1832.
9. *letter marked B:* letter missing.
10. *letter with which you lately favoured me:* see letter 2548.
11. *Rio:* Rio de Janeiro, Brazil, where Margaret Ginty lived.

To Alfred Russel Wallace [8 February 1867][1] 2558

My dear Wallace

I promised to read the tract[2] which you were kind enough to send me, and to give you my frank opinion upon it.

I read it with disappointment. I opened it with the idea that it contained a record of your new experiments & observations. There is nothing of the kind from the beginning of the book to the end.

I was acquainted with most if not all that you adduce; and in many cases more fully & closely acquainted with it than you appear to be. I know Rechenbach,[3] and in 1856 paid an oficial[4] visit to his house near Vienna.[5] I saw him in the midst of his apparatus—his magnets & his crystals—and he described to me his methods and results. I did not want to hear Riechenbach but <u>you</u>.

I also am the person who addressed Thackeray[6] on the occasion referred to by Weld.[7] The report that you give of the affair does not rightly represent it. Poor Thackeray was staggered and abashed by the earnestness of my remonstrance regarding the lending the authority of his name to "Stranger than fiction"[8]—(my great respect for Thackeray rendered my remonstrance earnest) In about a year afterward he poured scorn upon Spirit rapping in the Round About papers of the Cornhill Magazine.[9]

You speak of phenomena "such as no human being could possibly have produced". I have heard this statement only too often. I have sat myself at Spiritual Seances, and produced motions myself, which grave and quasi-scientific men immediately adjacent to me pronounced impossible to *[make]*. I have heard men talk in excited astonishment about the ardently supernatural character of vibrations which I was producing myself. And had they described <u>their impressions</u> to you you would in all probability have agreed with them. The "facts" of such people are <u>not</u> facts but <u>feelings</u>.

I see the usual keen powers of your mind displayed in the treatment of this question. But mental power may show itself whether its materials be facts or fictions. It is not lack of logic that I see in your book, but a willingness that I deplore to accept data which are unworthy of your attention. This is frank—is it not.

Yours very faithfully | John Tyndall

BL Add MS 46439, ff. 13–4

---

1. The date is written is written at the top right of the page, presumably at a later time and probably in Tyndall's hand. At the top left are the words 'Opinion of book.' in the same hand.

2. *tract:* A. R. Wallace, *The Scientific Aspect of the Supernatural: Indicating the Desirableness of an Experimental Inquiry by Men of Science into the Alleged Powers of Clairvoyants and Mediums* (London: F. Farrah, 1866). Wallace noted in his autobiography that while he had this work published in a secularist magazine in 1866, he 'had a hundred copies printed separately, which I distributed among my friends' (A. R. Wallace, *My Life: A Record of Events and Opinions,* 2 vols (London: Chapman & Hall, Ltd., 1905), vol. 2, p. 280). On Wallace and spiritualism, see I. Hesketh, 'The First Darwinian: Alfred Russel Wallace and the Meaning of Darwinism', *Journal of Victorian Culture,* 25 (2020), pp. 171–84; E.

Sera-Shriar, 'Credible Witnessing: A. R. Wallace, Spiritualism, and a "New Branch of Anthropology"', *Modern Intellectual History,* 17 (2020), pp. 357–84; and C. H. Smith, 'Wallace and the "Preter-normal"', in C. H. Smith, J. T. Costa, and D. A. Collard (eds), *An Alfred Russel Wallace Companion* (Chicago: Chicago University Press, 2019), pp. 41–66.

3. *Rechenbach:* Carl Reichenbach (1788–1869), a German chemist, geologist, metallurgist, and industrialist who attributed the ability of mediums to detect hidden magnets and crystals in darkened rooms to their possession of an 'odylic force' (*ADB;* C. Reichenbach, *Physikalisch-physiologische Untersuchungen über die Dynamide des Magnetismus,* 2 vols (Braunschweig: Vieweg, 1850).
4. *oficial:* misspelt by Tyndall.
5. *in 1856 paid an oficial visit to his house near Vienna:* Tyndall visited Reichenbach in September 1856, writing to Edward Frankland that he 'saw all his paraphernalia, his magnets his crystals and his electrical apparatus'. Tyndall continued in more detail about the visit; see John Tyndall to Edward Frankland, [6 October?] 1856, letter 1282, *Tyndall Correspondence,* vol. 5.
6. *Thackeray:* William Makepeace Thackeray (1811–63), an English novelist and satirist (*ODNB*).
7. *occasion referred to by Weld:* see Wallace, *The Scientific Aspect of the Supernatural,* pp. 39–40. Charles Richard Weld (1813–69), an English author known for his travel writings, and a historian of the RS (*ODNB*).
8. *"Stranger than fiction":* as editor of *Cornhill Magazine,* Thackeray allowed the publication of an article in support of spiritualist phenomena ([R. Bell], 'Stranger than Fiction', *Cornhill Magazine,* 2 (August 1860), pp. 211–24). Thackeray included a note with the article stating that readers should think for themselves regarding the claims in the article, but that the author, who went unnamed, was a long-time friend 'of good faith and honourable character' (this friend was Robert Bell). See G. Dawson, 'Stranger than Fiction: Spiritualism, Intertextuality, and William Makepeace Thackeray's Editorship of the *Cornhill Magazine,* 1860–62', *Journal of Victorian Culture,* 7 (2002), pp. 220–38.
9. *he poured scorn upon Spirit rapping in the Round About papers of the Cornhill Magazine:* Thackeray wrote a column in the *Cornhill Magazine* titled 'Roundabout Papers' (he became editor of the magazine in 1860). Tyndall was probably referring to an ironic ghost story in which Thackeray seemed to mock spiritualism: 'Roundabout Papers.—No. XX. The Notch on the Axe.—A Story À La Mode', *Cornhill Magazine,* 5 (May 1862), pp. 508–12, 634–40. Also see 'Superstition', *Cornhill Magazine,* 5 (May 1862), pp. 537–49.

To Henry Acland 9 February 1867 2559

9th February 1867.

My dear Dr. Acland.

Many times since I received your "oration"[1] I have thought of writing to thank you for it. But I have been hard pressed by other matters[2] and thus rendered tardy in the performance of my duty to you. I read the discourse with great interest, coming as it does from a high heart and a cultivated head. For my own part when I look over human history and see how the belief for which you contend helped men to the performance of deeds which are even classed among the greatest and most sublime achievements of humanity. I am tempted to ask is there no way of preserving this force intact? Must it go with the forces to which in special ages it linked itself? I cannot I confess look at the extinction of this grand motive power without regret, and I can hardly imagine it doomed to destruction.

Perhaps the most significant phrase of the whole lecture is that where you speak of a supreme will "whom the reason cannot comprehend but whom the heart can approach". What you here translate into the approach of the heart is the thing which gives the discussion value in your eyes. As a purely intellectual problem in which your "heart" had no share, the question of a supreme will would possess comparatively little interest for you. You find this feeling to be of an enabling character and you would therefore defend not only it, but the conceptions with which it has been associated from all assault. But is it necessary that this test of heart and height of soul, should express themselves in theological forms? I hope not, for if so they will come to grief. Judging of theological inquiry by its tendencies and achievements hitherto the idea of entrusting any department of investigation of general interest to humanity to the exclusive treatment of theologians is I confess exceedingly uncomfortable to me. If they knew their vocation the case might be different. But taking them as they are, they appear to me to be specially endowed with the capacity of bringing into contempt the noblest aspiration of the human soul.

Excuse this rather random production & believe me

Yours faithfully | John Tyndall

Oxford, MSS.Acland d.65 f.189

---

1. *your 'oration':* H. Acland, *The Harveian Oration, 1865* (London: Macmillan and Co., 1865). The Harveian Oration is an annual lecture at the Royal College of Physicians of London established by William Harvey in 1656. Acland's was the first to be delivered in English rather than Latin. If there was a letter from Acland to Tyndall, it is missing.
2. *hard pressed by other matters:* in letter 2556, Tyndall mentioned an 'accident' (which has not been identified) that prevented him from replying in a timely manner to that letter. This probably refers, however, to his lecturing duties at the RI; he gave a Friday Evening Discourse on 18 January (see letter 2543, n. 2) and began a series of lectures on 22 January (see letter 2545, n. 8).

To Alfred Russel Wallace [9 February 1867][1] 2560

My dear Wallace.

Could you & I work together at this question?[2] Can the experiments & observations to which you refer be made by us two without the introduction of a third party?

Yours very truly | John Tyndall

Mss.B.W15a. American Philosophical Society

1. *[9 February 1867]:* the date is written is written at the top right of the page, presumably at a later time and probably in Tyndall's hand. At the top left are the words 'Reply to note.' in the same hand.
2. *this question:* most likely a reference to spiritualist séances. This letter was probably a reply to Wallace's missing response to letter 2558.

To Joseph Dalton Hooker [11 February 1867][1] 2561

Royal Institution. Monday.

Dear Hooker

You once gave me your opinion of the negroes both in Jamaica and in western Africa.[2] Would you object to my stating this as your opinion? Or would you permit me to state it as that of a friend of mine who never gave a shilling to the Eyre Fund, who would not hurt the meanest creature, and who is regarded with the most profound respect by members of the Jamaica Committee itself—All this without mentioning your name?[3]

Give my kind regards to Mrs Hooker.[4] I hope she will encourage you to give me this permission.

Yours ever | John Tyndall

RI MS JT/1/TYP/8/2587[5]
LT Typescript Only

1. *[11 February 1867]:* the date is given by relation to letter 2567 (dated 15 February 1867, a Friday), which is Hooker's second reply to this letter and a second from Tyndall on 15 February, letter 2565. Hooker's first reply from 14 February, a Thursday, is letter 2563. The Monday preceding was 11 February. An additional letter from Tyndall on 15 February is letter 2566.
2. *your opinion of the negroes both in Jamaica and in western Africa:* see letter 2505, Joseph Hooker to John Tyndall, 13 November 1866, *Tyndall Correspondence,* vol. 9.

3. *without mentioning your name:* Tyndall spoke at a meeting of the Eyre Defence Committee, held at Willis's Rooms, on 18 February 1867, and indeed shared Hooker's views (see letters 2567 and 2572, n. 2). A transcript of the meeting, including Tyndall's speech, was published as Appendix C in H. Hume, *The Life of Edward John Eyre, Late Governor of Jamaica* (London: Richard Bentley, 1867), pp. 277–88, Tyndall's speech on pp. 278–85. See also letters 2571, 2572, and 2576.
4. *Mrs Hooker:* Frances Hooker.
5. This typescript also includes, above this letter, another that we have not determined if it is part of this letter, or a separate letter that for some reason LT typed up with this one. It is very short, and reads: 'Dear Hooker | I have passed it on'. It is unclear as to the date of this short letter, or to what 'passed it on' refers, but it may be referring to a letter of Hooker's that Tyndall passed on to the Jamaica Committee (see letters 2563 and 2567).

From Rudolf Clausius 13 February 1867 2562

ZÜRICH, 13/2 67.

Lieber Tyndall,

Ich habe Deine Vorlesung wieder mit vielem Vergnügen gelesen. Die Art wie Du durch Versuche alles zur klaren Anschauung zu bringen, und das Interesse am Gegenstande zu erhöhen weisst, ist wirklich bewunderungswürdig. Der Versuch mit dem durch den galvanischen Strom erhitzten und dann in Schwingungen versetzten Platindraht hat mir durch seine Neuheit und Eleganz besonders gefallen.

Ich habe nur wenige Bemerkungen zu machen, die zum Theil ganz unbedeutend sind, und will die betreffenden Stellen der Reihe nach anführen.

S.88 Zeile 3 von unten scheinen mir die Bechstehen AB unrichtig zu stehen.

S.91 liessen sich die beiden letzten Sätze wohl noch zweckmässig so zusammen fassen. Die Anzahl der Schwingungen ist umgekehrt proportional dem Quadratwurzel des Gewichtes der Saite, oder noch vollständiger ausgedrückt: <u>dem Quadratwurzel des Gewichtes der Längeneinheit der Saite</u>. Daran schliesst sich dann das, was auf S.98 über die überspannenen Saiten gesagt ist sehr bequem an.

S.96. Die in Italics gedruckte Stelle: "and from it, just as if it were a fixed point, the pulse bb' will be reflected while the segment b'c will oscillate as an independent string", würde ich etwas anders ausdrücken. Z.B. so: "es hat den Anschein als ob .. . . . reflectirt würde."

S.95. Die Stelle: "The tension at the present moment is such, that the whole string swings to and fro in a vertical plane" könnte so verstanden werden, als ob die Ebene der Schwingungen auch von der Spannung abhinge. Ich würde daher statt der Worte "in a vertical plane" etwa setzen: "und zwar finden die Schwingungen in verticaler Ebene statt," oder in irgend einer ähnlichen Weise die Worte "in a vertical plane" von dem Übrigen trennen.

S.113 Auch hier scheint es mir zweckmässiger statt des Durchmessers des Stranges das Gewicht zu erwähnen, weil 4 Stränge das 4fache Gewicht haben, aber nicht gerade einen Strang von vierfachem Durchmesser bilden.

S.122 ganz oben steht "$\frac{3}{7}$ ths of a second" und gleich nachher folgen diese Worte bei $\frac{3}{20}$ noch einmal. Helmholtz sagt aber $\frac{3}{7}$ und $\frac{3}{20}$ von der Schwingungsdauer des Grundtones.

Auf derselben Seite sprichst Du vom 7ten Oberton, wo Helmholtz vom 7ten Partialton spricht. Das, was Helmholtz den 7ten Partialton nennt, ist aber der 6te Oberton, weil Helmholtz bei den Partialtönen den Grundton mitzählt.

S.121 würde ich noch etwas bestimmter hervorheben dass die Angabe von Helmholtz sich auf den Fall bezieht, wo die Saite auf $\frac{1}{7}$ ihrer Länge abgeschlagen wird.

S.129 wird gesagt bei der Theilung b müsse man 225 sinuosities haben, eben so bei den Theilungen c, d, e, 625, 1225 und 2025 sinuosities. Das verstehe ich nicht, da ja nicht 225 Schwingungen auf einen Umlauf sondern auf 36 Umläufe kommen.

S.131. In Fig. 56 ist zu beachten, dass die Endstücke nur etwa $\frac{1}{4}$ so gross sein müssen als die Stücke zwischen zwei Knotenpuncten.

Dasselbe ist eigentlich auch bei Figur 63 auf S.138 zu bemerken; aber bei dieser wird es sich wohl nicht ändern lassen, weil dann die folgenden auf die Wheatstone'sche Erklärung bezüglichen Figuren auch nicht anwendbar wären. Vielleicht kannst Du indessen auf S.138 im Texte sagen, dass die Theilungsart in Fig. 63 der folgenden Constructionen wegen etwas anders gemacht ist, als sie bei einem schwingenden Stabe stattfindet. Die beiden letzten unter 63 stehenden Figuren b und c werden aber wohl geändert werden müssen:

Statt: 

etwa so: 

Die bei p und q in der Figur gezeichneten Krümmungen können bei einer solchen Schwingung nicht entstehen, da ja die Enden frei sind, und daher keine Kraft wirkt, welche das herabschwingende Ende nach oben drücken könnte.

S.132 Zeile 7 von unten sind zwei Buchstaben m und n angeführt, welche in der Figur auf S.133 nicht vorkommen.

Bei den Verhältnissen der Schwingungszahlen elastischer Stäbe, wie sie sich auf den Seiten 128, 131 etc. befinden, würde ich ein Wort wie "angenähert" anbringen, weil diese Verhältnisse doch nur näherungsweise richtig sind.

Ich hoffe dass Deine kurze Reise nach Italien Dir gut bekommen ist, und Du Dich wohl befindest. Uns geht es ganz gut. Von Johnny werden wir Dir nächstens wieder ein Bildchen schicken.

Mit freundlichen Grüssen von meiner Frau und mir an Dich und Hirst

Dein | Clausius.

ZURICH, 13/2/67.

Dear Tyndall,

I have read your lecture[1] again with much pleasure. The way in which you know how to clarify everything through experiments and to heighten interest in the subject is truly admirable. The experiment with the platinum wire that was heated by the galvanic current and then made to oscillate was especially pleasing to me through its novelty and elegance.

I have just a few comments to make, which in part are quite minor, and shall cite the relevant passages in order.

P. 88 line 3 from the bottom, the letters AB seem to be incorrect.

P. 91. The last two sentences could probably still be summarised usefully thus. The number of oscillations is inversely proportional to the square root of the weight of the string, or expressed even more completely: <u>the square root of the weight of the unit of length of the string</u>. What is said on p. 98 about the over-stretched strings then follows this very conveniently.

P. 96. The passage printed in italics: "and from it, just as if it were a fixed point, the pulse bb' will be reflected while the segment b'c will oscillate as an independent string", I would express somewhat differently. E.g., thus: "it has the appearance as if. . . . . was reflected."

P. 95. The passage "The tension at the present moment is such, that the whole string swings to and fro in a vertical plane" could be understood as if the plane of the oscillations also depended on the tension. I would therefore put instead of the words "in a vertical plane" with something like: "and indeed, the oscillations take place in a vertical plane" or in some similar manner or other separate the words "in a vertical plane" from the rest.

P. 113 Here too it seems more useful to me to mention the weight of the string instead of its diameter, because four strings have four times the weight, but do not exactly form one string of four times the diameter.

P. 122, at the very top it says "$\frac{3}{7}$ths of a second" and then immediately afterwards, these words follow at $\frac{3}{20}$ one more time. However, Helmholtz says:[2] $\frac{3}{7}$ and $\frac{3}{20}$ <u>of the duration of oscillation of the fundamental tone</u>.

On the same page, you speak of the 7th overtone, where Helmholtz speaks of the 7th partial tone. What Helmholtz calls the 7th partial tone is however the 6th overtone, because Helmholtz counts the fundamental tone with the partial tones.

P. 121 I would emphasise somewhat more precisely that Helmholtz's statement refers to the case where the string is being plucked at a point of $\frac{1}{7}$ of its length.

P. 129 It is said that with the division b one would have to have 225 sinuosities, just as with the divisions c, d, e, 625, 1225, and 2025 sinuosities. I do not understand this, as 225 oscillations of course do not come to one revolution, but rather to 36 circulations.

P. 131. in fig. 56 it should be noted that the end segments only have to be roughly $\frac{1}{4}$ as large as the segment between two nodal points.

The same should really also be noted with figure 63 on p. 138; but with this one it will probably not be able to be changed, because then the subsequent figures referring to Wheatstone's explanation[3] would also not be applicable either. Perhaps, however, you can say on p. 138 in the text that the manner of division in fig. 63 has been done somewhat differently because of the following constructions than it takes place with an oscillating rod. The last two figures b and c under 63, however, will probably have to be changed:

Instead of: 

something like this: 

The curves marked at p and q in the figure, cannot occur with such an oscillation, as the ends of course are free, and therefore no force is acting which could push up the end that is swinging down.

P. 132 line 7 from the bottom two letters m and n are mentioned, which do not occur in the figure on p. 133.

With the proportions of the oscillation frequency of elastic rods, as they are found on pages 128, 131, etc., I would add a word like "approximate", because these proportions after all are only approximately correct.

I hope that your short trip to Italy[4] did you good, and you feel well. We are doing quite well. We shall send you a little picture of Johnny[5] again before long.

With kind regards from my wife[6] and myself to you and Hirst

Your | Clausius.

RI MS JT/1/TYP/7/2298–300
LT Typescript Only

1. *your lecture:* presumably draft pages for lectures 3 and 4 in Tyndall's *Sound* (1867), which would be published on Friday, 5 July 1867 (*Athenaeum,* 29 June 1867, p. 843).
2. *Helmholtz says:* H. Helmholtz, *Die Lehre von den Tonempfindungen als physiologische Grundlage für die Theorie der Musik* (Braunschweig, Vieweg und Sohn, 1863), p. 135.
3. *Wheatstone's explanation:* see C. Wheatstone, 'On the Figures obtained by strewing Sand on Vibrating Surfaces, commonly called Acoustic Figures', *Phil. Trans.,* 123 (1833), pp. 593–633.

4. *your short trip to Italy:* possibly a reference to Tyndall's time in Italy the previous summer. When traveling in the Alps in July 1866 and confronted with undesirable weather, he decided to venture into Italy with the reverend and mountaineer Arthur Girdlestone (1842–1908), whom he met on the Eggishorn (see *The Ascent of John Tyndall,* pp. 236–7; and *Hours of Exercise in the Alps,* pp. 263–6). Tyndall had mentioned going to Italy in two previous letters to Clausius (see letters 2464 and 2490, *Tyndall Correspondence,* vol. 9).
5. *Johnny:* Rudolf John Clausius.
6. *my wife:* Adelheid Clausius.

From Joseph Dalton Hooker [14 February 1867][1] 2563

Athenaeum.[2] Thursday

My dear T.

I hoped to have caught you here today and explained why I have not earlier answered your note.[3]

My opinion of the character of the negro is that he is as a rule eminently fickle, impulsive and cruel, not systematically cruel, like the Red Indian, or a rejoicer in cold-blooded deeds, but one who is easily roused through fear, superstition or other motive to the acts of revolting cruelty for his own benefit. Whatever his mental qualities may be in the abstract, as compared with our own races, they desert him altogether in cases of difficulty: at the same time he is capable of an immense amount of passive endurance. Above all he is almost incapable of joint action for a continuous purpose: and in cases of insurrection, meeting, or emeute,[4] you never know what you are about with him, he has no "guard" and so you can hardly tell when you have him at an advantage if you judge him by yourself.

Under discipline or pressure whether artificial, of others, or engendered by the struggle for existence, he will work well and may I doubt not, prove a good and useful member of any community, but the circumstances under which his good qualities for self-government have hitherto been called out are altogether so exceptional that I do not think one can judge much from them.

I think you are mistaken as to the experience I have had of the Negro. I have met him on the African coast and Brazil,[5] but never resided with him or had him under me.

Most sincerely yours | J. D. Hooker.

RI MS JT/1/TYP/8/2588
LT Typescript Only

1. *[14 February 1867]:* the date is given by relation to letter 2567, dated 15 February 1867 (a Friday), which is Hooker's second reply to Tyndall's letters of 11 February (2561) and 15 February (2565). This present letter is Hooker's first reply to those letters. An additional letter from Tyndall on 15 February is letter 2566. Tyndall was preparing his speech before the Eyre Defence Committee on 18 February 1867 (see letter 2561, n. 3, and 2572, n. 2).
2. *Athenaeum:* see letter 2537, n. 8.
3. *your note:* letter 2561.
4. emeute: a popular rising or disturbance (*OED*).
5. *I have met him on the African coast and Brazil:* during his time on HMS *Erebus* (1839–43), Hooker visited the Brazilian island of Trindade and the Cape of Good Hope in South Africa before sailing to the Antarctic.

To James Croll 15 February 1867[1] 2564

Thursday[2]

My dear Sir,

Your reasoning seems to me to be quite fair.[3] But I confess I should feel a greater interest in the practical question if I were sure of the observations.

Is there any necessity for saying that the surface of the moon exposed to the sun's radiation is colder than the earth? Is it necessary for your argument to assume that the planetary envelopes are more pervious to the radiation from the planet than the radiation from the sun? If not necessary I would not if I were you bring these points so definitely forward. Your general principle is perfectly secure, that an atmosphere forms a local dam which elevates the temperature of the planet. I have said a few words on this subject in par[agraph]. 546 page 437 of the 2nd edition of my book on Heat.[4] Also in the tract[5] which I now send you at p.p. 45 & 47.

Yours very truly | John Tyndall

BL Add MS 41077, ff. 150–50a

J. C. Irons (ed), *Autobiographical Sketch of James Croll with Memoir of his Life and Work* (London: Edward Stanford, 1896), p. 153

1. *15 February 1867:* the date is given in the published letter, but is not included in the handwritten letter.
2. *Thursday:* 'Thursday' is given in the handwritten letter, but is not included in the published letter.
3. *Your reasoning seems to me to be quite fair:* Croll's letter to Tyndall is missing, but it appears that Croll sent him a paper that was to be published in the *Phil. Mag.* in March 1867: J.

Croll, 'On the Reason why the Difference of reading between a Thermometer exposed to direct Sunshine and one Shaded Diminishes as we ascend in the Atmosphere', *Phil. Mag.*, 33 (1867), pp. 213–6.

4. *2nd edition of my book on Heat: Heat* (1865).
5. *tract:* J. Tyndall, *On Radiation: the "Rede" lecture delivered in the Senate House before the University of Cambridge* (London: Longman, Green, Longman, Roberts & Green, 1865). In the published paper (see n. 3), Croll added a footnote with these references that Tyndall gave him (p. 216).

To Joseph Dalton Hooker [15 February 1867][1] 2565

Friday.

My dear Hooker

The question which I wished to ask you[2] was prompted by the manifest desire of the Negrophobists[3] to place Negroes and Englishmen on the same level.

The note you wrote to me some time ago[4] shows me that this is not your view. You said in that note

> "that the Negro in Jamaica, and even in the free towns of [W.] Africa is pestilential I have no hesitation in declaring, and that he is a most dangerous savage at the best."

It would strengthen my hands mightily if I could quote this in public[5] as your opinion, or even without mentioning your name as the opinion of a disinterested friend. Have you any objection to this? Do not allow me to bring your name up in the matter if you object to it:—Could I not say that the opinion was expressed in a private note?

Private

I have to move a resolution and in doing this I intend to say a few words temperately but strongly, and this is what causes me to ask you the foregoing question.

Yours affectionately | John Tyndall

RI MS JT/1/TYP/8/2589
LT Typescript Only

1. *[15 February 1867]:* the date is given by relation to letter 2567, dated 15 February 1867 (a Friday), which is Hooker's second reply to Tyndall's letters of 11 February (2561) and this present letter. Hooker's first reply from 14 February is letter 2563. An additional letter from Tyndall on 15 February is letter 2566. Tyndall was preparing his speech before the Eyre Defence Committee on 18 February 1867 (see letters 2561, n. 3, and 2572, n. 2).
2. *question which I wished to ask you:* Tyndall first asked Hooker about 'negroes' in letter 2561.
3. *Negrophobists:* Tyndall must have intended to write 'negrophiles'—white persons who are friendly or well-disposed towards black people (*OED*)—rather than 'negrophobists', white persons with an intense dislike or hatred of black people (*OED*).
4. *note you wrote to me some time ago:* see letter 2561, n. 3.
5. *quote this in public:* see letter 2561, n. 4, and letter 2572.

To Joseph Dalton Hooker [15 February 1867][1] 2566

Friday.

My dear Hooker

I feel somehow that I ought not to have written that 2nd note to you,[2] as I think I ought to have inferred from the first[3] that you did not wish to have yourself mixed up directly or indirectly with the matter in hand.[4] If therefore you have not already replied to me,[5] let the matter drop.

Yours ever | John Tyndall

RI MS JT/1/TYP/8/2590
LT Typescript Only

1. *[15 February 1867]:* the date is given by relation to letters 2561 (11 February 1867), 2563 (14 February 1867), 2565 (15 February 1867), and 2567 (15 February 1867).
2. *2nd note to you:* letter 2565.
3. *the first:* 2561.
4. *the matter in hand:* in letter 2561, Tyndall first asked Hooker about his views on "negroes' and whether or not he could quote him in his speech before the Eyre Defence Committee on 18 February 1867 (see letter 2561 n. 3, and 2572, n. 2). In his reply (2563), Hooker shared his views but did not give permission or not. Following a second letter from Tyndall (2565), Hooker again replied (2567), stating that 'I leave it entirely to you to do so or not as you please and how you see fit'.
5. *replied to me:* Hooker's reply is letter 2567.

From Joseph Dalton Hooker 15 February 1867 2567

Kew,[1] February 15th 1867.

Dear old Tyndall

I rather misunderstood your first note[2] or I should have been more explicit. In answer to your second,[3] (1) I consider The Negro far below the level of the Englishman. (2) From all I have read and heard of him, I consider him a savage, and a most dangerous savage too. I believe the power and position given to him in the free towns of W. Africa to have had a most pestilential influence, and the liberty given to him in Jamaica to have proved equally detrimental to the prosperity of that island. Now as regards (1) of course there are many Englishmen worse than any negro, because the worst classes of the best races are worse than the worst of any and all the inferior races; but I do not see how that affects the truth of my proposition.

As regards (2) it depends on the definition of the term "Savage". Johnson[4] defined savage as "a man untaught, uncivilised";[5] in general parlance, the world now superadds cruelty to the above. Now I hold that the negro in W. Africa and Jamaica is untaught, uncivilised and cruel too, not cruel in the same way as the N. American Indian, who is cruel for cruelty's sake; but essentially cruel, inasmuch as when his blood is up very cruel acts are his first acts and many of them.

1000 thanks for your second considerate note. I had no intention of hauling out; I only wished and wish to remind you that my opportunities of personally judging of the Negro are few and small, and that my opinion is formed on extensive reading and inquiry from travellers and settlers and philanthropists. I fear that if you quote me[6] it will be said "what can J.D.H know", and in fact that if I am quoted (I leave it entirely to you to do so or not as you please and how you see fit) it should be coupled with the statement of how I came to my conclusions and this would defeat your objects I fear. W. Greig[7] is the very man with whom I agree entirely in estimate of the negro.

Of course I am no advocate for slavery, but I believe that except the Negro is so situated that he must work fairly hard for his bread he will ever remain what he is, a dangerous savage, and I do not believe that he will work fairly hard except under compulsion and control.

I only yesterday got a horrid piece of work off my hands—taking an inventory of my father's scientific effects for the Gov[ernmen]t,[8] like cataloguing ye Brit[ish]. Mus[eum]! it wearied my life out. I was puzzled to know why it was called for, and only guess from this evening's paper (that announces a "supplementary grant of £7000 for Kew"[9]) that they are going to purchase all at once. I hope indeed they are as my Bankers are grumbling sorely!

Ever your affectionate | J. Hooker—

RI MS JT/1/TYP/8/2591
LT Typescript Only

1. *Kew:* see letter 2554, n. 2.
2. *your first note:* letter 2561.
3. *your second:* letter 2565.
4. *Johnson:* Samuel Johnson (1709–84), writer, lexicographer, and author of *A Dictionary of the English Language*, 2 vols (London: W. Strahan, 1755) (*ODNB*).
5. *"a man untaught, uncivilised":* Johnson's definition of the noun 'savage' in the first edition of his *Dictionary of the English Language* (1755) was 'A man untaught and uncivilized; a barbarian'.
6. *if you quote me:* see letter 2561, n. 4.
7. *W. Greig:* possibly Woronzow Greig (1805–65), barrister, FRS, and member of the Statistical Society of London. He was a son of Mary Somerville from her first marriage (see letter 2803, n. 1 and n. 6; *ODNB*). We have been unable to locate any sources which show Greig's views on 'negros'.
8. *taking an inventory of my father's scientific effects for the Gov[ernmen]t:* William Jackson Hooker (1785–1865), botanist, director of the Royal Botanic Gardens, Kew (1841–65), and father of Joseph Hooker. After his death his herbarium and library were sold to the government to be added to the permanent collection of Kew. The collection, which was purchased for £7000, contained more than 1 million plant specimens (*ODNB*; *CDSB*).
9. *"supplementary grant of £7000 for Kew":* Hooker wrote to Darwin that the £7000 was for 'Herbarium £5000 | Choice of books £1000 | Portraits, mss, | Correspondence &c £1000' (see Joseph Hooker to Charles Darwin, 19 October 1866, F. Burkhardt, et al. (eds), *The Correspondence of Charles Darwin,* 24 vols (New York: Cambridge University Press, 2005), vol. 14 (1866), pp. 354–5; this is letter 5247 in the online edition of the Darwin correspondence).

To James Croll 19 February 1867 2568

Tuesday.

Dear Sir,

I certainly think that the view you have expressed[1] is "more in harmony with existing physics" than that expressed by M^r. Wilson.[2]—

Yours very truly, | John Tyndall

BL Add MS 41077, F. 152
J. C. Irons (ed), *Autobiographical Sketch of James Croll with Memoir of his Life and Work* (London: Edward Stanford, 1896), p. 153

1. *view you have expressed:* see letter 2564.

2. *that expressed by Mr. Wilson:* James Maurice Wilson (1836–1931), an Anglican clergyman, schoolteacher, and astronomer, who, along with Tyndall and Thomas Huxley, served on a committee of the BAAS to study the question of promoting science education in England. In addition to being a vocal advocate of science education in Britain, Wilson was also the mathematics and science master at Rugby (1859–79) and headmaster of Clifton College (1879–90). This possibly refers to J. M. Wilson, 'Some Remarks on an observation of Mr. Glaisher's', *Phil. Mag.,* 31 (1866), pp. 104–6. Wilson was responding to Glaisher's paper, 'An Account of Meteorological and Physical Observations in Three Balloon Ascents made in the years 1864 and 1865', presented at the thirty-fifth annual meeting of the BAAS at Birmingham (*Brit. Assoc. Report 1865,* pp. 145–91). On Glaisher's observation, according to Wilson: 'That there is always a difference of reading of two thermometers, one with blackened bulb exposed to direct sunshine, and the other with bulb concealed, is well known. The latter makes the temperature of the air surrounding it; the former marks the sum of this and of the direct radiated heat which it receives from the sun. Mr. Glaisher has shown that the difference of reading of the two thermometers, when taken to great heights, instead of increasing or remaining constant, as would seem at first sight to be expected, actually diminishes, and appears as if it would diminish without limit' (p. 104).

From Charles Mackay[1] 19 February 1867 2569

Reform Club[2] | Feb. 19. 1867.

My dear Dr. Tyndall,

I thank you very cordially for all you have done, and all the trouble you have taken. It is of course unpleasant to be rejected by a Club,[3] but if the Athenaeum[4] had no other objection to me than that I was a public writer, I shall not take their rejection at all to heart. Perhaps too it may be recorded hereafter that I lost but little pleasure and no credit, by my non-admission. I belong already to two Clubs[5] to both of which I was elected unanimously, and can bear this rebuff with perfect equanimity.

Believe me | Ever truly yours | Chas. Mackay

Professor Tyndall, F.R.S.

RI MS JT/1/TYP/2/785

1. *Charles Mackay:* Charles Mackay (1814–89), a Scottish poet and journalist, mostly known for his study on crowd psychology in *Extraordinary Popular Delusions and the Madness of Crowds* (London: Robert Bentley, 1841), which covered such subjects as economic bubbles, the Crusades, witch mania, haunted houses, alchemy, and fortune telling, among others (*ODNB*). Mackay's son attended Queenwood College when Tyndall taught there.
2. *Reform Club:* a private gentleman's club in London, founded in 1836, whose members espoused progressive political ideas.

3. *rejected by a Club:* Mackay had been entered in the Candidates Book for the Athenaeum on 22 October 1855: 'Charles Mackay LLD Author of Studies from the Antique; Sketches from Nature; The Salamandrine; Ballad and Songs &c &c'. His nomination went to the ballot on 18 February 1867, with Sir John Bowring (1792–1872), an English political economist and writer (*ODNB*), as his proposer and Tyndall as his seconder. There were 25 signatures of support on his ballot sheet, including Charles Dickens, Wilkie Collins, and John Murray. However, in the ballot he received 80 yes votes and 25 no votes, and so was not elected; 8 no votes would have been sufficient to exclude him (Jennie de Protani, Archivist, the Athenaeum Club, personal communication).
4. *Athenaeum:* see letter 2537, n. 8.
5. *two Clubs:* Mackay belonged to the Reform Club and, for a period, to the Garrick Club. See C. Mackay, *Through the Long Day: Or, Memorials of a Literary Life During Half a Century,* 2 vols (London: W. H. Allen & Co., 1887), vol. 2, pp. 57, 66.

From Margaret Ginty 22 February 1867 2570

Rio de Janeiro | Feb. 22nd 67

Dear Dr. Tyndall,

I believe that Mr. James[1] in writing to you by the last English Mail forgot to say where the boys[2] were at school also where the accounts were to be paid—Georgie and Fred are at Linton House School[3] care of W. Brightwell. Colchester.[4] and the bill to be paid is for Mr. *[1 word illeg]*[5] | Exchange Arcade | Manchester.[6]

I am very sorry to trouble you so much but I hope soon to be free of all this kind of trouble.

I hope you will write when you have time to spare.

I remain dear Dr. Tyndall | yours sincerely | M. Ginty.

RI MS JT/1/TYP/11/3658
LT Typescript Only

1. *Mr. James:* Bartlett James.
2. *the boys:* George and Fred Ginty. The Gintys also had another son and two daughters (see letter 2539, n. 3).
3. *Linton House School:* a boy's secondary school in Colchester, Essex.
4. *W. Brightwell. Colchester:* William D. B. Brightwell (*c.* 1841–91), principal of Linton House School, Colchester ('Sad Suicide of William Brightwell, of Clacton-on-Sea', *Essex Standard,* 31 January 1891, p. 2).
5. *Mr. [1 word illeg]:* not identified. The last name is handwritten on the typescript, but is not legible.
6. *Exchange Arcade | Manchester:* located in the Royal Exchange, a building in Manchester.

From William Gomm[1] 23 February 1867 2571

33 Brunswick Terrace | Brighton.[2] | 23rd Feb. 1867.

My dear Sir,

I have only this morning obtained a sight of your speech of Monday last[3] through the good offices of the "Morning Post";[4] and cannot forego the pleasure I feel in telling you that I find it a worthy compeer of your flogging letter of reply[5] to the attacking Committee,[6] and more I cannot say for it, although I may feel tempted to particularise portions which have struck me as most telling.

Your dissection of the two parties[7] which compose the entire lump of aggression I think especially happy and strikingly just and true.

Mr. Mill[8] too, pointed out as the theoretic navigator through all this storm-broil Jamaican: the Acherontic[9] outburst of which at Morant Bay[10] was only a first and premature puff from the mine scientifically underlaid along the whole length and breadth of the island by "Head Centre" Gordon himself,[11] &c.

No! I will not believe that this England of ours is hopelessly peopled with "deaf adders",[12] and proof against what you so vividly place before them.

I feel much obliged by the kind reception you have given to my two Jamaica brochures[13] which so unceremoniously obtruded themselves upon your notice:—begging you to believe me,

Very faithfully yours | W M Gomm

RI MS JT/1/TYP/8/2584
LT Typescript Only

---

1. *William Gomm:* William Maynard Gomm (1784–1875), a British Army officer, born in Barbados, who served in the Napoleonic Wars, commanded troops in Jamaica in the 1830s, and was governor of Mauritius (1842–9), and commander-in-chief in India (1851–6) (*ODNB*).
2. *Brighton:* a resort town on the southern England coast, close to Hastings.
3. *your speech of Monday last:* see letter 2561, n. 3.
4. *'Morning Post':* 'The Eyre Defence and Aid Fund', *Morning Post,* 21 February 1867, p. 2.
5. *your flogging letter of reply:* see Tyndall's reply to the Jamaica Committee dated 7 November 1866, letter 2498, *Tyndall Correspondence,* vol. 9 (also in H. Hume, *The Life of Edward John Eyre, Late Governor of Jamaica* (London: Richard Bentley, 1867), Appendix C).
6. *the attacking Committee:* the Jamaica Committee (active 1866–9) was formed in 1866 in opposition to the violent suppression of the uprising of black peasants at Morant Bay, Jamaica. By 1867, with John Stuart Mill as chairman, the committee advocated prosecuting

Edward Eyre, Governor of Jamaica, for his direction of the execution of the Jamaican rebels. Eyre would be unsuccessfully tried for private criminal charges brought by the Jamaica Committee. Other notable members of the committee included Thomas Huxley, Charles Lyell, Herbert Spencer, and Charles Darwin (*ODNB*).

7. *two parties:* the Jamaica Committee (see n. 6), and the Governor Eyre Defence and Aid Committee. The latter was established and chaired by Thomas Carlyle. Other supporters of Eyre included Charles Dickens, Alfred Tennyson, John Ruskin, and John Tyndall (*ODNB*).
8. *Mr. Mill:* John Stuart Mill (1806–73), a philosopher, economist, and women's rights advocate. He was the chairman of the Jamaica Committee from 1866 (*ODNB*).
9. *Acherontic:* of or relating to Acheron; infernal, hellish; dark, gloomy (*OED*).
10. *Morant Bay:* a town in southeastern Jamaica, location of the violent suppression of the uprising of black peasants in October 1865.
11. *'Head Centre' Gordon himself:* George William Gordon (1820–65), a wealthy Jamaican businessman and politician, who was arrested, tried, and executed, at the direction of Edward Eyre, on suspicion of his having planned the Morant Bay rebellion (*ODNB*).
12. *'deaf adders':* in Biblical translation or allusion: an asp or other snake, regarded as deaf or wilfully unhearing (*OED*).
13. *my two Jamaica brochures:* not identified.

From Joseph Dalton Hooker 25 February 1867 2572

Kew[1] | Feb. 25th 1867.

Dear Tyndall

Thanks for the Post.[2] I do not deserve half the feelingly kind utterances in regard of me, that it contains. I only hope that my irrefragably true pictures of the negro character[3] may lead some persons to pause ere they judge too harshly of the conduct of those who have to do with that race <u>when their own masters</u>. I have just glanced at little du Chaillu's[4] book,[5] contemptible it is as far as giving any exact information, but what a picture of negro character it gives! and he is evidently a kindly little body full of love to the black, and without an atom of prejudice.

True enough, the negro has been a valuable servant, ally if you like, of the European, so has the horse, are we therefore to give the latter its head and let him kick us to pieces? but this is just what we have done with the negro.

Ever your aff[ectiona]te | J. D. Hooker.

RI MS JT/1/TYP/8/2594
LT Typescript Only

1. *Kew:* see letter 2554, n. 2.
2. *Post:* 'The Eyre Defence and Aid Fund', *Morning Post,* 21 February 1867, p. 2. The *Morning Post* published the speeches from the meeting of 18 February, including Tyndall's (see letter 2561, n. 3): 'let me now give you the opinion regarding the negro of Dr. Joseph Dalton Hooker—a man whose righteousness, mercy, and truth are blended with the highest intellectual excellence, and with knowledge gathered from all corners of the earth. Neither he nor I, let me say, ever had the slightest sympathy with negro slavery. Both of us equally abhor cruelty; both of us are ready to join in denouncing it, but neither of us is willing to commit himself to that "falsehood of extremes" which is the vice of the position of the Jamaica committee. This, then, in the opinion which extensive reading, much questioning of travellers, and some personal contact have forced upon Dr. Hooker: "That the negro in Jamaica," he writes to me, "and even in the free towns of Western Africa, is pestilential, I have no hesitation in declaring; nor that he is a most dangerous savage at the best." But my friend Hooker, if he had a word to say in favour of the Fiend himself, would be careful not to repress it, and he takes great care to explain that though the negro is cruel, he is not so in the sense of the North American Indian, cruel for cruelty's sake. Still he regards the negro as essentially cruel, "inasmuch as when his blood is up, very cruel acts are his first acts, and these in great number." In a second note to me he thus repeats the conviction in his first: "I consider him a savage, and a most dangerous savage too. I believe the power and position given to him in the free towns of Western Africa, to have had a pestilential influence; and the liberty given to him in Jamaica to have proved equally detrimental to the prosperity of that island." With testimony of which this is but a sample before me, I decline accepting the negro as the equal of the Englishman, nor will I commit myself to the position that a negro insurrection and an English insurrection ought to be treated in the same way' (Tyndall quoted Hooker from letter 0327, *Tyndall Correspondence,* vol. 9, and letter 2567 in this volume).
3. *my irrefragably true pictures of the negro character:* see letter 2567.
4. *du Chaillu's:* Paul Belloni Du Chaillu (*c.* 1835–1903), a French-American explorer and writer who claimed to have been the first European to observe gorillas in their natural habitat. From the 1860s he began writing numerous and highly popular books describing his travels in Africa and the peoples and animals he encountered (*ANB*).
5. *book:* probably P. Du Chaillu, *A Journey to Ashango-Land: and further Penetration into Equatorial Africa* (London: John Murray, 1867).

To Augusta Stanley[1] 3 March 1867 2573

Athenaeum Club[2] | Pall Mall S.W. | 3rd March | 1867

Dear Lady Augusta.

When your most kind note[3] reached the Royal Institution yesterday I was absent in Birmingham[4]—this accounts for the tardiness of my reply.

Any wish expressed by you must carry with it the weight of a command to me. I shall therefore loyally render myself at the Deanery[5] on Friday next.

Most truly yours | John Tyndall

Royal Archives/VIC/ADDA17/1785/118 by the permission of Her Majesty Queen Elizabeth II

---

1. *Augusta Stanley:* Augusta Elizabeth Frederica Stanley (née Bruce, 1822–76), a courtier who served as lady in waiting to Queen Victoria before marrying Arthur Stanley in 1863 (*ODNB*). She died on 1 March 1876, the day after Stanley officiated at Tyndall's wedding.
2. *Athenaeum Club:* see letter 2537, n. 8.
3. *your most kind note:* letter missing.
4. *absent in Birmingham:* possibly to visit the Chance Brothers glassworks in Birmingham, who produced lighthouse lamps for Trinity House. Although Tyndall was not yet officially adviser to Trinity House, he had been carrying out work for them while Faraday's health was in decline.
5. *Deanery:* Stanley's residence; her husband Arthur Stanley was Dean of Westminster.

From Wilhelm Weber 3 March 1867 2574

Göttingen 1867. März 3.

Hochverehrter Freund!

Beiliegend sende ich Ihnen ein Autograph von Chladni, auf einem kleinen Blatte, was leider! mit der Zeit schadhaft geworden ist. Es enthält Bemerkungen zu „Rudolphis Physiologie" Berlin 1821.

Unter diesen Bemerkungen dürfte wohl die letzte, die Vocale betreffende, zu §358, besonders in Betracht kommen.

Sie lautet:

> „Über die Selbstlauter (Vokale) deren es eigentlich 9 giebt, habe ich in meinen neuen Beyträgen zur Akustik, 3. Abth. mehreres gesagt. Diphthongen giebt es so viele, als es Möglichkeiten giebt, 2 Vocale in einer Sylbe auszusprechen. Die Lehre von den Consonanten hat Olivier in seinem orthoepographischen Lehrbuche (welches nicht angeführt ist) am besten vorgetragen. Chladni."

Ich hoffe dass dieses Autograph (wenngleich in deutschen Lettern) seinem Inhalte nach und wegen der Unterschrift des Namens Ihrem Zwecke am meisten entsprechen wird.

Die auf demselben Blatte vorhergehenden Bemerkungen sind zum Theil wohl für Chladni sehr charakteristisch, haben aber keine Beziehung auf die Akustik und würden sich auch aus anderen Gründen, wie mir scheint, sich weniger eignen. Doch legt mein Bruder, von dem ich sie erhalten, einigen

Werth darauf und wünscht dieses Blatt von Chladnis Hand wieder zurückzuerhalten wenn von demselben Gebrauch gemacht worden ist. Um die Erfüllung dieses Wunsches erlaube ich mir daher Sie noch besonders zu ersuchen.

Ihr | ganz ergebenster | Wilhelm Weber.

Göttingen[1] 1867. March 3.

Esteemed Friend!

I am sending you enclosed an autograph of Chladni's,[2] on a small piece of paper, which unfortunately! has been damaged over time. It contains remarks on "Rudolphis Physiologie" Berlin 1821.[3]

Among these remarks, the final one concerning the vowels, at §358, might well come into particular consideration.

It reads:

"I have said several things about vowels, of which there are actually 9, in my neuen Beyträgen zur Akustik,[4] 3rd section. There are as many diphthongs as there are possibilities of pronouncing 2 vowels in one syllable. Olivier[5] has best presented forward the theory of consonants in his orthoepographical textbook[6] (which is not cited). Chladni."

I hope that this autograph (although in German script) will suit your purpose most of all according to its content and because of the signature of the name.

The preceding remarks on the same sheet of paper are in part probably very characteristic of Chladni, but have no relation to Acoustics[7] and would also, as it seems to me, be less suitable for other reasons. However, my brother,[8] from whom I obtained them, sets some store by them, and wishes to have this sheet of paper from Chladni's hand returned again when use has been made of it. I am therefore taking the liberty of especially requesting you to fulfill this wish.

Your | most very devoted | Wilhelm Weber.

RI MS JT/1/W/16

---

1. *Göttingen:* see letter 2551, n. 1.
2. *Chladni's:* see letter 2551, n. 2.
3. *'Rudolphis Physiologie' Berlin 1821:* probably *Grundriss der Physiologie* (Berlin: Dümmler, 1821), written by Karl Asmund Rudolphi (1771–1832), a Swedish naturalist who made contributions to botany, helminthology, and comparative anatomy (*CDSB*).
4. *neuen Beyträgen zur Akustik:* E. Chladni, *Neue Beyträge zur Akustik* (Leipzig: Breitkopf und Härtel, 1817).
5. *Olivier:* Ludwig Heinrich Ferdinand Olivier (1759–1815), a German-Swiss educator who developed his own method of teaching reading and writing and wrote a number of textbooks (*ADB*).

6. *orthoepographical textbook:* L. Olivier, *Ortho-epo-graphisches Elementarwerk oder: Lehrbuch über die in jeder Sprache anwendbare Kunst rechtsprechen, lesen und rechtschreiben zu lehren* (Dessau, 1804).
7. *Acoustics:* see n. 4.
8. *my brother:* either Wilhelm's younger brother, Eduard Friedrich Weber (1806–71), or more likely his older brother Ernst Heinrich Weber (1795–1878), who was heavily influenced by Chladni in his scientific work (*CDSB*; *ADB*).

To George Granville Bradley[1] 7 March [1867][2] 2575

7th March

My dear Mr. Bradley

Through Mr. Farrar[3] I have this moment heard of the illness of Hallam Tennyson.[4] You judged rightly when you thought that the intelligence would interest me. It has shaken me more than I can describe. You would render me your grateful debtor if you would tell me yourself how the matter really stands. I cannot bear to think of the noble boy being taken away.

Believe me | Yours ever | John Tyndall

TRC Letters 6243

---

1. *George Granville Bradley:* George Granville Bradley (1821–1903), an Anglican clergyman and schoolmaster, who was headmaster of Marlborough College (1861–70), Master of University College, Oxford (1870–81), and Dean of Westminster (1881–1902). In the 1860s Bradley bought a house near Farringford and soon developed a close friendship with Tennyson (*ODNB*).
2. *[1867]:* the year is given based on the mention of the illness of Hallam Tennyson (see n. 4).
3. *Mr. Farrar:* Frederic William Farrar.
4. *illness of Hallam Tennyson:* Alfred Tennyson's son Hallam, who was a pupil at Marlborough College in Marlborough, Wiltshire, England, fell ill in late winter 1867, a 'serious attack on his lungs' (H. Tennyson, *Alfred Lord Tennyson: A Memoir by His Son,* 2 vols (London: Macmillan and Co., 1897), vol. 2, p. 42). Tyndall mentioned this illness again when he wrote to Hallam in February 1868 (letter 2766).

From Warren de la Rue[1] 11 March 1867 2576

The Observatory, Cranford[2]—| Middlesex. | March 11th /67.

My dear Tyndall

The perusal of your speech on the Eyre Defence Fund[3] incites me to fulfil an oft-recurring impulse to send a small contribution to that object[4] which your eloquent and thrilling words must have greatly aided. I look upon Governor Eyre[5] as one of the noblest examples of what England's sons can be when placed in positions of great danger and perplexity: one single minute of hesitation on his part would have caused the sacrifice not only of hundreds of valuable lives of men, but also the dishonour and cruel destruction of countless tenderly nurtured women. I will never admit that the negro can be placed on the same level as the European and that a negro insurrection can be suppressed by mild palliatives. Nothing but the display of overwhelming power can have any effect on the limited intelligence of the negro, he is and always will be a savage kept within bounds only by external influences.

Yours most sincerely | Warren de la Rue.

RI MS JT/1/TYP/8/2585
LT Typescript Only

---

1. *Warren de la Rue:* Warren de la Rue (1815–89), a British astronomer and chemist, and partner in the printing firm De La Rue founded by his father (*ODNB*).
2. *The Observatory, Cranford:* de la Rue's own observatory at his home in Cranford, Middlesex near Kew Observatory, where he moved in 1857 from Canonbury, where he also had an observatory but despised the poor skies there. He focused on lunar photography, and received medals from the RS (1864) and Royal Astronomical Society (1862) for this work. The observatory was dismantled in 1873—de la Rue's vision was failing and he sold his house—and the reflecting telescope given to Oxford in 1874 (M. L. Huggins, 'Obituary: Warren de la Rue', *The Observatory,* 12 (1889), pp. 245–50; M. E. W. Williams, 'Astronomy in London: 1860–1900', *Quarterly Journal of the Royal Astronomical Society,* 28 (1987), pp. 10–26).
3. *your speech on the Eyre Defence Fund:* see letter 2561, n. 3.
4. *that object:* the defence of Edward Eyre from the charges brought against him by the Jamaica Committee.
5. *Governor Eyre:* Edward Eyre.

From James Wilson[1] 14 March 1867 2577

Rugby[2] | March 14/67

Dear Dr Tyndall,

You were not present at the[3] Committee[4] meeting on March 2, & so do not know how irrationally we spent our time. There was one point, of very great importance, that[5] I intended then to bring before the Committee, but wh[ich]. was excluded by want of time. It is not too late to remedy it at the Council meeting on March 23.

If you will refer to the report[6] p. 3 you will see that we consider that

Exp[erimenta][l]. Physics, Chemistry, & Botany have paramount claims as subjects for school teaching. Now we devote a paragraph to the explanation of the merits of each of these subjects, & yet we seem to consider Chemistry only as subordinate to Physics. If you examine that paragraph on Chemistry, for wh[ich]. I believe you are responsible, I think you will see that it scarcely gives it so prominent a position as was intended. I quite agree that many parts of Chemical Physics may well be taught under the head of Chemistry; but a mere knowledge of chemical substances, of the laws of combination &c. scarcely represents what the gain of teaching in Chemistry is. Nor w[oul][d] a short course of lessons, illustrated by simple experiments, give a proper knowledge even of these things. I think the paragraph on Chemistry ought to have a numbered section to itself, & be entirely rewritten: & I have written to Griffith[7] to request him to propose this at the meeting of the Council.

Anything that looks crotchety or exclusive in this report will diminish its value; & it cannot fail to attract attention that we have so inadequately put forward the claims of Chemistry. After writing to Griffith, I took the report to D[r] Temple,[8] & he entirely agrees with what I have said: so also does Hutchinson.[9] On our Sub-Committee there was no chemist, or we should have had as eloquent a paragraph in the praise of Chemistry as the one on Physics.

We all agreed, if I remember right, that Exp[erimenta][l]. physics ought to precede Chemistry, but that Chemistry ought to be taught at schools, in the higher classes: & if so it must be taught a means of training, thoroughly, & be supplemented by work in the laboratory. Here boys learn themselves to experiment: in the study of Physics that master experiments.

Oxford[10] requires Chemistry as well as Mechanical Philosophy for the scholarships & studentships for Nat[ural]. Sc[ience]. & schools therefore must teach it; & I quite think Oxford is right.

I have therefore sent a paragraph, for proposal, to Griffith, but sh[oul]d be very glad if you w[oul]d write one, & submit it to the Council[11] for substitution for the existing paragraph.

I wish all this c[oul]d have been done before; but it is not too late, & is, as I said before, really important that the report sh[oul]d not be so onesided as the Universities, & Chemists generally will think it is.

Do not trouble to answer this letter, but if you can, attend the meeting of the Council, & see that the thing is properly done.

You will be glad that one of our Rugby boys[12] has just won the Nat[ural]. Sc[ience]. Studentship at Xt Church Oxford,[13] by his Chemistry & Exp[erimenta][l]. Physics.

Believe me | very truly yours | J M Wilson

I can if you like send you a copy of the paragraph I shall propose, & then if you cannot attend in person, you may signify y[ou][r] approval. Otherwise there will be an unwillingness to alter what you have written. I have not said what you will not approve of I am sure.

RI MS JT/1/W/77
RI MS JT/1/TYP/5/1785–6

1. *James Wilson:* see letter 2568, n. 2.
2. *Rugby:* Rugby School, an independent public school founded in 1567 in Warwickshire, England.
3. *the:* in the handwritten letter, Wilson wrote this and all other instances of the word 'the' using the thorn letter 'y', with 'the' written as 'y[e]'. Here we have used 'the' throughout this transcription.
4. *Committee:* the BAAS Committee on Scientific Education in Schools included Tyndall, Frederick Farrar, Thomas Hutchinson (see n. 9 below), Thomas Huxley, Joseph Payne (1808–76), and James Wilson (*Brit. Assoc. Rep. 1867,* p. xxxix). According to a letter from Farrar to Charles Darwin, George Griffith was also a member of the committee though he was not acknowledged in the published report (Frederick Farrar to Charles Darwin, 7 March 1867, F. Burkhardt, et al. (eds), *The Correspondence of Charles Darwin,* 24 vols (New York: Cambridge University Press, 2005), vol. 15 (1867), pp. 131–2; this is letter 5434 in the online edition of the Darwin correspondence).
5. *that:* in the handwritten letter, Wilson wrote this and all other instances of the word 'that' using the thorn letter 'y', with 'that' written as 'y[t]'. Here we have used 'that' throughout this transcription.
6. *the report:* 'Report of the Committee appointed by the Council of the British Association for the Advancement of Science to Consider the best means for promoting Scientific Education in Schools', *Brit. Assoc. Rep. 1867,* pp. xxxix–liv. This was submitted at the annual meeting of the BAAS at Dundee in 1867.
7. *Griffith:* George Griffith.
8. *D[r] Temple:* Frederick Temple (1821–1902), an Anglican clergyman and educator, who became Archbishop of Canterbury (1896–1902). From 1858–69 Temple was headmaster of Rugby School. He also contributed an article to the controversial *Essays and Reviews* (London: John W. Parker and Son., 1860) entitled, 'The Education of the World' (*ODNB*).
9. *Hutchinson:* Thomas Neville Hutchinson (1826–99), an Anglican clergyman and science educator. From 1865 Hutchinson was Natural Science Master at Rugby School, until he became Vicar of Broad Chalke in 1882. In 1898 he became Canon of Salisbury (see 'Wilts Obituary', *Wiltshire Archaeological and Natural History Magazine,* 30 (1899), p. 275).
10. *Oxford:* the University of Oxford.
11. *Council:* the Council of the BAAS, which managed affairs of the association in between the annual meetings (*Brit. Assoc. Rep. 1861,* p. xix, Council members at this time listed on p. xxxvi).

12. *one of our Rugby boys:* Frederick William Fison (1847–1927), 1st Baronet, an English mill owner and Member of Parliament for Doncaster (1895–1906). He was educated at Rugby School and Christ Church, Oxford ('Sir Frederick Fison', *Times,* 21 December 1927, p. 14; *A Companion to the Rugby School Register from 1675 to 1870* (Rugby: W. Billington, 1871), p. 110).
13. *Xt Church Oxford:* Christ Church, Oxford.

From Heinrich Gustav Magnus 15 March 1867 2578

Berlin 15 März 1867

Verehrter Freund!

Sie werden eine kurze Abhandlung von mir erhalten haben und ich kann nicht unterlassen Ihnen noch besonders auszusprechen wie sehr ich mich gefreut habe daß endlich die Ursache aufgefunden ist weßhalb die Resultate unserer Versuche in den nicht verschlossenen Röhren nicht übereinstimmen wollten. Ich bin überzeugt daß wenn Sie die Versuche mit Röhren aus verschiedenem Material wiederholen <Sie> bestätigt finden werden wie groß der Einfluß der Röhrenwand ist.

Ich gebe mich der Hoffnung hin daß ich Sie in Paris sehen werde und freue mich dann einmal wieder auf längere Zeit mit Ihnen verkehren zu können.

In alter Freundschaft | Ihr | G Magnus.

Berlin 15 March 1867

Dear friend!

You will have received a short paper[1] from me and I cannot refrain from particularly expressing to you how much I was pleased that the reason has finally been found why the results of our experiments in the unsealed tubes would not tally.[2] I am convinced that if you repeat the experiments with tubes from different material, you will find confirmed how great the influence of the tube wall is.

I cherish the hope that I shall see you in Paris[3] and look forward to being able to socialise with you then once again for a lengthy period of time.

In old friendship | Your | G Magnus.

RI MS JT/1/M/33

---

1. *a short paper:* presumably H. G. Magnus, 'On the Influence of the Adhesion of Vapour in Experiments on the Absorption of Heat', *Phil. Mag.*, 33 (1867), pp. 413–25. This was published along with Tyndall's remarks to the editor of the *Phil. Mag.* about the paper (see letter 2598). The original paper is: G. Magnus, 'Ueber den Einflufs der Vaporhäsion bei Versuchen über Absorption der Wärme', *Poggendorff's Annal.*, 130 (1867), pp. 207–26).

2. *why the results of our experiments in the unsealed tubes would not tally:* Magnus still believed that Tyndall's results were due to condensation of water on the sides of his tubes, a point Tyndall probably considered he had disproved some time previously, and had published in his 'Third Memoir'. For reference to a detailed account of their differences, see I. M. McCabe, *Second Best as a Researcher, Second to None as a Populariser? The Atmospheric Science of John Tyndall* (PhD dissertation, University College London, 2012), pp. 146–65. Tyndall's reply to this letter is missing, but Magnus wrote again (letter 2580), and Tyndall responded in his comments in the *Phil. Mag.* (letter 2598). Tyndall published no further direct response, and the challenge faded on Magnus's death in 1870. Even so, as late as the 1880s, Tyndall felt the need to demonstrate further evidence for the absorption of heat by water vapour, given its importance for meteorology, as described in *The Ascent of John Tyndall*, pp. 393–4.
3. *I will see you in Paris:* Tyndall went to Paris from 19–26 April for the International Exposition (or Exposition Universelle) of 1867, held from 1 April to 3 November (Brock & MacLeod, *Hirst Journals*, 19–26 April 1867, p. 1801), but was able to see Magnus only briefly while there. Tyndall served as a juror for Apparatus and Processes for Heating and Lighting (*Reports on the Paris Universal Exhibition, 1867*, 7 vols (London: Eyre and Spottiswoode, 1868), vol. 1, p. 30).

From James Wilson[1] 15 March 1867 2579

Rugby[2] | March 15/67

Dear Dr Tyndall,

I enclose a copy of the[3] paragraph on chemistry:[4] wh[ich]. I sh[oul]d like to see adopted, at least in spirit if not in words. If you approve of it, pray write a line to Griffith,[5] 5 Park Villas Oxford to say so, as he is sure to be present at the Council[6] meeting.

It will shew what my views are at least.

We never entered happily on a discussion of an ideal Public School System, or I fear that[7] I sh[oul]d differ as widely from you as I do from the head master of Eton.[8] I do not <u>want</u> more than four hours a week for ordinary Nat[ural]. Science teaching, & the opportunity of a laboratory for a few promising pupils. And I believe that three hours a week will be soon obtained in most of the public schools, & will be found to be enough. In five years boys w[oul]^d^ go well through Botany, Mechanic Hydrostatics & Pneumatics, Heat, Acoustics, Geom[etrical]. Optics, and Electricity & Magnetism, & have probably attended some series of lectures on Chemistry besides.

I am quite sure that an education principally based on Nat[ural]. Science w[oul]d be found as defective as an exclusively classical one. At present science is only taught by clever & enthusiastic men, who c[oul]d teach anything. But when it is taught by everybody there will be another story.

Our report[9] at present strikes me as very grave & weighty & considerate, & shews no revolutionary principles, the tip of the tail of wh[ich]. w[oul]d ruin your cause. And only in this matter of chemistry does it seem to be strikingly deficient.

With many thanks for your note,[10]

I remain | very truly y[ou]rs | J M Wilson

Do you see that Croll,[11] in the Phil. Mag,[12] has come to the same conclusion that I did[13] from Glaishers.[14] I wish you c[oul]d again take the question up.

I understood at the meeting of Committee[15] that the report was limited to Public Schools. We voted on it, & it was carried that the word Public remain. I must write about this to Griffith.

RI MS JT/1/W/78
RI MS JT/1/TYP/5/1787

---

1. *James Wilson:* see letter 2568, n. 2.
2. *Rugby:* see letter 2577, n. 2.
3. *the:* in the handwritten letter, Wilson wrote this and all other instances of the word 'the' using the thorn letter 'y', with 'the' written as 'y[e]'. Here we have used 'the' throughout this transcription.
4. *I enclose a copy of the paragraph on chemistry:* enclosure missing. In letter 2577, Wilson had written to Tyndall, 'I can if you like send you a copy of the paragraph I shall propose, & then if you cannot attend in person, you may signify y[ou][r] approval. Otherwise there will be an unwillingness to alter what you have written'.
5. *Griffith:* George Griffith.
6. *Council:* see letter 2577, n. 11.
7. *that:* in the handwritten letter, Wilson wrote this and all other instances of the word 'that' using the thorn letter 'y', with 'that' written as 'y[t]'. Here we have used 'that' throughout this transcription.
8. *head master of Eton:* Edward Balston (1817–91), an Anglican clergyman, and educator who served as head master of Eton College (1862–8) (*Alum. Cantab.*).
9. *Our report:* see letter 2577, n. 6.
10. *your note:* letter missing.
11. *Croll:* James Croll.
12. *in the Phil. Mag:* J. Croll, 'On the Reason why the Difference of reading between a Thermometer exposed to direct Sunshine and one Shaded Diminishes as we ascend in the Atmosphere', *Phil. Mag.*, 33 (1867), pp. 213–16.
13. *the same conclusion that I did:* J. M. Wilson, 'Some Remarks on an observation of Mr. Glaisher's', *Phil. Mag.,* 31 (1866), pp. 104–6. Tyndall commented on Wilson's paper in the following edition of the *Phil. Mag.*: J. Tyndall, 'On the Black-bulb Thermometer', *Phil. Mag.*, 31 (1866), pp. 191–3.

14. *Glaishers:* James Glaisher (1809–1903), an English astronomer and meteorologist, who worked at the Royal Observatory Greenwich (1835–74), managing the meteorological department there from 1841 until his retirement. Glaisher was perhaps best known for his pioneering balloon ascents during the 1860s with Henry Tracey Coxwell (1819–1900). The observations they made during these ascents—to which Wilson refers—were published in the BAAS *Reports* (1863–6) (*ODNB*).
15. *Committee:* see letter 2577, n. 4.

From Heinrich Gustav Magnus 23 March 1867 2580

Berlin 23 März 1867.

Sie wissen, theurer Freund daß ich stets gern bereit bin Ihre Aufträge auszuführen und so soll auch der den ich so eben von Ihnen erhalten so schnell und so gut als möglich ausgeführt werden, allein zuvor bedarf ich noch einiger Bestimmungen die ich mich beeile einzuholen. Sie schreiben zwar nicht wozu Sie die Apparate haben wollen aber doch ohne Zweifel um meine Versuche zu wiederholen. Da fragt es sich was Sie dazu wünschen. Eine Thermosäule sandte ich Ihnen vor längerer Zeit von hier. Sie waren anfangs nicht recht zufrieden aber später hat sie Ihnen M^r^. Becker, glaube ich, vollkommen *[eingestellt]*. Es kann sich also wohl nur um das Galvanometer handeln. Das würde Ihnen Sauerwald in kurzer Zeit liefern können, wobei ich voraussetze daß Sie Fernrohr und Scale dort sich beschaffen werden.

Ich gehe in den allernächsten Tagen nach Paris, wo ich Sie zu sehen hoffe, dann können wir alles besprechen. Für den Fall Sie aber jetzt nicht nach Paris kommen, bitte ich Sie mir nach dort zu schreiben. Ich hinterlasse hier solche Orders daß dann alles, was Sie wünschen, auf das Schnellste ausgeführt wird. Am sichersten wird es sein wenn Sie den Brief für mich richten an

M^r^. Herzog | Commissaire de la Prusse à l'Exposition | 29 Rue Marignan Champs Elysées dort werden Sie auch meine Adresse sogleich nach meiner Ankunft erfahren.

In alter Freundschaft | Ihr | G Magnus.

Berlin 23 March 1867.

You know, dear friend, that I am always happy and ready to carry out your instructions and so the one that I have just received from you[1] shall also be carried out as quickly and well as possible. However, I still require some details beforehand that I hasten to obtain. You do not write what you want to have the apparatus for, but it is without a doubt to repeat my experiments. The question then is what you wish for them. I sent you a thermopile[2] from here a long time ago. You were initially not really satisfied but Mr Becker,[3] I believe, *[adjusted]* it completed for you. It can therefore probably only be something to do with the galvanometer.[4] Sauerwald[5] would be able to deliver it to you within a short time, in which case I am assuming that you will obtain a telescope and a scale from there.

I am going to Paris in the very next few days, where I hope to see you;[6] we can then discuss everything. However, in case you are not coming to Paris now, I ask you to write to me there. I shall such leave orders here that everything that you wish will then be carried out as quickly as possible. It will be safest if you direct the letter for me addressed to

Mr Herzog[7] | Commissaire de la Prusse à l'Exposition | 29 Rue Marignan Champs Elysées

there, you will also find out my address immediately after my arrival.

In old friendship | Your | G Magnus.

RI MS JT/1/M/34

---

1. *the one that I have just received from you:* letter missing.
2. *thermopile:* a thermo-electric battery, used in connection with a galvanometer, for measuring minute quantities of radiant heat (*OED*).
3. *Mr Becker:* possibly Carl Ludwig Christian Becker (1821–75), a German instrument maker who moved to London (1849), where he became the manager of scientific instrument manufacturing firm Elliott Brothers (1858), and later (1873) a partner in the firm ('Obituary', *Monthly Notices of the Royal Astronomical Society,* 36 (1876), pp. 136–7).
4. *Galvanometer:* an apparatus for detecting the existence and determining the direction and intensity of a galvanic current (*OED*).
5. *Sauerwald:* Ferdinand Sauerwald, a German scientific instrument maker whose workshop was located in Berlin. He constructed apparatuses for Gustav Magnus, John Tyndall, Emil du Bois-Reymond, and Hermann Helmholtz, among others (*Helmholtz,* pp. 186, 264).
6. *hope to see you:* see letter 2578, n. 3.
7. *Mr Herzog:* 'M. Herzog' represented Prussia and Northern German States at the 1867 Paris International Exhibition (see letter 2578, n. 3). He was listed as 'conseiller intime au ministere du commerce, commissaire royal' and 'Privy Councillor for Commerce, Industry, and Public Works' in *Reports on the Paris Universal Exhibition, 1867,* 2 vols (London: Eyre and Spottiswoode, 1868), vol. 1, p. 22, and vol. 6, p. 9.

To Rudolf Clausius 27 March 1867 2581

27th March. 1867

My dear Clausius,

Your letter[1] reached me in proper time, but before replying to it I waited until I should be able to send you some more proofs[2] and also in the hope of being able to find somebody to take charge of the watch[3] for me from Zürich. But I dare say it will be best to send it direct if the duty is not too high—If very high I would wait a little in the hope of being able to find somebody to take charge of it.

I send you a cheque for twelve pounds sterling or 300 francs. Such cheques are every-where current in Switzerland and the watch-maker will therefore experience no difficulty in getting it cashed. It is by such cheques that I pay all my continental bills.

I am exceedingly glad to hear that you have found a position better than the one you occupy in Zürich.[4] In fact you ought to be in one of the first positions in Germany. No doubt dear Zürich will be very sorry for your departure. It will tempt me to make the Bavarian Highlands[5] occasionally my home when I think that you are at Wurzburgh.[6]

I hope you will excuse my having delayed so long in answering your last letter.

Believe me dear Clausius | ever yours truly | John Tyndall

If any further charges have been incurred be good enough to inform me.

I have lately been writing the concluding portion of my book[7] and have of course consulted Helmholtz regarding the theory of consonance and dissonance.[8] He lays great weight upon the Obertöne;[9] and of course they come very much into play, but I hardly think he puts the action of the primary tones in a clear light. For example, From page 309, where he speaks of the Gedackten Orgelpfeifen.[10] One would almost conclude that after the beats of the grand tone with the tones immediately adjacent to it there was no further dissonance between the two primaries. He says I think several times that whenever you press the interval of a third, the dissonance is due to the beats of the Obertöne. Now if I take two tuning forks which are a pure fifth [assuredly], and load one of them with wax I obtain beats: and these are the beats not of the Obertöne, but of the two primaries. I take also a fork $\underline{\text{III}}_2$ and sound it and $\underline{\text{III}}_3$ together. Loading either of them with wax I obtain loud beats. They appear beats of the deepest tone, and are powerfully heard when the higher of the two forks makes no sensible impression upon the ear. So also when I compare III_3 & III_4 the sound itself is obtained. So that if the Obertöne were entirely absent we should still have dissonance. I may not have properly understood Helmholtz, but it seems to me that he throws the interference of the primaries in the case of the wider interval too much into the shade.

RI MS JT/1/T/229

---

1. *Your letter:* letter 2562.
2. *some more proofs:* for Tyndall's book on sound (see n. 7).

3. *watch:* a watch that Tyndall presumably commissioned and was arranging to be sent to him. See letter 2585 of this volume, and letters 2464 and 2494, *Tyndall Correspondence,* vol. 9.
4. *a position better than the one you occupy in Zürich:* Clausius was professor of mathematical physics at the Polytechnic Institute (Eidgenössische Polytechnische Schule) in Zürich from 1855 until he took a professorship at the University of Würzburg in 1867. He remained at Würzburg for just two years before leaving for Bonn (*CDSB*).
5. *Bavarian Highlands:* the highland region or Bavaria in Germany (also known as the Bayerisches Oberland).
6. *Wurzburgh:* Würzburg, a city in northern Bavaria, Germany.
7. *my book:* Tyndall's *Sound* (1867), which would be published on Friday, 5 July 1867 (*Athenaeum,* 29 June 1867, p. 843).
8. *consulted Helmholtz regarding the theory of consonance and dissonance:* H. Helmholtz, *Die Lehre von den Tonempfindungen als physiologische Grundlage für die Theorie der Musik* (Braunschweig, Viewg and Son, 1863). An English translation was not published until 1875 (see n. 9).
9. *<u>Obertöne</u>:* upper partial tone (German). Translation found in the first English edition of Helmholtz's work referred to in n. 8: H. L. F. Helmholtz, *On the Sensations of Tone as a Physiological Basis for the Theory of Music* (London: Longmans, Green, and Co., 1875), p. 6.
10. *Gedackten Orgelpfeifen:* wide-stopped organ pipes (German). Translation found in H. L. F. Helmholtz, *On the Sensations of Tone as a Physiological Basis for the Theory of Music,* p. 172–3.

To Heinrich Debus [29][1] March 1867 2582

1867

My dear Heins

It would give me great pleasure to lend a helping hand to Tribe.[2] I know many Indian men, but the only persons in your list personally known to me are Hermann Merivale[3] & Sir James Fergusson.[4] Is it not Sir James Fergusson the eminent surgeon?[5] There is another James Fergusson, not a "<u>Sir</u>",[6] personally known to me—He is also an Indian—Please inform me if I am right in supposing that your Sir James is the surgeon.

When is the election[7] to take place? I shall be dining at Merivales on Monday next[8] and could then speak to him. If haste however be necessary I could write to him previously.

It is a mad world dear Heins. Perhaps a world with so many mad people in it has a right to be mad; and that it would be wrong if the few sane people held sway. This turbulence[9] will subside, I hope, and honest people will be again permitted to pursue their work in peace.

Yours ever, | John Tyndall

RI MS JT/1/T/307

---

1. *[29]:* '29' appears to be the day as written on the letter.
2. *Tribe:* Walter Harry Tribe (1832–1909), an English clergyman ordained in 1857 at which time he became curate of Broughton, Hampshire. He was Rector of nearby Stockbridge (1860–7), after which he served in several locations in India, before becoming the Archdeacon of Lahore from 1885–92 ('The Rev. W. H. Tribe', *Times,* 19 May 1909, p. 13).
3. *Hermann Merivale:* Herman Merivale (1806–74), an English economist and civil servant who was Permanent Under-Secretary of State for India, at the India Office (1860–74) (*ODNB*).
4. *Sir James Fergusson:* the 'Sir James Fergusson' to which Debus referred was probably James Fergusson (1832–1907), sixth baronet, a politician and colonial governor. He had inherited the baronetcy in 1849 and when this letter was written was Parliamentary Under-Secretary of State for India (1866–7). He was later the Governor of South Australia (1869–73), and New Zealand (1873–4) (*ODNB*).
5. *Sir James Fergusson the eminent surgeon:* not identified. It is possible Tyndall was thinking of William Fergusson (1808–77), first baronet, a Scottish surgeon, who was professor of surgery at King's College, London (1840–70). He was made a baronet in 1866 and was appointed sergeant-surgeon to the Queen in 1867 (*ODNB*).
6. *another James Fergusson, not a 'Sir':* Tyndall was probably referring to James Fergusson (1808–86), an architectural historian born in Scotland, who amassed a fortune as an indigo planter in Bengal. After retiring from this work he wrote several works on art and architecture including on the ancient architecture of India (*ODNB*).
7. *the election:* possibly having to do with Tribe's relocation to India in 1867.
8. *Monday next:* 29 March 1867 was a Friday, making 'Monday next' 8 April 1867.
9. *This turbulence:* a reference to the Eyre controversy in which Tyndall was involved; see letters 2567, 2571, 2572, and 2576.

To Heinrich Debus [March 1867][1] 2583

Tuesday

My dear Heinrich.

I fear I shall have to go to Paris at Easter.[2] I committed myself to a jurorship, and fear I must carry it out. I should prefer a thousand times a ramble round the island[3] with you.

And now for a matter of "business"—An examination occurs in May.[4] Tom will be unable to look over the papers; so I fear this heavy work will fall upon you. By the way it just occurs to me to ask whether you were ever paid for your last work in this line. Surely I cannot have been so stupid, or Tom cannot have been so stupid as to neglect this!

I want you to lend me; or to copy for me that passage of Lessing in which he refers to the Shield of Homer.[5]

What a curious imbroglio the political world is in!

If I do not go to Paris, then certainly we must have a walk together.

Yours ever | John Tyndall

RI MS JT/1/T/334

---

1. *[March 1867]:* although LT wrote '1866' on the letter, the month and year are given by reference to Tyndall going to Paris in April 1867 for the International Exposition (see. n. 2). He wrote 'at Easter' rather than 'next year' (which would be more likely if it was written in 1866). His reference to examinations in May (see n. 4) also indicates that this letter was written not long before the months of April and May.
2. *go to Paris at Easter:* see letter 2578, n. 3.
3. *island:* the Isle of Wight, where Tyndall would often escape London and enjoy long walks with friends.
4. *examination occurs in May:* probably referring to Tyndall's role beginning in 1859 as an Examiner for Science for the Department of Science and Art (sometimes Science and Art Department) at the South Kensington Museum. He had arranged for Hirst and Debus to grade examinations for him. On the South Kensington Museum, see B. Robertson, 'The South Kensington Museum in context: an alternative history', *Museum and Society,* 2 (2004), pp. 1–14.
5. *passage of Lessing . . . Shield of Homer:* a reference to the German philosopher Gotthold Ephraim Lessing's (1729–81) discussion of the shield of Achilles in his 1766 book *Laokoon: oder über die Grenzen der Malerei und Poesie,* or *Laocoon: or, The Limits of Poetry and Painting,* trans. W. Ross (London: Ridgeway, 1836).

To Frederic Farrar [March 1867][1] 2584

Saturday

My dear Mr. Farrar.

I agree to all your proposed alterations.[2]

I am also quite willing that what I have written regarding chemistry should be modified to suit the views of Mr Griffith[3] & the state of matters at Rugby.[4] It would perhaps be well before publishing finally to have the sanction of a chemist to this part of the Report.[5]

In the regard to the other matter I would briefly say that in my opinion any man in England might feel justly proud in having his name associated with your admirable Lecture.[6]

Yours ever | John Tyndall

Wellcome MS. 7777/101

---

1. *[March 1867]:* the month and year are given based on relation to Farrar's lecture in February (see n. 6) and that this letter very likely soon follows letter 2577 (dated 14 March 1867), in which James Wilson asked Tyndall to write a paragraph about chemistry for the education committee report: 'I have therefore sent a paragraph, for proposal, to Griffith, but sh[oul]d be very glad if you w[oul]d write one, & submit it to the Council for substitution for the existing paragraph'.
2. *your proposed alterations:* letter missing.
3. *Mr Griffith:* George Griffith.
4. *Rugby:* see letter 2577, n. 2.
5. *Report:* see letter 2577, n. 6.
6. *your admirable Lecture:* on 8 February 1867, Farrar delivered a Friday Evening Discourse at the RI, 'On Some Defects in Public School Education' (*Roy. Inst. Proc.,* 5 (1866–9), pp. 26–44).

From Rudolf Clausius 4 April 1867 2585

ZÜRICH, 4/4/67.

Lieber Tyndall,

Erst heute konnte ich zum Uhrmacher gehen, um wegen der Absendung Deiner Uhr mit ihm zu sprechen, da ich einige Tage unwohl war, und zu Hause bleiben musste. Der Uhrmacher konnte mir nicht sagen, wieviel der Zoll betrage, meinte aber dass er jedenfalls nicht so hoch sein würde, um deshalb die Absendung noch länger zu verzögern. Er hat es nun übernommen die Uhr gut zu verpacken, und sie auf die Post zu befördern. Ich hoffe Du wirst mit der Uhr zufrieden sein. Der von Dir gesandte Wechsel ist vom Banquier zu 301 ½ frs angenommen, und da die Uhr nur 300 fr kostet, so habe ich den Uhrmacher beauftragt die Sendung zu frankiren, wodurch der Unterschied sich ungefähr ausgleichen wird. Die Quittung des Uhrmachers lege ich hier bei.

Es würde mir lieb sein, wenn Du mir gelegentlich mittheilen wolltest, wie es mit der englischen Uebersetzung meines Buches steht, von der ich lange nichts gehört habe, ob sie schon erschienen ist, oder bald erscheinen wird.

Ich möchte Dich noch in einer anderen Sache um Deinen Rath bitten. Hirst hat mit meinem Buche so viel Mühe gehabt, dass ich ihm gern als kleines Zeichen meiner Dankbarkeit ein Geschenk schicken möchte, und da würde es mir sehr lieb sein, wenn Du mir etwas angeben könntest was ihm Freude machen würde. Ich habe daran gedacht, ihm eine solche Uhr machen zu lassen, wie Du sie Dir jetzt angeschafft hast. Aber dazu muss ich wissen, ob er sie auch gebrauchen kann, oder ob er vielleicht schon eine so schöne Uhr hat, dass ihm eine zweite nichts nützen kann. Ferner könnte man an einen Ring mit einem Edelstein denken, oder etwa an ein gemaltes Porzellanservice oder an einen physikalischen Messapparat. Da ich in diesem Sommer wahrscheinlich nach Paris reise, so könnte ich dort etwas aussuchen. Wenn Du mir irgend eine Art von Gegenstand bezeichnen wolltest, von der Du glaubst dass sie Hirst Freude machen könnte, so würdest Du mir einen grossen Gefallen thun. Natürlich wünsche ich, dass Hirst nichts davon erfährt.

Die Bogen Deines Buches von denen Du in Deinem Briefe schreibst, sind noch nicht angekommen. Wenn sie noch morgen oder übermorgen ankommen, so würde ich sie gern noch schnell durchlesen und sie ohne Aufenthalt zurücksenden; am Sonntag aber muss ich auf etwa eine bis zwei Wochen verreisen, indem ich versprochen habe, nach München und Würzburg zu kommen. In Würzburg muss ich unter anderem eine Wohnung für uns aussuchen und miethen, und das physikalische Cabinet durchsehen, um zu sehen, was etwa bis zum Herbst noch angeschafft werden muss, damit es für die Vorlesungen genügend ausgerüstet ist.

Was Du in Deinem Briefe über die Schwingungen einfacher Töne sagst, scheint mir in der That sehr geeignet, zur Vervollständigung der Helmholtz'-schen Ansichten über Dissonanzen zu dienen.

Deine Ankündigung, dass Du künftig auch die Baierischen Hochlande zu Deinen Sommerausflügen wählen willst, um uns dann in Würzburg zu besuchen, hat uns grosse Freude gemacht. Ich schicke Dir diesmal Bilder von der ganzen Familie: 1 Meine Frau mit dem kleinsten Mädchen (der kleinen Hedwig). 2 Ich mit Johnny. 3 die beiden ältesten Mädchen Mathilde und Helene.

Mit herzlichsten Grüssen von meiner Frau und mir an Dich und Hirst

Dein | Clausius.

ZURICH, 4/4 /67.

Dear Tyndall,

It was not until today that I was able to go to the watchmaker[1] only today to speak to him about the dispatch of your watch,[2] as I was unwell for a few days and had to stay at home. The watchmaker was unable to tell me how much the customs duty would amount to, but thought that in any case it would not be so high that it would therefore delay dispatch any longer. He has now undertaken to pack the watch well and to post it. I hope you will be satisfied with the watch. The bill of exchange sent by you has been accepted by the banker[3] at 301½ francs, and as the watch costs only 300 francs, I have instructed the watchmaker pay the postage for shipping it, whereby the difference will roughly balance itself out. I am enclosing the watchmaker's receipt here.

I would be glad if you could tell me sometime or other how things stand with the English translation of my book,[4] about which I have not heard anything for a long time as to whether it has already appeared, or will appear soon.

I would also like to ask you for your advice in one other matter. Hirst has had so much trouble with my book that I would really like to send him a gift as a token of my gratitude, and here I would be very glad if you could indicate something to me that would give him pleasure. I have thought about having such a watch made for him as you have just acquired now for yourself. However, for this I need to know whether he can do with it also, or whether perhaps he already has such a fine watch that a second one can be of no use to him. One could also consider a ring with a precious stone, or perhaps a painted porcelain set or a physical measuring apparatus. As I shall probably be travelling to Paris this summer, I could pick out something there. If you would indicate some kind of item or other to me that you believe would give Hirst pleasure, you would do me a great favour. Naturally, I do not wish Hirst to learn anything about it.

The sheets of your book[5] of which you write in your letter[6] have still not arrived. If they still arrive tomorrow or the day after, then I would still happily look through them quickly and send them back without delay; on Sunday, however, I have to travel for perhaps one to two weeks, for I have promised to go to Munich and Würzburg.[7] In Würzburg, among other things, I have to pick out and rent an apartment for us, and to inspect the physical laboratory in order to see what has to be purchased by about autumn so that it will be sufficiently fitted out for the lectures.[8]

What you say in your letter about the vibrations of simple tones indeed seems to me to be very suitable to serve to round off Helmholtz's views about dissonances.[9]

Your announcement that in future you will also choose the Bavarian highlands[10] for your summer excursions in order to visit us then in Würzburg has given us great pleasure. I am sending you this time pictures of the whole family as well: 1. My wife[11] with our youngest daughter (little Hedwig);[12] 2. Me with Johnny;[13] 3. the two oldest girls, Mathilde and Helene.[14]

With warmest regards from my wife and myself to you and Hirst

Your | Clausius.

RI MS JT/1/TYP/7/2301–2
LT Typescript Only

1. *watchmaker:* see letter 2581, n. 3.
2. *your watch:* see letter 2581, n. 3.
3. *the banker:* not identified.
4. *English translation of my book:* R. Clausius, *The Mechanical Theory of Heat, with its*

Applications to the Steam-Engine and to the Physical Properties of Bodies (London: John van Voorst, 1867). This work was edited by Thomas Hirst with an introduction by Tyndall.

5. *sheets of your book:* see letter 2581, n. 3.
6. *your letter:* letter 2581.
7. *Würzburg:* see letter 2581, n. 4 and n. 6.
8. *the lectures:* his lecturing duties at University of Würzburg, where Clausius took up a professorship in 1867.
9. *Helmholtz's views about dissonances:* see letter 2581, n. 8.
10. *Bavarian highlands:* see letter 2581, n. 5.
11. *my wife:* Adelheid Clausius.
12. *youngest daughter (little Hedwig):* Hedwig Clausius (1866–1946).
13. *Johnny:* Rudolf John Clausius.
14. *Mathilde and Helene:* Mathilde Clausius (1861–1907), who married Friedrich Zimmer, and Helene Clausius (1863–1919), who married Karl Budde.

From James Anderson Rose[1] 5 April 1867 2586

11 Salisbury Street, Strand.[2]

My dear Sir,

I am sure you were delighted at our success at Market Drayton.[3] Oh I would that you had heard Giffard's[4] speech—to enchain an audience for an hour, and to conclude still with everyone wishing you to go on, is the triumph of eloquence. It was perfect in law, in the facts, in discretion, sound judgment and good taste. But what I wish to say to you is this, to inform you of your share in the victory. We sat up till 4.0 on Friday morning and I then proposed to wind up by reading to Giffard your letter and speech[5] which I said I considered far away the best that had been said in Eyre's[6] defence.

I read them both to Giffard who agreed with me. He said he would read them both again before he went to sleep and again as soon as he woke, and I believe he did. You will not be offended will you if I say that your thoughts had every effect given to them by Mr. Giffard? Further I gave Mr. Eyre's life[7] by Hume[8] to a particular Baptist—a very clever young lady[9] of my acquaintance—and I hear she has said that no Englishwoman could read your speech and not be on Mr. Eyre's side.

Believe me to be | Yours very truly | J. Anderson Rose. | 5th April 1867.

RI MS JT/1/TYP/8/2586
LT Typescript Only

1. *James Anderson Rose:* James Anderson Rose (1819–90), a British solicitor and avid art collector, who defended Edward Eyre in the newspapers and in court ('Fine-Art Gossip', *Athenaeum,* 27 September 1890, p. 424; J. A. Rose, 'To the Editor', *Standard,* 14 February 1867, p. 6).
2. *11 Salisbury Street, Strand:* location of Rose's law firm in London.
3. *Market Drayton:* Market Drayton, Shropshire, where private charges were brought against Edward Eyre after he declined to appear alongside Colonel Nelson and Lieutenant Brand, at the Bow Street Magistrates' Court, London. On 29 March 1867 the magistrates at Market Drayton discharged Eyre on the grounds that the evidence brought before them would have resulted in an acquittal if the case were heard by a jury ('The Jamaica Prosecutions', *Times,* 30 March 1867, p. 12; see also reports with the same title appearing on 26, 28, and 29 March 1867).
4. *Giffard's:* Hardinge Stanley Giffard (1823–1921), first earl of Halsbury, a journalist, solicitor, and judge, who served as Eyre's counsel in the case brought against him at Market Drayton in 1867 (*ODNB*).
5. *your letter and speech:* see letter 2561, n. 4 (his speech), and letter 2571, n. 5 (his letter).
6. *Eyre's:* Edward Eyre.
7. *Mr. Eyre's life:* H. Hume, *The Life of Edward John Eyre, late Governor of Jamaica* (London: Richard Bentley, 1867).
8. *Hume:* Hamilton Hume (1797–1873), an Australian explorer and fellow of the Royal Geographical Society who helped to organize the Eyre Defence and Aid Fund Committee, serving as its secretary (*ODNB*).
9. *a very clever young lady:* not identified.

From James John Trayer 9 April 1867 2587

Bagenalstown[1] | 29th, April, 1867.

My dear Dr Tyndall

You, I fear, have not received my letter[2] directed to the Grand Hotel St James 211 Rue St Honoré Paris. I wrote the next post after I had made out your probable "where abouts" in Paris, i.e. on Wednesday the 24th inst[ant], at your mother's[3] request, to tell you of her very grave illness—

I have been in regular attendance on her since the 23rd—the day on which it became known that she had this terrible sore on her back—

I am very sorry to tell you that her strength has run down <u>very much</u> and thus lots of nourishment and fair amount of stimulants not-withstanding; I endeavoured in my former letter to explain to you that tho' she felt (and still today feels) as if she may come through this trial, that it was very doubtful if she could, and I grieve to say this doubt is still stronger to day. I do not think there is anything being left undone for her and as you seem to wish she should have a water bed—tho' there are particular circumstances which render it not so necessary as it often is—still I am writing to Dublin for a large water cushion for her which will answer your wish and not be expensive. I hope you will approve of this. She told me before and today again that she very much wished you to know how ill and suffering she is but that she did not want you to come over as she knew how this would put you about. It is not possible for any one to speak with certainty as to the result of, or with precision as to the duration to be expected in the case of such an illness. Still it is my duty to say to you plainly that I fear the worst and with no very long delay. At the same time my energies shall be directed for the best. Pray let me have a line by return to say if you are not coming over if it is your wish that I should in any emergency send you a telegram.

I hope I need not say that such plain speaking is not due to any want of feeling for the distress I inflict but is dictated by a desire to enable you to see your way clearly.

Your poor sister[4] is pretty much as usual but of course greatly agitated at her mother's weakness and suffering.

Yours very faithfully | J. J. Trayer

RI MS JT/5/16a
LT Typescript Only

1. *Bagenalstown:* a small town on the River Barrow in County Carlow, Ireland, about two miles south of Leighlinbridge.
2. *my letter:* letter missing.
3. *Mother's:* Sarah Tyndall (née Macassey).
4. *Your poor sister:* Emma Tyndall.

From Matthew Arnold[1] [before May 1867][2] 2588

ATHENAEUM[3] | Saturday—

My dear Tyndal

You do me a great deal too much honour,[4] and I should really like to lecture at the Royal Institution, but I hold on at Oxford till May, and with finishing there, and other things, I have my hands quite full for the present. But in 1868,[5] if all goes well and you will ask me again, I promise to try and give a lecture at the Institution.

Ever sincerely yours, | Matthew Arnold.—

Wellcome MS. 7777/102

---

1. *Matthew Arnold:* Matthew Arnold (1822–88), an English poet who was professor of poetry at Oxford from 1857–67 (*ODNB*).
2. *[before May 1867]:* the month and year are given by mention in the letter of the following year, which is 1868, and that May was still forthcoming.
3. *ATHENAEUM:* see letter 2537, n. 8. This is written on Athenaeum letterhead.
4. *You do me a great deal too much honour:* letter missing.
5. *But in 1868 . . . give a lecture at the Institution:* Arnold did not end up delivering a Friday Evening Discourse at the RI in 1868. He did, however, deliver discourses at the RI on 8 February 1878 (on 'Equality') and on 21 March 1884 (on 'Emerson').

From James John Trayer 1 May 1867 2589

TELEGRAM

From | Dr Trayer[1] | Bagenalstown[2]

To | Professor Tyndall | Royal Institution | May 1st, 1867.

Your mother[3] died last night telegraph at once when and where you wish the interment. Say will you come over[4]

RI MS JT/5/16a
LT Typescript Only

---

1. *Dr Trayer:* LT's typescript of the telegram shows 'Dr Fraser' but we believe this to be a mistranscription of 'Dr Trayer'.
2. *Bagenalstown:* see letter 2587, n. 1.
3. *Your mother:* Sarah Tyndall (née Macassey).
4. *come over:* Tyndall travelled to Ireland from 2–5 May for his mother's funeral. Thomas Hirst wrote in his journal for 2 May 1867: 'Went to see Hopwood. Tyndall ought to have accompanied me but he left suddenly to attend his mother's funeral in Ireland. Poor fellow the news was sudden and broke him down utterly. When my mother died it was he who broke the news to me in Marburg and did his best to sustain me under the calamity. It was he who put me into the diligence on my sad journey homewards. Now I tried to comfort him, we dined together at home previous to his departure and he tried to talk to me about his mother; abandoning his habitual reticence on such subjects he spoke with earnestness and deep affection of her rare qualities. I could not speak in return but shook his hand silently and bid him good bye. I fear the effect of this blow now when he had much work to do before he can leave for Switzerland' (Brock & MacLeod, *Hirst Journals*, p. 1802).

From Thomas Carlyle 5 May 1867 2590

Chelsea,[1] 5 May 1867.

Dear Tyndall,

I mean to call for you to-morrow (Monday) at Albermarle Street,[2] between 3 and 4. p.m.;—wishing 5 minutes of speech with you about a certain Mr Larkin[3] (of the "Magnesium Light"[4]), whose fate with the Trinity House people[5] is understood to be in your hands. He was my amanuensis,[6] map-maker, Index-maker, and Factotum[7] in all mechanical things for many years; and it is proper I should testify to you, in few words—, what kind of man and mechanician I found him.

Yours ever truly | T. Carlyle.

RI MS JT/1/TYP/1/203
LT Typescript Only

---

1. *Chelsea:* see letter 2546, n. 11.
2. *Albemarle Street:* location of the RI.
3. *Mr Larkin:* Henry Larkin (1820–99), a British inventor, employee of the Chelsea Steamboat Co., and assistant to Thomas Carlyle. He was his indexer from 1856, and later wrote *Carlyle and the Open Secret of His Life* (London: Paul, Trench, and Co., 1886) (*CLO*).
4. *"Magnesium Light":* H. Larkin, 'On a Magnesium Lamp', *Brit. Assoc. Report 1866,* p. 40: 'The distinguishing peculiarity of this lamp is that it burns magnesium in the form of powder, instead of riband [ribbon] or wire, and does not depend on clockwork or any similar extraneous motive power for its action. The stream of the metal powder is mixed with a small portion of gas and fine sand in its progress through the tube; they escape together at its mouth, where they are ignited and continue burning with a brilliant flame'. From the *Journal of the Society of Arts* (September 7, 1866), p. 519: 'At the *soirées* of the British Association, at Nottingham, the large refreshment annexe was lighted by two magnesium lamps of a novel character, the invention of Mr. H. Larkin'.
5. *Trinity House people:* while Tyndall had assisted Michael Faraday since May 1866 in lighthouse matters (see letter 2423, *Tyndall Correspondence,* vol. 9), Tyndall would in July 1867 officially take over for Michael Faraday as Scientific Adviser to the Board of Trade (a British government department concerned with making policy and instituting regulations regarding economic activity in Britain and its empire) and Trinity House (the lighthouse authority for England and Wales, established by Royal Charter in 1514) (see letters 2614, 2624, 2727, 2791, and 2829). In this role, Tyndall investigated and gave recommendations to Trinity House on matters of lighthouse illumination (magnesium lamps, it turned out, were not suitable), and later, sound signaling. On Tyndall and his role with the Board of Trade and Trinity House, see R. M. MacLeod, 'Science and Government in Victorian England: Lighthouse Illumination and the Board of Trade, 1866–1886', *Isis,* 60 (1969),

pp. 4–38; and *The Ascent of John Tyndall*, pp. 282–5, 316–7, 345–6, 389, 406–13, 420–3, 433–4, and 437. For a history of Trinity House, see A. Adams and R. Woodman, *Light Upon the Waters: The History of Trinity House 1514–2014* (London: The Corporation of Trinity House, 2013).

6. *amanuensis:* one who copies or writes from the dictation of another (*OED*).
7. *Factotum:* an employee who does all kinds of work (*OED*).

From George Gabriel Stokes 14 May 1867 2591

Cambridge 14th May 1867

My dear Tyndall,

By the regulations now and for many years in force (including the year 1853) the exam[inatio]$^{\underline{n}}$ for math & honours[1] is divided into two parts, 3 days for the easier subjects and after an interval 5 more for the more advanced.

The result of the first 3 days determines whether a man gets honours or not; if he does, the place he gets is determined by the result of all the 8 days.

The papers in the first 3 days are set according to a schedule which you will find in the Cambridge calendar at the Athenaeum.[2] One paper is devoted to hydrostatics and optics; acoustics and heat don't come in. The question of passing or not is decided by the total marks got in the 3 days. As optics occupy the half of one paper out of only 5 (for one paper is for problems only) men would not venture to leave it out.

In the last 5 days, several questions are put in the higher parts of common optics & the undulatory theory. Questions on sound are set but rather sparingly: in hastily looking over the papers for 1853 I found only one: I may possibly have missed one. Heat does not come in*. You now I think have got your data.

*except so far is is[3] found in books on hydrostatics such as questions on thermometers, heat developed by compression &c.

Yours very truly | G. G. Stokes

RI MS JT/1/S/243b
RI MS JT/1/TYP/4/1404

---

1. *exam[inatio]$^{\underline{n}}$ for math & honours:* the Mathematical Tripos, a written examination of undergraduate students of the University of Cambridge. Tyndall probably had asked Stokes about this (in a missing letter) in relation to his participation in the BAAS Committee on Scientific Education (see letters 2577 and 2579).
2. *Athenaeum:* see letter 2537, n. 8.
3. *is:* here Stokes repeated the word 'is'.

From Thomas Henry Huxley [24][1] May 1867 2592

Athenaeum Club.[2] | May 1867.

My dear Tyndall,

A conversation I had with Bence Jones yesterday reminded me that I ought to have communicated with you. But we do not meet so often as we used to do, being I suppose both very busy, and I forget to write.

You recollect that the last time we talked together you mentioned a notion of Bence Jones's to make the Fullerian Professorship of Physiology[3] a practically permanent appointment, and that I was quite inclined to stick by that (if such arrangement could be carried out) and give up other things.

But since I have been engaged in the present course of lectures[4] I have found reason to change my views. It is very hard work, and takes up every atom of my time to make the lectures what they should be; and I find that at this time of year, being more or less used up I suppose with the winter work, I stand the worry and excitement of the actual lectures very badly. Add to this that it is six weeks clean gone out of the only time I have disposable for real scientific progress, and you will understand how it is that I have made up my mind to resign.

I put all this clearly before Bence Jones yesterday, with the proviso that I could and would do nothing that should embarrass the Institution[5] or himself.

If there is the least difficulty about supplying my place or if the Managers[6] think I shall deal shabbily with them by resigning before the expiration of my term, of course I go on. And I hope you will all understand that I would do any thing rather than put even the appearance of a slight upon those who were kind enough to elect me.

Ever yours | T. H. Huxley

IC HP 8.53
Typed Transcript Only

---

1. *[24]:* the day is given based on relation to letter 2593.
2. *Athenaeum Club:* see letter 2537, n. 8.
3. *Fullerian Professorship of Physiology:* the RI Fullerian Professorship of Physiology was established in 1834 by an endowment from John Fuller (1757–1834). Thomas Huxley held the professorship twice, from 1855–8 and 1865–9.
4. *present course of lectures:* Huxley delivered twelve lectures at the RI, 'On Ethnology', on Thursdays and Saturdays from 2 May to 8 June 1867 (*Roy. Inst. Proc.,* 5 (1866–9), p. 25).
5. *the Institution:* the RI.

6. *the Managers:* the following were elected Managers of the RI on 1 May 1867: Henry Wollaston Blake (1815–99), George Busk (1807–86), Edward Cardwell (1813–86), Robert Gascoyne-Cecil (1830–1903), Warren de la Rue (1815–89), William Erle (1793–1880), John Peter Gassiot (1797–1877), John Hall Gladstone (1827–1902), Caesar H. Hawkins (1798–1884), John Carrick Moore (1805–98), Roderick Murchison (1792–1871), Algernon Percy (1810–99), William Pole (1814–1900), William Frederick Pollock (1815–88), and Philip James Yorke (1799–1874). Henry Holland (1788–1873) was elected President, William Spottiswoode (1823–83) was elected Treasurer, and Henry Bence Jones (1813–73) was elected Secretary ('Annual Meeting', *Roy. Inst. Proc.,* 5 (1866–9), p. 145; *ODNB*).

To Thomas Henry Huxley 25 May 1867 2593

25th. May, 1867.

My dear Huxley,

Had I known that your lectures at the Royal Institution[1] would take up six weeks of your time annually, instead of being gladdened by the prospect of having you there permanently I should have exhorted you to have nothing to do with them.

I do not think the men who elected you[2] will ever dream of associating with your present act[3] any other motive than those which you assign for it.

Give my love to the "pretty chickens and their dam."[4]

Yours ever | John Tyndall.

IC HP 8.54
RI MS JT/1/TYP/9/2911
Typed Transcript Only

---

1. *your lectures at the Royal Institution:* see letter 2592, n. 4.
2. *the men who elected you:* the Managers of the RI who elected Huxley Fullerian Professor of Physiology on 3 July 1865 ('General Monthly Meeting', *Roy. Inst. Proc.,* 4 (1862–6), p. 468; see p. 434 of the same volume for the names of the Managers at the time of Huxley's appointment).
3. *your present act:* Huxley considered resigning from the Fullerian Professorship of Physiology at the RI (see letter 2592). He previously held it from 1855–8, was elected again in 1865, and would remain in the position until 1869.
4. *"pretty chickens and their dam.":* Tyndall was quoting Macduff in Shakespeare's *Macbeth* IV.iii.258, referring in this case to Huxley's wife, Henrietta Huxley, and their children, Jessie, Marian, Leonard, Rachel, Henrietta, Henry, and Ethel.

From Rudolf Clausius 27 May 1867 2594

ZÜRICH, 27/5/67.

Lieber Tyndall,

Es sind nur wenige unbedeutende Puncte die ich zur VIII Vorlesung welche ich gleichzeitig mit diesem Briefe abschicke, zu erwähnen habe.

S.307 in der diatonischen Tonleiter ist der 7te Ton mit B bezeichnet; in der deutschen Bezeichnung heisst aber der 7te Ton der diatonischen Tonleiter, H.

S.308 in der untersten Zeile und S. 309, in der Zeile 18 von oben steht: one second statt half a second.

S.313. In der Aufzählung der verschiedenen Figuren welche entstehen, während die eine Stimmgabel der anderen um eine ganze Schwingung voreilt, ist nur eine gerade Linie erwähnt, während doch wohl die beiden auf einander senkrechten geraden Linien, welche auf S.319 in der obersten Figurenreihe vorkommen erwähnt werden müssen.

Ausserdem habe ich nur ein Paar Druckfehler bemerkt, die ich am Rande markirt habe, z.B. S.302: 8 statt 6 und S. 305: c' statt c''.

Ich habe alle Vorlesungen mit grossem Vergnügen gelesen und ich glaube gewiss, dass das Buch einen sehr guten Erfolg haben wird.

Mit herzlichen Grüssen

Dein | Clausius.

ZURICH, 27/5/67

Dear Tyndall,

There are only a few minor points that I have to mention on the VIIIth lecture,[1] which I am posting at the same time as this letter.

On p. 307 in the diatonic scale the 7th tone is noted with B; in the German notation, however, the 7th tone of the diatonic scale is called H.

On p. 308 in the bottommost line and on p. 309, in line 18 from the top it says: one second instead of half a second.

On p. 313. In the list of the different figures which arise while the one tuning fork precedes the other one by one full vibration, only one straight line is mentioned, while both of the straight lines perpendicular to each other which occur on p. 319 in the uppermost row of figures probably have to be mentioned.

Apart from that, I have noticed just a couple of printing errors, which I have marked on the margin, e.g. p. 302: 8 instead of 6 and on p. 305: c' instead of c''.

I have read all the lectures with great pleasure and I believe for certain that the book[2] will enjoy very good success.

With warm regards
Your | Clausius.

RI MS JT/1/TYP/7/2303
LT Typescript Only

---

1. *VIIIth lecture:* 'Lecture VIII' of *Sound* (1867).
2. *the book: Sound* (1867) would be published on Friday, 5 July 1867 (*Athenaeum,* 29 June 1867, p. 843).

To [Thomas Milner Gibson][1] [28 May 1867][2] 2595

Royal Institution of Great Britain | Tuesday

Dear Sir,

If you name a time my assistant,[3] to whom I have given the necessary instructions, will subject the needle in your daughter's[4] foot to more powerful magnetic traction than is possible with the magnet I gave you yesterday. It is worth while deciding the value of this method.

Yours truly | John Tyndall

Mss.Ms.C011.200. American Philosophical Society

---

1. *[Thomas Milner Gibson]:* Thomas Milner Gibson (1806–84), an English politician who served as President of the Board of Trade 1859–66. In 1832 he married (Susannah) Arethusa Cullum (1814–85), the noted political activist and hostess (*ODNB*). See n. 2.
2. *[28 May 1867]:* the date is given by relation to a journal entry of Thomas Hirst: 'Dr Percy introduced Milner Gibson to me at the Athenaeum. His daughter had trodden on a needle which had broken. He wished to have a powerful magnet to try and pull out the broken portion. I went with him to the Royal Institution and put him in Tyndall's hands. The magnet was sent but didn't pull out the needle' (Brock & MacLeod, *Hirst Journals,* 27 May 1867 p. 1803).
3. *my assistant:* probably John Cottrell.
4. *daughter's:* either Alice Mary Robertson (née Gibson, 1843–79) or Sydney Isabella Gibson (1849–80), daughters of Thomas Milner Gibson and Arethusa Gibson (née Cullum) (see J. J. Howard and F. A. Crisp (eds), *Visitation of England and Wales,* 33 vols (London: Privately Printed, 1893–1921), vol. 1, pp. 150–1).

To Augusta Huth[1] [28 May 1867][2] 2596

Tuesday

My dear Mrs. Huth

Before answering your note[3] I have been trying whether there is any reasonable possibility of my abandoning the Institution[4] on Friday the 7th.

To my deep regret I do not see it. It is the last evening of the year,[5] and a peculiar evening as to the lecturer.

Dr. Bence Jones[6] is not able to attend, Mr. Faraday of course is out of the question[7] and the Institution has only myself to look to that all shall go right. I do not add an iota to the truth when I say that there are few if any tables in London that have a greater attraction for me than yours, and when Hirst[8] first spoke of your entertaining the idea of feasting us I heard of it with delight. If Ruskin[9] does not keep us too long, and if I have shaken off the sore throat which now bothers me I will drive down after the lecture[10] and hear your plans about Switzerland.

Yours ever truly | John Tyndall

RI MS JT/1/TYP/2/651
LT Typescript Only

---

1. *Mrs. Huth:* probably Augusta Huth (née von Westenholz, 1823–99) who married Henry Huth (1815–78), a noted book collector. She was elected a member of the RI on 2 March 1868 (*Roy. Inst. Proc.*, 5 (1866–9), p. 308).
2. *[28 May 1867]:* the date is given based on mention of a Friday Evening Discourse on 7 June and the suggestion that Michael Faraday's health is in decline, both indicating 1867, with Tuesdays before the lecture being 28 May or 4 June. Since this precedes Tyndall's letter to Huth of 3 June (2600), this letter would then be 28 May.
3. *answering your note:* while Huth's letter is missing, Tyndall's reply to it following this letter is letter 2600.
4. *Institution:* the RI.
5. *last evening of the year:* on 7 June 1867, the art critic John Ruskin gave a Friday Evening Discourse at the RI, 'On the Present State of Modern Art with Reference to the Advisable Arrangements of a National Gallery'. Tyndall gave a Friday Evening Discourse on 21 June 1867 that was not previously scheduled (see letter 2603, n. 4).
6. *Dr. Bence Jones:* Henry Bence Jones.
7. *Mr. Faraday of course is out of the question:* at this time, Michael Faraday was in decline and would die on 25 August 1867.
8. *Hirst:* Thomas Hirst.
9. *Ruskin:* John Ruskin (1819–1900), an English art and social critic. While one of the

leading art critics of his generation Ruskin also wrote and lectured on many subjects including geology, literature, botany, and economics (*ODNB*).

10. *I will drive down after the lecture:* Hirst noted in his journal on 7 June 1867 that Tyndall arrived after dinner (Brock & MacLeod, *Hirst Journals,* p. 1804).

To unidentified[1] 31 May [1867][2] 2597

Royal Institution of Great Britain | 31st May

Sir

Please send me twelve copies of the Fortnightly Review containing my article on "Miracles & Special Providences"[3]

John Tyndall

Private Collection

1. *unidentified:* possibly John Morley (1838–1923), who succeeded George Henry Lewes (1817–78) as editor of the *Fortnightly Review* in January 1867 (until 1882) (*ODNB*); or more likely an assistant with the publication.
2. *[1867]:* the year is given by reference to the article mentioned in the letter (see n. 3).
3. *"Miracles & Special Providences":* J. Tyndall, 'Miracles and Special Providences', *Fortnightly Review,* 7 (1867), pp. 645–60.

To William Francis May 1867 2598

Note on the preceding Paper. | By Professor TYNDALL, *F.R.S.*

DEAR FRANCIS,

It gives me pleasure to learn from you that a translation of Professor Magnus's[1] last paper on the Action of Aqueous Vapour upon Radiant Heat[2] will be published in the forthcoming Number of the Philosophical Magazine.[3] In 1862, while walking with Professor Magnus through the streets of London, he drew my attention to the untransparent state of the atmosphere, and expressed the opinion that London smoke had much to do with the results that I had ascribed to aqueous vapour. That remark caused me to make a journey to the middle of the Isle of Wight,[4] during the prevalence of a south-west wind, so as to obtain a quantity of air which should be free from the taint of the metropolis. I cite this fact to illustrate the spirit in which I have met the objections which Professor Magnus has raised from time to time against my experiments.[5] Those objections demand, have received, and will continue to receive the utmost consideration from me. In all their forms (and these have been very various) I have endeavoured to meet them by the production of clear and irrefragable facts. This course it is my intention to pursue[6] with reference to Professor Magnus's last paper.

Yours very truly, | JOHN TYNDALL.

Royal Institution, May 1867.

Phil. Mag., 33 (1867), p. 425

1. *Professor Magnus's:* Heinrich Gustav Magnus.
2. *last paper on the Action of Aqueous Vapour upon Radiant Heat:* G. Magnus, 'Ueber den Einfluss der Vaporhäsion bei Versuchen über Absorption der Wärme', *Poggend. Annal.*, 130 (1867), pp. 207–26.
3. *will be published in the forthcoming Number of the Philosophical Magazine:* G. Magnus, 'On the Influence of the Adhesion of Vapour in Experiments on the Absorption of Heat', *Phil. Mag.*, 33 (1867), pp. 413–25. This paper and Tyndall's 'Note' appeared in the June 1867 number of the *Phil. Mag.*
4. *a journey to the middle of the Isle of Wight:* Tyndall visited the Isle of Wight in October 1862.
5. *the objections which Professor Magnus has raised from time to time against my experiments:* see letter 2578, n. 2, and letter 2580.
6. *my intention to pursue:* see letter 2578, n. 2.

From [Mary Egerton][1,2] [May 1867][3] 2599

it to you as to what part would be most level to our capacity, (if you grant my request.) We have the book on Heat,[4] so perhaps we c[oul][d] pick up a little of that, even in the middle of a series. Or if you are going to speak of the grand subject of Forces in general & their Co-relation it w[oul][d] be most interesting, though I fear too ambitious a subject for us.

Professor Miller[5] does not love his spectra as you do, lets them fall on any corner of the screen & sometimes eclipses them with his own body, & the lamps do not know him & obey him as they would their legitimate master. If you happened ever to be walking this way late any afternoon, I am almost always at home towards 6 o'clock. I know you can have no time for visiting, so ought not to propose any abridgement of y[ou][r] walk, but in these Parliament days[6] one never has an evening to ask anybody. We have changed our

RI MS JT/1/E/991

1. *[Mary Egerton]:* the recipient is given based on the handwriting, which is a match for Mary Egerton.
2. Letter incomplete; fragment only.
3. *[May 1867]:* the month and year are given by reference to William Allen Miller's 1867 course of four lectures at the RI on 'Spectrum Analysis, including its application to Astronomy' (*Roy. Soc. Proc.,* 5 (1866–9), p. 25).
4. *book on Heat: Heat* (1863) or *Heat* (1865).
5. *Professor Miller:* William Allen Miller (1817–70), an English chemist who held the chair of Chemistry at King's College London and was treasurer of the RS from 1861–70 (*ODNB*).
6. *these Parliament days:* Mary's husband Edward Egerton served as Parliamentary Under-Secretary of State for Foreign Affairs from 6 July 1866–1 December 1868.

To Augusta Huth[1] 3 June [1867][2] 2600

3rd. June[3]

My dear Mrs Huth

If you go to Chamounix[4] I would recommend you to go up to the Montanvert[5]—if equal to the task go to the Jardin[6]—but this is a very heavy day's work. At all events cross the glacier from the Montanvert; and go down by the Chapeau,[7] hav[in]g your mules to meet you at the other side.

The view from the Flegère[8] is also worth seeing.

From Chamounix you may cross the Tete noir,[9] or the Col de Balme[10] to Martigny[11]—thence by rail to Sion,[12] whence I would advise you to take a carriage to Visp.[13] From Visp you ride to Zermatt, where you may put up at the Hotel du Mont Rose.[14] At the Riffelberg[15] 3000 feet higher up you can get quarters. Go to the Görner Grat,[16] and look over the vast glaciers of Monte Rosa.[17] See the grand Matterhorn.[18] From the Riffelberg you will also see my beloved Weishorn.[19] If you wish to linger go to the Hörnli.[20] The view from it is very fine & the exercise exhilarating

From Zermatt return to Visp: thence to Brieg,[21] thence to Viesch,[22] and thence ascend to the Hotel Jungfrau[23] on the slopes of the AEggischhorn.[24] You may well stay in this region a couple of days.

From the AEggischhorn proceed up the Valley of the Rhone[25] to the Rhone Glacier[26]—go upon the ice up as close as you can to the ice fall, and drink my health when you are unable to go any further.

From the good hotel (kept by the Zermatt people)[27] at the end of the Rhone Glacier, you cross the Grimsel Pass.[28] Halting at the Grimsel you can ascend the Siedelhorn:[29] view very grand.

I do not think the end of the Aar Glacier[30] would repay you. It is so covered with debris that you cannot get upon the ice.

From the Grimsel go down the Valley of Hasli,[31] looking at the fall of Handeck[32] on your way to Meyringen.[33] There is a nice little hotel at Imhof[34] also, if you like to see a fine gorge you can look into the Finsteraarschlucht[35]—but it is not quite easy to get at.

From Meyrigen you cross the great Scheideck[36] to Grindelwald[37]—a noble mountain walk or ride—lots of hotels at Grindelwald.

From Grindelwald cross the Wengern Alp,[38] and admire the Jungfrau[39] and give her my love.

You then descend to Lauterbrunnen[40] and go down the valley to Interlaken. From Lauterbrunnen you may if you like go up to Mürren[41]—and ascend the Schilthorn[42] (where poor Mrs Arbuthnot was killed[43])—views noble.

From Interlaken if your time be limited you can go straight to Ball[44] & then to Paris.

But you can also go up the Brienzersee,[45] see the fall of the Giessbach,[46] and cross the Bruning[47] to Lucerne, whence if you are ignobly inclined you can ascend the Rigi.[48] A less ignoble inspiration will carry you up Pilatus.[49]

From Lucerne there are railways in all directions. See while there Thorwaldsen's Lion.[50]

Instead of crossing the Tete Noir from Chamounix you can if you like cross the Col de Bon Homme[51] and go through the Allé Blanche[52] to Courmayeur.[53] This is a fine excursion but then you are in Italy. Excuse this![54]

Good bye—A pleasant journey, good weather and glorious thoughts!

Yours | J. Tyndall

RI MS JT/1/T/1025
Transcript Only[55]

---

1. *Augusta Huth:* see letter 2596, n. 1.
2. *[1867]:* the year is given by relation to letter 2596, dated [28 May 1867].
3. *June:* at this point in the handwritten transcript, '1866 or 1867' has been added.
4. *Chamounix:* Chamonix, a town in the valley north of Mont Blanc in south-eastern France.
5. *Montanvert:* a lookout point above the Mer de Glace, a large glacier on the north side of Mont Blanc.
6. *Jardin:* a rocky spot in the middle of the Talèfre glacier frequently visited by tourists to the glacier, containing a profusion of plants at high altitude.
7. *Chapeau:* a lookout point above the Mer de Glace, a large glacier on the north side of Mont Blanc.
8. *Flegère:* a hill in the valley of Chamonix.

9. *Tete noir:* Tête Noir, a pass on the route between Chamonix and Martigny.
10. *Col de Balme:* an alternative pass between Chamonix and Martigny, between the summits of the Tête de Balme and Les Grandes Otanes.
11. *Martigny:* a town in the Rhône valley in the Swiss canton of Valais.
12. *Sion:* a town town in the Rhône valley, and the capital of the Swiss canton of Valais.
13. *Visp:* a town in the Rhône valley in the Swiss canton of Valais, at the junction of the valley to Zermatt.
14. *Hotel du Mont Rose:* the Monte Rosa Hotel.
15. *Riffelberg:* referring to the hotel at Riffelberg, above Zermatt.
16. *Görner Grat:* a rocky ridge, located southeast of Zermatt, Switzerland, and which overlooks the Gorner glacier.
17. *Monte Rosa:* the second highest mountain in the Alps located between Switzerland and Italy near the town of Zermatt.
18. *Matterhorn:* one of the high peaks of the Alps, located on the border between Switzerland and Italy. Tyndall attempted to climb the Matterhorn in August 1860 and July 1862 (*Hours of Exercise in the Alps,* pp. 27–52, 153–65), and his party was the first to traverse the mountain from the Italian to Swiss side from 26–8 July 1868 (see letter 2869; and *Hours of Exercise in the Alps,* pp. 271–94). Tyndall had wished to climb the Matterhorn in 1867 but the weather held him back (see letter 2651).
19. *Weishorn:* the Weisshorn, one of the high peaks of the Alps. On 19 August 1861, Tyndall and his guides were the first to summit the Weisshorn. See *Mountaineering in 1861,* pp. 41–58.
20. *Hörnli:* a mountain below the Matterhorn.
21. *Brieg:* Brig, a town in the Rhone Valley in the Swiss canton of Valais.
22. *Viesch:* Fiesch, a town in the Rhone Valley in the Swiss canton of Valais.
23. *Hotel Jungfrau:* a hotel halfway up the Eggishorn, opened in 1841. Tyndall stayed there often in his alpine travels.
24. *AEggischhorn:* Eggishorn, a mountain of the Bernese Alps in the Swiss canton of Valais.
25. *Valley of the Rhone:* the valley formed by the Rhône river, which starts in Switzerland and passes through southeastern France.
26. *Rhone Glacier:* a glacier in Switzerland that forms the river Rhône.
27. *good hotel (kept by the Zermatt people):* not identified.
28. *Grimsel Pass:* the mountain pass separating the upper valley of the river Aare from the upper valley of the Rhône.
29. *Siedelhorn:* the Sidelhorn, a mountain of the Bernese Alps, located west of the Grimsel Pass.
30. *Aar Glacier:* a system of glaciers located at the sources of the river Aare in the Bernese Alps, consisting of the Unteraargletscher (formed by the confluence of the Lauteraargletscher and Finsteraargletscher) and the Oberaargletscher.
31. *Valley of Hasli:* or Haslithal (now Haslital),
32. *fall of Handeck:* the waterfall above Handegg, a village in the Swiss canton of Bern.

33. *Meyringen:* Meiringen, a town in the Swiss canton of Bern.
34. *hotel at Imhof:* not identified.
35. *Finsteraarschlucht:* a gorge in the Haslital (see n. 31), which Tyndall described in a paper: J. Tyndall, 'On the Conformation of the Alps', *Phil. Mag.*, 28 (1864), pp. 255–71, on p. 259.
36. *Scheideck:* Gross Scheidegg, a mountain pass between the Schwarzhorn and the Wetterhorn in the Bernese Alps of Switzerland. One would cross this pass when traveling between Meiringen and Grindelwald.
37. *Grindelwald:* a town in the Swiss canton of Bern.
38. *Wengern Alp:* the slopes on the Lauterbrunnen side of the pass over Kleine Scheidegg towards Grindelwald.
39. *Jungfrau:* a mountain of the Bernese Alps.
40. *Lauterbrunnen:* a village in the Swiss canton of Bern.
41. *Mürren:* a village in the Lauterbrunnen valley.
42. *Schilthorn:* an alpine peak in the Swiss canton of Bern, above Lauterbrunnen.
43. *where poor Mrs Arbuthnot was killed:* on 21 June 1865, Alice Charlotte Arbuthnot (née Pitt-Rivers, b. 1841), first wife of William Arbuthnot (1838–93), was killed by lightning on the Schilthorn during a sudden thunderstorm ('Obituary', *Gentleman's Magazine, Or Monthly Intelligencer,* 80 (August 1865), p. 257).
44. *Bale:* Basel, a city in northwestern Switzerland, located on the Rhine river (Bale is the French name for the city).
45. *Brienzersee:* a lake in the Swiss canton of Berne, north of the Alps.
46. *fall of the Giessbach:* Giessbach Falls is located near the Brienzersee, Switzerland.
47. *Bruning:* a mountain pass connecting Meiringen to Rosenlaui to the south.
48. *Rigi:* a mountain massif of the Swiss Alps surrounded by several lakes.
49. *Pilatus:* a mountain massif in the Swiss Alps overlooking Lucerne.
50. *Thorwaldsen's Lion:* the Lion Monument, carved in sandstone overlooking a pond in Lucerne, Switzerland. Designed by the Danish sculptor Bertel Thorvaldsen (1770–1840) and carved by the German stonemason Lukas Ahorn in 1820–1, the monument commemorates members of the Swiss Guard that died while protecting the French royal family when a revolutionary mob attacked the Tuileries Palace in Paris on 10 August 1792.
51. *Col de Bon Homme:* a mountain pass in the Vosges range of France.
52. *Allé Blanche:* Vallée Blanche, a glacier leading down to Courmayeur.
53. *Courmayeur:* a town in the Aosta valley of northern Italy, located at the bottom of the southern side of Mont Blanc.
54. *Excuse this!:* following 'you are in Italy', the following words have been crossed out, with 'Excuse this!' immediately after: 'and to get to the north side of the Alps you will have either to retrace your steps or to cross one of the passes'.
55. LT included a note at the end of the letter that reads: '3/4 an hour to copy, what must it have taken to think out & write! L. C. T.'.

To Henry John Roby[1] 3 June 1867 2601

2. From JOHN TYNDALL, Esq., F.R.S.,
Professor of Physics in the Royal School of Mines.[2]

June 3, 1867.

SIR,

I HARDLY think that an Exhibition in Paris[3] furnishes the means of accurately testing the comparative merits of English and French education.

The simple inconvenience of transport tends to render England worse represented than France.

Still on other grounds I would express a general concurrence in the views of Dr. Playfair.[4] The facilities for scientific education are far greater on the Continent than in England, and where such differences exist, England is sure to fall behind as regards those industries into which the scientific element enters.

In fact, I have long entertained the opinion, that in virtue of the better education provided by continental nations, England must one day—and that no distant one—find herself outstripped by those nations, both in the arts of peace and war. As sure as knowledge is power this must be the result.

I am, &c. | JOHN TYNDALL.

H. J. Roby, Esq. | &c. &c.

'Report Relative to Technical Education' (Schools Inquiry Commission), C 3898, Parliamentary Papers (1867), p. 10

---

1. *Henry John Roby:* Henry John Roby (1830–1915), an English classical scholar, author, and Member of Parliament, who was involved in educational reform. He served as secretary of the Schools Inquiry Commission (1864–7), and later served on the endowed schools commission (*ODNB*). The Schools Inquiry Commission was established in 1864 by Henry Labouchere (Lord Taunton, 1798–1869) to 'examine those schools for boys of the middle class that had escaped scrutiny in the earlier Commissions on Popular Education (1858) and Public Schools (1861)' (R. M. Tullberg, *Women at Cambridge* (Cambridge: Cambridge University Press, 1998 [1975]), p. 24). Pedagogical and administrative aspects of secondary education were investigated, and a twenty-one volume report issued in 1868–9 (K. H. Parshall, *James Joseph Sylvester: Jewish Mathematician in a Victorian World* (Baltimore: John Hopkins University Press, 2006), p. 193).
2. *Royal School of Mines:* the Government School of Mines and Science Applied to the Arts, established in 1851 (from 1863 known as the Royal School of Mines). Tyndall held the

Chair of Natural Philosophy at the Royal School of Mines from 1859–68, and in this role lectured to students during the week and in some years gave evening lectures for working men. On the Royal School of Mines, see R. M. Clary, 'The Royal School of Mines: Henry de la Bache's Convergence of Professionalization and Public Advocacy', *Earth Sciences History*, 39 (2020), pp. 291–304.

3. *Exhibition in Paris:* see letter 2578, n. 3.
4. *Dr. Playfair:* Lyon Playfair (1818–98), first Baron Playfair, a chemist and politician. Playfair assisted Prince Albert with the planning of the Great Exhibition of 1851 and from 1853 served as secretary of science in the Department of Science and Art, in which role he sought to increase the quality of science education in Britain (*ODNB*).

From George Campbell 8 June 1867 2602

Argyll Lodge,[1] | Kensington. | June 8/67

My Dear Professor Tyndall

Can you give us the pleasure of y[ou]r. company at dinner on Saturday the 15th at ¼ to 8?

Y[ou]rs | Argyll

RI MS JT/1/A/97
RI MS JT/TYP/1/68

1. *Argyll Lodge:* previously Bedford Lodge and later Cam House, a home of Argyll's in Kensington, London from 1852 until his death in 1900.

To Jane Barnard 11 June [1867][1] 2603

11th June

My dear Miss Barnard.

I am anxious to make an experiment[2] of your uncle's[3] on my Friday evening.[4] He used to lend me in former days a piece of heavy glass for the experiment—Could you lay your hand upon a piece of the glass and lend it to me? After the lecture I would return it to you.

The experiment requires time & care for its preparation, if therefore you could let me have the piece of glass sometime before this week ends I should be extremely obliged to you.

yours most truly | John Tyndall[5]

RI MS JT/1/T/121
RI MS JT/1/TYP/12/4193

1. *[1867]:* the year is given based on mention of Tyndall's Friday Evening Discourse of 21 June 1867 (see n. 4).
2. *an experiment:* a reference to Faraday's discovery of the rotation of plane polarised light passing through his heavy glass in a magnetic field, and his subsequent discovery of diamagnetism. See M. Faraday, 'Experimental Researches in Electricity. Nineteenth Series.', *Phil. Trans.*, (1846), pp. 1–20, M. Faraday, 'Experimental Researches in Electricity. Twentieth Series.', *Phil. Trans.*, (1846), pp. 21–40, and F. A. J. L James, *Michael Faraday: A Very Short Introduction* (Oxford: Oxford University Press, 2010), pp. 76–82. Faraday's 'heavy glass', a lead borosilicate, had a much greater ability to rotate plane polarised light than ordinary glass, making the effect readily demonstrable.
3. *your uncle's:* Michael Faraday.
4. *my Friday evening:* on 21 June 1867, Tyndall gave a Friday Evening Discourse at the RI, 'On Some Experiments of Faraday, Biot and Savart'. This was not originally scheduled, and was requested by the President of the RI, Henry Holland (*Roy. Inst. Proc.,* 5 (1869), pp. 188–90).
5. *Tyndall:* LT wrote on the typescript: 'c/o Prof Faraday | Hampton C^t Green | Kingston on Thames | Hampton C^t' (Michael Faraday's house in London since 1858).

From Mary Egerton 11 June [1867][1] 2604

June 11th

Dear M^r. Tyndal[2],

I have been reading with intense interest an article of yours in the Fortnightly Review.[3] You have contrived to Smash the arguments of your opponent, without ever departing from that tone of moderation, dignity, & reverence, which is so suitable to those who "follow truth along her star-paved way",[4] but which is too often forgotten on both sides. But in the inference you imply, I cannot at once acquiesce. The subject has haunted me, and I feel as if I must talk to you about it, though I don't know what possible right I have to bother you with my thoughts upon such things! But it is your own fault for being so kind & good natured to us, that I cannot help looking upon you as a friend. Did you ever read an article by the Duke of Argyll "on the Supernatural"?[5] It forms now the first chapter of his book on the "Reign of Law",[6] and is, it seems to me, the most striking part of the book. I should so much like to know what you think of it? His view, that what we term miracles, or Special Providences, are not interference with the laws of nature, but adaptations of those laws, amalgams to those by which man has been enabled in a degree to rule nature, & to make it work his own Ends, seems so much the most satisfactory. Look at the illustration you afforded us of that the other day.[7] Was it not by the laws of nature that that geyser exploded? but you arranged the pipe, the basin, the heating apparatus—there was Providence—then you kindly thought we should enjoy seeing it, so you forestalled its place in your lectures for our benefit; might not that be compared to a Special Providence?—And may we not suppose that He to whom we cannot but attribute, at the least a perfect knowledge of all the laws of nature, may thus from an infinitely higher point control their springs?—but I know one is trembling here on the verge of an Abyss of mystery.

How truly you describe those flashes of truth that come upon the philosopher's mind, as "inspirations"! I believe they come, as surely as moral truth, from the one great Source of Light. It is fanciful, but it always seems to me that the Sun's mingled rays of heat & light, figure to us moral goodness, and intellectual illuminations, and as some instances in the physical world are transparent to one set of rays & not to the other, so it is in the Spiritual. Happy those whose Spirits can receive and transmit both! I suppose only one has ever done that perfectly on Earth, but some of you may be made like unto Him hereafter, for is He not the perfect man? Can Prof[esso]r Huxley persuade us that a Spirit full of Philosophy of Poetry of admiration for all that is glorious and beautiful, was developed from the grey or white tissues of a monkey's brain?[8] No—How when or where-so-ever he originated, I must believe that the Adam was as to his Spiritual nature, a "son of God".

Now after writing this I feel strongly disposed to know it, but I will leave it, for if you have such patience to read so far, it will first serve to show you that your kindness is not thrown away, that I do however ignorantly take a real interest in these subjects, an interest far deeper than I can ever express by word of mouth. They make one's heart glow like your wine when you pass the current through it![9] Forgive me for this scrawl—

Y[ou]rs truly & gratefully | M F Egerton

RI MS JT/1/E/40

1. *[1867]:* the year is given based on the mention of Tyndall's article in the *Fortnightly Review* (see n. 2).
2. *Tyndal*: misspelled by Egerton.
3. *an article of yours in the Fortnightly Review:* see letter 2597, n. 3.
4. *"follow truth along her star-paved way":* from the poem 'The Angel Song' by British poet John Keble (1792–1866): 'Thee, on the bosom laid | Of a pure virgin mind, | In quiet ever, and in shade, | Shepherd and sage may find; | They, who have bow'd untaught to Nature's sway, | And they, who follow Truth along her star-paved way' (lines 37–42).

5. *an article by the Duke of Argyll "on the Supernatural":* George Campbell, eighth Duke of Argyll, wrote the unattributed article 'The Supernatural', *Edinburgh Review,* 116 (1862), pp. 378–97.
6. *his book on the "Reign of Law":* G. Campbell, 'The Supernatural', in *The Reign of Law* (London: Alexander Strahan, 1867), pp. 1–52.
7. *the illustration you afforded us of that the other day:* presumably one of Tyndall's lectures at the School of Mines. Tyndall began a course of thirty-two lectures on 'Magnetism, Electricity, Sound, Light, and Heat' on 29 April 1867 (*PMG,* 27 April 1867, p. 14). Tyndall appears to have given Egerton fifty passes to the course (see letter 2804).
8. *a monkey's brain:* probably a reference to Huxley's work on comparing the brains of humans and apes, which got him into a row with anatomist Richard Owen (see letter 2647, n. 10). While Huxley saw the similarity between the brains of humans and apes as evidence of evolution, Owen disagreed. Egerton was probably familiar with the subject from Huxley's book, *Evidence as to Man's Place in Nature* (London: Williams and Norgate, 1863).
9. *your wine when you pass the current through it:* probably a reference to a lecture of Tyndall's she attended.

To Jane Barnard 13 June [1867][1] 2605

Thursday | June 13

My dear Miss Barnard.

I am very grateful to you for your note[2]—Your uncle[3] <u>did</u> give me a bit of the glass,[4] but since then the electric lamp enables us to project the thing upon a screen & show it to a large audience. A bigger bit of glass is therefore useful. Saturday, Monday, Tuesday or even Wednesday would suit my purpose.

Yours ever | John <u>Tyndall</u>

I was so glad to hear from the cook that your uncle was brighter & better.[5,6]

RI MS JT/1/T/132
RI MS JT/1/TYP/1/4193

---

1. *[1867]:* the year is given based on relation to letter 2603.
2. *your note:* Barnard's reply to Tyndall's previous letter (2603), which is missing.
3. *Your uncle:* Michael Faraday.
4. *a bit of the glass:* see letter 2603, n. 2.
5. *your uncle was brighter & better:* at this time Faraday was ill, and would die two months later on 25 August.
6. *better:* LT wrote on the typescript, presumably from the envelope: 'Hampton C^{t} Green | Hampton C^{t} | Kingston on Thames | SW' (Michael Faraday's house in London since 1858).

From John Haas[1] 21 June 1867 2606

Morat, Suisse,[2] | June 21st. 1867

Dear Sir,

The meteor,[3] about which I wrote you a few words,[4] having been seen at Paris, but to the South East, I infer that it was not noticed in England, and that consequently there will be no letters in the Times about it.

I think it would be a pity if this curious phenomenon were not recorded in the proper place and by proper persons, in the interest of science, and so I have the liberty of addressing the enclosed scraps & cuttings from newspapers concerning it,[5] to yourself as one of the Main Pillars of the Headquarters of Science, to do therewith that you think best.

The first article[6] is in my opinion the best and entirely agrees and coincides with all I wrote you last; it was sent in from here, I don't have by whom, but not by me. I was much vexed to find that an English "lady" residing at Morat,[7] of more fancy than judgment, had sent in an a/c[8] to the "Times",[9] in which she spoke of fiery swords, pointing to France, mystic & cabalistic signs, crosses & crescents & the like rubbish! (This is entre nous,[10] as the writer is a lady), but if unfortunately her contribution should see the light, I should be sorry if people were to think us Moratois[11] generally superstitious and silly.

With kind regards I am, Dear Sir, | Yours truly John Haas

RI MS JT/1/H/5

1. *John Haas:* John Jacob Haas (1828–94), a Swiss born teacher of modern languages at Queenwood College, and author of the textbook *Easy and Practical Introduction to the French Language* (London: Darton and Co., 1850) ('Deaths', *Manchester Courier,* 7 July 1894, p. 7).
2. *Morat, Suisse:* a town in the Swiss canton of Fribourg, between Bern and Lausanne.
3. *The meteor:* this meteor of 11 June 1867 was observed in Paris, France and Basel, Switzerland, among other places (see 'A Catalogue of Observations of Luminous Meteors, 1866–67', *Brit. Assoc. Report 1867,* pp. 288–430, on pp. 378–81).
4. *I wrote you a few words:* letter missing.
5. *the enclosed scraps & cuttings from newspapers concerning it:* Haas included five clippings from regional Swiss newspapers, all in German (one from the *Berner Intelligenzblatt,* and four from *Der Bund*).
6. *The first article:* a note with this article reads 'The meteor of June 11th. 1867. | (from the Berner Intelligenzblatt)'.
7. *English "lady" residing at Morat:* not identified.
8. *a/c:* account.

9. *to the "Times":* the account does not appear to have been published by the *Times*.
10. *entre nous:* between us (French).
11. *Moratois:* residents of Morat, Switzerland.

From Gustav Wiedemann 22 June 1867 2607

Carlsruhe, 22 Juni 1867. | Hirschstr. 54.

Lieber Tyndall!

Da meine Frau die zweite Hälfte Ihres Werkes übersetzt hat, so hat mir Helmholtz Ihre Correcturen für §547 und §604 (S. 483) geschickt; sie sollen richtig besorgt werden.—Sie meinen, dem Appendix zu Cap XII sollte Ihre vollständige Abhandlung über Calorescenz *[sistiert]* werden; ich fürchte mir, daß dadurch die Grenzen ein wenig überschritten würden, die für ein, einem größeren Leserkreise bestimmtes Werk zu stecken wären. Würde es nicht vielleicht genügen, wenn Sie mir in Ihrer Abhandlung die wichtigsten Punkte bezeichneten, welche in dem Anhang nicht angeführt sind *[&]* wenn ich dieselben gehörigen Orts *[einfügte]?* Bitte, überlegen Sie dies *[&]* schreiben Sie mir die Antwort. Um ein Exemplar Ihrer Arbeit über Calorescenz darf ich Sie dann wohl freundlichst bitten—

Meine Frau ist jetzt mit der Uebersetzung der zweiten Hälfte Ihres Werkes ganz fertig; das Manuscript ist schon in Viewegs Händen. Es hat ihr sehr große Freude gemacht, das Werk zu übersetzen und ebenso habe ich mit vielem Vergnügen die Uebersetzung mit dem Original verglichen. Ich hoffe, dieselbe wird auch Ihren Ansprüchen genügen.—Daß von der ersten, von Fr Helmholtz übersetzten Hälfte schon 8 Bogen gedruckt sind, wißen Sie wohl *[&]* so wird das Werk schnell beendet sein.

Unsere Damen bitten bei dieser Gelegenheit um Ihre Photographie, selbstverständlich in der sicheren Voraussetzung, daß sie mit uns Männern die Freude haben werden, Sie im kommenden Sommer auch persönlich bei uns zu begrüßen. In unserem Hause steht stets ein Zimmer für Sie bereit *[&]* Sie sollen uns jederzeit herzlichst willkommen sein.

In aufrichtiger Ergebenheit | Ihr | G. Wiedemann.

Karlsruhe,[1] 22 June 1867. | Hirschstr. 54.

Dear Tyndall!

As my wife[2] has translated the second half of your work,[3] Helmholtz[4] has sent me your corrections for §547 and §604 (p. 483); they shall be taken care of properly.—You think that your entire paper on calorescence should be *[held back]* for the appendix to Ch. XII; I fear that through this, the limits that would have to be set for a work meant for a larger readership would be exceeded a little. Wouldn't it perhaps be enough if you were to note for me the most important points in your paper which are not cited in the appendix *[&]* if I *[were to insert]* them in the appropriate place? Please, think about this *[&]* write me your reply. In that case, I may ask you kindly for a copy of your work on calorescence.—

My wife has now completely finished the translation of the second half of your work; the manuscript is already in Vieweg's[5] hands. It has given her very great pleasure to translate the work and I, too, have compared the translation with the original with great pleasure. I hope that it will meet your requirements too.—You probably know that of the first half, translated by Frau Helmholtz,[6] 8 sheets have already been printed *[&]* so the work will soon be completed.

Our ladies ask on this occasion for your photograph, naturally in the certain assumption that along with us menfolk they will have the pleasure of welcoming you in person also at our place in the coming summer. There is always a room available for you in our house *[&]* you shall be most warmly welcome any time.

In sincere devotion | Your | G. Wiedemann.

RI MS JT/1/W/45

1. *Karlsruhe:* University of Karlsruhe.
2. *my wife:* Clara Wiedemann.
3. *your work:* Clara Wiedemann and Anna Helmholtz translated the first German edition of Tyndall's *Heat* (1863), which appeared as *Die Wärme, betrachtet als eine Art der Bewegung,* trans. H. Helmholtz and G. Wiedemann (Braunschweig: Friedrich Vieweg und Sohn, 1867). While Clara and Anna translated the work, only their husbands were credited (*HLS*; B. Meijer, et. al. (eds), *Nordisk familjebok,* 2nd edn (Stockholm: Nordisk familjeboks förlags, 1904–26)). The first German edition was based on the second English edition of *Heat* (1865).
4. *Helmholtz:* Hermann Helmholtz.
5. *Vieweg's:* Heinrich Rudolf Vieweg.
6. *Frau Helmholtz:* Anna Helmholtz.

To Peter Riess[1] 23 June [1867][2] 2608

23rd. June.

Dear Friend.

I beg to introduce to you M[r]. Gore,[3] a fellow of the Royal Society, who desires to make your acquaintance. M[r]. Gore is the author of interesting papers in the Philosophical Magazine[4] in the Philosophical Transactions.[5]

Yours every truly, | John Tyndall

Professor Riess | Berlin

Staatsbibliothek zu Berlin (Stabi) Slg Darmstaedter 1855 Tyndall Be. 1–58, 52

---

1. *Peter Riess:* Peter Theophil Riess (1804–83), a German physicist who befriended Tyndall during his studies in Berlin in 1851 and 1853. He became professor of physics at the Akademie der Wissenschaften in Berlin (1842–83) and, in 1842, he was the first Jew to be elected to the Royal Prussian Academy of Sciences (*ADB*).
2. *[1867]:* the year is given based on a note written on the letter, '1867', which seems reasonable as Gore was elected FRS on 1 June 1865 ('Annual Meeting', *Roy. Soc. Proc.,* 14 (1865), p. 299).
3. *Mr. Gore:* George Gore (1826–1908), an English electrochemist who published dozens of papers on chemistry and electrometallurgy. He was a highly sought after scientific consultant in Birmingham and would later (1870–80) teach at King Edward's School (*CDSB*; *ODNB*).
4. *interesting papers in the Philosophical Magazine:* for example, the following papers from the 1860s: 'Note on the Specific Gravity of Electro-Deposited Amorphous Antimony', *Phil. Mag.,* 19 (1860), p. 403; 'Description of an Apparatus for Generating Hydrogen, Carbonic Acid, and Sulphurated Hydrogen', *Phil. Mag.,* 20 (1860), pp. 405–6; 'Experiments on the Adhesion of Liquids to Mercury', *Phil. Mag.,* 26 (1863), pp. 142–3; 'On the Electrical Relations of Metals &c. in Fused Substances', *Phil. Mag.,* 27 (1864), pp. 446–51; and 'On Chemical Reactions obtained by employing Anhydrous Liquids as Solvents', *Phil. Mag.,* 30 (1865), pp. 414–9;
5. *in the Philosophical Transactions:* for example, the following papers from the late 1850s and 1860s: 'On the Properties of Electro-deposited Antimony', *Phil. Trans.,* 148 (1858), pp. 185–98, 149 (1859), pp. 797–808, 152 (1862), pp. 323–31; 'On the Production of Vibrations and Sounds by Electrolysis', *Phil. Trans.,* 11 (1860–62), pp. 177–9, 491–3; 12 (1862–63), pp. 217–36; 'On the Properties of Liquid Carbonic Acid', *Phil. Trans.,* 11 (1861), pp. 85–6; 'Preliminary Note on the Nature and Qualities of Voltaic Currents', *Phil. Trans.,* 11 (1862), pp. 304–6; and 'On the Properties of Liquefied Hydrochloric Acid Gas', *Phil. Trans.,* 14 (1865), pp. 204–13.

From Mary Egerton 28 June 1867 2609

45, Eaton Place, | S. W.[1] | Friday | June 28/1867

Dear M^r^. Tyndal,[2]

To speak the honest truth I am rather glad you did not come the other night.[3] I have come to the conclusion that a party at one's own house is the stupidest thing in the world. One cannot have two minutes talk with any one for the constant disruption of "How d'ya dos", and one sees ones friends across the room, while one has to "civility" comparative strangers. Besides, the people I should much have liked you to meet[4] were prevented coming. But what does rise me, is that you should have come all the way down here, to find me out! A little chat in peace & quietness I should have enjoyed above all things. It was the only day this week (except Saturday) that I should have been out late, but we drove down to Roehampton.[5] Let me wish you "God speed" for your Alpine tour.[6] Bring back plenty of strength, and don't go crossing glaciers alone. Your life is of more value to the world than that of some members of the Alpine Club,[7] whose sole delight seems to be to go as near destruction as possible!

Y[ou]rs. very truly | M F Egerton

RI MS JT/1/E/80

1. *45, Eaton Place, | S. W.:* in London.
2. *Tyndal:* misspelled by Egerton.
3. *the other night:* not identified.
4. *the people I should much have liked you to meet:* not identified.
5. *Roehampton:* a district in southwest London.
6. *your Alpine tour:* Tyndall traveled in the Swiss Alps for about a month in the summer of 1867, departing London on 5 July and returning on 8 August. He had hoped to climb the Matterhorn, but the weather was not in his favor (see letter 2651).
7. *Alpine Club:* founded in London in December 1857, Tyndall was elected a member on 27 November 1858.

To Elizabeth Dawson Steuart [May to June 1867][1] 2610

Royal Institution | Saturday.

My dear Mrs Steuart,

I have halted to know my own possible movements before writing to you. I intended to leave London to-day, but a day or two ago I received a letter from the Trinity house[2] requesting me to meet the Deputy Master[3] at Birmingham on Monday. I am scientific adviser to the Trinity Corporation[4] and hence did not wish to neglect the request. Afterwards it was found that a gentleman[5] whose presence was essential could not be there on Monday. I then resolved to go to Ireland on that day. But this morning just before leaving home a messenger came to inform me that it was essential that I should go to Birmingham on Monday. This therefore has been the final settlement.

I start from Birmingham by the night train at 10.30 on Monday, reach Dublin at 7.35; on Tuesday morning and endeavour to catch the 8.35 train at King's Bridge. If I am successful I shall be in Bagnalstown[6] at 11.2. If not successful I reach Bagnalstown by the following train at 3.46.

I shall write a note to Dr. Trayer[7] asking him to give me a few minutes conversation in Bagnalstown, and then I will drive on to you. I thankfully accept your offer of a night's lodging: more than one night however it will be beyond my power to remain, for I must return to London immediately.

With regard to consulting persons in Leighlin,[8] the best plan will be for me to hear all and then form my own judgment. I do not intend to restore matters to their former footing, because I am sure it would never do. My main object will be to make certain whether Emma[9] is carefully attended to: on this score I have been somewhat alarmed of late.

The financial matters are also not satisfactory to me: I hope you will help me to make some definite arrangement regarding them when I come.

Yours ever truly | John Tyndall.

RI MS JT/1/TYP/10/3355
LT Typescript Only

---

1. *[May to June 1867]:* this letter is dated to these months since it deals with Tyndall sorting out his sister Emma's affairs following their mother's death (30 April 1867, and Tyndall was in Ireland 2–5 May for the funeral), before Emma is living in Gorey (she was there as early as August 1867, as Tyndall wrote to Hector Tyndale in letter 2627, that 'I have been able to place her under the medical supervision & indeed in the family of a very worthy fellow—a Dr. John Tyndall of Gorey in the County Wexford. He has an exceedingly nice house and garden, and his attention appears to be unremitting'), and before he goes to the Alps on 5 July. While Tyndall officially became scientific adviser to Trinity House on 10 July 1867 (see letters 2614 and 2624), he started to assist Faraday in May 1866 (see letter 2423, *Tyndall Correspondence,* vol. 9), and took over as Faraday's health declined.
2. *letter from the Trinity house:* letter missing.
3. *Deputy Master:* the Deputy Master of Trinity House at this time was Frederick Arrow (1818–75), an English merchant and navy commander. He became Deputy Master in 1865 ('Obituary of Eminent Person', *Illustrated London News,* 24 July 1875, p. 95).
4. *scientific adviser to the Trinity Corporation:* see letter 2590, n. 5, and letter 2614.
5. *gentlemen:* not identified.
6. *Bagnalstown:* a small town on the River Barrow in County Carlow, Ireland, about two miles south of Leighlinbridge.
7. *a note to Dr. Trayer:* James John Trayer. If Tyndall wrote a note, it is missing.
8. *Leighlin:* Leighlinbridge, County Carlow, Ireland, where Tyndall was born and his mother lived.
9. *Emma:* Emma Tyndall.

From Juliet Pollock [May or June 1867][1] 2611

59 Montagu Square[2] | *[Friday night]* W.

My dear M^r. Tyndall

It was impossible for me to go to the R.I. to day[3]—nor can I go tomorrow to hear Huxley[4]—so I must write to you and say how much I hope you may be able to dine with us on Sunday next at 7. You will find Henry Taylor[5] here. don't imagine that he wants you to be vivacious—he does however want to meet you; to shake hands with the man whose letter about Eyre[6] gratified him so very much, and the writing of which he admired so much. I rather think he has no knowledge of science at all; but he has a great deal of thought on whatever subject his mind is brought to bear.

He has funds of tenderness and benevolence which seem inexhaustible, and it is for these qualities that you will be fond of him. I am so certain of your sympathies with him and of the interest you will find in the society of his wife[7] and daughters,[8] when you come to know them, that I am anxious you should not lose this chance of meeting him.

They are at Bournemouth he is at a lodging | 16 South Audley S^t | for a few days

We have no party on Sunday.—Come if you can.

I am ever truly & affect[tionate]^ly Your Friend Juliet Pollock

RI MS JT/1/P/222

---

1. *[May or June 1867]:* the month and year are given by mention of RI lectures (see n. 3 and n. 4), and Tyndall's letter about Edward Eyre (see n. 6).
2. *Montagu Square:* the Pollocks' residence in Marylebone, London.
3. *R.I. to day:* given that Huxley's lecture was likely on a Saturday (see n. 4), Pollock would be referring to a Friday Evening Discourse. Those that fell between the dates of Huxley's lecture course were: John Stuart Blackie on 3 May with 'On the Music of Speech in the Greek and Latin Languages'; Alexander Bain on 10 May with 'On the Doctrine of the Correlation of Force in its Bearing on Mind'; William Odling on 17 May with 'On the Occlusion of Gases by Metals'; Alexander Stewart Herschel on 24 May with 'On the Shooting-Stars of the Years 1866–7, and on the Probable Source of Certain Luminous Meteors in the Material Substance of the Zodiacal Light'; Thomas Sterry Hunt on 31 May with 'On the Chemistry of the Primeval Earth'; and John Ruskin on 7 June with 'On the Present State of Modern Art with Reference to the Advisable Arrangements of a National Gallery' (*Roy. Inst. Proc.*, 5 (1866–9), p. iv).
4. *tomorrow to hear Huxley:* Huxley delivered twelve lectures at the RI, 'On Ethnology', on Thursdays and Saturdays from 2 May to 8 June 1867 (*Roy. Inst. Proc.*, 5 (1866–9), p. 25).
5. *Henry Taylor:* Henry Taylor (1800–86), a British poet, playwright, and friend of the Pollocks. He was a diplomat who spent much of his career working in the Colonial Office

(*ODNB*). Beginning in 1861, Taylor and his family spent summers at The Roost, his home in Bournemouth on the southeastern English coast (U. Taylor, *Guests and Memories: Annals of a Seaside Villa* (London: H. Milford, Oxford University Press, 1924), p. 14).

6. *letter about Eyre:* see letter 2571, n. 5.
7. *wife:* Theodosia Alice Taylor (née Spring-Rice, 1818–91), daughter of Thomas Spring-Rice (1790–1866), first Baron Monteagle of Brandon in Kerry. They married on 17 October 1839.
8. *daughters:* the Taylors had five children: Aubrey Taylor (1846–70), who went to Balliol College, Oxford and died young; Eleanor Ashworth Towle (née Taylor, 1847–1911), a novelist; Ida Ashworth Taylor (1850–1929), a novelist and biographer (*ODNB*); Harry Ashworth Taylor (1855–1907), a captain in the British Army; and Una Ashworth Taylor (1857–1922), a novelist and author of *Guests and Memories: Annals of a Seaside Villa,* her memoir about her family's summer home and the guests they entertained there (*ODNB;* H. Taylor, *Autobiography of Henry Taylor, 1844 to 1875* (London: Harrison and Sons, 1877); U. Taylor, *Guests and Memories: Annals of a Seaside Villa*).

From Annette Roche[1] 1 July 1867 2612

Leighlin Bridge.[2] | July 1st. 1867

My dear Doctor Tyndall

I have only just returned to my old house, and I am anxious to delay no longer expressing to you my sincere condolence and sympathy in the bereavement you have sustained in the loss of your dear Mother.[3] I can feel from experience, that such a loss leaves a blank which never can be filled up, and even though you were so much separated from your Mother, I am sure you can feel something of this.—

I have often regretted that I was never able to assist you, in the way I would have wished to do here, but you know now, the difficulties which lay in the way.—

However I will not dwell on painful subjects, but wish you a very happy & beneficial tour this Summer,[4] & I sincerely hope the clear cold of the Alps will, as heretofore—, strengthen your health & brace your nerves for future work.—

I daresay you have been much interested with the Paris Exhibition[5] as I believe most intelligent minds are.—

I am only at home for a few weeks, & am likely to join my niece, Mrs Haire,[6] again in England before Winter.—

Our weather is lovely now, not too hot in the country, where the air is laden with the perfume of new mown hay, & flowers.—

Believe me dear Dr Tyndall, with very kind regards, yours
most sincerely | Annette Roche

RI MS JT/1/R/35

1. *Annette Roche:* Annette Roche, wife of James Roche (1770–1853), an active supporter of literary, philanthropic, and mercantile movements in Cork, Ireland. She was a good friend of Tyndall's and cared for his mother and sister by managing money, arranging accommodations, and choosing suitable clothing. In 1864 she had arranged for the erection of a tombstone over the grave of Tyndall's father in Ireland.
2. *Leighlin Bridge:* Leighlinbridge, County Carlow, Ireland, where Tyndall was born and his mother lived.
3. *the loss of your dear Mother:* Sarah Tyndall (née Macassey) died on 30 April 1867 (see letters 2587 and 2589).
4. *tour this Summer:* see letter 2609, n. 6.
5. *Paris Exhibition:* see letter 2578, n. 3.
6. *Mrs Haire:* possibly Jane King Haire (née Roche), of Leighlinbridge and daughter of Benjamin Roche MD, who married Andrew Haire in 1866 (*Irish Genealogy*).

To Juliet Pollock [before 5 July 1867][1] 2613

St. Laurence Hotel near Ventnor.[2] | Wednesday.

My dear M^rs^. Pollock

I ran down here today to get rid of lumbago;[3] I got a chill last week; but I broke the back of my assailant in a Turkish Bath; and now I am here for two days for the purpose of picking up a little general vigour.

I too was exceedingly pleased with my visit to St. Julians:[4] it had been a household word with me so long. I was glad to find it more beautiful than I had anticipated. and then the music and the songs half intoxicated me. Kindest regards to the two syrens.[5]

I shall begin to think myself wrong in an opinion which I always hitherto entertained:—an opinion which I repeatedly expressed in public ages before I ever wrote a book. That opinion is that none but a real master is fit to write an elementary work. I have often said, & written, that it would be an unspeakable boon to humanity if our best men could be prevailed upon to write our schoolbooks. This has been always my opinion hitherto.

Besides you know I must lecture, though I have been thinking lately of kicking against too much of this work; I must get up my experiments, devise new ones & repeat old ones in a new manner. The only question with me was having gone to the very great trouble of preparing these lectures for our audience in Albemarle Street,[6] was it wise to let the labour be practically lost by confining it to an audience of two or three hundred persons, when it really might be made the source of solid instruction for tens of thousands?

Longman[7] writes to me[8] for another edition of my book on Heat.[9] In

Paris a gentleman introduced himself to me in one of the stalls of the exhibition:[10] He told me the book had been translated more than once into Russian,[11] that they had quite recently made the most perfect translation of it that ever appeared, & talked much of its influence for good. This personage was the Duke of Leuchtenberg,[12] the nephew of the Emperor of Russia,[13] a most accomplished man. The book is at this moment translated into German[14] by two of the most distinguished scientific men of Germany.[15] It ran like wildfire through America,[16] and both in Italy and France its sale has been very great.[17] These books reach as I have said tens of thousands: is this a thing to be despised?

However my hands are too full to permit of the hope of being able to return to such work for years to come. But if I withdraw from such work it is because I must, & not that I entertain a doubt of its importance.

Yours ever | John Tyndall.

Sie haben mir Rath gegeben, Sehen Sie hier das Resultat.[18] | J. T.

M^rs^. Pollock

RI MS JT/1/T/1200
RI MS JT/1/TYP/6/2069

---

1. *[before 5 July 1867]:* this letter appears to have been written before Tyndall traveled to the Alps for that year; he departed London on 5 July 1867. Letter 2607 from Gustav Wiedemann shows that the translation of *Heat* was active in June 1867.
2. *Ventnor:* a seaside resort town located on the south-east coast of the Isle of Wight.
3. *lumbago:* a rheumatic affection in the lumbar region of the body (*OED*); i.e., lower back pain.
4. *St. Julians:* the Herries family home in the town of Sevenoaks in Kent, England, southeast of London.
5. *syrens:* one who, or that which, sings sweetly, charms, allures, or deceives, like the Sirens (*OED*). There must have been a performance of some kind at St. Julians.
6. *our audience in Albemarle Street:* the RI.
7. *Longman:* see letter 2617, n. 5.
8. *writes to me:* letter missing.
9. *another edition of my book on Heat: Heat* (1868).
10. *the exhibition:* see letter 2578, n. 3.
11. *translated more than once into Russian:* any record of *Heat* having been translated into Russian has not been found, although Leo Tolstoy mentions it in his 1878 novel *Anna Karenina* (see *The Ascent of John Tyndall*, p. 207).
12. *Duke of Leuchtenberg:* Nicholas Maximilianovich (1843–91), fourth Duke of Leuchtenberg (1852–91).
13. *Emperor of Russia:* Alexander II (1818–81), Emperor of Russia (1855–81).
14. *translated into German:* see letter 2607, n. 3.

15. *two of the most distinguished scientific men of Germany:* Hermann Helmholtz and Gustav Wiedemann. Tyndall, however, knew that the book was being translated by their wives, Anna Helmholtz and Clara Wiedemann (see letter 2607, n. 3), thus it seems odd that he would write this to Juliet.
16. *ran like wildfire through America:* D. Appleton and Company, New York, had printed editions of Tyndall's *Heat* in 1863, 1865, and 1867.
17. *in Italy and France its sale has been very great:* J. Tyndall, *La Chaleur considérée come un mode de mouvement,* trans. F. Moigno (Paris: Gauthier-Villars, 1864). Any record of *Heat* having been translated into Italian has not been found.
18. *Sie haben mir Rath gegeben, Sehen Sie hier das Resultat:* You gave me advice. Here is the result (German).

From C. Cecil Trevor[1],[2] 10 July 1867 2614

(H. 2320.) | Board of Trade[3] to Commissioners of Northern Lighthouses.[4] | Board of Trade (Harbour Department), 10 July 1867.

Sir,

I am directed by the Board of Trade to acquaint you, for the information of the Commissioners of Northern Lighthouses, that the continued illness of Professor Faraday renders it impossible that he can any longer give his valuable services to the Elder Brethren of the Trinity House,[5] and to the public; and with that, having regard to the eminence of those services, and the devotion with which they have been performed, it has been arranged that the small salary Professor Faraday has received heretofore shall be continued to be paid to him during the term of his life.

It has consequently become necessary that another Scientific Adviser should be appointed to advise the Elder Brethren of the Trinity House; and, having regard to the constant need of scientific advice, the necessity for which, especially with the electric light and other experiments, has of late greatly increased, and to the fact that that need is not necessarily confined to the England, it has been considered advisable that the services of Professor Tyndall, who has been associated with, and is now appointed as successor to, Professor Faraday, shall, under certain circumstances, be at the disposal of the Commissioners of Northern Lighthouses, and the Commissioners of Irish Lights,[6] when required by them, as well as the disposal of the Trinity House and the Board of Trade.

The following arrangements have accordingly been made with the Trinity House relative to their Scientific Adviser.

That Professor Faraday should continue to receive during his lifetime* the 200l. per annum hitherto paid by the Elder Brethren, as well as 100l. per annum, paid by this Board out of the Vote for Colonial Lighthouses.[7]

That Professor Tyndall should undertake the inspection of the lighthouse apparatus of colonial lights, and should also, if required be ready to give advice to the Commissioners of Northern Lighthouses, and to the Commissioners of Irish Lights, at a salary of 400l. per annum.

That during Professor Faraday's life the annual salary of 400l. should be paid in the following manner, viz., 200l. to be paid by the Elder Brethren of the Trinity House, and 200l. by this Board, out of the Mercantile Marine Fund.[8]

That upon the cessation of the payment to Professor Faraday, the 400l. per annum to

be paid as follows:—

£. 200 by the Elder Brethren.

£. 100 by the Board of Trade, out of the Mercantile Marine Fund.

£. 100 by the Board of Trade, out of the Vote for the Colonial Lighthouses.

That the Board of Trade, and with their approval, the Commissioners of Northern Lighthouses and the Commissioners of Irish Lights, shall have the power of applying directly to Professor Tyndall for his services when required.

The Board of Trade trust that these arrangements will be satisfactory to the Commissioners of Northern Lighthouses. So far as they are concerned, the effect is that upon any special application from the Commissioners on matters concerning which Professor Tyndall has special experience and knowledge, the Board of Trade will be able to place his services at their disposal.[9]

I and, &c. | [(signed)] *C. Cecil Trevor.* | The Secretary, | Commissioners of Northern Lighthouses.

*Professor Faraday died on the 25th day of August 1867.

'Correspondence on Mode of Remuneration to Scientific Adviser to Board of Trade and Trinity House', HC 258, Parliamentary Papers (1872), pp. 3–4

---

1. *C. Cecil Trevor:* Charles Cecil Trevor (1830–1921), an English civil servant who at the time of this letter was the assistant secretary in the Harbour Department of the Board of Trade ('Sir Cecil Trevor', *Times,* 20 September 1921, p. 20).
2. According to the Parliamentary papers, similar letters to this one were sent to both Tyndall and the Commissioners of Irish Lights (*Correspondence on Mode of Remuneration to Scientific Adviser to Board of Trade and Trinity House,* 24 July 1872, HC 258 1872, pp. 3–4).
3. *Board of Trade:* see letter 2590, n. 5.
4. *Commissioners of Northern Lighthouses:* the lighthouse authority for Scotland.
5. *Elder Brethren of the Trinity House:* Trinity House (see letter 2590, n. 5) is governed by a council of thirty-one Elder Brethren who are drawn from the 400 Younger Brethren.
6. *Commissioners of Irish Lights:* the lighthouse authority for Ireland.
7. *Vote for Colonial Lighthouses:* the budget for Colonial Lighthouses.

8. *Mercantile Marine Fund:* established in 1854, a common fund for a variety of expenses related to merchant shipping, including those of the lighthouse authorities (Merchant Shipping Act, Part VII, Mercantile Marine Fund (London: Eyre and Spottiswoode, 1854), pp. 994–5).
9. *to place his services at their disposal:* Tyndall accepted the position; see letter 2624.

From Mary Adair 12 July 1867 2615

Heatherton Park.[1] | July 12th. 1867

Dear Mr. Tyndall,

Miss Hamilton[2] has just forwarded your very kind present[3] to me. I have not had time to read much of it yet; but it seems thoroughly interesting and I am so glad to have it now that we are home again, when there is some chance of having leisure to read and enjoy it. Pray accept my very best thanks.

I hope you are having fine weather for Switzerland.[4] Are you making a long stay there this summer? It is not impossible that we may be spending three weeks or so from the middle of August in Switzerland. Our plans are by no means settled, but there is a chance of our going, and I promised to tell you if there was any likelihood of our visiting your favourite mountains.[5]

My Mother[6] joins me in kind regards and believe me very truly yours

H. Mary Adair.

RI MS JT/1/A/4

1. *Heatherton Park:* see letter 2540, n. 1.
2. *Miss Hamilton:* see letter 2540, n. 4.
3. *your very kind present:* probably *Sound* (1867), which Tyndall had just published.
4. *for Switzerland:* see letter 2609, n. 6.
5. *any likelihood of our visiting your favourite mountains:* Adair later informed Tyndall that they would not be able to meet him in Switzerland (see letter 2628).
6. *My mother:* Harriet Adair.

From George Campbell 12 July 1867 2616

Argyll Lodge,[1] | Kensington. | July 12/67

My Dear Sir,

Many thanks for the information about Spectroscopes[2]—

Can you give us the pleasure of y[ou]r. company at dinner on Thursday next the 18th. at ¼ to 8 o'c[loc]k?

Y[ou]rs t[ru]ly | Argyll

RI MS JT/1/A/98
RI MS JT/TYP/1/69

1. *Argyll Lodge:* see letter 2602, n. 1.
2. *information about Spectroscopes:* letter missing.

From Jean-Albert Gauthier-Villars 12 July 1867 2617

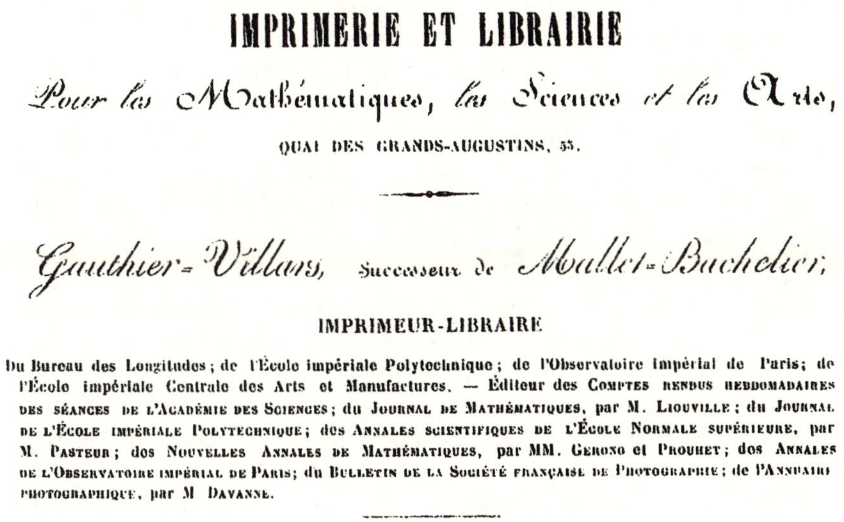

Imprimerie et Librairie | *Pour les* Mathématiques, *les* Sciences *et les* Arts | Quai des Grands-Augustins, 55. | *Gauthier-Villars*, successeur de *Mallet-Bachelier*, | Imprimeur-libraire | Du Bureau des Longitudes; de l'École impériale-Polytechnique; de l'Observatoire Impérial de Paris; de l'École impériale Centrale des Arts et Manufactures.—Éditeur des Comptes Rendus Hebdomadaires des Séances de l'Académie des Sciences; du Journal de Mathématiques, par M. Liouville; du Journal de l'École Impériale Polytechnique; des Annales scientifiques de l'École Normale Supérieure, par M. Pasteur; des Nouvelles Annales de Mathématiques, par MM. Gerono et Prouhet; des Annales de l'Observatoire Impérial de Paris; du Bulletin de la Société Française de Photographie; de l'Annuaire photographique, par M. Davanne.

N° 3573 | Paris, le 12 Juillet 1867

Monsieur,

J'ai l'honneur de vous remercier de l'exemplaire de votre ouvrage sur le son, que vous avez bien voulu m'envoyer. Je suis toujours disposé à user de votre autorisation, et à éditer une traduction française de ce livre.

Je résume par écrit les conventions verbales que nous avions échangées en présence de Mr. Jamin.

Mr. Tyndall donne à Mr. Gauthier-Villars qui l'accepte, l'autorisation de publier la traduction française d'un ouvrage dont il est auteur et qui a paru sous le titre de: Sound a course of Eight Lectures delivered at the royal institution of great Britain, aux clauses et conditions suivantes:

Art. 1er—L'ouvrage sera traduit par Mr. l'abbé Moigno et la 1ère édition sera tirée à 2400 exempl. avec mains de passe d'usage.

Article 2e—Mr. Gauthier-Villars fera à ses risques et périls tous les frais d'impression de reproduction des gravures et de traduction.

Art. 3.—Pour couvrir M. Gauthier-Villars de ses déboursés, la vente se fera d'abord à son profit jusqu'à *[due]* concurrence et ensuite pour la compte des deux parties contractantes qui se partageront les bénéfices par moitié.

Il sera accordé à Mr Gauthier-Villars sur le prix fort des exemplaires une remise de 33 pour cent et du 1/13 exemplaire sur 12, pour remises à faire au commerce de la librairie et frais *[divers]*.

Art. 4.—Le compte de cette opération sera établie tous les ans le 1er Mars par M. Gauthier-Villars, et la moitié des bénéfices acquis sera tenue à la disposition de Mr. Tyndall.

Art. 5.—Les conditions ci-dessus énoncées seront applicables aux éditions subséquentes qui pourraient être faites.

J'ai l'honneur de vous demander, Monsieur, de vouloir bien me faire savoir si vous acceptez mes propositions. J'écris à Mr. Longmann pour lui demander à quel prix il peut me céder les galvanotypes des figures de l'ouvrage. J'espère qu'il ne me demandera pas un prix trop élevé par galvanotypes, car vu le nombre des figures, cette dépense pourrait rendre l'opération difficile. Je vous prie du reste de vouloir bien ne considérer comme définitif mon engagement de publier l'édition française dont il s'agit, que lorsque Mr. Longmann m'aura adressé ses propositions et que je les aurai acceptées.

En vous remerciant de nouveau, Monsieur, de vouloir bien me donner les moyens de publier la traduction de votre remarquable ouvrage, je vous prie d'agréer l'assurance de ma haute considération.

Gauthier-Villars

P.S. M. l'abbé Moigno demande 1000*[f]* pour la traduction. J'ai accepté ce prix qui me parait raisonnable, à condition que Mr. Moigno ne vienne pas demander à participer aux bénéfices et à s'immiscer dans notre mutuelle opération.

Je vous prie de me dire combien d'exemplaires vous désirez avoir à votre discrétion, pour faire des hommages.

G. V.

M^r^. Tyndal, Royal institution | Albemarle Street W. Londres.

Printing and Bookstore | *For* Mathematics, Science *and* Arts | Quai des Grands-Augustins, 55. | *Gauthier-Villars*, successor of *Mallet-Bachelier*, | Printer-Seller | Of the Bureau of Longitude; of the Imperial Polytechnic School; the Imperial Observatory of Paris; the Central Imperial School of the Arts and Manufacturing.—Editor of the Comptes Rendus Hebdomadaires des Séances de l'Académie des Sciences; the Journal de Mathématiques' by Mr. Liouville; the Journal de l'École Impériale Polytechnique; the Annales scientifiques de l'École Normale Supérieure by Mr. Pasteur; the Nouvelles Annales de Mathématiques by Mr. Gerono and Mr. Prouhet; the Annales de l'Observatoire Impérial de Paris; the Bulletin de la Société Française de Photographie; the Annuaire photographique by Mr. Davanne.

N° 3573 | Paris, July 12 1867

Sir,

I have the honour of thanking you for the copy of your book on sound,[1] which you have kindly sent me. I am always willing to use your permission, and to publish a French translation of this book.[2]

I summarize in writing the verbal agreements that we exchanged in the presence of M^r^. Jamin.[3]

M^r^. Tyndall gives M^r^. Gauthier-Villars, who accepts, the permission to publish the French translation of a book of which he is the author and which appeared under the title: Sound, a Course of Eight Lectures Delivered at the Royal Institution of Great Britain, with the following clauses and conditions:

1st Article: The book will be translated by Father Moigno[4] and the 1st edition will be printed at 2400 copies with the usual surplus sheets.

2nd Article: M^r^. Gauthier-Villars will take, at his own risk, all of the costs of printing a reproduction of the engravings and translation.

3rd Article: To cover Mr. Gauthier-Villars' expenses, the selling of the books will first be to his profit up until the amount of debt, and then to count towards the two contracting parties that will share the profits by half.

M^r^. Gauthier-Villars will be granted a discount of 33 percent on high-priced copies, and of 1/13 copies on 12, for discounts to do at the bookstore and for various fees.

4th Article: An account of this project will be taken every year on the 1st of March by Mr. Gauthier-Villars, and half of the profits acquired will be held at Mr. Tyndall's disposal.

5th Article: The conditions set out above will be applicable to subsequent editions that could be produced.

I have the honour to ask you, Sir, to kindly tell me if you accept my propositions. I am writing to Mr. Longmann[5] to ask him at what price he would be able to offer me electrotypes[6] of the book's figures. I hope that he will not place too high of a price on the electrotypes, since given the number of figures, this expense would make the project difficult. Please be kind enough not to consider as definitive my commitment to publish the French edition in question until I have received the propositions from Mr. Longmann and have accepted them.

I thank you again, Sir, for wishing to give me the means to publish the translation of your remarkable book. I beg you to accept my highest regards.

Gauthier-Villars

P.S. Father Moigno asks for 1000[*f*][7] for the translation. I accepted this price, as it seems reasonable to me, as long as Mr. Moigno does not ask to take part in the profits and does not interfere in our mutual project.

Please tell me how many copies you want to have at your discretion, to make tribute.

G. V.

Mr. Tyndall, Royal Institution | Albemarle Street W., London.

RI MS JT/1/V/16

---

1. *your book on sound: Sound* (1867).
2. *French translation of this book:* J. Tyndall, *Le Son, Cours Expérimental fait à l'Institution Royale,* trans. F. Moigno (Paris: Gauthier-Villars, 1869).
3. *Mr. Jamin:* Jules Jamin.
4. *Father Moigno:* François Moigno.
5. *Mr. Longmann:* either Thomas Longman (1804–79) or William Longman (1813–77), brothers who ran the publishing firm Longmans, Green, and Co., publisher of Tyndall's *Sound* (1867).
6. *electrotypes:* a printing plate, made by a process involving the electrolytic deposition of copper or other metal (*OED*).
7. *[f]:* possibly a superscripted 'f' for francs.

From Heinrich Gustav Magnus 12 July 1867 2618

Berlin 12 July 1867.

Verehrter Freund!

Vorgestern habe ich Ihr neues Werk On Sound erhalten wofür ich Ihnen meinen besten Dank ausspreche. Ich bewundere Ihren Fleiß und Ihre Kraft zu arbeiten und wünsche daß der Himmel Sie noch recht lange so thätig und rüstig erhalten möge um ferner die Wissenschaft zu fördern.

Sehr habe ich bedauert daß Ihr Aufenthalt in Paris so kurz gewesen und daß wir uns in dem großen Wirwarr kaum gesehen haben.

Meine Frau und meine Töchter so wie Paul tragen mir die freundlichsten Grüße auf, und ich bitte Sie mich Herrn Frankland sowie allen andern Freunden bestens zu empfehlen.

Stets Ihr aufrichtig ergebener | G Magnus.

Berlin 12 July 1867.

Dear friend!

The day before yesterday I received your new work On Sound,[1] for which I give you my best thanks. I admire your industriousness and your energy to work and wish that Heaven may keep you so active and spry for quite a long time still, so that you may promote science further.

I very much regretted that your stay in Paris[2] had been so short and that we hardly saw each other in all that great confusion.

My wife and my daughters,[3] as well as Paul,[4] have asked me to give their kindest regards, and I ask you to give my best compliment to Herr Frankland[5] as well as all our other friends.

Always your sincerely devoted | G Magnus.

RI MS JT/1/M/35

1. *On Sound: Sound* (1867).
2. *your stay in Paris:* see letter 2578, n. 3.
3. *My wife and daughters:* Bertha Magnus (née Humblot, 1820–1910), who married Gustav Magnus in 1840, and their daughters Anna Magnus (1841–75) and Christine (1842–1936), who married her cousin Viktor von Magnus in 1866.
4. *Paul:* Paul Magnus (1845–1930), son of Gustav and Bertha Magnus.
5. *Herr Frankland:* Edward Frankland.

To Henry Acland [14 July 1867][1] 2619

Sunday
The Bel Alp[2] | near Brieg | Canton de Valais | Switzerland

My dear D^r^. Acland.

It would have given me great pleasure to make the acquaintance of your friend D^r^. Bowditch.[3] But I am here in the heart of the glaciers, endeavouring to mend the injuries physical, intellectual and emotional, which have been inflicted on me[4] during the last several months in London. I have written to D^r^. Bowditch at his London address & hope my note[5] may reach him.

Yours faithfully | John Tyndall

Oxford, MSS.Acland d.65 f.191

1. *[14 July 1867]:* the date is given by relation to letter 2622, in which it can be deduced that Tyndall arrived at Bel Alp on 13 July 1867 and 'rested a day or two'. As this letter was written on a Sunday, the date would be 14 July.
2. *The Bel Alp:* see letter 2622, n. 30.
3. *D^r^. Bowditch:* Henry Ingersoll Bowditch (1808–92), an American physician and abolitionist. In the summer of 1867, he took a trip abroad spending time in England in July and August before continuing on to continental Europe (*ANB*; V. Y. Bowditch (ed.), *Life and Correspondence of Henry Ingersoll Bowditch,* 2 vols (Boston: Houghton, Mifflin and Co., 1902), vol. 2, pp. 103–48.
4. *injuries physical, intellectual and emotional, which have been inflicted on me:* probably a reference to the disputes over Governor Eyre (see letters 2567, 2571, 2572, and 2576).
5. *my note:* letter missing.

From Hector Tyndale 19 July 1867 2620

Philadelphia July 19. 1867

My dear John

Emerson[1] says that

"Pride ruined the angels | Their shame them restores | And joys that are sweetest | Lurk in stings of remorse".[2]

Now both "shame" and "remorse" I have for my lazy and laissez faire neglect in keeping silent to you so long—thus cheating my own pleasure. But then, too, I have some "joys" in believing that you like me well enough to forgive my silence, for you have shown me your remembrance by sending at times papers of your own and letters of kindness and comfort. The last paper I had from you was your "Miracles and Special Providences",[3] which has been read here very widely, having been reprinted in several periodicals. Several persons have spoken to me of it and of you and always with admiration. It gave me much pleasure—have read it three times; it is clear, strong & earnest. While it must set Mr. Mozely[4] to rethinking without anger (unless the last sentence of it should stir him up)[5] it gives me a great respect for him too—rather unusual in a review. You are, I think, without Superstition—I am not. My Reason (logical) cannot accept Miracles for then they would be more to me, but it is possible and it may become essential to Me that they be worked hereafter. I say to Me, but I don't believe in them for "You" or Others except through my sympathies with You. That is my latent force contains a Miracle, which always shrinks into that hidden force, refusing to come out to my sight and understanding. So it shall always be a Superstition which I must deny and exclude from Knowledge or Science, if I would have a Science at all.

Your papers and works always give me much pleasure. Do not weary of sending me any of them you can spare.

More than a year ago I last wrote you[6] about Mr Peter Lesley[7] and received your reply.[8] A short time after I sent you a large photograph (half length) of myself (I do not mean half size) (rolled up) through Mr Lesley and the American Phil[osophical]: So[ciety]:[9] who were then about to send a box to London. Did you get it?

Last Autumn I sent you a paper containing notice of the death of my eldest Sister Mrs Elizabeth Mitchell,[10] on the 31st October—She of whom you have often heard me speak and of whom I gave you a small portrait (photograph) in 1861. By request of Mr Mitchell[11] I now enclose another from the same negative but enlarged. She was nearly 54 years old. With much knowledge and deep culture she was as innocent and pure as if her birth really was "but a sleep and a forgetting". I think she had better balance, more steadiness and courage—of all qualities—and with that she was more spiritual than any other I have known. Her life was not merely blameless; but it was continuously praiseworthy, good & lovely. To have known her is to believe in immortality, for her life is an element in our lives, as concrete and imperishable as the Universe, we know better and love more through her. Not given to many friends those who were close to her hinged the doors of every day upon her love. I cannot recall an injustice nor an ill temper nor a remorseful word or deed of hers. But in my memory of her there are more than forty years of beautiful affection and truth and of even-balanced good doing and self-forgetting. Too even and Catholic to be a martyr, through opposition to events, she bore her own great sufferings and supported those of others with courage and constancy—and Fate or Fortune never heard her complaints. For many years her lungs had been weakened and her death came very suddenly and not very painfully, her last suffering, of a few minutes duration, was from a quick and unpremonished hemorrhage coming while she seemed in comparative good health. But she had needed no premonition, for her whole life was prayerful, loving and true—she believed, and so lived, in the Omnipresence. Her death seems like a translation. She was a very beautiful woman of a high and rare order of beauty. There is no good likeness of her either painted or photographed, though we have several. My Brother, her husband, is desolate and very lonely and his future stay must be saddened. Such a loss, however, enlarges while it pains the heart, it refines the intellect and soul it pierces. As said he asks me to enclose the photograph with his respects. We often speak of you, as I did, too, with my Sister, and my friends know you—intellectually through your own and personally through my speech.

I send you also a newspaper with letter[12] from ___ Hughes[13] (the school boy at "Rugby", an M.P. and correspondent of the New York Tribune) in which he speaks of you, of Huxley[14] and others. I thought it might interest you.

How do you get on now—your own self I mean, or rather your own life? How is "Tom" Hirst? Give him my best regards always. Mrs. Tyndale,[15] who is well, and myself very often speak of both of you. Some time or other we may meet again when we can learn mutually the effect of life, circumstance and thought upon our selves—upon the Immortals—God's within us. But in anticipation of that meeting I will say that "the whirligig of time"[16] and the happenings of life have not made me wiser or better I fear. The wine of life has not fermented well and it tastes to me now like vinegar. But then vinegar hath it's uses and so must be accepted.

Some months ago I went down to Virginia to look at some lands and while there was taken very seriously ill, more painfully than ever before, with typhoid fever, inflammation of bowels supervening, the disease having lingered in me from the war time, it's exposure, privations &c., beyond what can be imagined of a war in Europe. I was in bed for two months and I have been two months more in recovering. But now I am well again and in better health than since I was broken down in the field; indeed from appearances it seems I shall grow stout and fat, more especially that I am not now engaged at any thing, am idle in action and thought.

Mrs Tyndale desires her regards to be sent you as to Mr Hirst. Commending my poor memory to you for an occasional and silence forgiving letter I am

Affectionately your Cousin | Hector Tyndale

John Tyndall Esq &c London

RI MS JT/1/T/59
RI MS JT/1/TYP/5/1667–9

1. *Emerson:* Ralph Waldo Emerson.

2. *"Pride ruined the angels . . . Lurk in stings of remorse":* from Emerson's poem 'The Sphinx' (1841), lines 89–92.
3. *your "Miracles and Special Providences":* see letter 2597, n. 3.
4. *Mr. Mozely:* James Bowling Mozley (1813–78), an English theologian and journalist (*ODNB*). Mozley's Bampton Lectures 'On Miracles' at the University of Oxford (1865) prompted Tyndall to write 'Miracles and Special Providences' (see n. 3).
5. *unless the last sentence of it should stir him up:* 'Here, in all frankness I would declare, that at present you are ill-informed, self-deluded, and likely to delude others. Farewell!'(J. Tyndall, 'Miracles and Special Providences', p. 660).
6. *I last wrote you:* on 15 May 1866; see letter 2428, *Tyndall Correspondence,* vol. 9.
7. *Mr Peter Lesley:* J. Peter Lesley (1819–1903), an American geologist (*CDSB*; *ANB*).
8. *your reply:* letter missing.
9. *American Phil[osophical]: So[ciety]:* American Philosophical Society, founded in 1743 in Philadelphia, where it remains an active institution.
10. *Mrs Elizabeth Mitchell:* Elizabeth Mitchell (née Tyndale, 1812–66), a sister of Hector Tyndale who married Edward Mitchell ('Died', *Philadelphia Inquirer,* 3 November 1866, p. 5).
11. *Mr Mitchell:* Edward Mitchell.
12. *letter:* not identified.
13. *___ Hughes:* Thomas Hughes (1822–96), an English politician, social reformer, and children's author who was educated at Rugby School (and wrote the 1857 novel *Tom Brown's School Days*) before attending Oxford and was elected to Parliament in 1865. Hughes was a vocal supporter of the North during the American Civil War and a member of the Jamaica Committee (*ODNB*).
14. *Huxley:* Thomas Huxley.
15. *Mrs. Tyndale:* Julia Tyndale.
16. *"the whirligig of time":* Tyndale was quoting Feste in Shakespeare's *Twelfth Night* V.i.2589, '. . . the whirligig of time brings in his revenges'.

From Richard Redgrave[1,2] [after 20 July 1867][3] 2621

Dear Mr Tyndall

I am flattered with your good opinion of my report & I assure you it annoys me to say "no" to you <u>and</u> your friend Dr Bence Jones proposal but when I recollect the severe labour before me[4] in April and May next, I cannot but decline undertaking lectures. But, were it otherwise, I should be most interested to [prepare].

[2 words illeg]

Wellcome MS. 7777/76

1. *Richard Redgrave:* Richard Redgrave (1804–88), an English painter who helped organize the South Kensington Museum and British contributions to international exhibitions. He was appointed surveyor of the Queen's pictures in 1857 (*ODNB*).
2. This is a draft of Redgrave's reply to Tyndall, which he wrote on Tyndall's letter (2626). The final letter is missing.
3. *[after 20 July 1867]:* the year is given by relation to letter 2626.
4. *the severe labour before me:* possibly a reference to the thirty-four volume 'Catalogue of Pictures the Property of Her Majesty Now at Windsor Castle' he produced while he was surveyor of the Queen's pictures (1857–80).

To Juliet Pollock 23 July 1867 2622

23rd July 1867

My dear Mrs. Pollock

I write this to you for the good of the public—that is, my public, viz. Mrs. Faraday,[1] Miss Moore[2] and Miss Julia Moore.[3] I started as you know on Friday the 5th, and reached Grindelwald[4] on the evening of Sunday the 7th. There I was in the heart of the mountains. There are two glaciers[5] at hand and I spent a day upon each of them. This was my training when the clouds dissolved and the mountains rose so gloriously in the sunlight that a longing to climb one of them beset me. I chose the most difficult—a splendid pyramid, which in point of steepness and real mountain character beats all others in the Oberland:[6] this mountain is called the peak Eiger;[7] what the meaning of the word is I know not. Well this Eiger or Ogre, if you like, is attacked from the summit of the Wengern Alp.[8] Thither I went on Wednesday the 10th. I went to bed early hoping to have a few hours sleep to prepare me for action. But a crowd of schoolboys rushed into the hotel[9] and made such a racket in my part of it that I was bound to transfer myself to a distant corner of it where I had the denied sleep. At 1½ A.M. we[10] started. The heavens were soon broad cast with stars. Not a heavy cloud was to be seen, and the stars twinkled—flashes of ruby and emerald darting from a thousand points. Just above the summit of the Wetterhorn,[11] a very noble mountain, hung the Pleiades,[12] forming a glorious diadem[13] for a most beautiful peak. We clambered over the green Alp: reached the snow, or rather mixed ice and snow which had been thrown to the very feet of the mountain from a broken glacier the cliffs of which fronted us for a time. Thence to the rocks. Onward & upward the steepness increasing at every step, varying from rock to snow and from snow to rock. The snow not yet shone upon by the morning sun was firm and afforded for a time good foothold. The rocks were[14] at some places covered by a thin crust of ice, and such places were avoided as plague spots[15] by the guides.[16] The east brightened, the stars gradually paled their ineffectual fires, and at length the topmost peak of the adjacent Jungfrau[17] was hit by the purple rays. Downward the beauty crept, steeping the snows lower and lower in that wonderful purple bloom. The whole of the mountains soon partook of the colour, and for a few minutes presented the most extraordinary aspect—Far more wonderful than their aspects when illuminated by the white light of the fully risen sun—After some seven hours of climbing we approached the upper slopes. They were for the most part ice and deep steps were required which might be used on our return. The whole duty of the climber in these high regions is to make these steps properly and to stand in them securely. There is no margin for a possible slip: no such thing must occur. At intervals we were roused by the cannon of the Jungfrau. From one of the highest glaciers of this mountain[18] five great avalanches were discharged before 8 oClock in the morning, and every one of them roared like a bark of artillery. They fell first as huge ice cliffs but the force with which they smote the base of the precipice from the summit of which they had leaked reduced them apparently to powder: clouds of ice dust rose in the air. While the main mass of the avalanche fell cascade-like from precipice to precipice through four or five thousand feet. The ice caused stiff work to the leading guide.[19] He had to cut the steps, and where ever it was possible he willingly escaped from the ice to seek a footing in the snow. To our left was a "cornice"; you know what a cornice is. It is a folding over of the snow of the mountain, which then forms an overhanging cave unsupported save by its own tenacity. We worked for a time along this cornice, with appalling precipices under our feet, and where the cornice became fretted or honeycombed the guide crept to its edge and looked over to see that the snow was of the requisite thickness. Where this was not the case we retreated from the cornice and took to ice-cutting once more. We reached the top before 9 oClock in the morning, and had an outlook over perhaps as glorious mountain scenery as this world affords.

The wind was wild and bitter and we did not spend more than half an hour on the top. The descent was more difficult and dangerous than the ascent. For the first time in my life I turned my face to the ice slopes and went backwards down them. This I did at the instance of my leading guide who thought this mode of locomotion prudent to our safety. I did not like it at first, but soon took kindly to it, though there seemed something ignoble about this mode of backing out of danger. Still the duty of the climber is to conform as much as possible to the wishes of his guide, because by opposing them you weaken his morale and self trust, which is not wise. We took the most anxious care of ourselves upon the snow slopes; at least the[20] amount of care taken by the guides in fixing the feet in the snow suggested anxiety on their part. I asked the leader whether if we slipped he thought we could stop ourselves: he thought decidely not. Careful treading in soft snow hanging on such an incline was therefore necessary. For had we given way there were precipices enough to reduce us to some extent to the condition of the Jungfrau avalanches. But we knew our work and did it, and fairly reached our hotel[21] at 6 oClock in the afternoon.

I found the work rather hard in my then state of no-training; and next day my muscles were very stiff. I dropped down to Lauterbrunnen[22] and went to a little auberge[23] at the head of the valley intending to make that my starting point over a very grand glacier pass called the Petersgrat.[24] I was detained in that little auberge a day by bad weather. At length however we broke away, & in the face of biting wind which was colder than the ice itself, we reached the summit, and dropped down the snow slopes at the other side into a valley called Lötschthal,[25] and into an exceedingly pretty village called Platten.[26] There is no auberge in the said village. The house of the curé[27] is usually the traveller's refuge, but when we called upon his reverence we did not find an open door. He would not take us in, but sent us to the house, or den, if you prefer the word of a neighbouring bauer.[28] The reason assigned for his reverence's inhospitality was that he had no wine to offer his guests: and the reason he had no wine was that when he purchased a supply he had the unconquerable habit of drinking it himself. Well I was obliged to manage as well as I could in this exceedingly dirty place & escaped from it gladly in the following morning. We then scaled another noble col,[29] and passed along glaciers to the Bel Alp[30] where we found good food and good beds, and where therefore I rested a day or two.

Thence down the mountain and through wild gullies to the town of Brieg:[31] thence to Visp[32] and thence to Zermatt; not halting there however but climbing to the hotel upon the Riffelberg[33] where I spent the night. There was the grim Matterhorn[34] in front of me. There was my own beloved mountain, the ever beautiful Weisshorn.[35] Nothing in the whole Alps approaches it in beauty and nobleness. But my thoughts were now turned towards the other monster. I crossed the Col de St. Theodule[36] on Saturday, remained on Sunday at Breuil:[37] tried to make some arrangements with the guides there, and left the place yesterday on the understanding that if the weather proved fine I was to go back, and see whether some crags and precipices are not to be subdued. On that point I do not entertain the shadow of a doubt provided the weather give us fair play.

I write this on the Riffelberg. I picked up a note[38] from Fred[39] here a few days ago. I think he purposes coming here; it would be a great pleasure to meet him; but I am the playball of the weather & am not in the least master of my own movements. I have not time to read this over, so excuse all its imperfections.

Yours ever | John Tyndall.

BL Add MS 63092, ff. 215–8
RI MS JT/1/TYP/6/2082–5

---

1. *Mrs. Faraday:* Sarah Faraday.
2. *Miss Moore:* Harriet Jane Moore (1801–84), a British watercolor artist who painted scenes of Faraday's apartment, study, and laboratory at the RI during the 1850s. Moore was the eldest child of surgeon James Carrick-Moore and Harriet Moore (née Henderson). She was elected a member of the RI in 1852. See F. A. J. L. James, 'Harriet Jane Moore, Michael Faraday, and Moore's Mid-Nineteenth-Century Watercolours of the Interior of the Royal Institution', in James Hamilton (ed), *Fields of Influence: Conjunctions of Artists and Scientists, 1815–1860* (University of Birmingham Press, 2001), pp. 111–28.
3. *Miss Julia Moore:* Julia Moore (1803–1904), the youngest of the three daughters of Harriet and James Carrick-Moore. Their other daughter was Louisa (1802–53).
4. *Grindelwald:* see letter 2600, n. 37.
5. *two glaciers:* probably the Lower Grindelwald glacier (Unterer Grindelwaldgletscher) and the Upper Grindelwald glacier (Oberer Grindelwaldgletscher) south of the town of Grindelwald, on the north side of the Bernese Alps.
6. *Oberland:* the geographically higher region of the Swiss canton of Bern.
7. *Eiger:* a mountain of the Bernese Alps.
8. *Wengern Alp:* see letter 2600, n. 38.
9. *hotel:* Hotel Adler. Tyndall mentioned the hotel and his guides in his published account in *Hours of Exercise in the Alps,* p. 267.
10. *we:* Tyndall was accompanied by the guides Christian Michel and Peter Baumann.
11. *Wetterhorn:* a peak in the Alps above Grindelwald.
12. *Pleiades:* a star cluster located in the constellation of Taurus.
13. *diadem:* a crown; an ornamental cincture or covering for the head, worn as a symbol of honour (*OED*).
14. *afforded for a time good foothold. The rocks were:* these words are transcribed from the typescript only.
15. *plague spots:* while literally a spot on the skin symptomatic of the bubonic plague or of another plague-like illness, or a place associated with bubonic plague (*OED*), the figurative use of *plague spot* would signify a place of peril or doom. Interestingly, the American historian, theologian, and mountaineer William Augustus Brevoort Coolidge (1850–1926) wrote in his *The Alps in Nature and History* (New York: E. P. Dutton and Company, 1908) that 'guideless climbing in the High Alps by incompetent persons . . . is the plague spot in Alpine matters at the present moment' (p. 248).
16. *the guides:* see n. 10.
17. *Jungfrau:* see letter 2600, n. 39.
18. *glaciers of this mountain:* these words are transcribed from the typescript only.
19. *leading guide:* Tyndall named Christian Michel first in his published account of the climb.

20. *of ourselves upon the snow slopes; at least the:* these words are transcribed from the typescript only.
21. *hotel:* see n. 9.
22. *Lauterbrunnen:* see letter 2600, n. 40.
23. *auberge:* inn or hotel (French).
24. *Petersgrat:* a glaciated mountain ridge, used as a pass, located on the border of the Swiss cantons of Bern and Valais.
25. *Lötschthal:* the Lötschental.
26. *Platten:* now Blatten, a village in the Lötschental.
27. *curé:* the curé of Platten. Not identified.
28. *bauer:* farmer (German), or perhaps a peasant (*OED*).
29. *another noble col:* possibly the Beich Pass.
30. *Bel Alp:* now Belalp, an alpine village where in 1877 Tyndall and his wife Louisa would build a home on Alp Lusgen, and spend their summers there.
31. *Brieg:* see letter 2600, n. 21.
32. *Visp:* see letter 2600, n. 13.
33. *hotel upon the Riffelberg:* see letter 2600, n. 15.
34. *Matterhorn:* see letter 2600, n. 18.
35. *Weisshorn:* see letter 2600, n. 19.
36. *Col de St. Theodule:* the summit of the Théodule Pass, connecting Zermatt and Breuil and running between the Matterhorn and the Breithorn.
37. *Breuil:* an alpine village in the Aosta Valley region of northwest Italy, underneath the Matterhorn.
38. *note:* letter missing.
39. *Fred:* Frederick Pollock.

To Thomas Archer Hirst [26 July][1] 1867 2623

Riffelberg.[2] Friday 1867.

My dear Tom

Previous to quitting London I asked you to look at my letters with a view of keeping back such as were of no immediate import. Up to this time no letters whatever have reached me from the Royal Institution, and as this is the only year where this has occurred I begin to fear that letters may have been sent and may have miscarried. Would you drop me a line to Zermatt stating whether you have posted any letters to me?

I went as I intended first to Grindelwald[3] and after two day's training went up rather formidable mountains, namely, the Eiger.[4] I found it rather stiff and it required caution which having duly bestowed the ascent was successful. I then dropped down to Lauterbrunnen[5] and afterwards crossed the Petersgrat[6] to a dirty village called Platten[7] where the curé[8] refused to take me in, and I was forced to take refuge in the house of a dirty bauer.[9] The reason assigned for the closing of the curé's door to all travellers this year is that he had no wine to place before his guests: and that whenever he has wine he has also the unconquerable habit of drinking it himself.

From Platten I crossed a snow pass to the Bel Alp,[10] and after resting there a day or two came on to Zermatt. I have been at the Riffel[11] for the last few days. I crossed the Theodule[12] with some hope that we were going to have fine weather and had a consultation with the Italian guides regarding our attempt on the Matterhorn.[13] It was arranged that I should ramble for a week and then return to Breuil,[14] when if the weather were favourable the snow would be sufficiently removed to justify a trial. Well I recrossed this Col,[15] but the weather since has been atrocious. Yesterday I climbed a little way up the Matterhorn but it was only the lowest cliffs that were clear of cloud. Today we have cloud wind & rain and also the stupidity of doing nothing.

I will remain here a few days, though I have hardly the shadow of a hope that the Matterhorn will be attackable this year.

Have the reviewers torn my little book[16] to pieces?

I shall probably be in London within the coming fortnight, certainly within the coming three weeks. I suppose when I arrive you will be in Scotland. This Dundee lecture[17] bothers me. I at first thought of giving a discourse without experiments more didactic than scientific; but I have lately thought a scientific lecture would be more in keeping with the general aims of the association.

Yours ever affect[ionate]ly | John Tyndall

BL Add MS 63092, ff. 33–4
RI MS JT/1/HTYP/541

---

1. *[26 July]:* the month and day are given as Tyndall is writing this letter during a twelve day stay at Riffelberg (*Hours of Exercise in the Alps,* p. 271), starting on 23 July (see letter 2622). In this letter he notes that he had 'been at the Riffel for the last few days', suggesting that this letter was written near the beginning of his twelve days there. Tyndall left Zermatt/Riffelberg for Visp (where he wrote letter 2624 on 4 August) and returned to London, arriving there on 8 August (see letter 2627).
2. *Riffelberg:* see letter 2600, n. 15.
3. *Grindelwald:* see letter 2600, n. 37.
4. *Eiger:* see letter 2622, n. 7.
5. *Lauterbrunnen:* see letter 2600, n. 40.
6. *Petersgrat:* see letter 2622, n. 24.
7. *Platten:* see letter 2622, n. 26.

8. *curé:* see letter 2622, n. 27.
9. *bauer:* a famer (German), or perhaps a peasant (*OED*).
10. *Bel Alp:* see letter 2622, n. 30.
11. *Riffel:* see letter 2600, n. 15.
12. *Theodule:* Col de St. Theodule (see letter 2622, n. 36).
13. *Matterhorn:* see letter 2600, n. 18.
14. *Breuil:* see letter 2622, n. 37.
15. *Col:* see n. 12.
16. *my little book: Sound* (1867).
17. *Dundee lecture:* see letter 2627, n. 31.

To C. Cecil Trevor[1] 4 August 1867 2624

Visp.[2] Switzerland 4th August 1867

Sir,

In reply to your letter (H. 2320),[3] dated the 10th of July 1867 I beg to state that I shall be glad to undertake the duty of scientific adviser to the Trinity House,[4] and also the other duty mentioned in your letter[5] on the terms there stated.

The highest object I can set before me in the discharge of these duties is to imitate as far as in me lies the example of the high-minded & celebrated man who has hitherto been your scientific adviser.[6]

I am Sir | Your obedient servant | John Tyndall

C. Cecil Trevor Esq. &c. &c. &c.

National Archives, MT 10—128, H2727/2
Correspondence on Mode of Remuneration to Scientific Adviser to Board of Trade and Trinity House, 24 July 1872, HC 258 1872, p. 5

1. *C. Cecil Trevor:* see letter 2614, n. 1.
2. *Visp:* see letter 2600, n. 13.
3. *your letter (H. 2320):* letter 2614.
4. *Trinity House:* see letter 2590, n. 5, and letter 2614.
5. *other duty mentioned in your letter:* Tyndall was also asked to advise the Commissioners of Northern Lighthouses (Scotland) and Commissioners of Irish Lights (see letter 2614).
6. *your scientific adviser:* Michael Faraday served as a scientific adviser to Trinity House from 1836 until his death in 1867, and although Tyndall had begun assisting Faraday in this role since May 1866 (see letter 2423, *Tyndall Correspondence,* vol. 9), it was not made official until July 1867. See also letter 2590, n. 5.

To Thomas Archer Hirst [9][1] August 1867 2625

Royal Institution Friday | Aug. 1867.

My dear Tom

In one of your letters[2] to me to[3] mentioned your willingness to take a few days trip with me before the meeting in Dundee.[4]

If you still entertain this truly christian thought I would put myself in your hands a clear week before the meeting.

I had a cold but have got practically over it, nevertheless a week's ramble prior to the meeting will be very grateful to me.

Do not imagine that I hold you bound by any pledge or promise: but if you are sufficiently loose it would do us both good to have a little trip.

Where am I to lodge in Dundee? I have declined two invitations for the purpose of preserving my independence.

Yours ever affect[ionatel]y | John Tyndall

RI MS JT/1/T/978
RI MS JT/1/HTYP/538

---

1. *[9]:* the day is given by relation to Tyndall's return to London from the Alps on 8 August. Fridays were 9, 16, and 23 August. Friday 23 August would have been too close to the annual meeting of the BAAS at Dundee for making these plans based on Tyndall's offer of putting himself in Hirst's 'hands a clear week before the meeting'. Letter 2635 suggests a missing reply to this letter informed Tyndall that Hirst would be on the Isle of Skye from 17–19 August, and thus 9 August seems an appropriate date.
2. *one of your letters:* letter missing.
3. *to:* Tyndall should have written 'you' here, and LT had corrected it in the typescript.
4. *meeting in Dundee:* see letter 2627, n. 22 and n. 31.

To Richard Redgrave[1] [9 August 1867][2] 2626

Royal Institution of Great Britain | Fri<u>day</u>.

Dear Mr Redgrave

I hope you may be induced to reply in the affirmative to the note of my friend Dr Bence Jones. I have been lately looking over your Report of the Paris Exhibition,[3] and from the tone & spirit of your writing I should infer that you would interest our audience[4] extremely. Do say "<u>yes</u>."

Yours very sincerely | <u>John Tyndall</u>[5]

Wellcome MS. 7777/76

1. *Richard Redgrave:* see letter 2621, n. 1.
2. *[9 August 1867]:* the year is given based on Redgrave's report on art at the Paris Universal Exhibition in 1867 (see n. 3) and mention of his being requested by Bence Jones and Tyndall to lecture in 1867 at the RI in his memoir compiled by his daughter (F. M. Redgrave, *Richard Redgrave, C.B., R.A.: A Memoir, Compiled from His Diary* (London: Cassell & Company, 1891)): 'The secretary of the Royal Institution, Dr. Bence Jones, asked him, about the year 1867, to give three lectures in Albermarle Street on the "Art of the Last Century," and Professor Tyndall wrote to him at the same time warmly pressing him to accede the request. He was, however, at that date so exceedingly busy, that he was obliged to decline what would have otherwise been to him a pleasant task' (p. 365). The letter must fall after 20 July since that was the date Redgrave's report Tyndall referred to was published (see n. 3). Moreover, this letter likely dates from just after he returned from the Alps on 8 August, with the first Friday being 9 August.
3. *Report of the Paris Exhibition:* Redgrave's report on 'Applications of Drawing and Modelling to the Common Arts' initially appeared in the *Illustrated London News,* 20 July 1867, pp. 69–71, and later was published with the other reports ordered by the Committee of Council on Education in *Reports on the Paris Universal Exhibition, 1867,* 7 vols (London: Eyre and Spottiswoode, 1868), vol. 2, pp. 151–68.
4. *interest our audience:* at the RI.
5. Redgrave wrote an unsigned draft reply on Tyndall's letter (see letter 2621). The final letter is missing.

To Hector Tyndale 9 August 1867 2627

9th August 1867.

My dear Hector

Last night on my return from the Alps[1] your letter[2] came into my hands, and I read it with comfort and pleasure before going to bed. Pray thank Mr. Mitchell[3] from me for the photograph which he has been kind enough to send me.[4] I can read the character which you give of your sister[5] in her sweet and earnest face. I shall put the photograph among those of my dearest friends and look at it now and then when my soul needs help and elevation.

I am glad to hear that the little shreds and patches of thought which I send to you at times reach you in safety. Previous to going to Switzerland I sent you a copy of my latest little book on Sound,[6] which I trust is by this time in your possession. The work was stereotyped[7] in London with a view to publication without delay in America. Appleton[8] is to be the publisher. An excellent and most useful man, Professor Youmans,[9] whom you doubtless know as a public lecturer, is the mediator in all such cases between the publisher and myself. Youmans has done excellent work in introducing and diffusing the writings of my friend Spencer[10] in America. He is altogether an exceedingly useful and exceedingly clever man.

He gave me by the way an account of Walt Whitman[11] which differs widely from that given by Moncure Conway[12] in the Fortnightly Review.[13] He told me of your mother[14] being greatly interested in Whitman, and giving herself some trouble to make his acquaintance.

I had the pleasure of meeting this year in Switzerland an American gentleman whose conversation interested me deeply. I was first struck by the intelligent and well bred deportment of his little boy,[15] and afterwards he came to ask me some question about the mountains. Until he handed me his card I did not know him to be an American. I afterwards learned that he was one of the Commissioners appointed to draw up a report of the probable working of emancipation in the slave states, and that he was Mr. Lincoln's[16] confidential friend. His name is McKaye[17] (Col. McKaye). He knew Emerson[18] and Lowel[19] and many others of that stamp. His conversation about America was full of life and interest:—in fact he impressed me altogether as a highminded, cultivated and most earnest man:—So earnest indeed as to excite by a kind of contagion in me an interest in subjects which under ordinary circumstances do not affect me deeply.

I am glad you liked the paper on miracles and special providences.[20] I do not object to <u>your</u> miracles: I believe with you that the universe is a standing miracle,—that we are all miracles, but feeling this I do not find it necessary to call in monstrosities to gratify my appetite for the miraculous. Far be it from me to say that science exhausts, or can exhaust, the problem of the universe: or that there are not a thousand things which the mind & heart of man might long to know, and regarding which science must for ever remain silent. But I take it to be the manlier way to push knowledge to its possible limits, and to have the courage to avow our ignorance regarding what lies beyond those limits. I know that in any really embracing and exhaustive system of philosophy the force of the human heart has as good a right to be taken into account as the force of gravitation. But our miracle mongers are the last men either to found such a system, or even to comprehend it should it be founded by others.

Tom Hirst is at the present moment in Scotland: he occupies an honorary position in the British Association,[21] and is now engaged on the business of the Association whose next meeting is to be in Dundee,[22] I will send him your letter which I know will give him delight. For we often speak of you, and I think neither he nor I is lacking in affection towards you.

The large photograph to which you refer[23] was duly received by me, and at this moment occupies the place of honour in my room. You stand between my friends Faraday[24] and Magnus,[25] right over my chimney-piece. Immediately under you is a crayon drawing of one of the most beautiful children in the world:—a little maiden[26] whom I know and love. The mother[27] of the child—a glorious woman—knowing my feeling drew the portrait herself for me; so that you divide the best space on my wall with this darling.

Tom Hirst is growing exceedingly popular in London. He now occupies the chair[28] in University College which was occupied for thirty years by Professor De Morgan.[29] The circumstances of his election to this post were in the highest degree gratifying both to him and me.

Debus[30] happens to be here at present. He desires his kindest remembrances. In fact Debus is a Hero-worshipper and <u>you</u> are one of his heroes. He never speaks of you without an expression of respect which rises at times to enthusiasm.

Though the weather has been far from satisfactory I had some capital exercise in the Alps and have returned from them in strong health. My intellectual work now begins. I am committed to a discourse to the working men of Dundee,[31] which will break sadly into my work of investigation. Had I been asked to lecture before the magnates of the Association I should have declined, but an audience of working men to whom I lecture sometimes in London[32] behave so exceedingly well that I could not continue to refuse when asked to address men of their order in the north.

Since you last heard from me[33] I have gone through some sorrow. Known properly to myself alone were the love, the fortitude, and the natural dignity of my dear mother.[34] Early last May I saw her lowered into the grave in that old Leighlin churchyard,[35] in which, I do not know how rests the dust of some of our ancestors. My sister's[36] health has been much deranged; but I have been able to place her under the medical supervision & indeed in the family of a very worthy fellow—a D^r^. John Tyndall of Gorey in the County Wexford. He has an exceedingly nice house and garden, and his attention appears to be unremitting.

Poor man, he lost a very promising young fellow—his eldest son[37]—who had been for some time surgeon to one of the Peninsular & Oriental Company's[38] vessels. Two other young Tyndalls[39] sons of a Wexford magistrate have been also lately in London. They have been educated at Trinity College Dublin, and are intelligent gentlemanly young fellows.

And now I will say good bye give my kindest regards to M^rs^. Tyndale;[40] and again thank my friend Mitchell for the photograph. It does me good to look at it.

Ever dear Hector affect[ionate]ly yours | John <u>Tyndall</u>

RI MS JT/1/T/1448
RI MS JT/1/TYP/4/1670–3

---

1. *my return from the Alps:* Tyndall traveled in the Swiss Alps for about a month in the summer of 1867, departing London on 5 July and returning on 8 August.
2. *your letter:* letter 2620.
3. *Mr. Mitchell:* Edward Mitchell.
4. *photograph which he has been kind enough to send me:* Edward Mitchell arranged for Hector to send a photograph of Elizabeth, Mitchell's recently deceased wife and Hector's sister.
5. *your sister:* Elizabeth Mitchell (see letter 2620, n. 10).
6. *my latest little book on Sound: Sound* (1867).
7. *stereotyped:* to prepare (literary matter) for printing by means of stereotypes (*OED*).
8. *Appleton:* William Henry Appleton (1814–99), an American publisher, who in 1848 became the president of the publishing firm D. Appleton & Co. founded by his father (*ANB*).
9. *Professor Youmans:* Edward Livingston Youmans (1821–87), an American editor and popularizer of science. He was the science editor for D. Appleton and Co. in New York, and founder of the publisher's famous International Scientific Series and *Popular Science Monthly* (in 1872). Also a renowned lecturer, Youmans was a prolific writer of popular science texts (*ANB*).
10. *Spencer:* Herbert Spencer.
11. *Walt Whitman:* Walt Whitman (1819–92), an American poet and journalist best known for his *Leaves of Grass* (1855) (*ANB*).
12. *Moncure Conway:* see letter 2552, n. 1.
13. *in the Fortnightly Review:* M. Conway, 'Walt Whitman', *Fortnightly Review,* 6 (1866), pp. 538–48.
14. *your mother:* Sarah Tyndale (née Thorn, 1792–1859), an American abolitionist born in Philadelphia who married Robinson Tyndale. Sarah became friends with Walt Whitman after visiting him in 1856 in the company of Henry David Thoreau and Bronson Alcott (D. Kohn, 'Tyndale, Sarah Thorn', in S. LeMaster and D. Kummings (eds), *Walt Whitman: An Encyclopedia* (New York: Routledge, 1998), pp. 749–50).
15. *his little boy:* possibly Henry Goodwin MacKaye (1856–1913), the youngest son of James Morrison MacKaye ('Dr. Henry Goodwin MacKaye', *New York Tribune,* 3 February 1913, p. 9).
16. *Mr Lincoln:* Abraham Lincoln (1809–65), sixteenth president of the United States of America (*ANB*).
17. *McKaye:* James Morrison MacKaye (1804–88), (sometimes McKay) an American lawyer and abolitionist who founded Wells Fargo and Western Union. He was made a colonel by Abraham Lincoln and was one of the three commissioners of the American Freedman's

Inquiry Commission (1863–4) (W. Ellis, *Norwich University, 1819–1911: History, Graduates, Roll of Honor,* 3 vols (Montpelier, VT: Capital City Press, 1911), vol. 3, p. 731).

18. *Emerson:* Ralph Waldo Emerson.
19. *Lowel:* probably James Russel Lowell (1819–91), an American writer and abolitionist who was the first editor of the *Atlantic Monthly* (1857–61) (*ANB*).
20. *paper on miracles and special providences:* see letter 2597, n. 3.
21. *an honorary position in the British Association:* Hirst served as General Secretary of the BAAS from 1866–70 (J. H. Gardner and R. J. Wilson, 'Thomas Archer Hirst–Mathematician Xtravagant VI. Years of Decline', *American Mathematical Monthly,* 100 (1993), pp. 907–15, on p. 907).
22. *next meeting is to be in Dundee:* the thirty-seventh annual meeting of the BAAS at Dundee from 4–11 September 1867.
23. *The large photograph to which you refer:* in letter 2620, Tyndale wrote, 'I sent you a large photograph (half length) of myself (I do not mean half size) (rolled up) through Mr Lesley and the American Phil[osophical]: So[ciety]: who were then about to send a box to London. Did you get it?'.
24. *Faraday:* Michael Faraday.
25. *Magnus:* Heinrich Gustav Magnus.
26. *a little maiden:* probably Amy Harriet Lubbock (1857–1929), eldest child of John and Ellen Lubbock ('Obituary', *Times,* 21 May 1929, p. 16).
27. *The mother:* Ellen Lubbock.
28. *the chair:* Hirst held the chair of mathematics at University College London from 1867–70.
29. *Professor De Morgan:* Augustus De Morgan (1806–71), an English mathematician who held the chair of mathematics at University College London from 1828–67 except for a short period between 1831–6 (*ODNB*).
30. *Debus:* Heinrich Debus.
31. *a discourse to the working men of Dundee:* on 5 September 1867, Tyndall gave a lecture on 'Matter and Force' to working-class men at Kinnaird Hall, Dundee, where the annual meeting of the BAAS was held (*Brit. Assoc. Rep. 1867,* p. lxxiii). See 'Lecture to Working Men by Professor Tyndall', *Dundee Advertiser,* 6 September 1867, pp. 2–3; and 'Professor Tyndall's Lecture to Working Men. Matter and Force', *Dundee Courier and Argus,* 6 September 1867, p. 3. Tyndall later published the lecture as 'Matter and Force', in *Fragments of Science,* vol. 2, pp. 53–75.
32. *an audience of working men to whom I lecture sometimes in London:* Tyndall gave evening lectures to working men in his position with the Royal School of Mines (see letter 2601, n. 2).
33. *Since you last heard from me:* the previous letter from Tyndall to Tyndale in the extant Tyndall correspondence is from 14 May 1864 (see letter 2095, *Tyndall Correspondence,* vol. 8). It is likely there were more recent letters that are missing.
34. *my dear mother:* Sarah Tyndall (née Macassey).

35. *old Leighlin churchyard:* in Leighlinbridge, County Carlow, Ireland, where Tyndall was born and his mother lived.
36. *My sister's:* Emma Tyndall.
37. *his eldest son:* not identified.
38. *Peninsular & Oriental Company's:* the Peninsular and Oriental Steam Navigation Company, a British shipping company that operated routes between England and the Iberian Peninsula (Spain and Portugal) and Alexandria in Egypt. See F. Harcourt, *Flagships of Imperialism: The P&O Company and the Politics of Empire From its Origins to 1867* (Manchester and New York: Manchester University Press, 2006).
39. *Two other young Tyndalls:* not identified.
40. *M^rs^. Tyndale:* Julia Tyndale.

From Mary Adair [13 August 1867][1] 2628

Heatherton Park[2] | Tuesday

My dear M^r^. Tyndall,

We are very sorry that there is no chance of our meeting you in Switzerland this season, but our plans were settled finally some time ago. Tomorrow we make our way to Folkestone[3] & sleep there, cross the Channel next day and hope to reach Geneva Saturday morning. On the Monday we go on to St Gervais[4] or Chamounix[5] and through next week make such excursion as are within the powers of ladies. Our cousin Miss Mitchell[6] will be with us. She knows Switzerland well and is an excellent travelling companion—but I do not know that we shall be able to attempt much mountaineering.

Would it be taxing your kindness too much if we asked you to give us one line to Geneva to tell us the name of a good and "prudent" guide? My Mother[7] would be much obliged if you would kindly do so, and do you think I should be able to get to the Jardin.[8] I am very anxious to accomplish the expedition if it is practicable. The Montanvert[9] and the Flégère[10] we think we can easily manage, and afterwards my cousin and I think of crossing the Tête Noir[11] and meeting my Mother (who would return via Geneva) at Morges[12] where we visit friends. After that we hope to get on to Interlaken and the Oberland[13]—but we fear Zermatt will be beyond our reach.

We spent all last autumn in the Highlands,[14] and thoroughly enjoyed the Scotch mountains. We saw all the finest scenery—the best of all is over the Spital of Glenshee between Dunkeld & Braemar.[15] I hope you will see the Highlands well and enjoy the heather.

I shall not be content with anything but Edelweiss![16]

With kind regards truly yours | H. Mary Adair

RI MS JT/1/A/10

1. *[13 August 1867]:* the date is given by reference to Adair having written in letter 2615 that 'we may be spending three weeks or so from the middle of August in Switzerland'. Given that this letter was written on a Tuesday, it could be either 13 August or 20 August, with 13 August more likely as she wrote 'from the middle of August'. Tyndall had returned to London on 8 August 1867.
2. *Heatherton Park:* see letter 2540, n. 1.
3. *Folkestone:* a town in Kent, England, on the English Channel.
4. *St Gervais:* a commune in the Rhône-Alpes region of southeastern France.
5. *Chamounix:* see letter 2600, n. 4.
6. *cousin Miss Mitchell:* not identified.
7. *My Mother:* Harriet Adair.
8. *Jardin:* see letter 2600, n. 6.
9. *Montanvert:* see letter 2600, n. 5.
10. *Flégère:* see letter 2600 n. 8.
11. *Tête Noir:* see letter 2600, n. 9.
12. *Morges:* a municipality in the Swiss canton of Vaud.
13. *Oberland:* the geographically higher region of the Swiss canton of Bern.
14. *Highlands:* the geographically northern and geologically more mountainous region of Scotland.
15. *Spital of Glenshee between Dunkeld & Braemar:* the distance between the Scottish towns of Dunkeld and Braemar is about 30 miles as the crow flies, with the village Spittal of Glenshee falling in between.
16. *Edelweiss:* an alpine flower in the daisy family.

From Frédéric Zurcher[1] 13 August 1867 2629

Toulon 13 août 1867.

Cher Monsieur,

Mon beau frère, M. Margollé & moi, avons à faire pour une collection de livres instructifs destinés à la jeunesse, un volume intitulé Ascensions célèbres aux plus hautes montagnes du globe. Nous venons vous prier de bien vouloir nous autoriser à traduire pour ce volume quelques fragments de votre beau récit: Ascent of the Finsteraarhorn, inséré dans le Tome IV des Matériaux pour l'étude des glaciers de M. Dollfus-Ausset, et tiré de votre livre The glaciers of the Alps. Nous serions heureux de pouvoir contribuer aussi à attirer l'attention sur des travaux si utiles et qui honorent à la fois la raison & l'humanité.

Veuillez agréer l'assurance de nos sentiments respectueux. | F. Zurcher
à l'Enclos, près du fort Lamalgue à Toulon (Var)

Toulon[2] 13 August 1867.

Dear Sir,

My step brother, Mr. Margollé[3] & I, have to do, for a collection of instruction books for youth, a volume entitled Ascensions célèbres aux plus hautes montagnes du globe.[4] We would like to ask you to allow us to translate for this volume some parts of your wonderful story: Ascent of the Finsteraarhorn,[5] inserted in Book IV of the Matériaux pour l'étude des glaciers[6] by Mr. Dollfus-Ausset,[7] and taken from your book The Glaciers of the Alps.[8] We would be delighted to take part in bringing more attention to such useful works, which both honour reason & humanity.

Please accept our best wishes. | F. Zurcher

At the Enclos,[9] near Fort Lamalgue in Toulon (Var)[10]

RI MS JT/1/Z/2

---

1. *Frédéric Zurcher:* Frédéric Zurcher (1816–90), a French popularizer of science who wrote, with Elie Margollé, numerous books on a variety of scientific subjects including *Les Glaciers* (Paris: L. Hachette, 1868) (Archives Nationales).
2. *Toulon:* a Mediterranean port city in France.
3. *Mr. Margollé:* Elie Phillipe Margollé (1816–84), a French popularizer of science who coauthored numerous books with Frédéric Zurcher (Archives Nationales).
4. *Ascensions célèbres aux plus hautes montagnes du globe:* F. Zurcher and E. Margollé, *Les ascensions célèbres aux plus hautes montagnes du globe* (Paris: L. Hachette, 1867).
5. *Ascent of the Finsteraarhorn:* chapter 16 in Tyndall's *Glaciers of the Alps.*
6. *Matériaux pour l'étude des glaciers:* D. Dollfus-Ausset, *Matériaux pour l'étude des glaciers,* 13 vols (Paris: F. Savy, 1864–72). Tyndall's account of his ascent of the Finsteraarhorn was printed in book four, *Ascensions* (1864), on pp. 225–37.
7. *Mr. Dollfus-Ausset:* Daniel Dollfus-Ausset (1797–1870), a wealthy Alsatian textile manufacturer, geologist, and glaciologist (*NDB*; *HLS*).
8. *The Glaciers of the Alps: Glaciers of the Alps.*
9. *Enclos:* l'Enclos; possibly the Domaine du Clos Lamalgue, a winery not far from Fort Lamalgue.
10. *Fort Lamalgue in Toulon (Var):* a military fort built in the mid-eighteenth century in the Mediterranean port city of Toulon in the department of Var.

To [Juliet Pollock][1] 14 August 1867 2630

14th Aug. 1867.

My dear friend,

Much to my regret I was absolutely unable to go to your house yestereven, and today the fear of wearying you keeps me from calling upon you. I hear with sorrow that you are not well; but I firmly hope that once rid of the air and association of London you will soon revive: this at all events is the earnest wish of

your friend | John Tyndall

RI MS/JT/1/TYP/6/2086
LT Typescript Only

1. *[Juliet Pollock]:* the recipient is identified as Juliet Pollock based on relation to letter 2637 (dated 21 August), in which Tyndall wrote: 'Had I not gathered from my conversation with your husband that you needed quiet, and that I might, as I said in my last note, really weary you, I should assuredly have gone to see you before you quitted London. Indeed my first thought on coming into it was to call upon you, and only an accident prevented me from doing so. I do sincerely trust that the change of air will banish that confounded fever'.

From Mary Egerton 14 August [1867][1] 2631

Mountfield Court | Hurst Green[2] | Sussex | Aug[us]t. 14th

Dear Mr. Tyndall,

I heard you were to be back about this time,[3] so as I think it most probable my letter to Grindelwald[4] would never reach you, I shall take the liberty of sending you a duplicate of thanks for the book[5] you so kindly sent me. If you did receive the former one, pray forgive me for boring you with a second, but I could not risk appearing ungrateful for what was such a real pleasure to me to receive. I hope you have had a better season for Switzerland than last year. Here the Summer seems to have just begun, and if we get all the fine weather for our tour, I shall consider it a just punishment to you for not staying long enough to meet us. We do not start till the 26th, so if you have a spare minute to bestow before then, do let us hear where you have been. We are devouring "Mountaineering",[6] (I, for about the 20th time), in the hope of seeing one or two of the places mentioned. If "Sound"[7] is "a bit of your mind", I think this may be called "a bit of your soul"—your descriptions are not pictures, they are realities. You have a (stereoscopic) way of seeing things from every point of view—Adventures—Poetry Science—& Worship! I think the book will go with me to the Aeggischhorn;[8] but Oh! if we should meet a heap of Yankees or Cockneys[9] there! Would that I could take refuge at the top of the Weisshorn![10] (but they must invent Air Navigation first!) Is "X" an "unknown", or a real friend? If you do write, would you tell me whether it makes much difference if one goes up or down from the Grimsel?[11] I mean for the views. My husband thought of taking in Lucerne & the St Gothard,[12] in which case it w[oul]^d. I suppose be the simplest way to go by the Furca,[13] which I see by the Guide book[14] professes to be now a carriage road! to the Grimsel road & then down by Obergesteln[15] to Viesch[16] for the Aeggischhorn, then back, & ride down Haslithal[17] to Meyringen[18] & by the Scheideck[19] to Grindelwald, reversing the line you gave us. This plan puts Zermatt still more out of the question I fear, for there is no other Easy way of getting back in the Rhone valley,[20] is there? Without going a very long way round. If I can get a glimpse of the Matterhorn[21] in the distance, I must be content. Will uninitiated Eyes be able to see the Curves & marks of lamination on any glacier from where we shall be? Forgive me for asking questions & don't answer them if you are busy; (but you always are!) I am forgetting that while I have been conversing with you, (in your book,) you have had almost time to forget my existence! Well! To return to my first reason for writing once more many thanks.

Y[ou]^rs. very truly | M F Egerton

I can hardly hold my pen in this heat! Do you know any small book for the Geology of the Alps?

RI MS JT/1/E/99c and E/99d

---

1. *[1867]:* the year is given by relation to letter 2645, in which Egerton discusses her travels in the Alps.
2. *Mountfield Court | Hurst Green:* the Egerton family estate.
3. *back about this time:* Tyndall returned to London from the Alps on 8 August.
4. *my letter to Grindelwald:* letter missing.
5. *book: Sound* (1867).
6. *"Mountaineering": Mountaineering in 1861.*
7. *"Sound":* see n. 5.
8. *Aeggischhorn:* see letter 2600, n. 24.
9. *Yankees or Cockneys:* Americans or someone from London.
10. *Weisshorn:* see letter 2600, n. 19.
11. *Grimsel:* see letter 2600, n. 28.
12. *St Gothard:* a mountain pass in the Alps connecting northern and southern Switzerland.
13. *Furca:* the Furka Pass, connecting Andermatt with Gletsch, and thence to the Grimsel Pass.
14. *Guide book:* probably J. Murray, *A Handbook for Travellers in Switzerland, and the Alps of Savoy and Piedmont,* 8th edn (London: John Murray, 1858).

15. *Obergesteln:* a village in the Swiss canton of Valais.
16. *Viesch:* see letter 2600, n. 22.
17. *Haslithal:* see letter 2600, n. 31.
18. *Meyringen:* see letter 2600, n. 33.
19. *Scheideck:* see letter 2600, n. 36.
20. *Rhone valley:* see letter 2600, n. 25.
21. *Matterhorn:* see letter 2600, n. 18.

From François-Napoléon-Marie Moigno [before 25 August 1867][1] 2632

Mon Cher M^r^. Tyndall

Merci des brochures que vous m'avez envoyées, merci plus encore de votre petit mot à M^r^. Sorby.

Ce que je voulais vous dire du livre des Glaciers, c'est qu'il n'avait pas paru encore; que si la Chaleur ne vous avait pas donné d'argent, elle ne s'était pas au moins fait attendre et qu'elle vous avait donné de la gloire.

Je trouve très bien que vous traitiez pour le Son avec une grande maison, Masson ou autre; mais je voudrais que vous me réserviez la traduction que je ferais aux frais du libraire ou à vos frais. Je tiens à la faire moi même, seul, sans l'intervention de M^r^. Raillard ou de qui que ce soit. Je veux en faire un chef d'œuvre accompli. La donner à un autre qui la fera moins bien que moi, ce serait m'infliger un petit déshonneur on croirait que j'aurais démérité de vous. Si vous m'envoyez l'ouvrage feuille par feuille chaque semaine, je vous renverrai la traduction feuille par feuille aussi chaque semaine, de sorte que les deux éditions pourraient paraitre même temps à huit jours près.

Rien n'est encore perdu pour la Chaleur; j'aurai je l'espère à vous donner un peu plus tard de bonnes nouvelles à ce sujet. Le libraire qui publiera Le Son pourra d'ailleurs publier en même temps une nouvelle édition de La Chaleur revue et refondue par moi sur la seconde édition anglaise.

Pardon de tous les détails; je ne vous reporterai plus de mon désir et vous laisse entièrement libre de faire ce que vous voudrez.

Lisez ma traduction d'Huggins et comparez là à la traduction de l'adresse de Grove faite par la Revue des cours publiés.

Voyez Hofmann ses leçons de chimie sont encore inconnues en France.

A vous de cœur | l'abbé F. Moigno

Je vous recommande instamment la compagnie L'Alliance et ses machines. Le temps de la livraison de la machine de Holmes va arriver prochainement.

My dear Mr Tyndall,

Thank you for the brochures that you sent me,[2] and thank you even more for your little note to Mr. Sorby.[3]

What I wish to tell you about the book on Glaciers, is that it has not appeared yet;[4] and that if la Chaleur[5] has not produced you money, at least it had not been delayed and had given you glory.

I think it very well that you deal with Sound[6] using a large publishing house, Masson[7] or another, but I would like you to reserve for me the translation of it,[8] which I will do either at the expense of the bookseller[9] or at your expense. I want to do it myself, alone, without the intervention of Mr. Raillard[10] or of anyone else. I want to create an accomplished masterpiece out of it. Giving it to another, who will do it worse than me, will inflict a little shame on me, it would be believed that I am not worthy of you. If you send me the book page by page every week, I will send you back the translations page by page every week as well, so that the two editions can get published at the same time, within eight days of each other.

All is not lost for Heat; I will give you, I hope, good news on this matter a bit later. The bookseller who will publish Le Son will also be able to publish a new edition of La Chaleur[11] at the same time, reviewed and reworked by me according to the second English edition.[12]

Forgive me for all the details, I will not be referring to my wishes anymore, and I leave you entirely free to do as you please.

Read my translation of Huggins[13] and compare it to the translation of Grove's[14] speech made by the Review of cour publiés.[15]

Look at Hofmann,[16] his chemistry lessons are still unknown in France.

Yours truly | Father Moigno

I strongly recommend the company L'Alliance[17] and their machines. The time for delivering Holmes'[18] machine[19] will arrive soon.

RI MS JT/1/M/138

---

1. *[before 25 August 1867]:* the date is given as this letter precedes the other three letters from Moigno in 1867, as they all discuss *Sound* during translation; the earliest of those is from 26 October. With no mention of Faraday's passing or Tyndall's work preparing his lectures on Faraday for the RI (see letter 2665, n. 3), this letter would not likely fall after the date of Faraday's death, 25 August 1867.
2. *brochures you sent me:* not identified.
3. *note to Mr. Sorby:* letter missing. Henry Clifton Sorby (1826–1908), an English geologist who pioneered the field of petrography through his application of the microscope to measure and determine the composition of rocks (*ODNB*).
4. *book on Glaciers . . . not appeared yet:* probably referring to Tyndall's *Glaciers of the Alps* (1860) being translated into French, but it never was.

5. *la Chaleur:* see letter 2680, n. 13.
6. *Sound: Sound* (1867).
7. *Masson:* Victor Masson and Sons, a French publishing firm.
8. *translation of it:* Moigno's translation of *Sound* appeared in 1869 (see letter 2617, n. 2).
9. *bookseller:* the translation of *Sound* would be published by Gauthier-Villars.
10. *M^r. Raillard:* l'abbé Raillard, not identified.
11. *new edition of La Chaleur:* the second French edition of *Heat* appeared in 1874 and was based on the fourth English edition: J. Tyndall, *La Chaleur: mode de mouvement,* trans. F. Moigno, 2nd edn (Paris: Gauthier-Villars, 1874).
12. *second English edition: Heat* (1865).
13. *my translation of Huggins:* Moigno's translation of William Huggins' *On the results of spectrum analysis applied to the heavenly bodies* (London: W. Ladd, 1866) appeared as W. Huggins, *Analyse spectrale des corps célestes,* trans. F. Moigno (Paris: Gauthier-Villars, 1866).
14. *Grove:* William Grove.
15. *translation of Grove's speech made by the Review of cour publiés:* probably the translation of William Grove's 1866 BAAS Presidential Address (W. Grove, 'La continuité dans la nature', *Revue des cours scientifiques,* 3 (1866), pp. 682–96).
16. *Hofmann:* August Wilhelm von Hofmann (1818–92), a German chemist (*CDSB*).
17. *L'Alliance:* a French company that made electric light machines.
18. *Holmes':* Frederick Hale Holmes (b. 1812), an English chemist and inventor who developed generators and electrical lighting devices for lighthouses. During the 1850s he was professor of chemistry at the short-lived Royal Panopticon of Science and Art (*Faraday Correspondence,* vol. 5, p. lii).
19. *machine:* not identified.

To François-Napoléon-Marie Moigno [before 25 August 1867][1] 2633

Monday

Dear M^r l'Abbé

I was in Paris last Friday[2] but was so extremely occupied that I had not time to call to inform you that I had made arrangements with M. Gauthier for the publication of a translation of my book on sound.[3] It will give me great pleasure to see the book translated by you, as indicated in your last note to me.[4] If you do it yourself no man can do it better. Having made this statement my connexion with the subject ends, all business matters are to be arranged not between you and me, but between you and M. Gauthier.

Yours very truly | John Tyndall

The Waller Manuscript Collection, Uppsala University, gb-01758

1. *[before 25 August 1867]:* this letter follows letter 2632 from Moigno, dated [before 25 August 1867].
2. *in Paris last Friday:* Tyndall was in Paris in 1867 on at least two occasions before this letter: he was in Paris from 19–26 April 1867 for the International Exposition (see letter 2578, n. 3), and he would have passed through Paris on his way to Switzerland in July 1867 (he departed London on 5 July).
3. *made arrangements with M. Gauthier for the publication of a translation of my book on sound:* Jean-Albert Gauthier-Villars published the French edition of Tyndall's *Sound* (1867), which appeared as: J. Tyndall, *Le Son, Cours Expérimental fait à l'Institution Royale,* trans. F. Moigno (Paris: Gauthier-Villars, 1869). In letter 2617, Gauthier-Villars wrote to Tyndall on 12 July 1867 that he had 'the honour of thanking you for the copy of your book on sound, which you have kindly sent me. I am always willing to use your permission, and to publish a French translation of this book'. Gauthier-Villars went on to clarify the terms of their agreement, including Moigno as translator.
4. *your last note to me:* letter 2632.

To Elizabeth Dawson Steuart [mid-August 1867][1] 2634

Saturday

My dear Mrs Steuart,

I am quite willing to conform to the advice of yourself and Dr Trayer[2] in all things regarding the outlay of money. I however send a copy of the note[3] I wrote to you in answer to the letter from the Liverpool minister.[4] This note will have been doubtless read to Sarah Tyndall's[5] brother,[6] and he will possibly have informed her of my intention. Perhaps it might be made known to Sarah Tyndall that whenever it is clearly made out to your satisfaction that the drawing of the money would be to her benefit she might consider herself at liberty to apply for it.

Caleb Tyndall's son[7] has again written to me, but I have thrown his letter[8] in the fire. He wants the means of carrying him home to Ireland. If his sister[9] thinks it right to devote a portion of the twenty pounds to this purpose she may do so. But I can have no direct communication whatever with this very troublesome and very stupid person.

With regards to the furniture I would willingly leave the management of the thing in your hands: Please bestow it in the names which you think best.[10]

I went down last Saturday to Chevening, the seat of Lord Stanhope.[11] I remained there till Monday. It is a beautiful place, and associated with many interesting memories of Lord Chatham[12] (the great premier). He lived there for a time and planned some of the improvements. Lady Stanhope[13] is full of kindness. I had been there before, and I like the people much. I run out of town to day also, and towards the end of the month go to Scotland.[14]

Yours ever truly | John Tyndall.

RI MS JT/1/TYP/10/3390
LT Typescript Only

1. *[mid-August 1867]:* the month and year are given based on mention of the status of his mother's furniture, implying that this letter was written after her death (on 30 April 1867), and that this letter appears to come before Michael Faraday's death on 25 August 1867, as he made no mention of it. For the reference to going to Scotland at the end of the month, Tyndall possibly intended to go to Scotland a few days earlier prior to the annual meeting of the BAAS at Dundee from 4–11 September 1867, and Faraday's death on 25 August changed the arrangements.
2. *Dr Trayer:* James John Trayer.
3. *note:* letter missing.
4. *Liverpool minister:* not identified.
5. *Sarah Tyndall's:* Sarah Tyndall (not Tyndall's mother; see Biographical Register).
6. *brother:* either Caleb, James, or William Tyndall, who were sons of Caleb Tyndall and brothers of Sarah Tyndall.
7. *Caleb Tyndall's son:* John Tyndall (see letter 2717, n. 6).
8. *his letter:* letter missing.
9. *his sister:* Sarah Tyndall.
10. *With regards to the furniture . . . which you think best:* the furniture of Tyndall's mother Sarah who died on 30 April 1867.
11. *Lord Stanhope:* Philip Stanhope (see letter 2707, n. 6). Chevening House, the country estate of the Earls Stanhope, is in the village of Chevening in Kent.
12. *Lord Chatham:* William Pitt (1708–78), first Earl of Chatham, an English politician and prime minister (1766–8) (*ODNB*).
13. *Lady Stanhope:* Emily Harriet Stanhope (née Kerrison, 1815–73), daughter of Edward Kerrison, she married Philip Stanhope in 1834.
14. *go to Scotland:* see n. 1.

From Thomas Archer Hirst [17 August 1867][1] 2635

Parkins and Gotto[2] | Saturday

My dear Tom

The letters were duly received at Visp: in fact I had forgotten that I had requested them to be sent to Visp. An accident revealed the fact to me.

I had but poor weather towards the end: the Matterhorn encased in ice, and swathed in cloud. Could I have calculated upon three days of fine weather I should have attacked it from the south without much fear of failure: but the weather seemed hopeless and I came away. There have been some glorious days since; but I hardly think the monster will be scaled this year.

I have had a very long letter[3] from Mr. Alexander Herschel.[4] He informs me that Mr Griffiths[5] has informed him that I should lend him my electric lamp at Dundee. Of course I would lend anything to Herschel that I could spare: but it is a risky thing for Mr Griffiths to entangle me in this way without ever hearing a word from me on the subject.

The lecture[6] is indeed a great bother to me—It has thus far entirely taken me off my work, and I fear for its success through the difficulty of getting the proper arrangements made. I did not know until yesterday that the Association meeting[7] and the working men's meeting are held in the same hall,[8] and then only incidentally from a passage in Herschel's letter. I have written to Griffiths asking him to inform me of the day and hour of lecturing to the working men: on these points I am still in the dark. If you could ascertain and inform me whether nitric and sulphuric acid can be obtained in Dundee I should be much obliged to you.

I want also to know whether a piece of clear ice could be obtained there—the so called Wenham Lake ice?[9]

You have hea[r]d of Lady Lubbock's[10] mishap: a son still-born. M^rs^. Busk[11] sends you a little note[12] by this post, in which doubtless you will find intelligence regarding High Elms.[13]

I fear you have had but poor weather in Skye:[14] I am sorry that you must return direct from Dundee, for I hoped we should go to Staffa[15] and to the parallel roads of Glenroy[16] together: goodbye. Yours affect[ionate]ly | John Tyndall

BL, Add MS 63092, ff. 35–6
RI MS JT/1/HTYP/538

---

1. *[17 August 1867]:* the date is given in relation to Tyndall's return to London on 8 August via Visp (where he wrote letter 2624 on 4 August) and in relation to letter 2636 (19 August) in which Tyndall asked Hirst to 'not say any thing to Griffiths about his remark to Herschel'.
2. *Parkins and Gotto:* embossed at upper left of first page; London stationers located on Oxford St.
3. *very long letter:* letter missing.
4. *Mr. Alexander Herschel:* Alexander Herschel.
5. *Mr Griffiths:* George Griffith.
6. *The lecture:* in Dundee; see letter 2627, n. 31.
7. *Association meeting:* the annual meeting of the BAAS in Dundee; see letter 2627, n. 22.
8. *hall:* Kinnaird Hall.
9. *Wenham Lake ice:* ice blocks were harvested from Wenham Lake in Massachusetts (United States) for shipment to countries across the Atlantic from the 1840s and into the twentieth century.

10. *Lady Lubbock's:* Ellen Lubbock.
11. *M^rs^. Busk:* Ellen Busk.
12. *little note:* not enclosed.
13. *High Elms:* the home of John Lubbock near Down House, the home of Charles Darwin in the English village of Downe.
14. *Skye:* or Isle of Skye, the largest island of the Inner Hebrides of Scotland. Hirst was in Skye from 17–19 August.
15. *Staffa:* an island of the Inner Hebrides of Scotland.
16. *parallel roads of Glenroy:* three parallel 'roads' along the canyon walls of Glen Roy, the valley of the River Roy in Scotland. The natural cause of these parallel 'roads'—in fact terraces formed by water levels in the valley—had long been a puzzle to geologists, including Charles Lyell, Charles Darwin, and Louis Agassiz (M. J. S. Rudwick, 'The origin of the Parallel Roads of Glen Roy: a review of 19th Century research', *Proceedings of the Geologists' Association,* 128 (2017), pp. 26–31). Tyndall would later lecture on the topic for a Friday Evening Discourse at the RI on 9 June 1876, stating: 'in 1867, with a view to self-instruction, I made a solitary pilgrimage to the place, and explored pretty thoroughly the roads of the principal glen' (J. Tyndall, 'The Parallel Roads of Glen Roy', *Roy. Inst. Proc.,* 8 (1876), pp. 233–46, on p. 240). In the lecture, he reviewed previous hypotheses and put forward his view that the action of ancient glaciers explained the parallel roads.

To Thomas Archer Hirst 19 August 1867 2636

19th Aug 67

My dear Tom.

Do not say any thing to Griffiths[1] about his remark[2] to Herschel.[3]

I hope the Lecture[4] will be got over early. I mean at an early portion of the meeting.[5] I should like to depart after the 1st. three days; I love the Highlands[6] better than the Sections.

It would also suit me if the Report on Schools[7] were introduced to the general Committee[8] at an early sitting. Three days are quite enough to demonstrate my loyalty to the Association.[9]

Yours affect[ionate]ly | John Tyndall

RI MS JT/1/T/918
RI MS JT/1/HTYP/539

---

1. *Griffiths:* George Griffith who was then the assistant secretary of the BAAS, while Hirst was one of two general secretaries. On 23 August, when Hirst arrived at Dundee for the BAAS annual meeting (see letter 2627, n. 22) he had dessert with Griffith (Brock & MacLeod, *Hirst Journals,* 23 August 1867, p. 1811).
2. *his remark:* see letter 2635.
3. *Herschel:* Alexander Herschel.

4. *lecture:* see letter 2627, n. 31.
5. *meeting:* see letter 2627, n. 22.
6. *Highlands:* the geographically northern and geologically more mountainous region of Scotland.
7. *Report on Schools:* see letter 2577, n. 4 and n. 6.
8. *General Committee:* the report was presented to the General Committee on 4 September 1867, the first day of the Dundee meeting (see letter 2577, n. 6). Tyndall was asking that Hirst, in his role as secretary, move the report to the start of the meeting so that he could leave the meeting early and explore the Scottish Highlands.
9. *Association:* the BAAS.

To Juliet Pollock [21 August 1867][1] 2637

Wednesday

My dear M[rs]. Pollock

Had I not gathered from my conversation with your husband[2] that you needed quiet, and that I might, as I said in my last note,[3] really weary you, I should assuredly have gone to see you before you quitted London. Indeed my first thought on coming into it was to call upon you, and only an accident[4] prevented me from doing so.

I do sincerely trust that the change of air will banish that confounded fever. It was in all probability the expression of the heat—and dissipation of London. Those causes are now removed and you will soon recover your strength.

I am glad you met Boxall.[5] He may grumble and complain—but there is a truth and beauty & kindliness in the man's eye that whisper a thousand pleasant things regarding the man's soul. I like him much, but do not tell him so. He is sure to find it out in other ways.

You describe the "rosebud"[6] accurately. I hardly know any body to whom the term is more applicable.—A most pleasant blessing truly. Give her my kind regards.

I was in Birmingham yesterday and got the dust of the line[7] into my eyes: this causes me to make this note shorter than it otherwise would be.

Pray present my respects to your aunt & cousin.[8] | Yours ever | John Tyndall

I am preparing for Dundee.[9] I have got entangled in a lecture there in connexion with the British Association. But I hope to be able to escape early from "the Sections", and have a run among the mountains.[10]

RI MS JT/1/T/1203
RI MS JT/1/TYP/6/1907

1. *[21 August 1867]:* the date for this 'Wednesday' letter is given by relation to letter 2630 (dated 14 August 1867, also a Wednesday) and reference to his upcoming trip to Dundee, in which he arrived on 4 September (another Wednesday). Wednesdays between 14 August and 4 September occurred on 21 and 28 August, and since this letter lacks any mention of Michael Faraday's death on 25 August, 21 August seems a more probable date.
2. *your husband:* William Frederick Pollock.
3. *my last note:* letter 2630.
4. *accident:* not identified.
5. *Boxall:* probably William Boxall.
6. *"rosebud":* a pretty young woman (OED); not identified.'
7. *dust of the line:* probably a reference to railway soot.
8. *your aunt & cousin:* probably Isabella Maria Herries (*c.* 1789–1870), a sister of Juliet Pollock's mother, Catherine Creed (née Herries) who raised Juliet Pollock after the death of her mother (see the Biographical Register in *Tyndall Correspondence,* vol. 6); and Isabella Ann Herries (d. 1897), a cousin of Juliet Pollock, daughter of John Charles Herries (1778–1855) and Sarah Dorington (d. 1821) (see letter 1550, *Tyndall Correspondence,* vol. 6).
9. *Dundee:* lecture to the Working Men of Dundee (see letter 2627, n. 31).
10. *run among the mountains:* see letter 2650, n. 2.

From Jane Barnard 26 August 1867 2638

26 Aug. 1867 | Hampton Court.[1]

My dear Dr Tyndall

We feel that we must not let you hear through public channels of our loss. My dear Uncle[2] is gone:[3] he passed quietly & calmly from this life yesterday afternoon. He had never taken to his bed & was sitting in his chair, in his own room at the time

I am yours | very sincerely | Jane Barnard

Dr Tyndall

Faraday Correspondence, 6:4602
RI MS JT/1/B/30

1. *Hampton Court:* Michael Faraday's house in London since 1858.
2. *my dear Uncle:* Michael Faraday.
3. *is gone:* Faraday's funeral, small and private to fellow Sandemanians, was held on 30 August at Highgate cemetery in north London.

To Jane Barnard 27 August [1867][1] 2639

27th Aug

My dear Miss Barnard.

Your little note,[2] announcing a great & sad event,[3] reached me this morning. I thank you for it. But why should I say "sad",? What could be better than to be released so peacefully after the conclusion of his glorious day's work:—I could almost envy him.

Yours ever | John Tyndall

RI MS JT/1/T/124
RI MS JT/1/TYP/12/4194

1. *[1867]:* the year is given based on mention of the death of Michael Faraday.
2. *Your little note:* letter 2638.
3. *event:* the death of Michael Faraday on 25 August 1867.

To [Juliet Pollock][1] 27 August 1867 2640

27th. August. 1867.

My dear Friend

Faraday[2] died on Sunday morning while sitting in his chair. He passed away as peacefully as if he had merely fallen asleep.

Yours ever | John Tyndall

RI MS JT/1/TYP/1/366
LT Typescript Only

1. *[Juliet Pollock]:* possibly Juliet Pollock, whom Tyndall often addressed his letters to with 'My dear friend'.
2. *Faraday:* Michael Faraday, who died on 25 August 1867.

To Thomas Archer Hirst 27 August 1867 2641

Royal Institution 27th. Aug. 1867.

My dear Tom

I hardly ever felt stronger than in Switzerland this year. I was fit for any work in the way of climbing[1] and came home not sunbrowned, but almost black, the image of health. I shut myself up to get this lecture for Dundee[2] ready, but went out of town on the Saturdays. Somehow or other this damp weather has struck rheumatism into my bones and it has brought me very low. I thought it might have been the kidneys that were wrong but Evans[3] declares the seat of the thing to be far away from the kidneys and moreover I must say the kidneys do their work perfectly. Lucky that it came down upon me so soon for had it occurred during the Dundee time I know not what I should have done. I hope to be able to meet you on Friday, but I will send you a telegram on Thursday stating yes or no.[4]

Glorious old Faraday[5] fell asleep for ever yesterday morning[6] while sitting quietly in his chair. There was no suffering of any kind.

I have been written to by the Editor of the Dundee Advertiser[7,8] asking for the manuscript of my lecture to the working men.[9] If you see no reason to the contrary I have no objection to giving it to him.

I send you some letters:—there are several large parcels which I do not think you need in Scotland. They are all books or pamphlets. I should so like to have two days scamper in Arran[10] with you.

Yours affectionately | John

I send this day by steamer addressed to me at the Kinnaird Hall[11] a large box of apparatus.[12] Will you see that it is received and properly housed.

Confound my stupid brain—I have promised Lord Wrottesley[13] to go down to him on Saturday. This quenches Arran.

RI MS JT/1/HTYP/539
LT Typescript Only

---

1. *in the way of climbing:* Tyndall traveled in the Swiss Alps for about a month in the summer of 1867, departing London on 5 July and returning on 8 August. He had hoped to climb the Matterhorn, but the weather was not in his favor (see letter 2651). He did climb the Eiger, went over the glacier pass Petersgrat, and crossed the Col de St. Theodule (see letter 2622).
2. *lecture for Dundee:* see letter 2627, n. 31.
3. *Evans:* not identified, but presumably a doctor.
4. *to meet you on Friday . . . yes or no:* according to Hirst's journals, Tyndall did not meet Hirst on Friday, 30 August (Brock & MacLeod, *Hirst Journals,* 27 August–15 September 1867, p. 1812).
5. *Faraday:* Michael Faraday, who died on 25 August.
6. *yesterday morning:* either Tyndall miswrote "yesterday morning", or he began writing this letter on 26 August.
7. *written to by the Editor of the Dundee Advertiser:* letter missing.

8. *Editor of the Dundee Advertiser:* John Leng (1828–1902), a politician and newspaper editor. He was the editor of numerous newspapers, including the *Dundee Advertiser* (1851–1901) (*ODNB*).
9. *my lecture to the working men:* 'Lecture to Working Men by Professor Tyndall', *Dundee Advertiser*, 6 September 1867, pp. 2–3. See also letter 2627, n. 31.
10. *Arran:* an island off the western coast of Scotland.
11. *Kinnaird Hall:* the location of Tyndall's lecture to the working men at Dundee (see 2627, n. 31).
12. *apparatus:* for Tyndall's lecture on 'Matter and Force' (see 2627, n. 31).
13. *Lord Wrottesley:* John Wrottesley (1798–1867), second baron Wrottesley, an English astronomer and landowner. He served as president of the Royal Astronomical Society (1841–3), the RS (1854–7), and the BAAS (1860) (*ODNB*).

From Jacob Siegfried Bürgi[1] 2 September 1867 2642

Hochgeehrter Herr.

Indem ich die Ehre habe Ihnen anbei die Einladung unserer festgebenden Section „Pilatus" zu übermitteln ist mir der Auftrag geworden der Bitte unseres Central-Comité Ausdruck zu geben um Sie zu bitten die Versammlung wo möglich mit Ihrer Gegenwart zu beehren.

In dieser Hoffnung zeichnet hochachtungsvoll

Im Namen des Central-Comité | Siegfried Bürgi | Central-Cassier des S.A.C

Zürich 2 Septbr. 1867.

Esteemed Sir.

In having the honour to send you enclosed the invitation of our organising Section "Pilatus",[2] I have been favoured with the task of expressing the request of our Central Committee[3] to request you to honour the meeting where possible with your presence.

In this hope I sign respectfully

On behalf of the Central Committee | Siegfried-Bürgi | Central Treasurer of the S.A.C.[4]

Zürich, 2 Sept[em]b[e]r. 1867.

RI MS JT/1/B/154

1. *Jacob Siegfried Bürgi:* His full name is found in *Jahrbuch des Schweizer Alpenclub. Vierter Jahrgang. 1867–1868* (Berne: 1868), p. 600.

2. *"Pilatus":* the name of a section of the Swiss Alpine Club, named for a mountain massif in the Swiss Alps overlooking Lucerne. There were eleven sections, among them Geneva, Monte Rosa, Bern, and Basle.
3. *Central Committee:* the Swiss Alpine Club's committee to decide on club rules and organize meetings.
4. *S.A.C.:* the Schweizer Alpen-Club (Swiss Alpine Club), founded in 1863.

From Emily Peel 2 September 1867 2643

Campagne Lammermoor | Sechéron | Genève[1]
September 2/67.

Dear Professor Tyndall

I hear that in all probability you are coming out again to Switzerland this Autumn,[2] & should Geneva fall in your route towards the mountains, I hope you will not forget to look in here & will give me an account of your excursions this year which I am told have been very remarkable ones. I remain

yours | truly | Emily Peel

RI MS JT/1/P/44

1. *Campagne Lammermoor | Sechéron | Genève:* probably an estate or house that the Peels took in the summer in the Sechéron neighborhood of Geneva, Switzerland.
2. *coming out again to Switzerland this Autumn:* in his reply (letter 2651), Tyndall wrote: 'It would give me true pleasure to be able to accept your invitation, but I have already accomplished my Swiss tour for the present year'.

To Elizabeth Dawson Steuart 11 September 1867 2644

11th Sep. 1867.

My dear Mrs Steuart,

I must have missed a letter of yours,[1] for there is an allusion in your last note[2] to me which I do not understand—you must have made some request which never reached me, or which escaped me and of which I have no memory. You say you thought I might have some friend with <u>time</u> on hand &c., and then there is a word between marks of quotation which I cannot quite make out it is written thus:—"The Linsner." Pray let me know what you want—it will go hard with me or your request shall be complied with.

I sent you some papers from Dundee which will tell you more of what I have been doing of late[3] than could be told in a letter. I am not in strong health and on that account shall remain in Scotland till the 20th. Afterwards

I shall spend a day or two with my friend Sir William Armstrong the great gun-man in Northumberland.[4] On the 23rd or 24th I shall be in London.

No doubt the account book is in London: I wish you had given yourself no trouble about it. I would not read over the accounts for a fee of 20 guineas.

Will you manage this for me in your wise way—Give Caleb Tyndall to understand that his daughters[5] shall not be left without a marriage portion provided their conduct is so good as to justify me in giving it. I shall take care that neither of them shall marry pennyless. Indeed if their behaviour were perfectly commendable I should not mind scraping together £100 to divide between them. It is perhaps as well that they should have some knowledge of this as it may influence their action.

Yours ever | John Tyndall.

RI MS JT/1/TYP/10/3362
LT Typescript Only

---

1. *letter of yours:* letter missing.
2. *your last note:* letter missing.
3. *papers from Dundee which will tell you more of what I have been doing of late:* see letter 2641, n. 9.
4. *Northumberland:* the most northern county in England, where Armstrong's home Cragside was located, near the town of Rothbury.
5. *his daughters:* probably Sarah Tyndall who married David Miller (1869) and Dora (or Dorah or Dorothea) Tyndall. Caleb's other daughters (and John Tyndall's first cousins) were Margaret who married Caleb Payne (1855), and Emma who married George Young (1841) and later William Hargrave (1865) (*Irish Genealogy*).

From Mary Egerton 11 September [1867][1] 2645

Terrace Hotel Jungfrau[2] | Eggischhorn[3] | Sept[embe]r 11th

Dear Mr. Tyndal,

We owe you an Eternal debt of gratitude, for it was you that first put it into our heads to come here! Everything has been glorious, from the first sights of the snowy peaks cut out against a cloudless sky which sent a thrill through one even in the hot railway going to Lucerne, to the present moment when I am just returned from that indescribable Marjelin See![4] Soft & lovely & blue in the midst of its almost mysterious wildness, one ought not to see those things merely for an hour or two! I clung to every last glimpse, & I could not I am sure have spoken the farewell that I seemed to give it when we finally had to turn away. They took us part of the way to the glacier[5] in a boat, & I must confess to my enjoyment having been slightly marred by some doubts of the stability of our craft. We were cracking through flakes of ice & occasionally jostling a small iceberg & did not appear to have the iron beak[6] one reads of in vessels prepared for the arctic regions! However when I found that by allowing the baling pan[7] to keep pace with every stroke of the oar, the water did not increase <u>more</u> than to wet our feet. I got tolerably /happy/ but even then we did not deem it wise to cross the deep part, so did not land on the glacier, as I could not venture the long walk round much to my eldest daughter's[8] disappointment, who wanted I believe to simulate your "English maiden".[9] She & her father[10] are now gone to the summit,[11] her sister[12] & I having left them to pursue their way & come down alone. It is a dreadful thing in these parts not being a first rate walker! You will despise us dreadfully as Alpine travellers when I tell you we all shirked the last peak of the Sparrenhorn,[13] except my girl,[14] who left us reposing on the ridge & went up with the guide. But how glorious were the views we did see. That Aletsch glacier[15] is wonderful! I only longed for you to be standing over it with me & explaining its curious lines of striations. One moraine especially puzzled us, it seems to begin suddenly in the midst of clear ice, without any apparent rock or function to account for it. Do you remember it from Bel Alp?[16] It appears this side of the central one. I hardly dare tell you how each crag & cliff of the Matterhorn's [17]stern pinnacle stood out against the sky, you w[d]. have been off to Zermatt that minute could you have seen it. & then the Weisshorn[18] with its graceful lines & sweeps of snow. It must indeed be something to remember having been the first to tread that stately summit![19] Is it that our ideas of beauty are formed upon the grandest objects of nature? but the lines in these mountains all seem so perfectly harmonious, & such endless variety, just what artists tell one to choose for a perfect picture; but who c[d] paint them. Bel Alp had been beyond our ambition, but I am so thankful we got there, & the ride here is so interesting; first the glacier scramble, which gave one such a much better idea of the scale it is built on, & then that lovely ride through the rocks & pine forests with the views of the glacier between the old riven stems; & the scent, & the lovely little oak ferns & Cystoptens[20] nestling under every stone. Alas! Alpen roses all over, but there are still many more flowers than I can give names to, the small blue Gentian[21] like Sapphire gems, & the ground alive with grasshoppers, brown, black & green & some with scarlet wings as well as hoppers, which was quite a novelty to me. After Rieder Alp,[22] where by the bye we eat nearly the last of the delicious strawberries & raspberries of the season we got into a mist & when we reached this place we could not see twenty yards before us, & felt as if our fortunes had turned; but no, we woke to a morning only the more lovely for flying clouds casting fitful shadows over the valley.—My husband & May have just returned proud of their atchievement & making us dreadfully jealous with their account of the glorious view they had. We have still the Grimsel[23] & Handeck[24] to see, but I could not resist writing to you from what I consider your special dominions. I hope you will not think it troublesome. You are probably now starting for a minor ramble[25] after the Dundee meeting[26] which I hope went off well. I have charged my friend M[rs]. Kingsley[27] to send me an account of it second hand as her husband[28] was to be there, & will I hope have heard your address too. We opened M[r] Wellig's[29] heart by talking to him of you, & he desires "bien des choses de sa part"[30] to you. These are wonderful places to be reached up at this height! As to Bel Alp, we had better dinners there, than at most of the grand hotels. We had a guide today by name Fritz[31] or something like it, who says he has been forty times up the Jungfrau.[32] He was with you there I think, or in some of your "grandes courses". We are to walk down to Viesch[33] by 8 o'clock tomorrow morning, hoping to get back to Rhône Glacier,[34] & thence on to Handek that night. With kind regards from us all, & many thanks for your kind suggestions about our journey, which has been, so far, a thing to look back to with thankfulness & delight all one's life! Believe me dear Mr Tyndal[35]

Y[ou][rs] very since[ere][ly] | M F Egerton

I never know which is the right way of addressing to you, but fall back upon the one most like your card, so I hope it is right.

(I stole the ink stand to write there, one cannot stay indoors.)

RI MS JT/1/E/48

---

1. *[1867]:* the year is given by mention of the annual meeting of the BAAS at Dundee (see letter 2627, n. 22 and n. 31; and 2650, n. 2).
2. *Terrace Hotel Jungfrau:* see letter 2600, n. 23.
3. *Eggischhorn:* see letter 2600, n. 24.
4. *Marjelin See:* the Märjelensee, a glacial lake that formed beside the Aletsch glacier, near the Eggishorn. It was a tourist attraction in the nineteenth century, and is now much smaller and further from the glacier.
5. *glacier:* see n. 15.
6. *iron beak:* a wedge-shaped structure on the front of a vessel for breaking through ice.
7. *baling pan:* a bucket to bale water out of a vessel.
8. *my eldest daughter's:* Mary Alice Egerton (1848–1924), 'May', the eldest daughter of Edward and Mary Egerton. She married Beauchamp Tower in 1902 ('Births', *Times,* 19 February 1848, p. 8; 'Wills and Bequests', *Times,* 11 June 1924, p. 15).
9. *your "English maiden":* not identified.
10. *her father:* Edward Egerton.
11. *summit:* presumably of the nearby Eggishorn.
12. *her sister:* probably Charlotte Egerton (1849–1926), second eldest daughter of Edward Egerton and Mary Egerton ('Deaths', *Times,* 17 May 1926, p. 1). Possibly also one of Egerton's younger daughters Emily Margaret (*c.* 1852–94) or Georgina Renira (1853–1930).

13. *Sparrenhorn:* now Sparrhorn, a peak above Belalp, with a wonderful view.
14. *my girl:* see n. 8.
15. *Aletsch glacier:* the longest glacier in the Alps, which flows past the Eggishorn (see 2600, n. 24).
16. *Bel Alp:* see letter 2622, n. 31.
17. *Matterhorn's:* see letter 2600, n. 18.
18. *Weisshorn:* see letter 2600, n. 19.
19. *first to tread that stately summit:* on 19 August 1861, Tyndall and his guides were the first to summit the Weisshorn. See *Mountaineering in 1861,* pp. 41–58.
20. *Cystoptens:* a genus of fern.
21. *Gentian:* referring to a genus of flower, *Gentiana.*
22. *Rieder Alp:* an alpine village in the Swiss canton of Valais.
23. *Grimsel:* see letter 2600, n. 28.
24. *Handeck:* see letter 2600, n. 32.
25. *minor ramble:* see letter 2650, n. 2.
26. *Dundee meeting:* see letter 2627, n. 22.
27. *M^rs. Kingsley:* Frances Eliza Kingsley (née Grenfell, 1814–91), married Charles Kingsley in 1844. She later wrote a biography of her husband, *Charles Kingsley: His Letters and Memories of His Life,* 2nd edn, 2 vols (London: Henry S. King & Co., 1877) (*ODNB*).
28. *her husband:* Charles Kingsley (1819–75), an Anglican clergyman and novelist who held the Regius professorship of modern history at Cambridge (1860–9) (*ODNB*).
29. *M^r. Wellig's:* Mr. Wellig was the operator of the Jungfrau Hotel, located on the slopes of the Eggischhorn (see *A Hand-book for Travellers in Switzerland and the Alps of Savoy and Piedmont* (London: John Murray, 1871), p. 108).
30. *"bien des choses de sa part":* wishes you all the best, or sends you his best regards (French).
31. *Fritz:* not identified.
32. *Jungfrau:* see letter 2600, n. 39.
33. *Viesch:* see letter 2600, n. 22.
34. *Rhône Glacier:* see letter 2600, n. 26.
35. *Tyndal:* misspelled by Egerton.

To Thomas Henry Huxley 12 September 1867 2646

12th September 1867.

My dear Huxley,

I wrote to Bence-Jones[1] about Hodgson[2] before I left London: I send you his reply.[3] You will notice and smile at that peculiarity of which we have so often spoken. The cause <u>now</u> however, is the row that we had about my article in the Fortnightly Review.[4] Poor fellow, he does not think hopefully of himself.

The Dundee meeting[5] was wonderfully successful. Apart from my lecture[6] I took next to no part in it, for I was by no means well. I scarcely entered any of the sections.

The first evening I went into the Kinnaird Hall[7] at the most distant point from the platform, to get some idea of the strength of voice necessary to fill it. Thomson[8] was standing in one of the passages, and I noticed him moving downwards and stopping just opposite to the end of the form on which Busk[9] and myself sat. Some ladies came in, and I, according to my nature, at once resigned my seat to them. I moved past Thomson further up the hall. Having waited till John Phillips[10] rose to speak I turned and walked towards the door. I was met by Thomson with a smile; he stretched out his hand and I took it, and passed on with the remark that it gave me pleasure to take it: next day we met, and he was cordial. I went to Saint Andrews[11] and there met Tait:[12] we shook hands: they both came to my hotel, and I went with them to the other club, and smoked and drank beer with them. The transformation was wonderful. We seemed like brothers together.

The lecture to the working men[13] was successful: the audience was exceedingly impressive. Such a mass of upturned faces I had rarely seen. Their behaviour was excellent. A small deputation of them headed by the editor of the Dundee Advertiser[14] came to me afterwards to request the redelivery of the lecture. This was impossible as the apparatus[15] had been sent away. They then put in a petition for a short course of three to be delivered in the same week at some future time. I did not give them much hope; but still left the question open.

I came away from Dundee on Tuesday, and put up at Pitlochry,[16] near the pass of Killiecrankie:[17] I had a splendid day yesterday, but to-day I am imprisoned by fog, drizzle, and rain. I need some exercise much.

With kindest regards to your wife[18] and love to the children[19] | Yours ever | John Tyndall.

IC HP 8.55
RI MS JT/1/TYP/9/2912
Typed Transcript Only

1. *Bence-Jones:* Henry Bence Jones.
2. *Hodgson:* possibly William Ballantyne Hodgson (1815–80), a Scottish education reformer and economist. In 1854 he lectured at the RI 'On the Importance of the Study of Economic Science as a Branch of Education for all Classes' (*Roy. Inst. Proc.,* 2 (1851–4), p. 420; published separately in 1860 by William Blackwood and Sons), and in 1858 he was appointed an assistant commissioner to the inquiry into primary education. From 1863–8, Hodgson was examiner in political economy at London University. Although he retired

in 1870, the following year he was appointed to the new chair of political economy and mercantile law at the University of Edinburgh (*ODNB*). See letter 2647, n. 4.

3. *his reply:* letter missing.
4. *row that we had about my article in the Fortnightly Review:* for the article, see letter 2597, n. 3. Tyndall wrote in his journal in June 1867: 'A few hot words with Bence Jones regarding the article in the Fortnightly Review. He had read it in manuscript and strongly opposed its publication. No good I knew could arise from my trammeling myself so I published. Bence Jones is a man of warm affections; but they are strangely easily deflected and converted into something very different from affection—They were so in the case of Faraday—They threatened to be so with me' (Journal, RI MS JT/2/13c/1333–4).
5. *Dundee meeting:* see letter 2627, n. 22.
6. *my lecture:* see letter 2627, n. 31.
7. *Kinnaird Hall:* see letter 2641, n. 11.
8. *Thomson:* William Thomson.
9. *Busk:* George Busk.
10. *John Phillips:* John Phillips (1800–74), an English geologist who was appointed professor of geology at Oxford University (1860) and was president of the BAAS in 1865 (*ODNB*).
11. *Saint Andrews:* University of St Andrews in Scotland. Tyndall wrote in his journal that following the Dundee meeting he 'Shook hands with Tait afterwards at St. Andrews' and 'Spent a most pleasant Sunday on the sea shore near St. Andrews' (RI MS JT/2/13c/1334).
12. *Tait:* Peter Guthrie Tait.
13. *lecture to the working men:* see n. 6.
14. *editor of the Dundee Advertiser:* John Leng (see letter 2641, n. 8).
15. *apparatus:* see letter 2641, n. 12.
16. *Pitlochry:* a town in Scotland on the River Tummel, to the northwest of Dundee.
17. *pass of Killiecrankie:* a river gorge just to the north of Pitlochry.
18. *your wife:* Henrietta Huxley.
19. *the children:* Jessie, Marian, Leonard, Rachel, Henrietta, Henry, and Ethel Huxley.

From Thomas Henry Huxley 18 September 1867 2647

Swanage, Dorset.[1] | Sept. 18th. 1867.

My dear Tyndall,

I suppose you will have got back to London by this time so here goes for an epistle.

Many thanks for Bence Jones's's[2] letter[3] which I enclose. If I see Hodgson[4] I will tell him that it is doubtful if existing arrangements will allow him to lecture. That will leave matters open.

I am consoled to find that Bence Jones "Dr's" you again but it is an odd notion. I fear his lease of life is not a long one; and it is a thousand pities, for the man is a good one notwithstanding his way of beginning a letter.

Thanks also for a copy of the "Dundee Advertiser" containing your lecture.[5] It seemed to me that the report must be a very good one, and the lecture reads exceedingly well. You have inaugurated the working men's lectures of the association in a way that cannot be improved. And it was worth the trouble, for I suspect they will become a great and noble feature in the meetings.

Everything seems to have gone well at the meetings[6]—the educational business carried and the Anthropologers making fools of themselves in a most effectual way.[7] So that I do not feel I have anything to reproach myself with for being absent.

I am very pleased to hear of the reconciliation with Thomson[8] and Tait.[9] The mode of it speaks well for them, and the fact will remove a certain source of friction from amongst the cogs of your mental machinery.

It might be better for me if the like could happen with Owen,[10] but it can't happen, for the man is base, and I could not trust him if I would.

You made a jolly speech about dear old Hooker.[11] We must have a good meeting for him at Norwich.[12]

We shall be in town again on the 26th or 27th. The place has suited all of us exceedingly well, and the chicks[13] are wonderfully flourishing. They and their mother[14] send their love to you.

Ever yours faithfully | T. H. Huxley

IC HP 8.56
Typed Transcript Only

---

1. *Swanage, Dorset:* a coastal town in southern England.
2. *Bence Jones's:* Henry Bence Jones.
3. *letter:* enclosed letter not identified.
4. *Hodgson:* see letter 2646, n. 2. While Hodgson did lecture at the RI in 1854, it does not appear he did so in 1867 or thereafter.
5. *copy of the "Dundee Advertiser" containing your lecture:* see letter 2641, n. 9.
6. *the meetings:* Huxley is referring to the annual meeting of the BAAS at Dundee (4–11 September), where the Committee on Scientific Education in Schools gave their report on 4 September (see letter 2577, n. 4 and n. 6), and anthropologists held a separate meeting of their own after the general meeting (see n. 7).
7. *Anthropologers making fools of themselves in a most effectual way:* because of the exclusion of 'Anthropology' as a section of its own at the BAAS annual meeting in 1867 (sections were divided into Mathematics and Physics, Chemistry, Geology, Biology, Geography and Ethnology, Economic Science and Statistics, and Mechanical Science), anthropologists held their own meeting after the BAAS meeting to allow papers to be presented. In the 1860s there was a division between the two disciplines of the sciences of human diversity: ethnology and anthropology (see E. Sera-Shriar, 'Observing Human Difference: James Hunt, Thomas Huxley and Competing Disciplinary Strategies in the 1860s', *Annals of Science,* 70

(2013), pp. 461–91). Leading the anthropologist camp and taking the role of president at this separate meeting was John Hunt (1833–69), a speech therapist and anthropologist, and co-founder in 1863 of the Anthropological Society of London. In 1864 and 1865, Hunt had tried unsuccessfully to get anthropology added to the sections, yet in 1866 Huxley, of the ethnology camp, helped to get anthropology added to the biology section (P. Sillitoe, 'The Role Of Section H At The British Association For The Advancement Of Science In The History Of Anthropology', *Durham Anthropology Journal,* 13 (2005), pp. 1–17, on pp. 5–6). Tensions between the disciplines, however, led to the 1867 exclusion of anthropology. For reports from the anthropology camp about their separate meeting, see C. W. Davis, 'Report on Anthropology at the British Association, 1867', *Journal of the Anthropological Society of London,* 6 (1868), pp. iii–xiii; and 'The Dundee Anthropological Conference', *The Anthropological Review,* 6 (1868), pp. 71–88.

8. *Thomson:* William Thomson.
9. *Tait:* Peter Guthrie Tait.
10. *Owen:* Richard Owen (1804–92), a highly influential English comparative anatomist with whom Huxley was engaged in a series of public conflicts over human evolution and Owen's professional conduct. He was superintendent of the natural history department of the British Museum, and later founded the Natural History Museum in Kensington, London (*ODNB*).
11. *Hooker:* Joseph Dalton Hooker.
12. *good meeting for him at Norwich:* Hooker had been elected the president of the BAAS for the 1868 annual meeting at Norwich (*Brit. Assoc. Rep. 1867,* p. xxxvi; see letter 2554).
13. *chicks:* Huxley's children, Jessie, Marian, Leonard, Rachel, Henrietta, Henry, and Ethel Huxley.
14. *their mother:* Henrietta Huxley.

From George Biddell Airy 19 September 1867 2648

Royal Observatory, Greenwich | London, S.E. |
1867 September 19

My dear Sir

Have you a Sirène[1] at the Royal Institution?

And, if so, could you give me a few minutes' lecture on it?

{My whole knowledge of Induced Galvanic Currents[2] is derived from a private lecture which you gave me several years ago.}[3]

I believe that you are in the country:[4]—but this can well wait.

I am, my dear Sir, | Yours very truly | G B Airy

Professor Tyndall | &c. &c. &c.

RI MS JT/1/A/41
RI MS JT/1/TYP/1/31

1. *Sirène:* a siren or syren (as Tyndall often spelled it); an acoustical instrument (invented by Cagniard de la Tour in 1819) for producing musical tones and used in numbering the vibrations in any note (*OED*).
2. *Induced Galvanic Currents:* galvanism, electricity developed by chemical action (*OED*).
3. *a private lecture which you gave me several years ago:* Tyndall demonstrated induction coils to Airy on 12 May 1857 (see letter 1393, *Tyndall Correspondence,* vol. 6).
4. *in the country:* Airy is referring to Tyndall having gone to Dundee, Scotland for the thirty-seventh annual meeting of the BAAS (4–11 September) and further travels in Scotland after the meeting. From letters 2650 and 2651, we know Tyndall was in Scotland until at least 27 September.

From Juliet Pollock 22 September [1867][1] 2649

Hotel de l'Aigle[2] | Grindelwald[3] | Sep. 22nd

My dear Mr Tyndall,

I am getting well at last, and should have been well, sooner, I believe, had we listened to the voice of wisdom (Yours) and got away from Lucerne as soon as possible. But thinking a long rest desirable after the great fatigue of the journey which took us there, and under which I twice collapsed; we stayed a whole week at Lucerne at the Pension [*1 word excised*][4] and during that week a slow fever took possession of me with a depression of spirits such as I rarely suffer from. This partly arose no doubt from the associations of the place; from the vivid recollection of an adored companion[5] who was with me when I was last at that beautiful place; before my marriage;[6] and who was taken from us by a sudden death at St. Gervais[7] in the year 1858. Physical weakness probably made me more entirely the subject of the past than I otherwise should have been; I saw a face and heard a voice in every rising mist and sighing wind; and my place seemed to be with the dead. And yet my husband[8] and my sons[9] were with me and I knew that I was happy; and yet felt often miserable: not always; for nothing could be more beautiful rowing on the lake at moonlight; listening to the music sounding from the government boat; so far from us, that we saw neither boat nor oars, nor men, only the lights they read their music by; so that the melody seemed to steal out of these floating lamps. Then too some pleasant times were past with Thorwaldsend Lion,[10] the perfection as I think of tragic sculpture, and some diversion was created by the passionate proposition which a Swiss lady[11] at the pension took for me. She was prepared to follow me wherever I went; and I was obliged at last to use some gentle remonstrance to prevent her from renouncing all her previous plans in order to join us here!! She told me in reply that she regretted that I did not desire her company but that she could not be angry "impossible de se jacher[12], <*10–15 words excised*>: and then she flung her arms round me and kissed me over and over again! She was a dear amusing woman and I could not but be grateful for her preference; but it became rather alarming as it increased. The journey from Lucerne to Brientz[13] knocked me up again; and so ill I was all night at Brientz; so entirely exhausted; that I really thought my career in this world was going to close. But I rallied there, in the middle of the day sufficiently to push on for this place. My husband rightly thinking that my best chance was to reach these altitudes as soon as possible. There was a pretty fair chambermaid at Brientz, who devoted herself to me as if she had been my own servant; and who was a wonderful comfort to me. Her name was Elise. Though it was raining I went to sit by the lake to breathe cool air, and while my husband was wrapping me up in cloaks &c. Elise suddenly appeared with a footstool. "Madame *<10–15 words excised>*". She objected to my travelling further that day; but she was wrong there; for I no sooner felt this light cool air than I felt better—and a daily improvement has ended now in a real recovery. We came here last Tuesday. I am not able to climb; as you may suppose; but chaises à porteurs[14] take me up the mountains into the heart of the glaciers, and then I feel a new existence springing up within me. We made a delightful expedition in this way to the ice the rest walking while I was carried and we enjoyed a primitive repast at the Wirsthaus.[15] The best part of the whole thing is that my husband enjoys Switzerland to the utmost; and is every where and always, one of the best travellers. I don't think he will ever again sit still in an English watering place. I think we shall always contrive a part of our vacation in foreign lands. Good bye now. I have seen and heard an avalanche this [morning] which was worth a good many sermons.

Your affectionate friend | Juliet Pollock.

RI MS JT/1/TYP/6/2088–9
LT Typescript Only

1. *[1867]:* the year is given by mention of this trip by Juliet's husband in his memoir: 'To Paris with Donne and his daughter Valentia for a little tour of Swizerland—Basle, Lucerne, Rigi, Pilatus, Gründelwald [Grindelwald], Berne, Chillon, Neufchâtel, Dijon' (W. F. Pollock, *Personal Remembrances of Sir Frederick Pollock: Second Baronet, Sometime Queen's Remembrancer,* 2 vols (London: MacMillan and Co., 1887), vol. 2, pp. 167–8).
2. *Hotel de l'Aigle:* the first hotel built in Grindelwald, in 1818 by Weibel Christian Bohren.
3. *Grindelwald:* see letter 2600, n. 37.
4. *Pension [1 word excised]:* LT left out the name of this hotel in her typescript, indicating that she probably could not make out the word in Pollock's letter.
5. *adored companion:* Maria Julia Herries (*c.* 1820–57), a cousin of Juliet Pollock with whom she was raised by her aunt Isabella Maria Herries after the deaths of their mothers. Maria Herries died at St. Gervais in the summer of 1857 (see letter 1418, *Tyndall Correspondence,* vol. 6).

6. *my marriage:* Juliet Creed and William Frederick Pollock were married in 1844.
7. *St. Gervais:* a commune in the Rhône-Alpes region of southeastern France.
8. *my husband:* William Frederick Pollock.
9. *my sons:* Frederick, Maurice, and Walter Pollock. Maurice Emilius Pollock (1857–1932) was the third son of Juliet and William Frederick Pollock. He became a sculptor and married Lydia Helen Roberts (1853–84) in 1880, and Mabel Mary McPherson (1860–1937) in 1889. Walter Herries Pollock (1850–1926) was their second son. He was educated at Eton College (1862–65) and was admitted to Trinity College, Cambridge, where he obtained a BA (1871) and MA (1875). He was admitted to the Inner Temple in 1870 and called to the bar in 1873. Walter became a writer, poet, and journalist, and was editor of the *Saturday Review* (1884–94). He married Emma Jane Pipon in 1876.
10. *Thorwaldsend Lion:* see letter 2600, n. 50.
11. *Swiss lady:* not identified.
12. *impossible de se jacher:* impossible to hide (French).
13. *Brientz:* now Brienz, a village on the shore of Lake Brienz in the Bernese Oberland region of Switzerland.
14. *chaises à porteurs:* sedan chairs; a closed vehicle to seat one person, borne on two poles by two bearers, one in front and one behind. In fashionable use during the 17th, 18th, and early 19th cent. (*OED*).
15. *Wirsthaus:* inn or tavern (German).

To Thomas Archer Hirst 24 September 1867 2650

Tuesday, 24th. Sep. /67 | Cragside[1] Morpeth

My dear Tom.

I came here on Saturday after my trip in the Highlands.[2] It is a fine windy moorland with bold heights rising out of it, the horizons being bounded northward by the Cheviots.[3] Nothing could be kinder than Sir Wm. & Lady Armstrong.[4] We have been every day upon the moors wandering knee deep in heather & indulging at times in a little climbing; for the moorland is bastioned by fine crags. Lubbock[5] has not yet put in an appearance though letters are lying for him here. I leave this place tomorrow (Wednesday) with Sir Wm. Armstrong & we travel both together to London. I shall reach home by about 9 ½ P.M.

Yours affectionately | John Tyndall

RI MS JT/1/T/962
RI MS JT/1/HTYP/540

1. *Cragside:* William Armstrong's home near Rothbury, and the larger town of Morpeth, Northumberland. It is now a National Trust site.

2. *my trip in the Highlands:* following the annal meeting of the BAAS at Dundee (4–11 September), where Tyndall lectured to the working men there (see letter 2627, n. 31), he spent more time in Scotland before returning to London, exploring in the Highlands with Scottish lawyer and folklorist John Stuart Stuart-Glennie (Journal, RI MS JT/2/13c/1334).
3. *Cheviots:* the Cheviot Hills, the hills straddling the border between Northumberland and the Scottish Borders.
4. *Sir Wm. & Lady Armstrong:* William Armstrong and his wife Margaret Armstrong (née Ramshaw, 1807–93). Margaret was the daughter of William Ramshaw, the owner of an engineering works that William Armstrong regularly visited while he attended a Grammar School in Bishop Auckland. Margaret and William were married in 1835 (*ODNB*).
5. *Lubbock:* John Lubbock.

To Emily Peel 27 September 1867 2651

27th Sept., 1867

Dear Lady Emily,

I have been for some time among the Scottish Highlands[1] without receiving letters, and your kind invitation, though dated on the 2nd of September,[2] did not reach me until my return from the North.

It would give me true pleasure to be able to accept your invitation, but I have already accomplished my Swiss tour for the present year. I spent three very uneventful weeks among the mountains this year. I climbed the Great Eiger,[3] which was very grand, but other mountaineers had done the same thing before me. I hung for nearly three weeks about the Matterhorn,[4] but the weather was obstinately bad and I came away not hoping for a favourable change. No sooner however did I leave Switzerland than the clouds vanished, the rain ceased, the weather brightened, and a gentleman[5] in whose company I intended to ascend the Matterhorn succeeded in getting to the top.

Altogether the mountain has treated me very badly. It pelted me down by a hailstorm in 1862, after I had succeeded in surmounting its principal difficulties.[6] The man[7] who accompanied me there as a porter succeeded in reaching the summit two years ago, and now a second ascent has been made, while nothing but ill-nature was my lot as long as I remained near the mountain.

The weather in Scotland was hardly more successful. Two or three days only could I call really good. One day indeed rose to magnificence, the mountain colouring beating even the indescribable purple of the Italian Alps.

I feel very grateful to you for your kind remembrance of me.

Most truly yours, | John Tyndall.

RI MS JT/1/TYP/3/955
LT Typescript Only

1. *some time among the Scottish Highlands:* see letter 2650, n. 2.
2. *your kind invitation, though dated on the 2nd of September:* letter 2643.
3. *Great Eiger:* see letter 2622, n. 7.
4. *Matterhorn:* see letter 2600, n. 18.
5. *gentleman:* Florence Crauford Grove (1838–1902), an English mountaineer and writer, who was one of the early members of the Alpine Club ('Founders of the Alpine Club', *ODNB*; *Hours of Exercise in the Alps*, pp. 269–70; A. L. Mumm, *The Alpine Club Register*, 3 vols (London: Edward Arnold & Co., 1923–8), vol. 1, pp. 125–6).
6. *It pelted me down . . . its principal difficulties:* in *Hours of Exercise in the Alps*, Tyndall described the hail storm as a 'malicious cannonade': 'A tempest of hail was here hurled against us; as if the Matterhorn, not content with shutting its door in our faces, meant to add an equivalent to the process of kicking us downstairs. The ice-pellets certainly hit us as bitterly as if they had been thrown in spite' (p. 165).
7. *man:* Jean-Antoine Carrel (1834–91), an Italian soldier, mountaineer, and alpine guide who Tyndall had hired as a porter during his failed attempt on the Matterhorn in 1862. Carrel made the second ascent of the Matterhorn on 17 July 1865 just three days after Edward Whymper (1840–1911) made the first ascent (*Hours of Exercise in the Alps*, pp. 157–65; C. D. Cunningham and W. W. Abney, *The Pioneers of the Alps*, 2nd edn (London: Sampson Low, Marston, Searle, and Rivington Ltd., 1888), pp. 126–31).

From George Biddell Airy 27 September 1867 2652

Royal Observatory, Greenwich, | London S.E.[1] |
1867 September 27

My dear Sir

In the matter of the University of London and Sir John Lubbock.[2]

Two applications have been made to me, and to both I have given the same answer:—that I was one of the original Members of Senate,[3] that I found things so uncomfortable that I retired; and that I considered myself by that step estopped[4] from interfering in the concerns of the University.

But your letter[5] has caused me to think over the matter again, and my scruples begin in some measure to relax. Considering that a great many years have elapsed since the occurrence to which I refer, and that my feeling towards the University is of the most friendly kind, I begin to think that I may act as another stranger would.

This being so, I have no hesitation in saying that I think very highly of Sir John Lubbock, and heartily wish him success: and I authorise you to place my name on the Committee List.[6]

I am, my dear Sir, | Yours very truly, | <u>G B Airy</u>
Professor Tyndall.

RI MS JT/1/A/42
RI MS JT/1/TYP/1/32

---

1. Included on this Royal Observatory letterhead are the following words, brackets included: '[*It is requested that all correspondence on the subject of this letter may be addressed to the Astronomer Royal.*]'.
2. *matter of the University of London and Sir John Lubbock:* in 1867 John Lubbock was put forward as a candidate for the first Member of Parliament to represent the University of London, the committee for which included a number of men of science including Tyndall (see n. 6). Lubbock was one among many Liberal candidates, but despite the support he received from colleagues, he later withdrew in June 1868, and the role would fall to Robert Lowe (1811–92) (F. M. G. Willson, *The University of London, 1858–1900: The Politics of Senate and Convocation* (Woodbridge: Boydell Press, 2004), pp. 160–8). See also, G. C. Foster and W. M. Odling, 'To the Members of the University of London', *The Laboratory: A Weekly Record of Scientific Research,* 3 August 1867, p. 307.
3. *original Members of Senate:* George Airy served as a member of the Senate of the University of London from 1835 until he resigned in 1847 (see 'Airy, Professor Sir George Biddell, 1835–1850', Senate House Library Archives, University of London, GB 96 MS 929). Airy's posthumously published autobiography described his repeated opposition to the introduction of religious examinations at the University (see W. Airy (ed), *Autobiography of Sir George Biddell Airy* (Cambridge: Cambridge University Press, 1896)).
4. *estopped:* to stop, bar, hinder, preclude (*OED*).
5. *your letter:* letter missing.
6. *the Committee List:* a partial list of members of the committee formed to support Lubbock's candidacy for Parliament is given in H. G. Hutchinson, *Life of Sir John Lubbock, Lord Avebury,* 2 vols (London: Macmillan and Co., 1914), vol. 1, pp. 94–5: George Airy, Benjamin Brodie (1817–80), David Brewster, George Busk, Charles Darwin, Charles Lyell (1797–1875), Max Müller (1823–1900), James Paget (1814–99), Andrew Ramsay (1814–91), Edward Frankland, Thomas Graham (1805–69), William Grove, John Herschel, Thomas Hirst, Joseph Hooker, Thomas Huxley, James Edwin Thorold Rogers (1823–90), James Joseph Sylvester, John Tyndall, William Spottiswoode, William Thomson, William Sharpey (1802–80), Herbert Spencer, Alfred Russel Wallace, and Charles Wheatstone. William Scovell Savory (1826–95), Frederic William Farrar, George Carey Foster (1835–1919), William Odling (see letter 2919, n. 1), and Philip Henry Pye-Smith (1839–1914) acted as Secretaries.

To Juliet Pollock [September 1867][1] 2653

Friday

If my eyes were waters & my head a fountain of tears I might hope to weep out the memory of my loss.[2]

I want you to help me to choose some jewelry for two ladies who have been wearing themselves out in doing justice to my work.[3]

Yours ever | John Tyndall

You quite won the blood, soul & divinity of my friend Wynne.[4]

RI MS JT/1/T/1207
RI MS JT/1/TYP/6/2217

---

1. *[September 1867]:* the month and year are given by relation to letters 2677 (18 November 1867) and 2691 (1 December 1867), and that it appears to have been written soon after when Michael Faraday died (25 August 1867).
2. *my loss:* Tyndall's friend and mentor at the RI Michael Faraday had died on 25 August 1867.
3. *two ladies who have been wearing themselves out in doing justice to my work:* Tyndall gifted brooches to both Anna Helmholtz and Clara Wiedemann for their work on translating his *Heat* (1863) into German. See letters 2677 and 2691.
4. *Wynne:* probably George Wynne (1804–90), an Irish surveyor, engineer, and military officer. He met Tyndall while employed in the Irish Ordnance Survey (1835–40). Wynne later served with distinction in Greece, China, and the West Indies, rising to general by the time he retired in 1877 ('The Late General Wynne', *Times,* 12 July 1890, p. 11). However, this 'Wynne' could also be one of George's three sons with his wife Anne (née Osborne, 1808–64): Henry Le Poer Wynne (d. 1874), a civil servant in India; Edward Toler Wynne (1837–89), a military officer and head of the Irish Ordnance Survey; and Frank George Wynne (dates unknown), a civil engineer.

From Auguste de la Rive 1 October 1867 2654

Mon cher Tyndall,

Je viens, suivant votre demande, vous donner le récit abrégé des circonstances particulières qui amenèrent dans le temps entre Faraday et mon père ces relations d'amitié que vous savez avoir été si intimes. Vous pourrez, suivant votre désir, rectifier par la connaissance exacte des faits, les bruits erronés qui ont couru sur ce sujet en Angleterre. Si je n'ai pas donné ces détails dans la notice biographique que j'ai publié sur Faraday, c'est par un sentiment de délicatesse que vous apprécierez facilement ; mais les considérations de ce genre doivent céder devant le devoir de rétablir la vérité. La voui doit telle que ma mémoire ne la retrace fidèlement.

C'était en 1813; il n'y avait pas longtemps que Faraday était placé à l'Institution Royale quand Davy, auprès duquel il était attaché en qualité de préparateur, se décida à faire un voyage sur le continent dont l'accès était interdit alors aux Anglais par l'Empereur Napoléon qui s'empressa toutefois de lever l'interdiction en faveur de l'illustre Chimiste. Poussé par cet esprit à la fois actif et curieux qui l'a toujours distingué, Faraday témoigna à Davy le vif désir qu'il avait d'être du voyage. Davy répondit que, voyageant avec sa femme et deux domestiques, il ne voyait pas le moyen de trouver une cinquième place dans sa voiture. Faraday insista et finit par demander en grâce à Davy de l'emmener avec lui en remplacement de son domestique homme. Davy refusa d'abord, mais finit par consentir, non sans répugnance, à cause de la position subalterne qui en résultait pour Faraday. Le voyage s'effectua dans ces conditions, & après un séjour à Paris, dans le midi de la France et en Italie, Davy arriva à Genève au milieu de l'été de 1814.

Mon père avait connu Davy en Angleterre en 1799 & avait assisté à ses premiers débuts dans la carrière scientifique; il n'avait dès lors cessé de soutenir avec lui une correspondance régulière & avait eu le privilège de faire connaître le premier au Continent par le moyen de la Bibliothèque Britannique, les belles découvertes qui ont immortalisé le nom du savant Anglais. Il l'accueillit donc avec joie quand il arriva à Genève & l'engagea à venir demeurer chez lui à la campagne.

Davy et mon père étaient chasseurs; un jour qu'ils chassaient ensembles, mon père resté en arrière pour charger son fusil se trouva à coté de ce soidisant domestique qui chargeait le fusil de Davy. Avec la bonhommie qui lui était naturelle il entre en conversation avec lui; il est bien vite frappé de l'intelligence remarquable de son interlocuteur; puis questionnant Davy à son sujet, il apprend qu'il se nomme Faraday & qu'il est attaché comme préparateur à l'Institution Royale. Davy ajouta que c'est sur sa demande & pour satisfaire au désir qu'il avait de voyager qu'il l'a emmené avec lui comme son domestique. Mon père comprenant alors ce qu'est Faraday, reproche à Davy de ne le lui avoir pas dit plus tôt & lui déclara que le préparateur mangera désormais à la table des maîtres & non à celle des domestiques. Davy s'y oppose disant que Faraday a accepté volontairement la position qu'il occupe; nouvelles instances de mon père suivies d'une transaction d'après laquelle il est entendu que Faraday ne sera plus considéré comme un domestique et qu'en particulier il mangera seul dans sa chambre et non plus à l'office. A partir de ce jour Davy traita Faraday avec plus de considération. Faraday de son côté n'oublia jamais la manière dont mon père l'avait accueilli à une époque où il n'était encore qu'un obscur garçon de laboratoire, & dont il continua à l'encourager & à le soutenir dans ses débuts scientifiques qui ne furent pas sans difficulté.

Voilà donc le commentaire du passage consacré à mon père dans la lettre que Faraday m'adressa en Octobre 1858, lettre dont j'ai inséré la traduction dans la notice biographique que j'ai publiée sur lui et que vous avez reproduite textuellement dans votre notice. Il me semble que je ne puis mieux terminer mon petit récit qu'en transcrivant ce passage. Après avoir rappelé dans sa lettre tout ce qu'il doit à Madame Marcet et la vénération qu'il a pour sa mémoire, Faraday ajoute:

"J'ai les mêmes sentiments pour la mémoire de votre père, car il fut, je puis le dire, le premier qui m'encouragea et me soutint, d'abord à Genève, quand j'eus le plaisir de l'y voir, puis ensuite par la correspondance que je soutins régulièrement avec lui."

En vous autorisant, très cher ami, à faire de ma lettre l'usage que vous voudrez, je vous prie d'agréer l'assurance de mes sentiments les plus affectueux.

Augte de la Rive. | Presinge, le 1 Octobre, 1867.

My dear Tyndall,

I am following up on your request and giving you a brief account of the particular circumstances that led to the intense friendship, as you know, between Faraday[1] and my father.[2] By knowing the exact facts, you may, if you desire, correct the false rumours that spread on this matter in England. If I did not give these details in the biographic note that I published on Faraday,[3] it is because of a sense of delicacy that you could easily appreciate; but considerations of this kind must yield to the obligation of restoring the truth. It must be such that my memory does not retrace it faithfully.

It happened in 1813. It had not been long since Faraday was placed in the Royal Institution when Davy,[4] to whom he was attached as an assistant, decided to take a trip across the continent at the time when Emperor Napoleon prohibited it for the English, but who hastened, nevertheless, to lift the ban in favour of the illustrious chemist. Driven by this active and curious state of mind, which always distinguished him, Faraday showed Davy his great desire to travel. Davy replied that since he traveled with his wife[5] and two servants, he could not see a way to fit a fifth person in his car. Faraday insisted and finally asked Davy to take him instead one of his man servants. At first Davy refused, though eventually he did agree, but not without reluctance, because of the subordinate position that Faraday would be placed in. The journey took place under these conditions, & after their stay in Paris, in Southern France and in Italy, Davy arrived in Geneva in the middle of the summer 1814.

My father was acquainted with Davy in England in 1799 & had attended his first steps in the scientific career; since then he had kept a regular correspondence with him & had the privilege of being the first to inform the Continent through the British Library[6] of the beautiful discoveries that immortalized the name of the English savant. Therefore, he welcomed him with joy when he arrived in Geneva & invited him to stay with him in the countryside.

Davy and my father were hunters; one day when they hunted together, my father stayed behind to load his gun and found himself next to the so-called servant who was loading Davy's gun. With his natural gentleness, he started conversing with him. He was soon struck by the remarkable intelligence of his interlocutor, and upon asking Davy about him, he learned that the man's name was Faraday & that he was attached as an assistant at the Royal Institution. Davy added that it was due to Faraday's own request & to satisfy his desire to travel, that he brought him along as a servant. Upon realizing who Faraday was, my father accused Davy of not telling him about it earlier & announced that the assistant would now eat at the masters' table & not at that of the servants'. Davy opposed him by saying that Faraday willingly accepted the position which he occupied. New instances of my father were followed by a deal in which it was understood that Faraday will no longer be considered a servant, and in particular that he would eat alone in his room rather than in the office. From that day on, Davy treated Faraday with more consideration. From his own end, Faraday never forgot the way in which my father welcomed him at a time when he was nothing but an obscure laboratory boy, & whom he continued to encourage & to support in his first scientific steps, which did not go without difficulty.

So here is the commentary of the passage dedicated to my father in the letter that Faraday wrote to me in October 1858,[7] the letter for which I inserted the translation in the biographic note that I published on him, and which you reproduced verbatim in your note.[8] It seems to me that I could not finish my little story better than with transcribing this passage. After recalling in his letter everything that he owes to Mrs Marcet[9] and the veneration for her memory, Faraday adds:

"I have the same feelings for the memory of your father, since he was, I may say, the first who encouraged me and supported me, first in Geneva when I had the pleasure to see him, and later by correspondence that I regularly maintained with him."

By allowing you, dear friend, to make use of my letter as you wish, I beg you to accept the assurance of my most affectionate sentiments.

Aug=[10] de la Rive. | Presinge,[11] 1 October, 1867

RI MS JT/1/D/100
RI MS JT/1/TYP/1/370–1

---

1. *Faraday:* Michael Faraday.
2. *my father:* Charles-Gaspard de la Rive (1770–1834), a Swiss physician, and father of Auguste de la Rive. He was educated at the University of Edinburgh, worked at the British

Library in London, and on his return to Switzerland studied electricity and chemistry (*CDSB*).

3. *biographic note that I published on Faraday:* see letter 2670, n. 3.
4. *Davy:* Humphry Davy.
5. *his wife:* Jane Davy (née Kerr, 1780–1855), a wealthy widow who married Humphry Davy in 1812 (*ODNB*).
6. *first to inform the Continent through the British Library:* Charles-Gaspard de la Rive published works by Davy in the French journal *Bibliothèque Britannique* (established 1796), which sought to publish works by British scientists for the use of scientists and other interested persons on the continent.
7. *letter that Faraday wrote to me on October 1858: Faraday Correspondence,* 5:3519.
8. *your note:* letter missing.
9. *Mrs Marcet:* Jane Marcet.
10. *Aug≡:* de la Rive used three parallel horizontal lines to abbreviate his first name.
11. *Presinge:* a village in the Swiss canton of Geneva where de la Rive lived.

From George Biddell Airy 4 October 1867 2655

Royal Observatory, Greenwich, | London, S.E.[1] |
1867 October 4

My dear Sir,

In your excellent lectures on Sound, Book[2] page 41, you speak of the experiments of Wertheim[3] and Chevandier[4] on the transmission of sound through wood in various directions.

Do you happen to have the precise reference at hand? and would you oblige me with it? I have no opportunity of ranging through the volumes of Poggendorf[5] or the Annales de Chemie,[6] though I can procure a special volume when the n°. is precisely known.

Some time ago (before your Lectures) I began writing a Tract on Sound[7] of very different character and with very different object. But I cannot tell you how much advantage for it I am deriving from your Lectures.

I am, my dear Sir, | Yours very truly | G B Airy

Professor Tyndall

RI MS JT/1/A/43
RI MS JT/1/TYP/1/33

---

1. Included on this Royal Observatory letterhead are the following words, brackets included: '[*It is requested that all correspondence on the subject of this letter may be addressed to the Astronomer Royal.*]'.

2. *lectures on Sound, Book: Sound* (1867).
3. *Wertheim:* Guillaume Wertheim (1815–61), a French physician, mathematician, and physicist who studied the mechanical properties of wood, bones, and tissue (Verdet, 'Notice sure les travaux scientifiques de M. Guillaume Wertheim', *Société philomathique de Paris: Extraits des procès-verbaux des séances,* (1861), pp. i-xvi).
4. *Chevandier:* Eugene Chevandier de Valdrome (1810–1878), a French politician and industrialist involved in forestry and glass manufacture. He published numerous accounts of investigations on the properties of wood and glass including several with Guillaume Wertheim (*DBF*). See E. Chevandier, 'Recherches sur la composition élémentaire des différents bois, et sur le rendement annuel d'un hectare de forêts', *Annales de Chimie et de Physique,* 10 (1844), pp. 129–62; E. Chevandier, 'Recherche expérimentale des propriétiés mécaniques du bois', *Paris, Compte Rendus,* 14 (1847), pp. 537–41; E. Chevandier, 'Sur la croissance et la composition chimique de divers bois', *Bibliothèque universelle de Genève,* 8 (1848), pp. 5–21 (later *Archives des sciences physiques et naturelles*); and E. Chevandier and G. Wertheim, 'Mémoire sur les propriétés mécaniques du bois', *Paris, Compte Rendus,* 23 (1846), pp. 663–74.
5. *Poggendorf: Poggendorff's Annal.,* in full *Annalen der Physik und Chemie,* was often referred to as just *Poggendorff's Annalen* while under the editorship of Johann Poggendorff from 1824–76.
6. *Annales de Chemie: Annales de Chimie et de Physique.*
7. *Tract on Sound:* G. B. Airy, *On Sound and Atmospheric Vibrations: with the mathematical elements of music* (London: Macmillan and Co., 1868).

From Henry Bence Jones 18 October [1867][1] 2656

From Dr. Bence Jones. | [Copy.]
Royal Institution of Great Britain

My dear friend

At first, as soon as I had read a few lines,[2] I thought I saw wrong. Then when I got more light I thought you had gone out of your mind, but as I read on I found that it was all serious, and that there was no sign of insanity in your expressions or in your hand-writing.

If you decide to go[3] I think we had better shut up the Institution. We have been going through a very provisional condition the last two years, and now that the change has come it will require far more from you than from any one else to make the place flourish & to keep it flourishing. On you its future depends, as long as your bodily and mental health and strength last, just as much as the Institution has depended upon Mr. Faraday[4] for half a century.

I could say a great deal more, but it would do no good and might do harm.

Ever yours truly | [Signed] H. Bence Jones. | Oct. 18.

From Dr. Bence Jones | Hon[orary]. Secretary | Royal Institution | No. I

BL Add MS 60631, ff. 119–20

1. *[1867]:* the year is given as this letter is Bence Jones' response to Tyndall's consideration of accepting a position in India. This letter is a copy made as an enclosure for letter 2663.
2. *read a few lines:* letter missing.
3. *decide to go:* Tyndall was offered a position to undertake meteorological research in British India, but declined on account of Faraday's death and his commitment to the RI (see letters 2658, 2661, and 2663).
4. *Mr. Faraday:* Michael Faraday.

From Frédéric Zurcher[1] and Elie Margollé[2] 18 October 1867 2657

Toulon 18 octobre 1867.

Cher Monsieur,

Nous recevons aujourd'hui une lettre de Madame Jean Reynaud qui nous annonce qu'elle vous adressera de Paris, le 23 de ce mois, les divers volumes composant les œuvres complètes de son mari. Permettez nous de vous recommander encore les belles études de cet *[*éminent*]* esprit, qui n'appartient pas seulement à la France, mais à toutes les nations affranchies par la science des erreurs du passé et qui cherchent aujourd'hui une idée religieuse plus en rapport avec la grandeur et la magnificence de la Création. Le culte de Madame Jean Reynaud pour la chère mémoire de son mari et tout ce qu'elle fait pour attirer le respect et la sympathie sur cette mémoire vous diront assez qu'elle était digne d'un tel compagnon.

Nous vous sommes bien reconnaissants, mon beau frère, M. Margollé et moi, de l'envoi de votre beau livre des Glaciers, qui selon nous, est aussi un livre de philosophie religieuse en ce sens qu'il ouvre à l'esprit de nombreuses & lumineuses perspectives sur l'œuvre d'écrire &

sur les grandes lois qu'en développent incessamment la beauté. Vous montrez aussi le calme courage du savant dans la recherche de la vérité et vous ennoblissez ainsi la science en mettant l'héroïsme à son service.

Nous nous unissons pour vous prier d'agréer, cher Monsieur, l'assurance de nos sentiments respectueux & dévoués.

F. Zurcher | Margollé

Toulon[3] October 18 1867.

Dear Sir,

Today we received a letter from Madame Jean Reynaud,[4] who told us that she will send you from Paris, on the 23rd of this month, the various volumes that form the complete work[5] of her husband.[6] Allow us to recommend again the superb studies of this eminent mind, which do not belong solely to France, but to all the nations freed by science from past mistakes, and which explores today a religious idea that is more in tune with the greatness and magnificence of Creation. The adoration of Madame Jean Reynaud for the dear memory of her husband and everything she has done to bring respect and sympathy for that memory, will tell you enough that she was worthy of such a partner.

We are very grateful, my step brother M. Margollé and I, of the sending of your wonderful book on Glaciers,[7] which in our opinion, is a book of religious philosophy in the sense that it brings to the mind numerous and luminous perspectives on the work of writing & on the great laws that incessantly develop beauty. You also show the calm courage of a savant in search for truth, and you also ennoble science by putting heroism in his service.

We both ask you to please accept, dear Sir, our deepest and respectful sentiments.

F. Zurcher | Margollé[8]

RI MS JT/1/Z/3

---

1. *Frédéric Zurcher:* see letter 2629, n. 1.
2. *M. Margollé:* Elie Margollé (see letter 2629, n. 3).
3. *Toulon:* see letter 2629, n. 2.
4. *Madame Jean Reynaud:* not identified; the wife of Jean Reynaud (see n. 6).
5. *complete work:* this would include his *Minéralogie des Gens du Monde, ou Notions Générales sur les Minéraux les Plus Utiles à la Société* (Paris: Moutardier, 1836) and *Philosophie Religieuse: Terre et Ciel* (Paris: Furne, 1854).
6. *her husband:* probably Jean Reynaud (1806–63), a French politician, philosopher, and mine engineer who founded the *Encyclopédie nouvelle* in 1833 (F. Hoefer (ed), *Nouvelle Biographie Générale,* 46 vols (Paris: Firmin Didot fréres, 1852–66), vol. 42, pp. 79–81).
7. *book on Glaciers: Glaciers of the Alps.*
8. *Margollé:* signed by Margollé himself.

To William Thomas Thornton 21 October 1867 2658

Royal Institution | To M[r]. Thornton 21st. Oct. 1867

Dear Sir

I have communicated with the President[1] and Hon[orary] Secretary[2] of the Royal Institution reg[ardin][g]. the proposal of the Indian govt. that I sh[oul][d]. visit that country for a year or 2 to investigate certain branches of its meteorology.

I greatly regret to say that my friends consider that the recent death of M[r]. Faraday[3] & other matters connected with the Royal Institution render it an unfortunate duty on my part to remain here. I am therefore compelled to decline an invitation which under other circumstances it would have given me great pleasure to accept.

Permit me however to say that you have at this moment an officer in India[4] who is eminently fitted to carry out investigations on the meteorology of the country. There is more sound sense & sound service in what he has written regarding the meteorology of India[5] than is to be found in the pages of far more voluminous writers. And were it my good fortune to be able to accept the invitation of the Indian govt., my first endeavour would be to secure the cooperation of the gentlemen to whom I now refer.

That gentleman is colonel Richard Strachey.

I am dear Sir | Yours very faithfully | John Tyndall.

To M^r^. Thornton. | N°. III.[6]

BL Add MS 60631, ff. 123–4

---

1. *President:* Henry Holland.
2. *Hon[orary] Secretary:* Henry Bence Jones.
3. *recent death of M^r^. Faraday:* Michael Faraday died on 25 August 1867 (see letter 2638).
4. *an officer in India:* Richard Strachey.
5. *what he has written regarding the meteorology of India:* possibly R. Strachey, 'On the Physical Geography of the Provinces of Kumáon and Garhwál in the Himálaya Mountains, and of the Adjoining Parts of Tibet', *Journal of the Royal Geographical Society of London,* 21 (1851), pp. 57–85. Strachey's findings on the 1848 expedition were so significant that the Royal Geographical Society published his full account fifty years later (*ODNB*; R. Strachey, 'Narrative of a Journey to the Lakes Rakas-Tal and Manasarowar, in Western Tibet, Undertaken in September, 1848', *Geographical Journal,* 15 (1900), pp. 150–70, 243–64, 394–415).
6. *To M^r^. Thornton. | N°. III.:* written on the outside page of the folded letter. In letter 2663 to Jane Maria Strachey, however, Tyndall wrote that this letter to Thornton was 'N° IV', so possibly writing 'N°. III' on this envelope was a mistake on his part.

From Henry Bence Jones [21 October 1867][1] 2659

Copped Hall | Epping[2] | Monday

Dear friend

Many thanks on the part of the Institution[3] for your decision.[4] I have nothing more to say. I only wish your *[role]* at the Institution may be a long & happy one to you.[5]

Ever yours | H B J

RI MS JT/1/J/95
RI MS JT/1/TYP/2/712

1. *[21 October 1867]:* the date is given in relation to letter 2656, also from Bence Jones. That letter, dated 18 October [1867], a Friday, would make this Monday letter 21 October.
2. *Copped Hall | Epping:* an estate in Epping, Essex where Bence Jones must have been staying.
3. *Institution:* the RI.
4. *your decision:* Tyndall was offered a position to undertake meteorological research in British India but declined on account of Faraday's death and his commitment to the RI (see letters 2658, 2661, and 2663).
5. *your [role] . . . happy one to you:* as Superintendent of the RI, a position which Tyndall would hold until 1887. The word 'role' may have also been 'rule', as LT transcribed it in her typescript.

From Henri Étienne Sainte-Claire Deville[1] 22 October 1867 2660

Laboratoire de Chimie | Ecole Normale Supérieure. |
Paris, le 22 Y^bre^ 1867

Cher ami

Permettez-moi d'introduire auprès de vous mon élève M. Rayet, astronome adjoint à l'observatoire. Il désire traduire & publier votre ouvrage sur le son—Si vous lui confiez ce travail, il le fera bien & vous serez content de lui.

Votre ami bien dévoué | H Sainteclair Deville

Chemistry Laboratory | Ecole Normale Supérieure |
Paris, 22 October 1867

Dear friend

Allow me to introduce to you my student, Mr. Rayet,[2] assistant astronomer at the observatory. He wishes to translate & publish your book on sound.[3] If you will entrust him this work, he will do it well & you will be happy.

Your very devoted friend | H Sainteclair Deville

RI MS JT/1/D/129

1. *Henri Étienne Sainte-Claire Deville:* Henri Étienne Sainte-Claire Deville (1818–81), a French chemist who worked at the Chemistry Laboratory, Ecole Normale Supérior, Paris. He developed a process for producing aluminium (*CDSB*).

2. *Mr. Rayet:* Georges Antoine Pons Rayet (1839–1906), a French astronomer educated at the École Normale Supérieure, who in 1867 was an observer at the Paris observatory (*CDSB*).
3. *book on sound: Sound* (1867). For the first French edition, see letter 2617, n. 2.

From Henry Holland 22 October 1867 2661

From Sir Henry Holland Bart.[1] | Brook St.[2] Oct. 22nd

My dear Tyndall,

Thank you heartily for your letter[3] which gave me heartfelt satisfaction.

The deep interest I feel in the fame and welfare of the R[oyal]. Institution, and the sense I entertain of your value as attached to it might reasonably have attached some suspicion to my opinion, before your own resolution was taken. But now I can fairly and honestly say that in my best judgment, and with reference to your own as well as other interests I feel sure you have resolved rightly. My ambition for you as well as for the R. Institution has been that you should be considered as the direct successor of Faraday,[4] taking his place in all ways. It is no mean succession, that of Davy[5] and Faraday. But it was a succession which would have been broken even by a years residence in India.[6] And let me add my conviction, on some knowledge of Indian matters, that a year would have been a very inadequate time (one third of it spent on going and returning) to fulfill the object which your own genius & resources would suggest, in addition to those which might enter into the official programme put before you.

I might say somewhat more on this matter, but I trust to have frequent opportunities of seeing you & repeating the satisfaction I feel in your decision. You will not doubt I am sure my most earnest desire, as far as I have any influence, to render your future position at the Royal Institution as consonant to your wishes & feelings as possible in every way.

Ever my dear Tyndall yours | most sinc[ere]ly | H. Holland. Signed.

From Sir Henry Holland Bart | President of the | Royal Institution | No. II[7]

BL Add MS 60631, ff. 121–2

---

1. *Bart:* Baronet.
2. *Brook St.:* in London.
3. *your letter:* letter missing.
4. *direct successor of Faraday:* in September 1867, Tyndall succeeded Michael Faraday as the Superintendent of the House and the Directory of the Laboratory at the RI, until he retired in 1887.
5. *Davy:* Humphry Davy.

6. *a years residence in India:* Tyndall was offered a position to undertake meteorological research in British India, but declined on account of Faraday's death and his commitment to the RI (see letter 2658 and 2663).
7. *From Sir Henry Holland Bart | President of the | Royal Institution | N°. II:* written on the outside page of the folded letter.

To Heinrich Rudolf Vieweg 23 October 1867 2662

The Royal Institution London | 23rd. October 1867

My dear Sir,

I should be extremely obliged to you if you would send me a copy of the translation[1] of my book "on Heat as a mode of motion".[2] Even if it should be incomplete I should like to see it, as in the 3rd. edition of the work[3] which is now in the hands of the printers I should like to refer to the translation

Yours very truly | John Tyndall

TU Braunschweig/Universitätsbibliothek/UABS V 1 T : 63

---

1. *translation:* for the German translation of *Heat* (1863), see letter 2607, n. 3.
2. *"on Heat as a mode of motion": Heat* (1863).
3. *3rd. edition of the work: Heat* (1868).

To Jane Maria Strachey 24 October 1867 2663

24th Oct. 1867

My dear Mrs. Strachey.

You know I received a note[1] from your excellent husband[2] regarding an invitation on the part of the Indian government that I should visit India.[3]

About a week ago I was visited by a gentleman from the India office,[4] who wished to ascertain my notion on the subject: he afterwards wrote to me[5] and then consulted with an authority at Albemarle Street[6] regarding the possibility of my getting away.

I send you

N° I. which is the reply of the Honorary Secretary to my note to him.[7]

I send you

N° II. The letter of the President,[8] from which you can gather his notions:

I send you N°. III.

A letter addressed by myself to the President,[9] whom I had seen personally, after I found that my departure from London would be so thoroughly deprecated.

I send you

N° IV. My letter to Mr Thornton of the India office,[10] who at the instance of Sir Stafford Northcote[11] called to see me.

These will put you in possession of the state of the question.

I shall meet you tomorrow at Kew:[12]—would you kindly bring the letters with you, as I wish to preserve them?

Would you also present my kindest regards to Lady Colvile?[13]

Yours ever | John Tyndall

BL Add MS 60631, ff. 113–4

---

1. *a note:* letter missing.
2. *husband:* Richard Strachey.
3. *invitation on the part of the Indian government that I should visit India:* Tyndall was offered a position to undertake meteorological research in British India, but declined on account of Faraday's death and his commitment to the RI (see letter 2658).
4. *a gentleman from the India office:* William Thornton.
5. *wrote to me:* letter missing.
6. *an authority at Albemarle Street:* probably Henry Bence Jones, but possibly Henry Holland.
7. *reply of the Honorary Secretary to my note to him:* Henry Bence Jones (see letter 2656).
8. *letter of the President:* Henry Holland (see letter 2661).
9. *letter addressed by myself to the President:* letter missing.
10. *My letter to Mr Thornton of the India office:* see letter 2658, though it is labelled N°. III.
11. *Sir Stafford Northcote:* Henry Stafford Northcote (1818–87), eighth baronet, an English politician who was Secretary of State for India (1867–8). He was later made first earl of Iddesleigh (*ODNB*).
12. *Kew:* probably the Royal Botanic Gardens at Kew in southwest London.
13. *Lady Colvile:* Frances Elinor Colvile (née Grant, 1838–1919), known as Lady Colvile after her 1857 marriage to James William Colvile (1810–80), a lawyer and judge who served in the courts of Bengal in the 1840s and 1850s, and was later appointed to the Judicial Committee of the Privy Council (*ODNB*).

From François-Napoléon-Marie Moigno — 26 October [1867][1] — 2664

Mon cher Mr Tyndall

Vous voulez que je vous dise la vérité toute entière sur la publication de la Chaleur; je vais le faire non sans chagrin.

Les mille francs que je me suis attribués pour la traduction, sont sortis de ma caisse des Mondes, je me les suis donnés à moi même.

Les frais d'impression montant à environ quatre mille francs, ont été payés par la caisse des Mondes, c'est à dire par moi. Les cinq cent francs donnés à votre éditeur anglais sont encore sortis de ma bourse.

La Chaleur m'a donc couté 5500 francs. L'édition est entièrement épuisée 2300 exemplaires sont aujourd'hui vendus, il n'en reste plus un seul, Mr Giraud a touché 11500, et il ne m'a pas donné un centime. Le bénéficie net fait par lui est de 7000 francs. Retranchant les 1200 francs, que vous nous avez accordés à chaçun, il reste 4600 francs à partager en trois, ce qui nous donnait à vous et à moi 1533 chiffre assez satisfaisant. En outre de ce qui concerne La Chaleur, Mr Giraud m'a fait perdre de 36 à 40 mille francs pour la gestion des Mondes. Il s'est laissé mettre dernièrement en faillite ou *[1–2 words illeg]*.

Mon procès avec lui n'est pas encore terminé, et je ne sais pas ce que je pourrai rattraper.

Il a avoué, il y a quelque temps que La Chaleur ne lui appartenait pas, mais il osé aussi lui proposer d'acheter les clichés.

C'est pour vous surtout que je regrette son indigne conduite. Et combien je serai heureux le jour ou je pourrai vous payer les 1533 francs qui vous appartiennent.

Mr Crookes m'avait envoyé la petite lettre sur la quelle vous exprimez votre surprise qu'on vous présentât la quittance des Mondes. Ce petit mot m'avait fait de la peine, en ce sens seulement que j'aurais voulu, pour tous au monde, qu'on ne vous réclamât rien; j'allais même donner ordre qu'on vous serve les Mondes gratuitement quand votre lettre m'est parvenue, et je vous prie de les accepter.

Vous voyez que je ne vous oublie pas; vous avez dû recevoir une petite brochure dans laquelle j'ai publié vos deux conférences sur la force, la matière et la force, avec une dissertation de moi sur la constitution de la matière que je recommande à votre plus sérieuse attention. En reproduisant votre conférence de Dundee je m'étais permis sur votre péroraison quelques petites remarques critiques que j'ai supprimées avec la péroraison dans la réimpression.

Je ne me contente pas de reproduire vos conférences, je les refais en public avec le plus grand succès, comme vous l'avez vu sans doute par le programme des Mondes. Demain je vous prépare une nouvelle ovation.

Je m'étais mis avec ardeur à la traduction des leçons sur le son, j'aurais tout donné pour qu'elles paraissent avant le petit volume de Radiation; la première leçon est prête à envoyer à l'impression; mais Mr Gauthier-Villars recule plus qu'il n'avance; il prétend n'avoir reçu aucune réponse aux lettres dans lesquelles il a demandé le prix d'une collection entière des clichés.

Si vous n'étiez pas sous le coup de la malhonnêteté de Giraud, si vous pouviez avoir encore en moi la confiance que je mérite à tous d'égards, je vous proposerais de faire moi même les frais de traduction et d'impression des leçons sur le son, aux mêmes conditions que Mr Gauthier-Villars. Du reste Mr Savy libraire rue Hautefeuille 24 demande à corps et à cris à donner une édition française de ces leçons en me payant convenablement la traduction. Si vous voulez en effet que la traduction française paraisse, il faut vous presser car le succès dépend de la rapidité de la publication. Votre édition anglaise est du reste si chère qu'il y aura toujours une bonne place à prendre par ma traduction; j'oserai dire en outre sans vanité, que le soin avec lequel je la ferai, et la notoriété de mon nom sont une excellente condition de succès.

En résumé et je ne le regrette que pour vous la gloire d'avoir attaché mon nom à votre œuvre vaut de l'argent; en résumé, j'ai dépensé pour La Chaleur 4500 francs, et elle ne m'a pas rendu un penny. Je vous devrais 1533 fr!

Acceptez comme hommage de reconnaissance et d'amitié l'abonnement gratuit aux Mondes.

Je les ai retirés des mains des libraires qui gardent tout pour eux, et je m'en trouve très bien; j'ai un petit bureau ou je publierai désormais les traductions que je ferai et mes grands ouvrages, donc j'aurai les bénéfices pour les partager avec les auteurs. J'ai douze cents abonnés, et ils augmentent tous les jours, l'année prochaine mon bénéfice sera de plus de dix mille francs, tandis que jusqu'ici je n'ai rien eu.

Je n'ai rien, cher Mr Tyndall, à me reprocher envers vous; personne plus que moi n'a *[1 word illeg]* de votre gloire et ne l'a plus *[1 word illeg]*. Je désire que le voile qui vous cache encore la vérité se déchire.

2 rue d'Erfurt | A vous de cœur | l'abbé F. Moigno | 26 octobre

Prenez le plutôt possible une résolution relativement aux leçons sur le son.

My dear Mr Tyndall,

You want me to tell you the whole truth about the publication of La Chaleur;[2] and I will do so, but not without distress.

The one thousand francs that I allocated for the translation came from my fund of Les Mondes,[3] so I gave them to myself.

The costs of the printing, amounting to around four thousand francs, were paid by the fund of Les Mondes, that is to say, by me. The five hundred francs given to your English editor[4] were again taken from my purse.

La Chaleur, therefore, cost me 5500 francs. The edition is entirely out of stock; 2300 copies were sold, and there is none left. Mr Giraud[5] received 11500 and he did not give me a penny. The net profit that he made is 7000 francs. Subtracting the 1200 francs that you allocated to each of us, there remains 4600 francs to divide by three, which gives to you and me 1533—a sufficient number. Moreover, regarding La Chaleur, Mr Giraud made me lose 36 to 40 thousand francs for the management of *Les Mondes*. He recently went bankrupt or *[1–2 words illeg]*.

My lawsuit with him has not yet finished, and I do not know what I could get back.

He confessed some time ago that La Chaleur did not belong to him, but he also dared to offer to buy the pictures.

It is especially in regards to you that I lament his unworthy conduct. And how happy I will be on the day when I can pay you the 1533 francs that belong to you.

Mr Crookes[6] sent me the little letter in which you express your surprise at being presented with the receipt for Les Mondes. That short message pained me, in the sense that all I ever wanted that nothing be collected from you; I even went to give the order to give you Les Mondes for free, when your letter reached me, and I beg you to accept it.

You see that I have not forgotten you. You should have received a small booklet in which I published your two lectures on force, matter and force,[7] with a dissertation from me on the constitution of matter,[8] to which I suggest you pay your most serious attention. By reproducing your Dundee lecture, I allowed myself to add several critical remarks to your peroration, which I removed with the reprinting of the peroration.

I am not satisfied with simply reproducing your lectures; I remake them publicly with great success, as you have doubtlessly seen in the program of Les Mondes. Tomorrow I will prepare a new ovation.

I worked arduously on the translation of the lessons on sound.[9] I would have given everything for them to appear before the small volume of Radiation.[10] The first lesson is ready to be sent to print, but Mr Gauthier-Villars[11] draws back more than he moves forward. He claims to have not received any response to the letters[12] in which he asked the price for a whole collection of pictures.

If you were not under the blow of Giraud's dishonesty, if you can still entrust me with the confidence that I deserve in any respect, I would suggest that I should myself do the costs of the translation and the printing of the lessons on sound, at the same conditions as Mr Gauthier-Villars. For the rest, Mr. Savy[13] bookseller 24 Hautefeuille Street, is asking to give a French edition of these lessons and to pay me properly for the translation. If, indeed, you wish for the French translation to appear, you must hurry because success depends on the swiftness of the publication. Your English edition is still so expensive[14] that there will always be a good place for my translation; and I dare to say without vanity, that the care with which I will do it, and the reputation of my name are excellent conditions for success.

In summary, and I regret that for you the glory of having my name attached to your work is worth money, in short, I spent 4500 francs for La Chaleur, and it did not bring me a penny. I owe you 1533 francs!

Accept as a tribute of gratitude and friendship, the free subscription to Les Mondes.

I took it back from the hands of the booksellers who wish to keep everything for themselves, and I find it very good. I have a small office where I will now publish the translations that I will do and my major works, so that I will have the profits to share with the authors. I have twelve hundred subscribers, and they increase every day. Next year my income will be more than ten thousand francs, while up until then, I have had nothing.

I have nothing, dear Mr Tyndall, to reproach myself to you. More than anyone, I have *[1 word illeg]* and *[1 word illeg]* your glory. I wish for you that the veil that hides you from the truth tears open.

2 Erfurt Street | Yours truly | Father F. Moigno | October 26

Take a decision as soon as possible in regards to the lectures on sound.

RI MS JT/1/M/133

---

1. *[1867]:* the year is given based on content of the letter.
2. *publication of Heat:* see letter 2664 in which Moigno discusses the issues following the publication of J. Tyndall, *La Chaleur considérée come un mode de mouvement,* trans. F. Moigno (Paris: Étienne Giraud, 1864).
3. *Les Mondes:* see letter 2706, n. 9.
4. *your English editor:* possibly a reference to Longman (see letter 2617, n. 5), who had published *Heat* in 1863.
5. *Mr Giraud:* Étienne Giraud.
6. *Mr Crookes:* probably William Crookes (1832–1919), an English chemist who was also the proprietor and editor of the *Chemical News* (1859–1906) (*ODNB*).
7. *lectures on force, matter and force:* J. Tyndall, *La matière et la force. La Force: deux conférences,* trans. F. Moigno (Paris: Gauthier-Villars, Les Mondes, 1867). The first lecture was Tyndall's 'Matter and Force' delivered on 5 September 1867 at the annual meeting of the BAAS at Dundee (see letter 2627, n. 31). The second lecture, 'On Force', Tyndall had given at the RI on 6 June 1862 (see J. Tyndall, 'On Force', *Roy. Inst. Proc.*, 3 (1862), pp. 527–36).'
8. *dissertation from me on the constitution of matter:* Moigno's paper 'Matière et force' appeared on pp. 43–71 following Tyndall's two lectures.
9. *lessons on sound:* see letter 2617, n. 2.
10. *Radiation:* J. Tyndall, *Radiation,* trans. F. Moigno (Paris: Gauthier-Villars, 1867). Tyndall had given the Rede lecture at Cambridge in 1865 (see letter 2564, n. 5).
11. *Mr Gauthier-Villars:* Jean-Albert Gauthier-Villars.
12. *the letters:* see letters 2617 and 2690.
13. *M^r^. Savy:* M. F. Savy (fl. 1860), a Parisian bookseller located at 24 rue Hautefeuille.
14. *English edition is still so expensive: Sound* (1867) was priced at 9s ('New Works', *Saturday Review of Politics, Literature, Science and Art,* 12 October 1867, p. 489).

From Mary Egerton 29 October [1867][1] 2665

Mountfield Court | Hurst Green[2] | Oct[obe]r 29th

Dear Mr. Tyndall,

We were greatly disappointed on hastening home from Church on Sunday to find you had not appeared, and all day we went on thinking you might have taken a long walk & would turn up at last, though the non-arrival of any luggage was rather against that notion. I am truly sorry for the reason that kept you away. An accidental illness would be bad enough, but such evident tokens of over-work, now, with months of labour before you, is what I still less like to hear of! You must have more rest, I know, at least I can conceive the difficulties, impossibilities if you will; but please to bear in mind that the most suspect piece of mechanism will wear out if you keep it incessantly at work without ailing, & you have not got discovered how to make a new body or brain when you regain one! I wish I had a clearer idea in what your actual daily work consists, & how poor Mr Faraday's death[3] has influenced it. It is touching to see how deeply he was loved & honoured by everyone who knew him.

But you are not going to India![4] At that I rejoice, though perhaps I ought not, for you write as if you would have liked to go; but independently of other considerations. I cannot think the climate would have suited you, have you not said yourself that "Intellect is in some sense a functioning temperature"? and what w[oul]d become of your brain with the therm[omete]r at 200 in the shade? Perhaps though, you were to go to some glorious point of observation up in the mountains, with a fine climate & scenery that w[oul]d dwarf the Alps! I hope not. I am determined to fancy you in the swamps of Calcutta or the burning plains of Madras[5] that I may indulge without scruple my gladness at your refusal—I wish you had mentioned what was the post offered you.[6]

It was very good of you to be willing to put up with so poor a "party"! Believe me I did not mean to regret not having a houseful of people,—as such,—to meet you; but I feel that when any one takes the trouble to come so far, we ought if possible to have something agreeable to make it worth while; especially as we can be to you but such slight acquaintances, whereas we feel to know you so well through holding converse with you in your books, as my scratchings & scorings in "mountaineering"[7] & other places would testify.

We are pretty sure to be back here before Saturday the 30th of November, so if you are able (& willing) at any time to make amends to us for your failure now, between that time & the session of Parliament in January, I shall trust to your writing to say so. If there was any doubt about our being returned, a note under cover to Mr. E.[8] at the Foreign Office, (marked "private",) w[oul]d be sure to be forwarded whether he was there or not.

We have been reading "Sound"[9] aloud with the greatest interest, but some of it is very hard, & I am afraid our "mental pictures" are of the haziest description. The most perfect Photographer could not, I suppose, produce a clear image on ill-prepared paper!

Forgive this long scrawl—take rest at all hazards—& believe me

Y[ou]rs very truly | M F Egerton

RI MS JT/1/E/56

---

1. *[1867]:* the year is given in relation to the death of Michael Faraday and Tyndall's refusal of the post in India (see n. 4).
2. *Mountfield Court | Hurst Green:* the Egerton family estate.
3. *Mr Faraday's death:* Michael Faraday died on 25 August 1867. After his return from Scotland to London, Tyndall on 25 September 1867 visited Henry Bence Jones who had asked him to give a lecture on Faraday. This request to give an account of Faraday also came from the managers of the RI (who are listed in letter 2592, n. 6), and Tyndall spent considerable time preparing the lectures, finding it very stressful. He wrote after the 17 and 24 January 1868 Friday Evening Discourses at the RI, 'On Faraday as a Discoverer', that he had 'suspended all other work: reread all Faraday's memoirs and gave two evenings' (Journal, RI MS JT/2/13c/1335). These were published in *Roy. Inst. Proc.,* 5 (1866–9) pp. 199–272; and in book form as J. Tyndall, *Faraday as a Discoverer* (London: Longmans, Green and Co., 1868).
4. *not going to India:* see letters 2656, 2658, 2661, 2663, and 2667.
5. *Madras:* the former name of both the Indian state of Tamil Nadu and Chennai, the state's capital city.
6. *post offered you:* Tyndall was offered a position to undertake meteorological research in British India, but declined on account of Faraday's death and his commitment to the RI (see letter 2658).
7. *"mountaineering": Mountaineering in 1861.*
8. *Mr. E.:* Edward Egerton.
9. *"Sound": Sound* (1867).

To John Herschel 30 October 1867 2666

30th Oct. 1867

My dear Sir John,

Would you kindly inform me whether the efforts of Faraday, with whom you & Mr. Holland,[1] were associated, to produce glass for optical purposes resulted in any great improvement of the manufacture? Or whether the chief scientific interest of "the heavy glass"[2] is not derived from the facts that it was by means of it that Faraday effected the rotation of the plane of polarization of a luminous ray, and discovered the phenomena of diamagnetism?

Yours ever faithfully | John Tyndall

Sir John Herschel Bart.[3] | &c. &c. &c.[4]

RS HS 17/401
RI MS JT/1/TYP/2/541

1. *M^r. Holland:* Henry Holland. While Tyndall clearly wrote 'Holland' in this letter, he may have actually meant to write 'Dolland', for George Dolland (see letter 2669, n. 12).'
2. *"the heavy glass":* see Herschel's reply (letter 2669), and F. A. J. L. James, 'Michael Faraday's work on optical glass', *Physics Education,* 26 (1991), pp. 296–300.
3. *Bart.:* Baronet.
4. Herschel wrote the following notes on this letter: 'ans[were]^d. Nov 3, I believe not as Guinand & Merz's processes have come into general use.—Mentioning my making Borate of lead for optic glass at Souths in 1822± Refractive index 1.866. (see Tables in "Light") also my specimen of *[perfect]* flint made on my suggestion in Com. R. S by melting fragments of flint &c. &c.'. For references in these notes, see Herschel's full response in letter 2669.

To Richard Strachey 1 November 1867 2667

1st. November 1867

My dear Strachey.

Your letter[1] served to put within my power the realization of a dream I had indulged in for years. Before replying to it I waited to see what action would be taken by the India Office,[2] and to consult with persons here who are interested in my movements. M^r. Thornton[3] came to me about a fortnight ago, being requested by Sir Stafford Northcote[4] to communicate unofficially with me on the subject. He afterwards wrote me a note;[5] and then I consulted the two persons chiefly interested in the affairs of the Royal Institution, namely its President Sir Henry Holland, and its Honorary Secretary D^r. Bence Jones.

Doubtless long ere this you have heard of the death of dear old Faraday.[6] You will I think sympathise with me when I say that I should not like to do any thing which would injure the Institution, which has had associated with it since its foundation the names of Young,[7] Davy,[8] & Faraday. And whether my going to India would injure it or not, was a question which I was compelled to submit to the decision of others. That opinion is emphatic, and it is such as to preclude all possibility, at all events for the present, of my going to India.

In order to put you in possession of all the circumstances I send you copies of a few of letters bearing on the question. After receiving M^r. Thornton's note I at once wrote to D^r. Bence Jones, who was then at Folkestone,[9] telling him how much I desired to supplement my knowledge by going to India, and asking him the definite question whether he thought my going at present would be of serious injury to the Royal Institution.

The enclosed note N° I, is a copy of his reply.[10]

The note N°. II is from our President,[11] and it will indicate how he regards the subject.

The note N° III is my reply to Mr. Thornton.[12]

I feel a kind of heartache at the thought of giving up the trip to India.

But you ought to undertake the matter—if they would only pay you adequately. Long before I received your note I chanced to meet Sir Stafford Northcote in a Railway carriage, both of us being bound for the same house. I caught your name in a copy of the Globe[13] which I chanced to be reading, and I let him know a bit of my mind regarding you.

Your kind wife[14] takes this out to you. Long days of health & happiness to both of you.

Yours ever | John Tyndall

BL Add MS 60631, ff. 116–8

---

1. *Your letter:* letter missing. See, however, Tyndall's letter to Jane Strachey, Richard Strachey's wife (letter 2663).
2. *India Office:* the India Office was a government department responsible for the administration of India during the British Raj (1858–1947). It was established in 1858 when the East India Company was dissolved by the Government of India Act 1858 (21& 22 Vict. c. 106), following the 1857 Sepoy Mutiny.
3. *Mr. Thornton:* William Thornton.
4. *Sir Stafford Northcote:* Henry Stafford Northcote (see letter 2663, n. 11).
5. *wrote me a note:* letter missing.
6. *dear old Faraday:* Michael Faraday, who died on 25 August 1867.
7. *Young:* Thomas Young (1773–1829), an English physician and natural philosopher whose work helped to establish the wave theory of light. Young had held the RI professorship of natural philosophy (1801–3) (*ODNB*).
8. *Davy:* Humphry Davy.
9. *I at once wrote to Dr. Bence Jones, who was then at Folkestone:* letter missing. Folkestone is a port town on the English Channel in Kent, where Bence Jones had a home at 5 Albion Villas since 1851.
10. *his reply:* letter 2656.
11. *from our President:* letter 2661.
12. *my reply to Mr. Thornton:* letter 2658.
13. *the Globe:* a London evening newspaper which ran from 1803–1921.
14. *your kind wife:* Jane Maria Strachey.

To Jane Maria Strachey 2 November 1867 2668

Royal Institution of Great Britain | 2nd Nov. 1867

My dear Mrs. Strachey

To my great regret there is not a single copy of the book on Heat[1] extant for love or money.

I have not even a separate copy of my own.

Again goodbye | Yours ever | John Tyndall

BL Add MS 60631, f. 125

---

1. *book on Heat:* either *Heat* (1863) or *Heat* (1865). The third edition of *Heat* would be published in 1868, and the book was presumably out of print at this point.

From John Herschel 3 November 1867 2669

Collingwood[1] Nov. 3 /67

My dear Sir

I am quite unable to say[2] with certainty whether Faradays heavy glass (Bora silicate of lead)[3] has been used in the manufacture of any considerable telescope—but I think, <u>not</u>, as the processes of Guinand[4] & Merz[5] having become generally used the manufacturers have no difficulty <u>now</u> at least in procuring optically perfect Flint-glass adequate to all their purposes. Possibly however for some scientific purposes as for the intermediate lens in the dialytic construction—or in Barlow's[6]—or for triple combinations to combine 3 different rays of the spectrum—it may have been employed.

I will take this opportunity to mention that I believe myself to have originated the suggestion of the employment of Borate of lead for <u>optical</u> purposes.[7] It was somewhere in the year 1822 as well as I can recollect, that I mentioned it to Sir James (then Mr) South[8] and in consequence the trial was made in his laboratory at Blackmans Street by precipitating & washing a large quantity of borate of lead and fusing it under a muffle[9] in a porcelain evaporating dish.—A very limpid (though slightly yellow) glass perfectly free from striae resulted—the refraction index 1.866! (which you will find set down in my table of refraction indices in my article "Light" Encyc[lopaedia] Metropol[itana][10])—It was however too soft for optical use as an object glass. This Faraday overcame at least to a certain considerable degree by the introduction of silica.

Being on the subject of optical glass I should notice that before the subject was handed over by the Council of the R.S. to Faraday—Davy[11] had proposed an open Committee of Members of the Council & others[12]—of which I was one—and divers[13] experiments were made at Pellatt & Green's glass house[14] with a view to producing veinless Flint glass. One of the earliest trials was on a principle suggested by myself—to remelt respectably good but still somewhat streaky flint—broken up into small fragments—applying the heat as much as possible from below upwards & so slowly as to let the lower & exterior parts melt first & expel the air. The trial perfectly succeeded and I still possess half of a thick circular disc about 3 in diameter of that exp[erimen]t. intended for an object glass and which was broken—whose purity from streaks & veins is absolutely perfect.

Believe me | My dear Sir | Yours v[er]y truly | JFW Herschel

P.S. Will nobody ever try to make low refraction glass by the use of Fluate[15] of Strontia or fluosilicates[16] with strontia[17] in them?

RI MS JT/1/H/107
RI MS JT/1/TYP/2/542–3

---

1. *Collingwood:* Herschel's home near Hawkhurst in Kent.
2. *I am quite unable to say:* this letter is a reply to letter 2666.
3. *Faradays heavy glass (Bora silicate of lead):* see F. A. J. L. James, 'Michael Faraday's work on optical glass', *Physics Education,* 26 (1991).
4. *Guinand:* Pierre Louis Guinand (1748–1824), a Swiss glass smelter, who worked at the Bavarian glassworks of Joseph von Utzschneider (1763–1840) and Joseph Fraunhofer (1787–1826). In 1805 he established a method for producing flint glass of a quality suitable for optical instruments (*HLS*).
5. *Merz:* Georg Merz (1792–1867), a Bavarian optical instrument maker, who had worked at the Bavarian glassworks of Joseph von Utzschneider (1763–1840) and Joseph Fraunhofer (1787–1826). In 1839 Merz, along with Joseph Mahler (1795–1845), purchased the business from Utzschneider and renamed it G. Merz & Mahler (*NDB*).
6. *Barlow's:* Peter Barlow (1776–1862), an English mathematician and physicist after whom the Barlow lens is named (*ODNB*).
7. *for optical purposes:* on optical glass, Herschel had published: J. Herschel, 'On the aberrations of compound lenses and object-glasses', *Phil. Trans.,* 111 (1821), pp. 222–67; and J. Herschel, 'Practical Rules for the Determination of the Radii of a Double Achromatic Object-glass', *Edinburgh Philosophical Journal,* 6 (1822), pp. 361–70.
8. *Sir James (then Mr) South:* James South (1785–1867), an English astronomer and controversialist who established a private observatory at his home on Blackman Street, London, where, from 1821–3, he and Herschel observed nearly 400 double stars (*ODNB*).
9. *muffle:* a cover which encloses an object that is to be heated in a furnace or kiln and shields it from the flames and the products of combustion (*OED*).
10. *my article "Light" Encyc[lopaedia] Metropol[itana]:* J. Herschel, 'Light', in E. Smedley et. al. (eds), *Encyclopaedia Metropolitana,* 29 vols (London, 1845), vol. 4, pp. 341–586.

11. *Davy:* Humphry Davy.
12. *Committee of Members of the Council & others:* a joint RS and Board of Longitude committee was established in 1824 to research improvements to the manufacture of optical glass for achromatic lenses, and in 1825 a subcommittee formed to investigate more thoroughly through experimentation. This subcommittee consisted of Faraday, Herschel, and the English optician and instrument maker George Dollond (1774–1852), and continued into 1829, with Faraday summarizing their efforts in a Bakerian Lecture given in three instalments on 19 November, and 3 and 10 December 1829. See F. A. J. L. James, 'Michael Faraday's work on optical glass'; F. A. J. L. James, 'The Military Context of Chemistry: The Case of Michael Faraday', *Bulletin for the History of Chemistry,* 11 (1991), pp. 36–40; and M. Faraday, 'The Bakerian Lecture.—On the manufacture of Glass for optical purposes', *Phil. Trans.,* 120 (1830), pp. 1–57.
13. *divers:* an obsolete form of *diverse* (*OED*).
14. *Pellatt & Green's glass house:* Pellatt and Green's was a premier glass manufacturer. Their shop was located in St. Paul's Churchyard, London.
15. *Fluate:* archaic form of *fluosilicate* (*OED*).
16. *fluosilicates:* a salt of fluosilicic acid; (also) a compound that is both a fluoride and a silicate (*OED*).
17. *Strontia:* one of the alkaline earths, the monoxide of strontium (*OED*).

To William Francis 5 November [1867][1] 2670

The Athenaeum[2] | 5th Nov

My dear Francis.

I have looked into the Bibliotheque Universelle—You cannot err in translating the memoir of De la Rive[3]—He was one of Faraday's most intimate friends & no man is better fitted to write about him—I say translate the article.[4]

Yours truly | John Tyndall

St Bride Library, Authors' letters

1. *[1867]:* the year is given by reference to de la Rive's memoir on Faraday, published in October 1867.
2. *The Athenaeum:* see letter 2537, n. 8.
3. *memoir of De la Rive:* A. de la Rive, 'Michael Faraday, sa vie et ses travaux', *Bibliothèque universelle et revue Suisse: Archives des sciences physiques et naturelles,* 30 (1867), pp. 131–76.
4. *translate the article:* the translation appeared in December as: A. de la Rive, 'Michael Faraday, his Life and Works', *Phil. Mag.,* 34 (1867), pp. 409–37.

To Mary Lyell[1] 5 November 1867 2671

5th Nov. 1867

My dear Lady Lyell,

It would give me extreme pleasure to join you but I am circumstanced thus:—

The managers[2] desire that I should give some image of the man Faraday:[3] I have undertaken to do so in the first two Friday evenings of the session.[4]

It is a heavy undertaking, it is in my brain day & night for having the Juvenile Lectures[5] to give this year the time I can devote to this memoir is but short.

A week ago, therefore, I registered a vow, not, save in one or two cases determined by old engagements—to dine out until the heart of the work was broken.

This vow deprives me of what would be a great pleasure.

Kind regards to Sir Charles,[6] | Ever yours, | John Tyndall

RI MS JT/1/T/1055
RI MS JT/1/TYP/3/840

---

1. *Mary Lyell:* Mary Elizabeth Lyell (née Horner, 1808–73), geologist, conchologist, and wife and scientific partner to Charles Lyell (see M. Ogilvie and J. Harvey (eds), *Biographical Dictionary of Women in Science,* 2 vols (London: Routledge, 2000), vol. 2, pp. 813–14).
2. *managers:* see letter 2592, n. 6.
3. *some image of the man Faraday:* Michael Faraday had died on 25 August 1867.
4. *first two Friday evenings of the session:* see letter 2665, n. 3.
5. *Juvenile Lectures:* the RI's annual juvenile Christmas Lectures began in 1827 and still continue. Tyndall delivered the Christmas Lectures a dozen times over his time with the RI, his first in 1861 (see F. A. J. L. James (ed), *Christmas at the Royal Institution: An Anthology of Lectures* (Hackensack, NJ: World Scientific, 2007)). His fourth time, in 1867–8, was on 'Heat and Cold', and the lectures were given on 26, 28, and 31 December 1867, and 2, 4, and 7 January 1868 (*Roy. Inst. Proc.,* 5 (1866–9), pp. 197). The lectures appeared in the *Chemical News,* 17 between January and March 1868, and were later published in French as J. Tyndall, *Chaleur et froid: six leçons faites devant un jeune auditoire pendant les vacances de Noël 1867,* trans. F. Moigno (Paris: Gauthier-Villars, 1868).
6. *Sir Charles:* Charles Lyell.

From John Herschel 5 November 1867 2672

Nov 5/67

My dear Sir

You are quite at liberty to quote anything you please from my note[1]—but as regards the extension of the use of Faraday's glass[2] among the object glass makers I speak only according to the best of my knowledge & belief but my means of information are not extensive

Yours very truly | JFW Herschel

RI MS JT/1/H/108
RI MS JT/1/TYP/2/544

1. *my note:* letter 2669, to which Tyndall must have replied (letter missing).
2. *Faraday's glass:* see letter 2666, n. 2.

From Jane Barnard 11 November 1867 2673

11 Nov. 1867

Dear Dr Tyndall

My Aunt[1] & I have gone through your paper[2] with much interest.

We like it all very much—my Aunt has made two or three slight corrections as to dates &c & she would like if you approve that the moving cause for his taking to the professional business in 1830–1; it should be mentioned, it was the ill health of his friend Richard Phillips[3] who requested him to take the excise office business for him while he was unable to do it.

If you would kindly come up & see us I should much like to shew you a page in his account book showing his income from that source for many years.

Since writing the above I have been reading some facts aloud & we see nothing we could wish altered—still if you can do spare us 5 minutes

Yours sincerely | Jane Barnard

RI MS JT/1/B/31

1. *My aunt:* Sarah Faraday.

2. *your paper:* Tyndall's draft for his Friday Evening Discourses on Faraday (see letter 2665, n. 3).
3. *Richard Phillips:* Richard Phillips (1778–1851), an English chemist and friend of Michael Faraday (*ODNB*).

From Thomas Henry Huxley [12 November 1867][1] 2674

My dear Tyndall,

Your reviewer in the Pall Mall to-night[2] wishing to be extra clear in the final paragraph has made a donkey of himself.[3]

Do you care to skewer him? If so I will write you a note which you can if you like send on to the Pall Mall.[4]

Ever yours faithfully | T. H. Huxley.

IC HP 8.59
Typed Transcript Only

---

1. *[12 November 1867]:* the date, given as 'probably' on the typescript, is corroborated by reference to a review in the *PMG* (see n. 2).
2. *reviewer in the Pall Mall to-night:* 'Tyndall on Sound', *PMG,* 12 November 1867, pp. 10–11. In letter 2679, Huxley suggested the reviewer was George Henry Lewes (1817–78), an English philosopher, literary critic, and physiologist (*ODNB*).
3. *has made a donkey of himself:* the final paragraph reads: 'The critical objection is this: The opening sentence of the volume quite needlessly repeats the popular error that "the various nerves of the human body have their origin in the brain;" the fact being that (with the questionable exception of the olfactory nerves) *none* have their origin in the brain, as Professor Tyndall of course knows perfectly well, though addressing a popular audience he chooses to use the old traditional language. Had he simply said that the nerves were in *connection,* more or less direct, with the brain, his object would have been secured, and he would have avoided the repetition of a stereotyped error which modern science has displaced' (p. 11).
4. *I will write you . . . to the Pall Mall:* Huxley responded to the reviewer. See T. H. Huxley, 'The Nerves and the Brain', *PMG,* 16 November 1867, pp. 4–5; T. H. Huxley, 'The Nerves and the Brain', *PMG,* 20 November 1867, pp. 4–5 (this includes a reply from the journal editor); and T. H. Huxley, 'The Nerves and the Brain', *PMG,* 22 November 1867, pp. 3–4.

From Abbey Doyle 13 November 1867 2675

Leighlin Bridge[1] | Nov. 13th 1867.

Professor Tyndall

Sir,

I presume you will think it rather presumptuous of me to address you. I would not attempt to do so only at the request of Mr Caleb Tyndall, he knows my destitute situation, he wrote in my behalf to you some time since[2] and as he got no reply he considered you were on the Continent. I have 11 children all depending on the wages of my husband,[3] provisions are now so dear that we are really in a destitute situation, I therefore humbly solicit your assistance the cold weather approaching and everything combined adds to our distress, knowing your humane and charitable disposition emboldens me to make my distress known to you, my prayers and those of my family shall be fervently offered for your long life and happiness here and hereafter.

Hoping you will pardon | your obedient Servant | Abbey Doyle.

RI MS JT/1/TYP/10/3364
LT Typescript Only

1. *Leighlin Bridge:* Leighlinbridge, County Carlow, Ireland, where Tyndall was born and his mother lived.
2. *he wrote in my behalf to you some time since:* letter missing. For further correspondence concerning Abbey Doyle, see letters 2681, 2718, and 2730.
3. *my husband:* not identified.

To Auguste de la Rive 18 November [1867][1] 2676

Royal Institution, 18th. Nov.

Excellent Friend!

I have lately received your admirable notice of Faraday,[2] for which I offer you my best thanks.

I intend to occupy the two first Friday evenings of our next session with some account of Faraday's scientific life.[3] Dare I ask you to be so good as to send me a copy of the letter he wrote to you[4] on the occasion of Mrs Marcet's[5] death? I could of course retranslate it, but then I might miss his actual words

Yours ever | John Tyndall

RI MS JT/1/TYP/1/366
LT Typescript Only

1. *[1867]:* the year is given based on the mention of Tyndall's lectures about Faraday (see n. 3).
2. *I have lately received your admirable notice of Faraday:* see letter 2670, n. 3. A translation of the text was published in the December 1867 number of the *Phil. Mag.* (see letter 2670, n. 4).
3. *two first Friday evenings . . . Faraday's scientific life:* see letter 2665, n. 3.
4. *letter he wrote to you:* Faraday had written to de la Rive on 2 October 1858 (*Faraday Correspondence,* 5:3519).
5. *Mrs Marcet's:* Jane Marcet.

From Anna Helmholtz 18 November 1867 2677

Heidelberg[1] Nov. 18th. 67.

Dear Sir,

Pray accept my most sincere thanks for your very kind way of reminding me of last winters dear work.[2] During three days I did not know to whom I might feel grateful for the beautiful little brooch,[3] which came to me in such a mysterious manner; now I know the friendly giver. I shall value it most highly considering it as a proof of your approbation of our alas but very insufficient efforts to do justice to your beautiful work. I feel but too well that I don't merit any such approbation, knowing how very imperfect the translation is. I need not tell you, that whatever ground for satisfaction you shall find in our part of the translation, is due to my husbands[4] help and corrections. But in spite of all the imperfections of our work, you will surely allow me, dear Sir, to thank you for the great pleasure we found in doing it; I felt quite proud of being the instrument to make such a book known to the great public in our country.

I hope you will fulfill your promise, given to my husband, of coming to see us one day; and please, do not forget, that there is always a room ready for you in our home, and that nothing would give us more pleasure than a visit from you.—Allow me once more to thank you for the charming souvenir—I have rather a weakness for such things and now I may even feel proud of the weakness.

My husband sends his best regards and thanks, believe me, dear Sir

Yours very truly | Anna Helmholtz

P.S. my husband wishes me reopen my letter, to tell you, that nothing would give us greater pleasure, than your confiding some other of your works in to our hands;[5] I know nothing would make <u>me</u> so happy, but I was afraid, you would think me indiscreet, if I said so of my own accord; with my husbands sanction, you will pardon me, I am sure

Yours again | A. H.

RI MS JT/H/31

1. *Heidelberg:* a town in southwestern Germany, location of the University of Heidelberg where Hermann Helmholtz was Professor of Physiology from 1858–71.
2. *last winters dear work:* Anna Helmholtz and Clara Wiedemann, along with their husbands had translated Tyndall's *Heat* into German (see letter 2607, n. 3; *Helmholtz,* pp. 311–2).
3. *little brooch:* Tyndall sent Anna Helmholtz 'a small brooch with inlaid gold and pearls' as thanks for her translating *Heat* (*Helmholtz,* p. 311).
4. *husbands:* Hermann Helmholtz.
5. *your works in to our hands:* Tyndall would forward his *Sound* (1867) to be translated into German as *Der Schall: acht Vorlesungen gehalten in der Royal Institution von Grossbritannien,* trans. H. Helmholtz and G. Wiedemann (Braunschweig: Friedrich Vieweg, 1869) (see letters 2680, 2687, 2691, and 2709). In practice, the book was not translated by H. Helmholtz and G. Wiedemann, but by their wives Anna Helmholtz and Clara Wiedemann (see letter 2691).

To Thomas Henry Huxley 19 November [1867][1] 2678

November 19th

Dear Hal,[2]

The point we were speaking about depends in some measure upon the mechanism of walking which I have thought about only in a general way. But since you were here I have made a posture-master of myself, throwing myself into numerous Belvederian[3] attitudes, which would have excited the same enthusiasm in an appreciative sculptor that the aspect of my teeth excites in Fletcher[4] when he scrapes the lamp-black[5] from their surfaces. The intellectual outcome of these enquiries you shall immediately hear.

When you first spoke I thought of the act of jumping, and here I do not think Haughton's statement[6] would apply. The case of jumping would be in great part that of a projectile, and this is why I referred to the case of the gun pointed obliquely and fired vertically upwards. The vertical component in the case of oblique projection would not express the work done.

But in walking the case is different, and to explain myself I must call upon my talent for drawing. Suppose then this figure to represent me

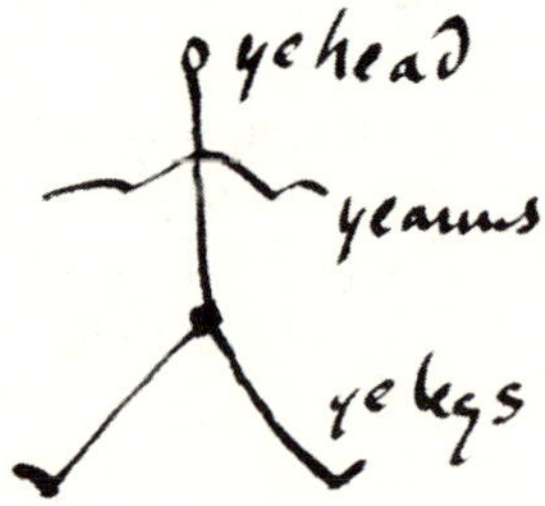

Then suppose me to be in this attitude walking from A towards B with one leg vertical

the muscles of which have just raised my centre of gravity, with the other leg just about to move forward. That other leg is a pendulum with the hip-joint for its point of suspension, and as a pendulum it will swing forward by gravity alone. At all events it would not be correct to say that its forward motion was muscular work. When the leg is brought forward you have your friend in this position,

the longitudinal propagation having invoked no work. He has simply to let his foot fall, and to recommence the process. Q.E.D.[7]

If the man spring forward when his foot reaches the ground then you would be to some extent right, but the poor devil cannot spring because you have placed him upon ice without crampons.

I would recommend the omission of the longitudinal push.

The other point is this:—suppose A to be a fellow's head, and AB a long nerve running down to his big toe: suppose the nerve to be 6 feet long and that it is divided into 6 equal parts, marked by the small letters, each a foot long:

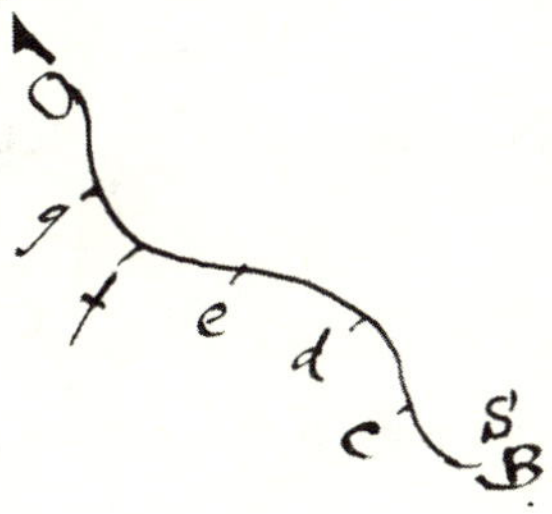

Suppose the points B, c, d, e, f, g, to be excited in succession. Then, calling the time occupied in the transmission the excitement from B to the seat of sensation, unity or 1; if the lines of transmission from the points, c, d, e, f, and g, be represented by the series of fractions 5/6, 2/3, 1/2, 1/3, 1/6, I say that this proves A to be the seat of sensation and entirely abolishes the claim of any damned ganglion that may stick itself between A and B.

Yours ever | John Tyndall.

IC HP 8.42
RI MS JT/1/TYP/9/2896
Typed Transcript Only

---

1. *[1867]:* the year is given by relation to letters 2674 and 2679.
2. *Hal:* nickname for Huxley.
3. *Belvederian:* a reference to the posture of the famous classical Roman marble sculpture *Apollo Belvedere* by Leochares, which depicts the Greek god Apollo.
4. *Fletcher:* not identified; possibly Tyndall's dentist.
5. *lamp-black:* a pigment consisting of almost pure carbon in a state of fine division; made by collecting the soot produced by burning oil or gas (*OED*).
6. *Haughton's statement:* Samuel Haughton (1821–97), an Irish mathematician, geologist, anatomist, and physiologist who studied the mechanics of limb movement in animals in the 1860s and early 70s, and published *The Principles of Animal Mechanics* (London: Longmans, Green, and Co., 1873) (*ODNB*). The statement is possibly: 'A young man, weighing 120 lbs., leaps horizontally 20 ft.; what is the work done? By the same method of calculating, we find the work done in this case to be 1200 lbs. lifted through one foot. This large amount of work is not given out in a single spring; for it represents the sum total of the single spring and of the velocity acquired in running up to the point of starting; and the whole art of long jumps resolves itself into jumping vertically with a velocity equal to the acquired horizontal velocity, and making both quantities a maximum' (S. Haughton, 'Notes on Animal Mechanics. No. III.—On the Muscular Mechanism of the Leg of the Ostrich', *Dublin Quarterly Journal of Science,* 5 (1864), pp. 204–17, on pp. 214–5.
7. *Q.E.D.: quod erat demonstrandum* (Latin), meaning 'which was to be demonstrated'; used to convey that a fact or situation demonstrates the truth of a person's theory or claim (*OED*).

From Thomas Henry Huxley [after 19 November 1867][1] 2679

Athenaeum Club.[2]

My dear Tyndall,

Your argument[3] is perfectly just—but would not meet the objections that would be raised by your grumbling reviewer.[4]

Suppose a ganglionic corpuscle at a, and that the nerve from the toe e a is traceable into it. Then anatomically a might be called the origin of the nerve, though neither structurally nor physically need there be any break between the nerve a e and the nerve a x which places the ganglionic corpuscle in communication with the brain. So that your excitement might be transmitted as if there were one continuous nerve fibre and yet anatomically speaking (or in a sense which might be employed by anatomists—though the practice is capricious and inconsequent) a might be the origin of the fibre. At present we really know very little with certainty about the origin of nerves in the anatomical sense. If I say that some nerve fibres pass continuously from the end of the great toe to the surface of the brain, I say more than I can prove, but not more than is possible or probable and nobody can disprove the assertion. If I say that all the sensory or afferent nerve fibres are continuous from the excitable surface to the brain without the interposition of a ganglionic cell, I say not only what I cannot prove but what is decidedly against probabilities as anatomy now stands.

If I say (as that blundering Pall Mall reviewer does) that no nerve fibres are continued from an excitable surface to the brain without the interposition of a ganglionic cell, I not only say what I cannot prove, but what can be disproved.

It must be Lewes,[5] nobody else could be so clever and so ignorant. In which case I have put my foot into his good graces.

Ever yours | T. H. H.

Explanation of the Figure.

The Beautiful, the Good, and the True attitudinizing after the manner described in this note.

a ganglionic corpuscle to which the nerve a e is attached: | a x, nerve continued from corpuscle to brain: b, arachnoid[6] fluid composed of pure alcohol (in the case of this subject), in which floats c, the brain (in the proper sense) here very small: d, appendage believed to exist by the clergy.

IC HP 8.57
Typed Transcript Only

1. *[after 19 November 1867]:* the date is given by relation to letters 2674 and 2678.
2. *Athenaeum Club:* see letter 2537, n. 8.
3. *Your argument:* see letter 2678.
4. *your grumbling reviewer:* see n. 5.
5. *Lewes:* George Henry Lewes (1817–78), an English philosopher, literary critic, and physiologist (*ODNB*).
6. *arachnoid:* the delicate serous membrane or membranous sac lining the *dura mater,* and enveloping the brain and spinal cord (*OED*).

To Hermann Helmholtz 20 November 1867 2680

20th. Nov. 1867

My dear Helmholtz

Quandt and Händel[1] of Leipzig have written[2] to ask my permission to publish a translation of my book on Sound.[3] Would it interfere in any way with the Tonempfindungen?[4] I have written[5] to Vieweg[6] asking him whether he had any thoughts of translating the work.

What interests me more at present is a memoir on Faraday[7] which I am working at night and day. I think it will be of public interest, and I wish to ask your advice about the translation of it.[8] Do you think it would be well to publish it in a separate form in Germany. It would probably be illustrated by two portraits.

I hope the Jeweller duly paid the postage for the little trinket[9] which I took the liberty of sending to Mrs. Helmholtz.[10] The Heat book[11] is a credit to every body connected with it. It looks far better than the English edition[12] and leaves the French nowhere.[13]

Yours ever | John Tyndall

What do you think of that Calculation of Faradays[14] where he makes out that the electricity of a single gram of water equal to that of a powerful flash of lightening? The utmost possible energy, to be got out of the formation of a gram of water would heat about 1/12th of an ounce Troy[15] of water from the freezing point to the boiling point.

BBAW, NL. Helmholtz, Nr. 477, BI_7–8
RI MS JT/1/T/484

1. *Quandt and Händel:* a publishing firm located in Leipzig, Germany, that specialized in scientific works.
2. *have written:* letter missing.
3. *translation of my book on Sound:* see letter 2677, n. 5.
4. *Tonempfindungen:* see letter 2581, n. 8.
5. *have written:* letter 2687.

6. *Vieweg:* Heinrich Rudolf Vieweg.
7. *memoir on Faraday: Faraday as a Discoverer* (see letter 2665, n. 3).
8. *translation of it:* see letter 2686, n. 4.
9. *little trinket:* see letter 2677, n. 3.
10. *Mrs Helmholtz:* Anna Helmholtz.
11. *Heat book:* see letter 2607, n. 3.
12. *English edition: Heat* (1865). The first German edition was based on the second English edition.
13. *the French nowhere:* J. Tyndall, *La Chaleur considérée come un mode de mouvement,* trans. F. Moigno (Paris: Gauthier-Villars, 1864).
14. *Calculation of Faraday's:* probably a reference to M. Faraday, 'Experimental Researches in Electricity. Seventh Series', *Phil. Trans.,* 124 (1834), pp. 77–122, see p. 116–17.
15. *Troy:* the standard system of weights used for the precious metals and precious stones (*OED*).

To [Elizabeth Dawson Steuart] 20 November [1867][1] 2681

Royal Institution | 20th. Nov.

My dear Mrs [Steuart],[2]

I have received from that wretched woman[3] who gave us so much trouble two applications for aid.

Would you hand her over two guineas for me. I send them in consideration of the fact that my poor mother[4] liked the woman.

Emma[5] has been very ill; and is still very low—She has every possible attention paid to her.

I am working under terrible pressure. The death of Faraday[6] has rendered this necessary. I take his place[7] and my work will be thereby augmented. At present however I am trying to give the world some image of the great discoverer.[8]

Yours ever | John Tyndall.

RI MS JT/1/TYP/10/3363
LT Typescript Only

---

1. *[1867]:* the year is given based on the year Michael Faraday died.
2. *Mrs [Steuart]:* although the typescript of this letter states 'My dear Mrs Faraday', we believe the recipient to be Elizabeth Dawson Steuart and was mis-transcribed by LT as Sarah Faraday. Steuart handled most of Tyndall's finances in Ireland, and it would seem odd that Tyndall would write 'The death of Faraday' to Faraday's wife.
3. *wretched woman:* Abbey Doyle, who wrote Tyndall on 13 November 1867 seeking financial assistance, after a previous request went unanswered (see letter 2675).
4. *my poor mother:* Sarah Tyndall (née Macassey).

5. *Emma:* Emma Tyndall.
6. *death of Faraday:* Michael Faraday died on 25 August 1867.
7. *his place:* see letter 2661, n. 4.
8. *I am trying to give the world some image of the great discoverer:* see letter 2665, n. 3.

To Heinrich Rudolf Vieweg[1] 20 November 1867 2682

20th Nov. 1867

Gentlemen

Messrs. Quandt and Händel of Leipzig[2] have written to me[3] with reference to a translation of my book on Sound,[4] but before giving them permission I wish to ask whether you entertain any thoughts of publishing the work.

Yours very truly | John Tyndall

Messrs. Vieweg & Son.

TU Braunschweig/Universitätsbibliothek/UABS V 1 T : 63

1. *Heinrich Rudolf Vieweg:* while the letter is addressed to 'Gentlemen', we believe the recipient is Heinrich Rudolf Vieweg, who took over the family publishing firm, Friedrich Vieweg und Sohns, from his father Eduard Vieweg (1797–1869). Eduard had a stroke in 1866 and was bedridden until his death. Heinrich would run the publishing house until his death in 1890, and control then went to his wife Helene Vieweg (née Brockhaus), the daughter of the Leipzig publisher Heinrich Brockhaus, whom he married in 1855 (*ADB*; *NDB*; P. Forman, 'The Archive of Friedr. Vieweg & Sohn, Braunschweig: Correspondence of Nineteenth-Century Physical Scientists, Especially German Chemists', *Isis*, 60 (1969), pp. 384–6).
2. *Messrs. Quandt and Händel of Leipzig:* see letter 2680, n. 1.
3. *written to me:* letter missing.
4. *translation of my book on Sound:* Tyndall's *Sound* (1867) was translated into German and published by Vieweg (see letter 2677, n. 5).

From Jane Barnard 21 November 1867 2683

21 Nov. 1867

Dear Dr Tyndall

Many thanks for the pamphlet.[1] I have read it through with very great interest & pleasure—it is beautiful. One sentence reminds me of one of my uncle's[2] sayings which I copy from a sheet of paper in his own hand, writing—["]criticize one's own views in every way by experiment—if possible no objection to be found by others"

The extracts I sent you yesterday[3] were from a sheet of paper on which my uncle had jotted down some reminiscences of his early days & which he did at our request.

I did not know where it had got to until I came upon it amongst other very interesting documents in one of his drawers.

It is here for you to see if you wish.

Yours ever sincerely | Jane Barnard

Dr Tyndall

RI MS JT/1/B/32

1. *pamphlet:* not identified.
2. *my uncle's:* Michael Faraday.
3. *The extracts I sent you yesterday*: not identified.

From Heinrich Rudolf Vieweg[1] 22 November 1867 2684

Braunschweig 22^n Novbr. 1867.

Verehrtester Herr!

Wir sind ihnen für Ihre freundliche Notiz vom 20 Novbr. sehr dankbar.

Bis jetzt kennen wir das fragl. Buch nicht aus eigener Anschauung, zweifeln aber keinen Augenblick, daß es sich für unsern Verlag eignen wird, und werden in diesem Falle Sorge tragen, eine wirklich gute und der wißenschaftlichen Bedeutung des Buches angemeßene Übersetzung davon zu veranstalten. Vielleicht können wir die Herren Helmholz und Wiedemann, deren Übersetzung allgemein als eine meisterhafte anerkannt wird, wiederum für die Herausgabe gewinnen. Sie würden uns sehr zu Dank verbinden, wenn Sie Ihren Verleger veranlaßen wollten, uns sogleich per bookpost ein Exemplar zu übersenden; zugleich bitten wir um Mittheilung der Bedingungen, welche Sie etwa für Überlaßung des Übersetzungsrechtes stellen würden.

Mit der größten Hochachtung | ganz ergebenst | Friedr[ich]. Vieweg & Sohn

Braunschweig[2] 22 Nov[em]b[e]r. 1867.

Most honoured Sir!

We are very grateful to you for your kind note of the 20 November.[3]

Up till now we are not familiar with the book in question[4] from our own experience, but do not doubt for a moment, that it will be suitable for our publishing house, and shall in this case attend to arranging a truly good translation[5] that is appropriate to the scientific significance of the book. Perhaps we can again enlist Mr Helmholtz[6] and Mr Wiedemann,[7] whose translating is generally acknowledged as masterful, for the publication. We would be very indebted to you, if you would get your publisher to send us a copy by bookpost[8] right away; at the same time we ask you to communicate the terms which you would more or less set for the transfer of the translation rights.

With the greatest respect | your most humble servants | Friedr[ich]. Vieweg & Sohn

RI MS JT/1/V/8

1. *Heinrich Rudolf Vieweg:* while the letter is signed 'Friedr[ich]. Vieweg & Sohn', we believe the sender is Heinrich Rudolf Vieweg, who took over the family publishing firm, Friedrich Vieweg und Sohns, from his father Eduard Vieweg (1797–1869). Eduard had a stroke in 1866 and was bedridden until his death. Heinrich would run the publishing house until his death in 1890, and control then went to his wife Helene Vieweg (née Brockhaus), the daughter of the Leipzig publisher Heinrich Brockhaus, whom he married in 1855 (*ADB; NDB;* P. Forman, 'The Archive of Friedr. Vieweg & Sohn, Braunschweig: Correspondence of Nineteenth-Century Physical Scientists, Especially German Chemists', *Isis,* 60 (1969), pp. 384–6).
2. *Braunschweig:* a city in Lower Saxony, Germany.
3. *your friendly notice of the 20th of November:* letter 2682.
4. *book in question: Sound* (1867).
5. *truly good translation:* Tyndall's *Sound* (1867) was translated into German and published by Vieweg (see letter 2677, n. 5).
6. *Herr Helmholtz:* Hermann Helmholtz.
7. *Herr Wiedemann:* Gustav Wiedemann.
8. *bookpost:* the system and regulations under which books and printed matter may be sent through the post, often at a reduced price compared to other items (*OED*).

From Auguste de la Rive 23 November 1867 2685

Genève le 23 Novembre 1867.

Mon cher ami,

Votre petit mot du 18 9bre m'a fait un grand plaisir; je suis très heureux que vous ayez été satisfait de ma notice sur Faraday. J'ai cherché de mon mieux à résumer & à classer ses belles et nombreuses recherches scientifiques de manière à en donner une idée aussi claire & complète que possible. Cela m'a donné assez de peine & je suis bien satisfait de voir que mon travail vous a intéressé.

Je vous envoie la copie textuelle de la lettre que Faraday m'a adressée à la mort de Madame Marcet; je n'en ai donné que quelques fragments dans ma notice; mais pour des Anglais comme pour des Genevois, vu les deux personnes dont il s'agit, la lettre entière a de l'intérêt, c'est pourquoi je vous en envoie une copie exacte, en ne supprimant que les passages qui ne sont pas relatifs au sujet principal. Vous ferez de cette lettre l'usage que vous voudrez; je serais extrêmement heureux si elle peut vous servir ainsi que ma notice dans les deux séances que vous comptez donner sur la vie scientifique de Faraday.

Recevez, très cher ami, l'assurance de mes sentiments bien affectueux.

Votre tout dévoué | Aug[uste]. de la Rive.

*Vous comprendrez facilement pourquoi je n'ai pas donné la traduction entière de la lettre dans ma notice qui était destinée aux savants en général.

Geneva, November 23 1867

My dear friend,

It was a pleasure to receive your small note of the 18th of November.[1] I am very happy that you were satisfied with my report on Faraday.[2] I tried my best to summarize & to classify his numerous, beautiful scientific works, in a way that would present the ideas as clearly & completely as possible. This gave me enough trouble, & I am glad to see that my work has interested you.

I am sending you an exact copy of the letter that Faraday sent me[3] when Madame Marcet[4] died. In my report I only gave a few fragments of it; but for the English as for the Genevans, given the two people whom it concerns, the entire letter is of interest. This is why I am sending you an exact copy, deleting only the passages that are not related to the main topic. You can use this letter for whatever you wish; I would be extremely happy if it can be of use to you, as well as my report, in the two sessions that you plan to give on the scientific life of Faraday.[5]

Please accept, my dear friend, the assurance of my very affectionate sentiments.

Truly yours | Aug[uste]. de la Rive.

You can easily understand why I did not provide the full translation of the letter in my report, which was intended for savants in general.

RI MS JT/1/TYP/1/367
LT Typescript Only

---

1. *your small note of the 18th of November:* letter 2676.
2. *my report on Faraday:* see letter 2676, n. 2.
3. *letter that Faraday sent me:* enclosure missing; see letter 2676, n. 4.
4. *Madame Marcet:* Jane Marcet.
5. *two sessions that you plan to give on the scientific life of Faraday:* see letter 2665, n. 3.

From Hermann Helmholtz 24 November 1867 2686[1]

Heidelberg. 24. Novb. 1867.

Lieber Freund,

wir haben Ihre beiden Briefe empfangen, und ich erlaube mir auf die darin enthaltenen Fragen folgendes zu antworten.

Wir würden mit dem grössten Vergnügen die Übersetzung von Faraday's Leben übernehmen, wenn Sie uns dasselbe anvertrauen wollen. Ich glaube auch, dass ein solches Buch meiner Frau viel leichter zu übersetzen sein wird als On Heat, und dass die Übersetzung besser gelingen wird. In letzterem Buche musste ich verhältungsmässig zu viel verbessern und das stört immer die gleichmässigen Fluss der Übersetzung, so dass wir nicht ganz so zufrieden mit derselben sind, wie Sie.

Nach Herrn Tyndall's Theorie muss die frei gewordene latente Wärme von der Stelle, wo zwei Eisstücke zusammenfrieren durch das benachbarte Eis fortgeleitet werden, und da die Differenz des Gefrierpuncts zwischen dem Innern und der Oberfläche des Eises jedenfalls sehr gering ist, muss sie sich über eine relativ beträchtliche Eismasse vertheilen, wenn nirgends das innere Eis schmelzen soll. Bei dem schlechten Wärmeleitungs vermögen des Eises und den äusserst kleinen Temperaturdifferenzen erscheint mir dies unwahrscheinlich in den Fällen, wo das Eis unter einen kräftigen Druck, ja selbst unter einem Hammerschlage, sich so schnell *[zusammenlöthet]*. Ich gebe aber zu, dass die für die Wärmeleitung *[hierbei]* *<possible page(s) missing>* wo an eine chemische Wirkung nicht zu denken ist. Aber diese Ladung ist zu erklären durch eine Anziehungskraft der Metalle von verschiedener Grösse für die beiden Elektricitäten, und strebt einem Gleichgewichtszustande zu; es kann dadurch kein continuirlicher Strom entstehen. Der Gleichgewichtszustand der sich berührenden Leiter kann gestört werden 1/ durch mechanische Kräfte in den Condensatorversuchen 2/ wenn der eine Leiter durch chemische Zersetzung zerfällt. Die rein chemische Theorie kann die Condensatorversuche gar nicht, und die Abhängigkeit des Stromes und der ihm begleitenden Zersetzung von der Anordnung der Metalle nur in sehr unbestimmter Weise erklären; während die Berücksichtigung der elektrischen Ladungen der Metalle eine verhältnissmässig einfache und bestimmte Erklärung giebt, *<possible page(s) missing>* Sie können vielleicht sagen, dass diese Durchführung von Faraday's Ideen Ahnungen tieferen Zusammenhangs aller Imponderabilien uns eröffnet. Ich glaube aber nicht dass man die Sache schon als abgeschlossen betrachten kann.

Ich meine hierbei nur Maxwell's letzte mathematische Darstellung in den Philos. Transct., nicht seinen künstlichen Mechanismus von Zellen mit rotiriendem Inhalt und Frictionsrollen.

Beste Grüsse; wir sehen der Beendigung Ihrer Arbeit mit grossem Interesse entgegen; für das deutsche Publicum werde ich, glaube ich, einige Zusätze als Anmerkungen machen müssen, die ich Ihnen später zur Billigung zusenden werde.

In treuer Freundschaft Ihr | H. Helmholtz.

Heidelberg.[2] 24th Nov. 1867.

Dear friend,

We have received both your letters,[3] and I am taking the liberty to make the following reply to the questions contained in them.

We would take on the translation[4] of Faraday's life[5] with the greatest pleasure if you want to entrust it to us. I believe also that such a book will be much easier for my wife[6] to translate than On Heat,[7] and that the translation will succeed better. In the latter book I had to improve too much, */relatively[8] speaking/*, and that always disrupts the[9] even flow of the translation, with the result that we are not always quite as satisfied with it as you are.

According to Mr. Tyndall's theory, the latent heat that has been released has to be transmitted from the spot where two pieces of ice freeze together through the neighbouring ice, and as the difference in the freezing point between the interior and the surface of the ice is very slight in every case, it has to be distributed over a relatively considerable mass of ice, if the inner ice is not supposed to melt anywhere. With the poor heat-conducting ability[10] of the ice and the extremely small differences in temperature, this appears improbable to me in cases where the ice */solders together/* so quickly under strong pressure, indeed, even under a hammer blow. I admit, though, that the heat conduction */in this connection/* *<possible page(s) missing in original>* where a chemical reaction is out of the question. However, this charge can be explained by an attractive force of metals of different size for both electricities, and strives towards a state of equilibrium; a continuous current cannot be produced through this. The state of equilibrium of the conductors in contact with each other can be disrupted 1/ through mechanical forces in the experiments with capacitors[11] 2/ if one conductor decays through chemical decomposition. The purely chemical theory cannot explain the experiments with capacitors at all, and only in a very vague fashion the dependence of the current and the decomposition accompanying it on the arrangement of the metals; while the consideration of the electrical charges of the metals provides a relatively simple and precise explanation, *<possible page(s) missing in original>* You can perhaps say that this implementation of Faraday's ideas opens up to us inklings of a deeper connection of all imponderables. I do not believe, though, that one can consider the matter already closed.

I mean in this connection only Maxwell's[12] latest mathematical presentation in the Philos[ophical]. Transact[ions].,[13] not his artificial mechanism of cells with rotating contents and rolling friction.

Best regards; we look forward to the completion of your work with great interest; for the German readership, I shall, I believe, have to make some additional remarks as footnotes, which I shall send you later for approval.

In loyal friendship Your | H. Helmholtz.

RI MS JT/1/H/45

---

1. Our German translator indicated that based on grammatical and contextual breaks, there may be pages missing from this letter, as we have noted in the text. We confirmed with the archive at the RI that what we have is all there is for this letter.
2. *Heidelberg:* see letter 2677, n. 1.
3. *both your letters:* letters missing.
4. *translation:* J. Tyndall, *Faraday und seine Entdeckungen,* trans. H. Helmholtz (Braunschweig: Friedrich Vieweg und Sohn, 1870). In practice, the book was not translated by Helmholtz, but by his wife Anna Helmholtz (see letter 2903).
5. *Faraday's life: Faraday as a Discoverer.*
6. *my wife:* Anna Helmholtz. See letter 2607, n. 3.
7. *On Heat: Heat* (1863). For the German translation, see letter 2607, n. 3.
8. *relatively:* in the German, Helmholtz used the word 'verhältungsmässig' which our German translator noted does not exist in German, and that Helmholtz probably intended to write 'verhältnismässig' (meaning 'relatively').
9. *the:* in the German, Helmholtz used the word 'die', for which our German translator noted that Helmholtz should have used 'den' (meaning 'the').
10. *heat-conducting ability:* in the German, Helmholtz used 'Wärmeleitungs vermögen', for which our German translator noted that he should have combined them as one word, 'Wärmeleitungsvermögen'.
11. *capacitors:* our translator noted that this, and the same instance in the following sentence, could also be translated as 'condensers'.
12. *Maxwell's:* James Clerk Maxwell.
13. *latest mathematical presentation in the Philos[ophical]. Transact[ions].:* J. C. Maxwell, 'On the Dynamical Theory of Gases', *Phil. Trans.,* 157 (1867), pp. 49–88.

To Heinrich Rudolf Vieweg[1] 27 November 1867 2687

27th Nov. 1867

Gentlemen

I have given directions to my publisher Mr. Longman[2] to forward you a copy of my book on Sound.[3]

With regard to the conditions of publication I am aware of what my friend Mr. Huxley & other friends of mine receive from French & German publishers. But I should prefer leaving it to you to offer what you consider fair and just. It is in my power to obtain liberal terms from other German publishers, but I would rather trust to your gentlemanly feeling in proposing terms than make any proposition of my own.

I am writing a memoir of Faraday[4] which will, I trust, be of interest to all who take an interest in science throughout the world. I have already written to Prof. Helmholtz[5] on the subject and he is willing to undertake the translation.[6] In this case also I should prefer leaving the question of terms in your hands.

I will write to Prof. Wiedemann about the book on Sound: from a letter received yesterday from Prof. Helmholtz I infer that one or both of these distinguished men will be willing to undertake the translation.[7]

Yours very truly | John Tyndall

TU Braunschweig/Universitätsbibliothek/UABS V 1 T : 63

1. *Heinrich Rudolf Vieweg:* while the letter is addressed to 'Gentlemen' and is a reply to a letter signed 'Friedr[ich]. Vieweg & Sohn' (2684), we believe the recipient is Heinrich Rudolf Vieweg, who took over the family publishing firm, Friedrich Vieweg und Sohns, from his father Eduard Vieweg (1797–1869). See letter 2682, n. 1.
2. *Mr. Longman:* see letter 2617, n. 5.
3. *my book on Sound: Sound* (1867).
4. *memoir of Faraday: Faraday as a Discoverer* (see letter 2665, n. 3).
5. *written to Prof. Helmholtz:* Hermann Helmholtz; letter missing.
6. *translation:* see letter 2686, n. 4.
7. *translation:* see letter 2677, n. 5.

From John Peter Gassiot[1] 18 November 1867 2688

"CLAPHAM COMMON, SURREY, | "28th *November,* 1867.

"MY DEAR TYNDALL,

"Sir H. Davy[2] was accustomed to call on the late Mr. Pepys[3] in the Poultry on his way to the London Institution,[4] of which Pepys was one of the original managers; the latter told me that on one occasion, Sir H. Davy, showing him a letter said, 'Pepys, what am I *to do,* here is a letter from a young man named Faraday;[5] he has been attending my lectures and wants me to give him employment at the Royal Institution, *what can I do?' 'Do,'* replied Pepys, 'put him to wash bottles; if he is good for anything, he will do it directly; if he refuses, he is good for nothing.' 'No, no,' replied Davy; 'we must try him with something better than that.' The result was, that Davy engaged him to assist in the Laboratory[6] at *weekly* wages.

Davy held the joint office of Professor of Chemistry, and Director of the Laboratory; he ultimately gave up the former to the late Professor Brande,[7] but he insisted that Faraday should be appointed Director of the Laboratory, and, as Faraday told me, this enabled him on subsequent occasions to hold a definite position in the Institution, in which he was always supported by Davy. I believe he held that office to the last.

"Believe me, my dear Tyndall, yours truly, | "J. P. GASSIOT. | "Dr. Tyndall."

Roy. Inst. Proc., 5 (1869), pp. 200–1
Faraday as a Discoverer, pp. 4–5

1. *John Peter Gassiot:* John Peter Gassiot (1797–1877), an English electrician and wealthy wine merchant. He was one of the founding members of the London Electrical Society (1837), and pursued electrical research in the laboratory he established in his home (*ODNB*).
2. *H. Davy:* Humphry Davy.
3. *Mr. Pepys:* William Hasledine Pepys (1775–1856), an English instrument maker and natural philosopher, who was one of the founders of the London Institution (*ODNB*).
4. *London Institution:* an educational institution established in 1806 aimed at promoting science, literature, and the arts.
5. *Faraday:* Michael Faraday.
6. *Laboratory:* at the RI.
7. *Professor Brande:* William Thomas Brande (1788–1866), an English chemist who succeeded Davy as professor of chemistry at the RI in 1813. He resigned the professorship in 1852 in order to become a superintendent at the Royal Mint (*ODNB*).

From François-Napoléon-Marie Moigno 28 November 1867 2689

Mon cher M^r^. Tyndall

J'avais répondu à cœur ouvert à la question que vous m'avez faite relativement à la chaleur, et je m'attendais à une petite lettre de vous. Je n'ai rien reçu, et j'en suis fort triste. Peut être que ma franchise vous aura blessé et que ce que je vous disais vous aura paru tellement extraordinaire qu'une réponse devenait pour vous par trop embarrassante.

Vous avez du recevoir aussi la petite brochure Force en matière, et il me tarde de savoir si vous avez lu la dissertation de la fin.

Le Son avance assez rapidement mais je ne sais pas encore quand l'impression commencera, parce que M^r^. Gauthier-Villars veut faire graver les figures à Paris au lieu d'acheter les clichés anglais.

J'ai lu avec beaucoup de plaisir le programme de vos conférences d'hyver à Royal Institution. Les leçons sur Faraday seront surtout intéressantes; je voudrai bien les entendre.

Daignez, je vous en prie, m'honorer d'une courte réponse, ne fut ce que pour me dire que vous ne me gardez pas rancune.

J'ai l'honneur d'être dans les sentiments d'une affection respectueuse et dévouée | Votre humble serviteur et ami | L'abbé F. Moigno | 2 rue d'Erfurth | 28 novembre 1867.

My dear Mr Tyndall

I had sincerely responded to the question that you asked me about Heat,[1] and I was expecting a small letter from you. I did not receive anything, and I am very saddened. Perhaps my frankness would have hurt you, and what I told you may have appeared so unusual that responding to it became too embarrassing for you.

You should also have received the small brochure Force en matière,[2] and I cannot wait to know if you have read the essay at the end.

Le Son[3] progresses quickly enough, but I don't know when printing will begin, because M^r^. Gauthier-Villars wants to etch the figures in Paris instead of buying English pictures.

I enjoyed reading the schedule of your winter lectures at the Royal Institution.[4] The lectures on Faraday[5] will be especially interesting; I would like to hear them.

Please condescend to honour me with a short response, simply to tell me that you do not hold a grudge against me.

I have the honour of sending my most affectionate and devoted regards | Your humble servant and friend | Father F. Moigno | 2 Erfurth Street[6] | 28 November 1867.

RI MS JT/1/M/110

1. *I had sincerely answered . . . about Heat:* see letter 2664 in which Moigno discusses the issues following the publication of J. Tyndall, *La Chaleur considérée come un mode de mouvement,* trans. F. Moigno (Paris: Étienne Giraud, 1864).
2. *small brochure Force en matière:* see letter 2664, n. 7.
3. *Le Son:* see letter 2617, n. 2.
4. *winter lectures at the Royal Institution:* see letter 2671, n. 5.
5. *lectures on Faraday:* on 17 January 1868, Tyndall delivered the first of two Friday Evening Discourses on Faraday (see letter 2665, n. 3).
6. *2 Erfurth Street:* in Paris.

From Jean-Albert Gauthier-Villars 30 November 1867 2690

IMPRIMERIE ET LIBRAIRIE

Pour les Mathématiques, les Sciences et les Arts,

QUAI DES GRANDS-AUGUSTINS, 55.

Gauthier-Villars, successeur de *Mallet-Bachelier;*

IMPRIMEUR-LIBRAIRE

Du Bureau des Longitudes; de l'École impériale Polytechnique; de l'Observatoire Impérial de Paris; de l'École impériale Centrale des Arts et Manufactures. — Éditeur des Comptes rendus hebdomadaires des séances de l'Académie des Sciences; du Journal de Mathématiques, par M. Liouville; du Journal de l'École impériale Polytechnique; des Annales scientifiques de l'École Normale supérieure, par M. Pasteur; des Nouvelles Annales de Mathématiques, par MM. Gerono et Prouhet; des Annales de l'Observatoire impérial de Paris; du Bulletin de la Société française de Photographie; de l'Annuaire photographique, par M. Davanne.

Imprimerie et Librairie | *Pour les* Mathématiques, *les* Sciences *et les* Arts | Quai des Grands-Augustins, 55. | *Gauthier-Villars*, successeur de *Mallet-Bachelier*, | Imprimeur-libraire | Du Bureau des Longitudes; de l'École impériale-Polytechnique; de l'Observatoire Impérial de Paris; de l'École impériale Centrale des Arts et Manufactures.—Éditeur des Comptes Rendus Hebdomadaires des Séances de l'Académie des Sciences; du Journal de Mathématiques, par M. Liouville; du Journal de l'École Impériale Polytechnique; des Annales scientifiques de l'École Normale Supérieure, par M. Pasteur; des Nouvelles Annales de Mathématiques, par MM. Gerono et Prouhet; des Annales de l'Observatoire Impérial de Paris; du Bulletin de la Société Française de Photographie; de l'Annuaire photographique, par M. Davanne.

N° 4148 | Paris, le 30 novembre | 1867

Monsieur

Je n'ai pas reçu la lettre que vous m'annoncez m'avoir adressée au sujet de la réédition de votre ouvrage sur la chaleur.

Je pense que vous voudrez bien m'excuser de vous signaler une cause de perte de vos lettres. Elles arrivent, avec un affranchissement insuffisant. Quand on me les présente, je recevrais votre écritures, et je les accepte de la poste; mais si je suis absent, mes commis, suivant la règle de la maison, peuvent être exposés à les refuser. Si je me permets de vous parler de cet affranchissement, c'est que je crains d'autres refus de lettres, malgré mes recommandations pour votre correspondance.

J'ai fait commencer les dessins des figures destinées à l'édition française de votre ouvrage sur le son, mais il serait encore temps de supprimer les figures que vous ne jugez pas indispensables. Je vous serai excessivement obligé de me donner les numéros de celles que vous m'autorisez à ne pas faire graver. Quant aux figures relatives aux expériences de Koenig, vous les trouverez dans l'Acoustique de M. Jamin (2e fascicule du cours de physique, tome II) qui va paraitre dans une huitaine de jours.

Je comprends bien que vous ayez désiré faire paraître une édition allemande de votre ouvrage sur le son; mais j'ai appris personnellement avec un certain regret, cette décision; car je comptais placer un bon nombre d'exemplaires de l'édition française en Allemagne.

Quoiqu'il en soit, je continuerai à activer le plus possible notre travail; car il serait bien important que l'édition allemande ne paraît pas la première.

Avant de vous répondre au sujet de votre mémoire sur Faraday, j'ai absolument besoin de savoir quel nombre de pages il contient environ en supposant par exemple des pages semblables à celles du Son.—Vous conviendrait-il, s'il m'était difficile de faire paraître séparément ce mémoire, que M. Babinet, membre de l'Institut, en fût paraître la traduction dans le prochain volume de ses études et lectures sur les sciences d'observation. M. Babinet connait et traduit l'anglais parfaitement bien.

En terminant, je vous renouvelle ma demande de vouloir bien m'indiquer, aussitôt que vous le pourrez les figures que vous ne jugez pas indispensables.

Veuillez agréer, Monsieur, l'assurance de ma considération la plus distinguée.

Gauthier-Villars

Mr Tyndall, Prof[esso][r]., Royal Institution | Albemarle Street W | London

Printing and Bookstore | *For* Mathematics, Science *and* Arts | Quai des Grands-Augustins, 55. | *Gauthier-Villars*, successor of *Mallet-Bachelier*, | Printer-Seller | Of the Bureau of Longitude; of the Imperial Polytechnic School; the Imperial Observatory of Paris; the Central Imperial School of the Arts and Manufacturing.—Editor of the Comptes Rendus Hebdomadaires des Séances de l'Académie des Sciences; the Journal de Mathématiques by Mr. Liouville; the Journal de l'École Impériale Polytechnique; the Annales scientifiques de l'École Normale Supérieure by Mr. Pasteur; the Nouvelles Annales de Mathématiques by Mr. Gerono and Mr. Prouhet; the Annales de l'Observatoire Impérial de Paris; the Bulletin de la Société Française de Photographie; the Annuaire photographique by Mr. Davanne.

N° 4148 | Paris, November 30 1867

Sir,

I have not received the letter,[1] which you said that you sent me, about the republishing of your book on heat.[2]

I think that you will be kind enough to excuse me for reporting the reason why the letters were lost. They arrived with insufficient postage. When they arrive, I can receive your writings and accept them from the post office; but if I am absent, my clerks, following the rules of the house, may be inclined to refuse. The reason why I am telling you about this postage is because I fear that other letters may have been refused, despite my recommendations for your correspondence.

I had begun the drawing of the figures meant for the French edition of your book on sound,[3] but there is still time to remove the figures that you do not think are essential. I would be extremely grateful to you if you would give me the numbers of those that you do not permit me to engrave. As for the figures in Koenig's[4] experiments, you will find them in Mr Jamin's Acoustics (2nd specification of the physics course, volume II)[5] that will appear in a week.

I understand that you wanted to release a German edition of your book on sound;[6] but I learned personally about this decision with some regret; since I intended to put a good number of the French edition copies into Germany.

In any case, I will continue to speed up our work as much as possible; because it is important that the German edition not appear first.

Before giving you an answer about your paper on Faraday,[7] I absolutely need to know how many pages there are, in order to be able to compare them, for example, to those of the Sound.—Would it be alright with you, if it was difficult for me to publish separately this paper, that Mr. Babinet, member of the Institute, would publish the translation of it in the next volume of his Studies and Lectures on Observational Science.[8] Mr Babinet knows and translates English perfectly well.

Finally, I renew my request for you to tell me kindly as soon as possible which figures you do not deem necessary.

Please accept, Sir, the assurance of my very best regards.

Gauthier-Villars

Mr. Tyndall, Prof[esso]r., Royal Institution | Albemarle Street W[est]. | London

RI MS JT/1/V/17

---

1. *the letter:* letter missing.
2. *book on heat: Heat* (1863). The first French edition (1864) was published by Étienne Giraud, and the second French edition (1874) was published by Gauthier Villars.
3. *French edition of your book on sound:* see letter 2617, n. 2.

4. *Koenig's:* Karl Rudolph Koenig (1832–1901), a German physicist and instrument maker whose main subject of study was acoustics (*CDSB*).
5. *Jamin's Acoustics (2nd specification of the physics course, volume [II]):* J. Jamin, *Cours de physique l'École Polytechnique,* 2nd edn, 3 vols (Paris: Gauthier-Villars, 1868), vol. 2.
6. *German edition of your book on sound:* see letter 2677, n. 5.
7. *your paper on Faraday:* a reference to Tyndall's forthcoming RI discourses on Faraday (17 and 24 January 1868) later published as *Faraday as a Discoverer* (1868).
8. *translation of it in the next volume of his Studies and Lectures on Observational Science:* J. Babinet, *Études et lectures sur les sciences d'observation et leurs application pratiques,* 8 vols (Paris: Gauthier-Villars, 1858–68), vol. 8 (1868). Tyndall's essay did not appear in the volume.

From Gustav Wiedemann[1] 1 December 1867 2691

Carlsruhe *[d.]* 1ten Decbr 1867

Mein lieber Tyndall!

Ihren freundlichen Brief habe ich gestern Mittag erhalten und sogleich an Helmholtz geschrieben. Wir sind schon lange der Ansicht, daß Ihr Werk „on sound", welches wir in der Schweiz mit großer Befriedigung gelesen haben, in deutscher Bearbeitung erscheinen müßte, sind aber in Folge unserer fortgesetzten Abwesenheit von Hause und mehrfachen Logirbesuche bisher noch nicht dazu gekommen, ernstlich die Ausführung davon zu betreiben. Meiner Frau würde es sehr große Freude machen, sich an der Uebersetzung zu betheiligen nur könnte sie dieselbe jetzt nicht gut allein übernehmen, da sie schon Ende Februar für einige Zeit nach Berlin reisen will. Wir wollen also hoffen, daß Frau Helmholtz die Arbeit mit ihr theilen will und in diesem Fall alsbald beginnen. Von Vieweg erhielt ich noch gestern Abend einen Brief, worin er ebenfalls den Wunsch ausspricht, daß wir die Bearbeitung Ihres Buches übernehmen sollten.

Meine Frau wird Ihnen schon vor einigen Tagen geschrieben haben, wie eine große Freude Sie durch Ihr Buch in unser Haus gebracht haben. Die Uebersetzung fiel gerade in die Zeit, wo meine Frau aus dem heimischen Norden hierher übergesiedelt war und mir die ernste Beschäftigung so manche schwere Stunde des Heimwehs verscheuchte. Wir beide sind Ihnen wirklich zu vielem Dank verpflichtet, daß Sie uns so eine schwere Zeit wesentlich erleichtert und sogar zu einer recht erfreulichen umgestaltet haben. Ihr so liebenswürdiges Geschenk, die schöne Broche, hat meiner Frau noch eine ganz besondere Freude gemacht; sie ist sehr glücklich darüber, daß es ihr gelungen ist, im Verein mit Frau Helmholtz, Ihren strengen Anforderungen an die Übersetzung zu genügen, die doch zugleich die Klarheit und elegante Form des Originals wiedergeben sollte. Auch von mir nehmen Sie dafür den besten Dank.

Mit besten Empfehlungen meiner Frau

Ihr | stets ergebener | G. Wiedemann.

Karlsruhe[2] 1st Dec. 1867

My dear Tyndall!

I received your kind letter[3] midday yesterday and wrote immediately to Helmholtz. We have long been of the view that your work "On Sound",[4] which we read with great satisfaction while in Switzerland, must appear in a German edition,[5] but as a consequence of our continued absence from home and several houseguests, have still not got around to seriously advancing its execution until now. It would be a very great pleasure for my wife to assist in the translation, only she could not take it on well by herself just now, as she will be travelling to Berlin for some time at the end of February. We shall therefore be hoping that Mrs. Helmholtz[6] will share the work with her and in that case begin straight away. I received a letter from Vieweg[7] only yesterday evening in which he also expresses the wish that we should take on the adapting of your book.

My wife[8] has already written to you some days ago[9] how you have brought great pleasure to our household through your book. The translating occurred exactly at the time when my wife shifted here from her home in the north and the serious work drove away so many a difficult hour of homesickness for me. We both owe a true debt of gratitude to you that you have made such a difficult time considerably easier for us and even turned it into a quite enjoyable one. Your such kind present, the beautiful brooch,[10] has given my wife further quite special pleasure; she is very happy that in conjunction with Frau Helmholtz, she has managed to meet your strict requirements for the translation, which was to simultaneously reproduce the clarity and elegant form of the original. Please accept my best thanks for this as well.

With the best compliments of my wife

Your | always devoted | G. Wiedemann

RI MS JT/1/W/46–46TRANS

---

1. An English language summary of the German letter, probably written by LT, is included with the letter at the RI.
2. *Karlsruhe:* University of Karlsruhe.
3. *your kind letter:* letter missing.
4. *"On Sound": Sound* (1867).
5. *German edition:* see letter 2677, n. 5.
6. *Mrs. Helmholtz:* Anna Helmholtz.
7. *Vieweg:* Heinrich Rudolf Vieweg.

8. *My wife:* Clara Wiedemann.
9. *written to you some days ago:* letter missing.
10. *beautiful brooch:* Tyndall gave a small brooch of gold and pearl to Anna Helmholtz as thanks for her work translating *Heat* (1863) into German, and must have done the same for Clara Wiedemann (see letter 2677, n. 3).

To Frances Colvile[1] 2 December 1867 2692

2nd Dec. 1867

My dear Lady Colville

It is pleasant to be remembered and I should much like to go: but some weeks ago I began a memoir of Faraday[2] which occupies me day & night, and compelled me to register a vow in heaven not to dine out until its heart was broken. For several weeks I have lived the life of a hermit, and I must continue to do so for several weeks more.

I hope to see you on Wednesday evening[3] & still I fear you will find me dull; for first of all I have absolutely no time to think of the lecture, (which is the fulfillment of our old engagement) and secondly Spottiswoode has forbidden all heresy which might have added spice to the occasion.[4]

Yours ever | John Tyndall

BL Add MS 60634, f. 121

---

1. *Frances Colvile:* Frances Colvile (see letter 2663, n. 13).
2. *memoir of Faraday: Faraday as a Discoverer* (see letter 2665, n. 3).
3. *Wednesday evening:* Thomas Hirst wrote in his journal for 4 December 1867: 'Dined at Spottiswoode's with Sir James and Lady Colville Sir John and Lady Lubbock and Edgar Lubbock and Greg, Afterwards went to the hall of the Whittington Club to hear Tyndall's lecture to the printers employed by the Spottiswoodes. The lecture was almost the same as the one at Dundee and although inferior to the latter (in consequence of Tyndall's time being absorbed at present by the preparation of his forthcoming lectures on Faraday at the Royal Institution) gave great pleasure to the men and to Spottiswoode's guests who were numerous and amongst whom was Miss Thackeray' (Brock & MacLeod, *Hirst Journals,* p. 1823). For Tyndall's Dundee lecture, see 2627, n. 31. For his lectures on Faraday, see letter 2665, n. 3.
4. *forbidden all heresy which might have added spice to the occasion:* probably a reference to a passage from Tyndall's lecture at Dundee, which he likely left out of the version he gave for Spottiswoode's employees (see n. 3): 'Depend upon it, if a chemist, by bringing the proper materials together in a retort or crucible, could make a baby, he would do it. There is no law, moral or physical, forbidding him to do it' (see 'Matter and Force', in *Fragments of Science,* vol. 2, pp. 53–75, on p. 72).

From Emily Peel 2 December 1867 2693

Drayton Manor | Tamworth[1] | Dec[embe]r 2/67.

Dear Professor Tyndall,

Should your numerous occupations admit of your leaving London towards the middle of the present month, would you do Sir Robert[2] & me the pleasure—to spend a few days with us[3] here, coming to us on the 17th.—

I hope that some mutual friends of ours, Lord & Lady Tankerville,[4] will be of our party, & I need scarcely say how glad we shall be, if you also will join our little gathering.

I remain yours | truly | Emily Peel

RI MS JT/1/P/45

---

1. *Drayton Manor | Tamworth:* a Peel family estate in Drayton Bassett, Staffordshire, England.
2. *Sir Robert:* Robert Peel (1822–95), third baronet, a British politician who married Emily Peel (née Hay) in 1856. He served as Chief Secretary for Ireland between 1861 and 1865 (*ODNB*).
3. *to spend a few days with us:* Tyndall was not able to go (see letter 2694).
4. *Lord & Lady Tankerville:* Charles Bennet (1810–99), sixth Earl of Tankerville, and Olivia Bennet (née Montagu) (G. E. Cokayne and G. H. White (eds), *The Complete Peerage,* 2nd edn, 14 vols (London: St. Catherine Press, 1910–98), vol. 12, pt. 1 (1953), pp. 635–6).

To Emily Peel 4 December 1867 2694

London, | 4th December, 1867

Dear Lady Emily,

I hardly know anything in the world which would give me greater pleasure than the liberty to accept your invitation.[1] It would gladden me and brighten me all over. But I am at present a chained and doomed individual. The death of that great and good man Faraday[2] has thrown duties upon me[3] which occupy my night and day and which fetter me to London. I was thinking when your letter came to me how Faraday's heart helped his head in his investigations:—the intellect saw and guided, but the heart acted. Everything that cheers and brightens his labour adds to the motive power of the worker. And if you care to do so you can take credit to yourself for having cheered and brightened my work and thus added to my power. I shall often look across the gloom and tangle of my thoughts at you and Lady Tankerville,[4] wishing myself near you. Be pleased to convey to her my faithful remembrances.

Good-bye, most kind Lady Emily, | and believe me | Most truly yours, | John Tyndall.

RI MS JT/1/TYP/3/956
LT Typescript Only

---

1. *your invitation:* see letter 2693.
2. *death of that great and good man Faraday:* Michael Faraday died on 25 August 1867.
3. *thrown duties upon me:* see letter 2665, n. 3. Tyndall wrote to Frances Colvile two days before that the work of writing his memoir of Faraday—two lectures at the RI that would be published as *Faraday as a Discoverer*—'occupies me day & night' (letter 2692).
4. *Lady Tankerville:* Olivia Bennet.

From Sarah Faraday 4 December [1867][1] 2695

Barnsbury Villa,[2] 320, Liverpool Road

Dear Dr. Tyndall

You are a troublesome man to make me write another note after I had told you that it was always a pleasure to see you and I meant what I said—so I shall expect to enjoy myself next Tuesday 6 o'clock

Yours sincerely | S. Faraday

Dec. 4th. | If you forget our high number just write Barnsbury Villa | Liverpool Road, N.

RI MS JT/1/TYP/12/4184
LT Typescript Only

---

1. *[1867]:* the year is given based on Sarah Faraday having moved to Barnsbury Villa in 1867 (see n. 2).
2. *Barnsbury Villa:* Sarah Faraday and Jane Barnard moved from Hampton Court to Barnsbury Villa shortly after Michael Faraday's death (Frank A. J. L. James, personal communication).

From Edward Sabine 7 December 1867 2696

Dec. 7. 1867

Dear Tyndall;

Unfortunately I have not heard M^r^. Farrar's lecture,[1] so that I am obliged to return the Testimonial[2] without my signature, which I am sorry to do because you have asked me to sign it.[3] The fact is I have been so deeply engaged with the more immediately pressing business connected with my Presidency[4] that I have been obliged to postpone every thing that admitted of postponement until the Christmas Holydays.[5] The latter part of the Autumn I was in Dublin, deep in the affairs of the Great Melbourne Telescope;[6] since I returned to London, the reorganisation of the Meteorological office,[7] & my own anniversary address[8] have occupied every instant of my time & thoughts—and now, I must work hard, night & day, for some weeks at the Southern Magnetic Survey,[9] which I hope to have ready for the Royal Society about the middle or towards the end of January.[10] I begin to fear that the short remainder of my life[11] will not suffice to enable me to complete the work which I have actually in hand. But, whatever happens, I am always

most sincerely yours | Edward Sabine.

Professor Tyndall.

RI MS JT/1/S/24
RI MS JT/1/TYP/4/1321

---

1. *lecture:* Frederic Farrar gave a lecture at the RI on 8 February 1867, 'On Some Defects in Public School Education', and he had given a lecture at the annual meeting of the BAAS at Nottingham in 1866 on 'The Teaching of Science at the Public Schools' (see *Roy. Inst. Proc.,* 5 (1866–9), pp. 26–44; *Brit. Assoc. Rep.* 1866, pp. 72–3).
2. *Testimonial:* Farrar probably sought testimonials as he had applied for the position of headmaster of Haileybury in 1867, but he was unsuccessful (R. Farrar, *The Life of Frederic William Farrar* (New York: Thomas Y. Crowell & Company, 1904), p. 141).
3. *you have asked me to sign it:* if this was done through a letter, then the letter is missing.
4. *my Presidency:* Sabine served as President of the RS from 1861–71.
5. *Holydays:* alternative spelling of 'holidays'.
6. *Great Melbourne Telescope:* a 48-inch-diameter telescope that would be built in 1868 by Thomas Grubb (1800–78) in Ireland, and in 1869 installed at the Melbourne Observatory (R. Gillespie, *The Great Melbourne Telescope* (Melbourne: Museum of Victoria, 2011).
7. *reorganisation of the Meteorological office:* the Meteorological Department of the Board of Trade, established by Robert Fitzroy (1805–65) in 1854, was renamed to the Meteorological Office in February 1867. When Fitzroy died in April 1865, new leadership was needed. Sabine put forward Robert Henry Scott (1833–1916) to fill the role of Director left vacant by Fitzroy's death, whom the Meteorological Committee of the Royal Society approved. Scott accepted the position and began work on 7 February 1867 (M. Walker, *History of the Meteorological Office* (Cambridge: Cambridge University Press, 2012), pp. 71–86).
8. *anniversary address:* Sabine delivered the presidential address at the anniversary meeting of the RS on 30 November 1867 (see 'Anniversary Meeting', *Roy. Soc. Proc.,* 16 (1867–8), pp. 163–83).

9. *Southern Magnetic Survey:* Sabine's report of the magnetic survey of the 'South Polar Regions of the Globe', or Antarctic (1840–5), published as: E. Sabine, 'Contributions to terrestrial magnetism.—No. XI', *Phil. Trans.,* 158 (1868), pp. 371–416. See also, J. Cawood, 'The Magnetic Crusade: Science and Politics in Early Victorian Britain', *Isis,* 70 (1979), pp. 492–518.
10. *ready for the Royal Society . . . end of January:* Sabine's paper was read to the RS on 18 June 1868.
11. *short remainder of my life:* at this time, Sabine was 79 years old, yet he would have more years left than he perhaps anticipated. He died in 1883 at the age of 94.

From Charles André[1] and Georges Rayet[2] — 10 December 1867 — 2697

Observatoire Impérial | Paris, le 10 décembre 1867.

Monsieur,

Nous avons lu avec un vif intérêt les remarquables leçons que vous venez de publier sur l'Acoustique. Il nous semble qu'une traduction française de votre livre rendrait service à l'enseignement élémentaire de cette partie de la Physique.

Il n'y a pas en France de traité d'acoustique; s'adressant aux personnes qui n'ont aucune culture scientifique spéciale; un grand nombre d'entre elles trouveraient donc dans cette traduction, l'occasion de s'initier aux progrès les plus récents de la Science.

Nous avons présent à la mémoire, le succés de la traduction française de vos leçons sur la chaleur; aussi nous espérons que, rencontrant dans ce nouvel ouvrage, la même clarté d'exposition le public français le recevrait avec le même empressement?

Quant à nous, Monsieur, nous serions heureux de contribuer pour quelque chose à la popularité qui entoure votre nom de ce coté du détroit.

Nous avons donc l'honneur de vous demander l'autorisation de traduire en français vos leçons d'Acoustique.

Dans le cas où vous n'auriez point encore autorisé une traduction française, et où vous accueilleriez favorablement notre demande, nous vous prions de vouloir bien nous dire, à quelles conditions nous pouvons disposer du texte et des figures de l'ouvrage. Ces détails nous sont nécessaires pour nous entendre avec un éditeur.

L'un de nous, Mr Rayet, publie depuis plusieurs années des articles scientifiques dans l'annuaire de Mr Dehérain, mais comme cela ne constitue pas un titre littéraire suffisant, nous vous serions très reconnaissants de vouloir bien lire les épreuves de notre traduction.

Nous vous prions, Monsieur, d'agréer à l'expression de notre profond respect.

Ch. André G. Rayet

Ancien élève de l'école normale supérieure. | Agrégée des sciences physiques. | Astronome adjoint à l'Observatoire de Paris.

Imperial Observatory[3] | Paris, 10 December 1867.

Sir,

We have read with great interest the impressive lectures you have just published on Acoustics.[4] It seems to us that a French translation of your book[5] would be useful for to the elementary teaching of this part of Physics.

There is in France no acoustic treatise intended for people who do not have any specific scientific culture; a great number of them would thus find in this translation, the chance to become initiated into the most recent progress of Science.

We have *[in]* memory, the success of your French translation of your lectures on Heat;[6] therefore we hope to fulfil, in this new work, the same clarity of exposition the French public would receive it with the same eagerness

As for us, Sir, we would be happy to contribute something to the popularity that surrounds your name on this side of the channel.

We thus have the honour to ask your permission to translate into French your lectures on Acoustics.

In case you have not yet authorized a French translation, and you would favourably welcome our request, we ask you, please, to give us the conditions under which we may manage the text and the figures in the book. These details are necessary to reach an agreement with an editor.

One of us, Mr Rayet, has been publishing, for some years, scientific articles in the yearbook of Mr Dehérain,[7] but because it does not constitute a sufficient literary title, we would be very grateful if you were kind enough to read the proofs of our translation.

Please accept, Sir, the assurance of our deep respect.

Ch. André G. Rayet

Graduate of the Normal Superior School.[8] | Professor of physical science. | Assistant astronomer at the Observatory of Paris.

RI MS JT/1/R/7

---

1. *Charles André:* Charles Louis Francois André (1842–1912), a French astronomer educated at the École Normale Supérieure, Paris. Around 1866 he began working under Urbain Le Verrier at the Paris Observatory (*CDSB*).
2. *Georges Rayet:* Georges Antoine Pons Rayet (1839–1906), a French astronomer educated at the École Normale Supérieure, who in 1867 was an observer at the Paris observatory (*CDSB*).

3. *Imperial Observatory:* under Louis XIV, construction of the Paris Observatory was completed in 1675.
4. *impressive lectures you have just published on Acoustics: Sound* (1867).
5. *French translation of your book:* see letter 2617, n. 2.
6. *French translation of your lecture on Heat:* see letter 2680, n. 13.
7. *yearbook of Mr Dehérain:* Pierre-Paul Dehérain (1830–1902), a French agricultural chemist. From 1862–70 he edited the *Annuaire scientifique* (*DBF*).
8. *Normal Superior School:* the École Normale Supérieure in Paris, founded in 1794, served as a teacher-training school.

From Louisa Baring [9 December 1867][1] 2698

32, Grosvenor St., | W[2] | Monday Ev[enin]g.

Dear Professor Tyndall,

I am spending a few days here with my friend Miss Davenport Bromley,[3] and she begs me to say how very happy she will be if you will give her the pleasure of seeing you here on Thursday next 12th at ¼ to 8 to dinner. I shall be so glad if you can come—as it is an age since Mentone.[4]

Ever yours very truly | L. Ashburton

Elton Hall, Ashburton Collection

RI MS JT/TYP/1/120

1. *[9 December 1867]:* the date is given based on a note on the handwritten letter: '[env. Dec 10.67]' (a Tuesday). Since 12 December 1867 was a Thursday, this letter written on a 'Monday Ev[enin]g' is dated as 9 December.
2. *32, Grosvenor St., | W:* in London.
3. *Miss Davenport Bromley:* Caroline Davenport Bromley (*c.* 1820–91), sister of William Bromley-Davenport (1821–84), and close friend of Louisa Baring and Jane Welsh Carlyle (1801–66) ('Deaths', *Times,* 5 March 1891, p. 1).
4. *Mentone:* it had been almost a year since Tyndall and Thomas Carlyle went to Mentone to visit Baring. See letter 2535, n. 4 and n. 5.

To Joseph Dalton Hooker 11 December [1867] 2699

11th Dec

My dear Hooker.

I take it for granted that you know Farrar:[1] if you do you will sign this[2] and return it to me.

I thank your kind wife[3] fifty times over for her most seasonable present.[4]

Yours ever | John Tyndall

Kew JDH, vol. 4, 1262

---

1. *Farrar:* Frederic William Farrar.
2. *this:* a testimonial; see letter 2696, n. 2.
3. *your kind wife:* Frances Hooker.
4. *seasonable present:* not identified.

From Jean-Albert Gauthier-Villars 12 December 1867 2700

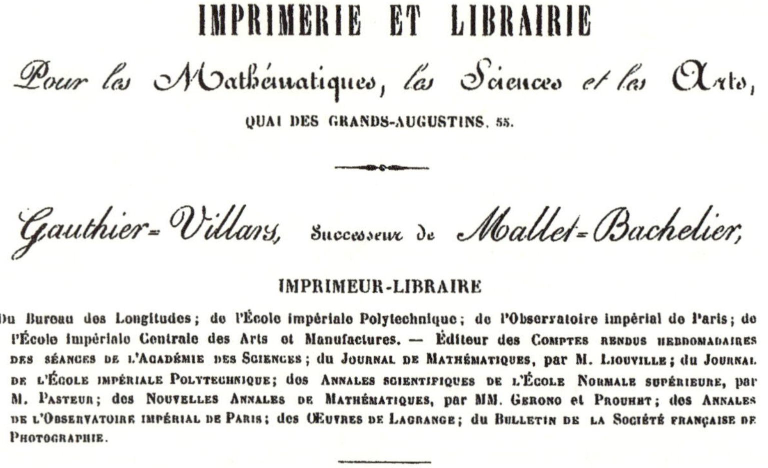

Imprimerie et Librairie | *Pour les* Mathématiques, *les* Sciences *et les* Arts | Quai des Grands-Augustins, 55. | *Gauthier-Villars,* successeur de *Mallet-Bachelier,* | Imprimeur-libraire | Du Bureau des Longitudes; de l'École impériale-Polytechnique; de l'Observatoire Impérial de Paris; de l'École impériale Centrale des Arts et Manufactures.—Éditeur des Comptes Rendus Hebdomadaires des Séances de l'Académie des Sciences; du Journal de Mathématiques, par M. Liouville; du Journal de l'École Impériale Polytechnique; des Annales scientifiques de l'École Normale Supérieure, par M. Pasteur; des Nouvelles Annales de Mathématiques, par MM. Gerono et Prouhet; des Annales de l'Observatoire Impérial de Paris; du Bulletin de la Société Française de Photographie.

N° 4180 | Paris, le 12 décembre 1867.

Monsieur,

Avant tout, je tiens à vous remercier de l'extrême bienveillance que vous voulez bien me témoigner. Il est certain que je n'avais fait aucune réserve au sujet de la traduction allemande de l'ouvrage sur le Son et que vous aviez toute liberté à cet égard. Mon tort est de ne vous avoir pas demandé vos instructions à cet égard.

Ceci posé, j'ajoute que si j'ai appris avec regret la prochaine apparition de cette traduction allemande, c'est que je comptais beaucoup, pour l'écoulement de l'édition française, sur le marché allemand qui avait pris un grand nombre d'exemplaires de l'édition française de la Chaleur.

Quoiqu'il en soit, j'accomplirai mes engagements, et je m'occupe activement de faire graver les figures sur le Son, en supprimant quelques-unes de celles que vous me laissez libre de ne pas conserver. Vous me dites très généreusement que vous désirez me couvrir des pertes que je pourrais subir. Je ne crois pas devoir accepter en entier cette proposition. Si vous le voulez bien, nous conviendrons seulement de ceci.

Si au bout de deux ans, après l'apparition de l'ouvrage, les frais ne sont pas soldés par la vente nette, vous prendrez à votre charge la moitié de la dépense qui restera non couverte. Ensuite, la vente se continuera, et je vous tiendrai compte de la moitié du prix net des exemplaires qui continueront à se vendre, de telle sorte que vous soyez successivement indemnisé et que vous arriviez à la fin de l'édition à avoir la moitié des bénéfices.

En résumé, je fais l'avance de tous les frais; nous partagerons par moitié la perte ou les bénéfices; mais en cas de perte, je n'ai rien à vous réclamer avant deux ans. Il est entendu comme je l'ai dit dans mes lettres précédentes, que suivant les habitudes de la librairie française, je prélèverai sur le prix fort 33 pour % et le 13e exemplaire, pour les remises que j'ai à faire moi-même à mes correspondants. Ainsi, le prix porté sur mes catalogues étant 6 francs, je ferai figurer dans nos comptes respectifs, l'ouvrage au prix de 4 francs.

Pour l'ouvrage sur la Chaleur, il y aura lieu, je pense, de faire une nouvelle édition française; mais je suis si encombré de travaux, que je désire ne l'entreprendre qu'après avoir imprimé l'ouvrage sur le Son.

J'arrive à votre Notice sur Faraday. Elle ne sera pas assez volumineuse pour former un volume entier de Mr Babinet, et je serai en conséquence obligé de réunir un autre mémoire de M. Babinet au vôtre, dans le même volume. Mais je pourrai faire tirer à part pour vous un certain nombre d'exemplaires de votre propre Notice, que vous aurez ainsi séparément et que vous pourrez distribuer.

J'attends votre réponse, avant de faire en ce qui me concerne, les démarches nécessaires auprès de M. Babinet.

J'ai eu l'honneur de vous adresser le fascicule de M. Jamin sur l'Acoustique, qui vient de paraître. Je pense que vous y trouverez plusieurs figures que vous pourrez introduire dans l'édition française de votre ouvrage sur le Son.

Veuillez agréer, Monsieur, l'assurance de ma considération la plus distinguée,

Gauthier Villars

P.S.—J'accepte bien volontiers votre proposition de faire exécuter à mes frais les clichés des figures 104 et 138. Je ne ferai pas graver ces figures et j'attendrai les clichés.

À Monsieur Tyndall, professeur, Royal Institution | Albemarle Street W. à Londres.

Printing and Bookstore | *For* Mathematics, Science *and* Arts | Quai des Grands-Augustins, 55. | *Gauthier-Villars*, successor of *Mallet-Bachelier*, | Printer-Seller | Of the Bureau of Longitude; of the Imperial Polytechnic School; the Imperial Observatory of Paris; the Central Imperial School of the Arts and Manufacturing.—Editor of the Comptes Rendus Hebdomadaires des Séances de l'Académie des Sciences; the Journal de Mathématiques by Mr. Liouville; the Journal de l'École Impériale Polytechnique; the Annales scientifiques de l'École Normale Supérieure by Mr. Pasteur; the Nouvelles Annales de Mathématiques by Mr. Gerono and Mr. Prouhet; the Annales de l'Observatoire Impérial de Paris; the Bulletin de la Société Française de Photographie; the Annuaire photographique by Mr. Davanne.

N°4180 | Paris, 12 December 1867.

Sir,

First of all I would like to thank you for the extreme kindness that you have shown me. It is certain that I had no restrictions regarding the German translation[1] of the book on Sound,[2] and that you had full freedom in that regard. It was my mistake not to ask you for your instructions on this matter.

This said, I will add that, if I learned with regret about the next appearance of this German translation, it is because I very much counted for the disposal of the French edition,[3] on the German market, which had taken a large number of copies of the French edition of Heat.[4]

Either way, I will fulfil my commitments, and I am actively engaged in engraving the figures on Sound, while removing some of the ones that you allowed me to discard. You were extremely generous in telling me that you would like to cover the losses that I may sustain. I do not think that I should accept this proposal in its entirety. If you wish it, we will agree only on this.

If by the end of two years, following the emergence of the book, the costs have not been balanced out by the net sales, you will take charge of half of the amount that has not yet been covered. Then, the selling will continue, and I will keep track of half of the net price of the copies that will continue to sell, so that you are in turn compensated and that you will end up with having half of the profits by the end of the edition.

In short, I am paying in advance for all the costs; we will share half the losses or the profits; but in case of a loss, I will have nothing to ask of you before two years have passed. It is understood, as I mentioned in previous letters,[5] that according to the procedures of the French bookstore, I will deduct from the price 33% and the 13th copy, for the discounts that I will have to make myself for my correspondents. <u>Thus, for the price that shows in my catalogues as 6 francs, I will note the book in our respective accounts at the price of 4 francs</u>.

For the book on Heat, there will be room, I think, to make a new French edition.[6] But I am so overwhelmed by work that I would like to take it up after the printing of the book on Sound.

Now I come to your note on Faraday.[7] It is not large enough to form an entire volume of Mr Babinet,[8] and therefore I would have to combine yours with another of Mr. Babinet's essays in the same volume. But I might be able to print for you several copies of only your own essay, which you will have separately and would be able to distribute.

I am waiting for your reply[9] before taking, as far as I am concerned, any necessary actions with Mr. Babinet.

I have had the honour of sending you Mr. Jamin's[10] instalment on Acoustics,[11] which had just appeared. I think that you will find there many figures that you could introduce into the French edition of your book on Sound.

Please accept, Sir, the assurance of my highest regards,

Gauthier Villars

P.S.—I gladly accept your proposal to produce the pictures of the figures 104 and 138 at my expense. I will not engrave those figures, and I will wait for the pictures.

To Mr. Tyndall, professor, Royal Institution | Albemarle Street W. in London.

RI MS JT/1/V/17

---

1. *German translation:* see letter 2677, n. 5.
2. *book on Sound: Sound* (1867).
3. *French edition:* see letter 2617, n. 2.
4. *French edition of Heat:* see letter 2680, n. 13.
5. *previous letters:* see letter 2690.
6. *book on Heat . . . new French edition:* the second French edition of *Heat* was published by Gauthier-Villars in 1874.
7. *your note on Faraday:* probably a reference to Tyndall's forthcoming RI discourses on Faraday or *Faraday as a Discoverer* (see letter 2665, n. 3).
8. *volume of Mr Babinet:* see letter 2690, n. 8.

9. *your reply:* letter missing.
10. *Mr. Jamin's:* Jules Jamin.
11. *instalment on Acoustics:* see letter 2690, n. 5.

To Frederic Farrar [13 December 1867][1] 2701

Friday

Dear M[r]. Farrar.

On seeing your note[2] yesterday I immediately telegraphed[3] to Kew.[4] Hooker[5] has sent me the testimonial[6] with his signature. Fearing I may be late I send it to you though it does not contain half the names that I intended to have attached to it.

Besides I find from a note of Griffith[7] to Hirst[8] that my crotchety friend Grove[9] has objected to the tone of this Testimonial.

—I can only say that I shall be prepared to sign any thing which you may prefer.

Yours ever | John Tyndall

Wellcome MS. 7777/100

---

1. *[13 December 1867]:* the date is given by mention of a testimonial (see n. 6) and relation to other letters that discuss it (2696 and 2699).
2. *your note:* letter missing.
3. *telegraphed:* see letter 2699.
4. *Kew:* see letter 2554, n. 2.
5. *Hooker:* Joseph Dalton Hooker.
6. *testimonial:* see letter 2696, n. 2.
7. *Griffith:* George Griffith.
8. *Hirst:* Thomas Hirst.
9. *Grove:* William Grove.

To Heinrich Rudolf Vieweg 14 December 1867 2702

Royal Institution of Great Britain | 14th. Dec. 1867

My dear Sir,

My agreement with M[r]. Longman[1] is this:—he purchased the work[2] edition by edition; and when the book is printed & its price fixed, the expenses are deducted and the profits divided into three equal parts. Two of these are paid to me on the day of publication.

I must be mainly guided by what you consider right and honourable: and therefore if you consider £30 to be so I do not refuse to accept it.

The electris[3] if furnished by M^{r}. Longman would come to about 30 pounds: some of the figures I think might be omitted & the expense thus reduced.

Yours very truly | John Tyndall

TU Braunschweig/Universitätsbibliothek/UABS V 1 T : 63

---

1. *Mr. Longman:* either Thomas Longman (1804–79) or William Longman (1813–77), brothers who ran the publishing firm Longmans, Green, and Co., publisher of Tyndall's books.
2. *the work:* probably Tyndall's *Heat,* which had editions in 1863 and 1865, and a third in the works (1868).
3. *electris:* probably German for *electrotype,* 'a copy or duplicate of an object, *esp.* a printing plate, made by a process involving the electrolytic deposition of copper or other metal; a relief printing plate made in this way' (*OED*).

From William Frederick Pollock 15 December 1867 2703

59 Montagu Square[1] | 15th Dec. 1867

My dear Tyndall

The lecture[2] which you have given me the great pleasure of reading, is admirable—and I must congratulate you upon so much eloquence—& a most successful condensation without loss of life and interest. In looking a second time over the sheets I have almost regretted that I ventured to pencil a few first impressions in the margin—but I allow them to remain—& most of them refer to small matters. I have also corrected one or two "literals" which had escaped your pen—and altered the spelling of Chromotrope—which from its Greek derivation should be Chromatrope[3]—& I think was so spelt when the word came out. Having now seen it, I can now well understand the trouble & anxiety the thing has given you—but will be amply repaid to you in the end—and I will only add one suggestion, that at the end of the lecture—if what I have represents one lecture—you should return to the man for a few sentences, & so, as it were, come home again, to the point from which you start—but no doubt you intended to do so. Juliet[4] is writing separately.[5]

Y[ou]^{rs}. ever | W. F. Pollock

RI MS JT/1/P/237
RI MS JT/1/TYP/6/2092

1. *59 Montagu Square:* the Pollocks' residence in Marylebone, London.
2. *lecture:* possibly a draft of the second part of Tyndall's RI lecture 'On Faraday as a Discoverer' that he would deliver on 24 January 1868 (see letter 2665, n. 3). Tyndall suggested Faraday was responsible for inventing Chromatrope on p. 206.
3. *Chromatrope:* a magic-lantern slide consisting of two superposed circular glasses, brilliantly coloured, one of which is made to rotate in front of the other (*OED*).
4. *Juliet:* Juliet Pollock.
5. *writing separately:* letter missing.

To George Henslow[1] 17 December 1867 2704

Royal Institution of Great Britain | 17th. Dec. 1867

My dear Sir,

I think I did not attend more than one of your lectures at the Royal Institution[2] and that lecture impressed me as being extremely clear & instructive.

Yours very sinc[er]ely | John Tyndall

The Rev[eren]d. Geo[rge]. Henslow

Private Collection #3

1. *George Henslow:* George Henslow (1835–1925), an English botanist and clergyman. He was the son of John Stevens Henslow (1796–1861) and brother of Frances Hooker (*ODNB*).
2. *your lectures at the Royal Institution:* Henslow gave 'Six Lectures on the Practical Study of Botany' at the RI from 5 March to 9 April 1867 (*Roy. Inst. Proc.,* 5 (1866–9), p. 4).

To William Frederick Pollock [18 December 1867][1] 2705

Wednesday

My dear Pollock

Much to my regret I find that I shall not be able to go to Zion College[2] tomorrow. This is a great disappointment to me for Farrar ought to be supported.[3] Huxley[4] will no doubt be here & you & he can cab it together.

Yours ever | John Tyndall

RI MS JT/1/T/1253
RI MS JT/1/TYP/6/2210

1. *[18 December 1867]:* the date, a Wednesday, is given based on mention of a lecture the next day (see n. 3).

2. *Zion College:* Sion College, a religious institution in London, founded in 1630.
3. *Farrar ought to be supported:* Fredric Farrar gave a lecture at Sion College on 19 December 1867 (see 'Sion College Lectures', *Standard,* 27 November 1867, p. 5). The lecture was published the following year as: F. W. Farrar, 'The Attitude of the Clergy towards Science', *Contemporary Review,* 9 (1868), pp. 600–20.
4. *Huxley:* Thomas Henry Huxley.

From François-Napoléon-Marie Moigno 18 December 1867 2706

Mon cher M[r]. Tyndall

Je comprends que vous n'ayez pas pu croire a tout ce que je vous disais dans mon avant dernier lettre. C'est cependant la verité toute entière. C'est moi qui me suis payé les mille francs de la traduction, et M[r]. Giraud ne ma pas donné un centime sur les produits de la vente; quoique les deux mille exemplaires de la première édition sont complètement écoulis.

J'ai prié M[r]. Gauthier-Villars qui a été en relation avec M[r]. Giraud, à qui M[r]. Giraud a avoué qu'il n'était nullement propriétaire de la chaleur, de réclamer ce qui vous est du, et par conséquent le compte de cette opération de librairie. Il m'a repondu qu'il le ferait volontiers, mais qu'il fallait que vous lui donnez une procuration. Envoyez la lui sur le champ, et il agira; c'est d'autant plus parce que, comme je vous l'ai dit M[r]. Giraud est en faillite.

La traduction du Son m'occupe tout entier en ce moment; c'est un rude travail qui exige une patience énorme. Cinq leçons seront prêtes avant le premier janvier. M. Gauthier-Villars fait faire les bons; je trouve que c'est bien long; et je ne comprends pas bien pourquoi il veut introduire d'autres figures de M. Jamin ou de Mr. Koenig.

Je suis prêt à commencer la traduction de vos leçons sur Faraday.

Je vous en prie, envoyez, courrier par courrier votre procuration à Monsieur Gauthier-Villars pour qu'il puisse agir contre M. Giraud.

Votre humble et dévoué serviteur | L'abbé F. Moigno | 2 rue d'Erfurt | 18 Décembre 1867

PS: je suis désolé de vous avoir effrayé de la métaphysique de mon appendice. Vos expériences Biot, Savart, Fresnel sont charmantes. Je tacherai de les répéter en public. En attendant, elles sont dans les mondes que vous ne recevez plus, je crois, et que je suis prêt à vous envoyer gratis.

My dear Mr Tyndall

I understand that you could not believe everything I told you in my penultimate letter. It is, however, the whole truth. It was I who paid the thousand francs for the translation,[1] and M^r^. Giraud[2] did not give me a penny on the proceeds of the sale; although the two thousand copies of the first edition are completely sold.

I have asked M^r^. Gauthier-Villars, who has been in contact with M^r^. Giraud, to whom M^r^. Giraud has confessed that he was not the owner of la chaleur, to reclaim what is yours, and therefore the account of this bookstore operation. He answered that he would do it gladly, but you had to give him a power of attorney. Send it to him immediately, and he will act; this is all the more because, as I told you, M^r^. Giraud is bankrupt.

The translation of Sound occupies me entirely at this moment; it's a difficult job that requires an enormous patience. Five lessons will be ready before the first of January. M. Gauthier-Villars is having the proofs made; I think it is really long; and I do not really understand why he wants to introduce other figures from M. Jamin[3] or from M. Koenig.[4]

I am ready to start the translation of your lessons on Faraday.

Please, send, one after the other your proxy to Mister Gauthier-Villars so that he can act against M. Giraud.

Humbly and devotedly yours, | Father Moigno | 2 rue d'Erfurt | 18 December 1867

PS: I am sorry that I frightened you with the metaphysics from my appendix. Your experiments[5] of Biot,[6] Savart,[7] Fresnel[8] are charming. I will try to repeat them in public. Meanwhile, they are in Les Mondes[9] which you no longer receive, I believe, and which I am ready to send to you[10] free of charge.

RI MS JT/1/M/111

---

1. *translation:* J. Tyndall, *La Chaleur considérée come un mode de mouvement,* trans. F. Moigno (Paris: Étienne Giraud, 1864).
2. *M. Giraud:* Étienne Giraud.
3. *M. Jamin:* Jules Jamin.
4. *M. Koenig:* Karl Koenig (see letter 2690, n. 4).
5. *Your experiments:* probably a reference to Tyndall's Friday Evening Discourse at the RI on 21 June 1867, 'On Some Experiments of Faraday, Biot and Savart' (see letter 2603, n. 4).
6. *Biot:* Jean-Baptiste Biot (1774–1862), a French physicist and mathematician who made significant contributions to the study of magnetism and the polarization of light. For the latter he was awarded the RS's Rumford medal in 1840 (*CDSB*).
7. *Savart:* Félix Savart (1791–1841), a French mathematician and physicist known for developing with Biot (see n. 6) the Biot-Savart law of electromagnetism which describes the magnetic field produced by electric currents. He also made numerous experiments on acoustics and vibration (*CDSB*).
8. *Fresnel:* while it appears that Moigno wrote 'Fresnel" instead of 'Faraday' by accident, Augustin-Jean Fresnel (1788–1827) was a French civil engineer and natural philosopher

whose work helped to establish the wave theory of light. He invented the stepped lens which bears his name for use in lighthouses (*CDSB*).

9. *Les Mondes:* after leaving the journal *Cosmos,* which he had edited since 1852, Moigno founded the popular science journal *Les Mondes: revue hebdomadaire des sciences* in 1863.
10. *to send to you:* 'Répétition par projection de quelques experiences de Faraday, Biot et Savart, par M. Tyndall', *Les Mondes,* 15 (1867), pp. 458–61.

To Elizabeth Dawson Steuart 19 December [1867][1] 2707

Royal Institution. | 19th. Dec.

My dear Mrs Steuart,

It seems a long time since I heard from you—I hope you are well and wish you a happy Christmas and many of them.

I have been working terribly hard: and now the Christmas lectures[2] are descending on me. But my health bears up well: Still I must have a day in the country before the lectures begin on the 26th.

This cold weather has put me in mind of asking you to let John Mcassey[3] have that bedding. It had escaped my memory.

I have had very pleasant invitations into the country, but have been unable to avail myself of them. Lady Emily Peel wrote me[4] an exceedingly kind letter wishing me to go down to Drayton Manor.[5] Lord Stanhope[6] wanted me to spend Christmas at Chevening,[7] but I am fettered here.

I am about changing my place of abode. Hitherto I have lived in St. John's Wood[8] but now, as Faraday's successor I shall have to come to the Royal Institution.[9]

What diabolical fellows those Fenians[10] are:—They are doing great injury to Ireland.

Yours very truly | John Tyndall.

Remember me to Trayer[11] when you see him. Emma[12] is recovered.

RI MS JT/1/TYP/10/3365
LT Typescript Only

---

1. *[1867]:* the year is given based on Tyndall having given the Christmas lectures at the RI that year (see n. 2) and mention of the Fenian Rising of 1867 (see n. 10).
2. *Christmas lectures:* letter 2671, n. 5.
3. *John Mcassey:* possibly John McAssey, Tyndall's mother Sarah's older brother.
4. *Lady Emily Peel wrote me:* see letter 2693.
5. *Drayton Manor:* see letter 2693, n. 1.
6. *Lord Stanhope:* Philip Henry Stanhope (1805–75), fifth Earl Stanhope, an English politician and historian (*ODNB*).

7. *Chevening:* Chevening House, the country estate of the Earls Stanhope, in the village of Chevening in Kent.
8. *St. John's Wood:* since June 1861 Tyndall lived with Hirst at 14 Waverley Place, St. John's Wood, London and moved out on 25 January 1868 to occupy the rooms at the RI that Faraday had lived in. Hirst noted in his journal for 25 January: 'Tyndall went to High Elms [home of John Lubbock] and left 14 Waverly Place for good. When I returned at night the house felt dismal enough. We have lived together here for 7 or 8 years, happily. At any moment we could see each other; henceforth it will necessarily not be so and I regret it deeply. Dear John my love and respect for him has gone on deepening from the day I first knew him—23 years ago' (Brock & MacLeod, *Hirst Journals*, pp. 1825–6).
9. *I shall have to come to the Royal Institution:* Tyndall would now occupy the rooms at the RI that Faraday had lived in. In letter 2770, Tyndall shared with Steuart that he had moved into the rooms, and that 'This I suppose will be my residence to the end of the Chapter'.
10. *Fenians:* the Fenians aimed to establish an independent Irish Republic by force of arms. There were unsuccessful uprisings in early 1867. On 13 December 1867 Fenians blew up a wheelbarrow laden with explosives against a wall of Clerkenwell Prison in London, to attempt (unsuccessfully) to release Ricard O'Sullivan Burke. The explosion killed several people across the street (N. Whelehan, *The Dynamiters: Irish Nationalism and Political Violence in the Wider World, 1867–1900* (Cambridge: Cambridge University Press, 2012), pp. 71–2).
11. *Trayer:* James John Trayer.
12. *Emma:* Emma Tyndall.

From Ughtred James Kay-Shuttleworth[1] 19 December 1867 2708

38 Gloucester Square. W.[2] | Dec. 19. 1867.

My dear Sir,

At length, after all sorts of unforeseen delays, I am able to ask you to accept a copy of my small book,[3] and—with it—my very cordial thanks for the assistance I have derived from your "Heat as a Mode of Motion"[4] as well as from such of your lectures as I have had the advantage of attending.

Though I have availed myself very freely of your permission to use passages of your book, I hope I have nowhere excluded proper bounds or failed to acknowledge my obligation.

I know that it would be too much to ask you to wade through the whole of a mere student's manual, but I hope you may be able to spare a few minutes some day in order to run your eye over the 2nd & 3rd Chapters, which are chiefly on Heat, and in great measure based on your book. If you are able to comply with this comparatively moderate request, I have no doubt you will find plenty to criticise, and you would do me a great favour if you would give me any suggestions or corrections for my own benefit & that of a possible second edition.

Pardon my thus troubling you and Believe me | Yours faithfully | U. J. Kay Shuttleworth Professor Tyndall L.L.D.[5] F.R.S.

RI MS JT/1/S/73

---

1. *Ughtred James Kay-Shuttleworth:* Ughtred James Kay-Shuttleworth (1844–1939), an English landowner and politician. He was the son of James Phillips Kay-Shuttleworth (1804–77).
2. *38 Gloucester Square. W.:* in London.
3. *my small book:* U. J. Kay-Shuttleworth, *First Principles of Modern Chemistry* (London: John Churchill & Sons, 1868).
4. *"Heat as a Mode of Motion": Heat* (1863).
5. *L.L.D.:* a Doctor of Law degree had been conferred upon Tyndall by the University of Cambridge on 30 May 1865 (see letter 2274, *Tyndall Correspondence,* vol. 9).

From Gustav Wiedemann[1] 21 December 1867 2709

Carlsruhe *[d.]* 21. Decbr 1867

Mein lieber Tyndall!

Helmholtz hat meinen gestrigen Brief umgehends beantwortet und ich freue mich, Ihnen nun mittheilen zu können, daß Frau Helmholtz gern bereit ist, gemeinschaftlich mit meiner Frau die Übersetzung Ihres Werkes on sound zu bearbeiten. Hoffentlich wird dieselbe recht gut ausfallen, so daß wir nach dem jetzt erschienenen Werk auch mit ihr Ehre einlegen können. Daß wir Männer das unsrige dazu nach Kräften beitragen werden, versteht sich von selbst. An Vieweg schreibe ich sogleich, daß die Arbeit alsbald begonnen werden kann.

Mit besten Empfehlungen meiner Frau

Ihr | treu ergebener | G. Wiedemann

Karlsruhe,[2] 21st Dec. 1867

My dear Tyndall!

Helmholtz[3] has answered my letter from yesterday by return mail and I am pleased to be able to inform you now that Mrs. Helmholtz,[4] is happy and willing to undertake, together with my wife,[5] the translation of your work of sound.[6] Hopefully it will turn out really well, so that we can make a good impression with it too, following on from the work that has just now appeared.[7] That we men shall contribute our bit to it to the best of our ability, goes without saying. I shall write to Vieweg[8] immediately that work can be begun straight away.

With the best compliments of my wife

Your | loyally devoted | G. Wiedemann

RI MS JT/1/W/47–47TRANS

1. An English language summary of the German letter, probably written by LT, is included with the letter at the RI.
2. *Karlsruhe:* University of Karlsruhe.
3. *Helmholtz:* Herman Helmholtz.
4. *Mrs. Helmholtz:* Anna Helmholtz.
5. *my wife:* Clara Wiedemann.
6. *translation your work on sound:* see letter 2677, n. 5.
7. *work that has just now appeared: Sound* (1867).
8. *Vieweg:* Heinrich Rudolf Vieweg.

To Hermann Helmholtz 24 December [1867][1] 2710

Royal Institution of Great Britain | 24th. Dec.

My dear Friend.

To expedite matters I send by this day's post a portion of the Memoir on Faraday:[2] It is not finally corrected, but the alterations if any will be very small.

Many happy returns of the season to you and M[rs]. Helmholtz.[3]

Give my greetings to Bunsen[4] and Kirchhoff.[5]

Yours ever | John Tyndall

I hope Bunsen will not be angry with me for the manner in which I have spoken of the Contact Theory.[6]

BBAW, NL Helmholtz, Nr. 477, BI_9
RI MS JT/1/T/507

1. *[1867]:* the year is given by reference to Tyndall's 'Memoir on Faraday' (see n. 2).
2. *Memoir on Faraday:* see letter 2665, n. 3.
3. *Mrs Helmholtz:* Anna Helmholtz.
4. *Bunsen:* Robert Bunsen.
5. *Kirchhoff:* Gustav Robert Kirchhoff (1824–87), a German physicist who made important contributions to the study of electricity and spectroscopy (*CDSB*).
6. *spoken of the Contact Theory:* for Tyndall's comments on contact theory see *Faraday as a*

Discoverer, pp. 59–66. For an overview of contact theory, see S. Hong, 'Controversy over Voltaic Contact Phenomena, 1862–1900', *Archive for History of Exact Sciences*, 47 (1994), pp. 233–89.

To Heinrich Rudolf Vieweg 24 December 1867 2711

24th Dec. 1867

Dear Sir,

I send you by this day's post a portion; by far the greater portion—of my memoir on Faraday.[1] I have also sent a copy of it to Prof. Helmholtz.[2]

The form in which it is here printed is that of our own Proceedings; but the memoir as published by Mr. Longman[3] will be in larger type, and in smaller form. If you have seen a little book of mine on "Radiation"[4] you have seen the size in which the memoir as published by Mr. Longman will appear. The number of words in a page is 250.

I think the letters which here produced in smaller type might appear in the same type as the general letter-press.

I am dear Sir | Yours very truly | John Tyndall

TU Braunschweig/Universitätsbibliothek/UABS V 1 T : 63

1. *memoir on Faraday:* see letter 2665, n. 3.
2. *sent a copy of it to Prof. Helmholtz:* see letter 2710.
3. *memoir as published by Mr. Longman: Faraday as a Discoverer.* For 'Mr. Longman', see letter 2617, n. 5.
4. *little book of mine on "Radiation":* J. Tyndall, *On Radiation* (Longman, Green, Longman, Roberts & Green, 1865).

From Rudolf Clausius[1] 26 December 1867 2712

Würzburg, 26/12 67.

Lieber Tyndall,

Dass ich solange Zeit habe vergehen laßen, ohne zu schreiben, hat seinen Grund nur darin, daß ich hier in der ersten Zeit meiner Thätigkeit sehr beschäftigt bin, indem das physikalische Cabinet in grosser Unordnung war, und ich daher alle Apparate, die ich brauchen will, erst in Stand setzen muß. Im Uebrigen gefällt es mir hier ganz gut, und ich hoffe daß wir uns, wenn erst die Unbequemlichkeiten, die mit einer Uebersiedelung nothwendig verbunden sind, überwunden sein werden, hier ganz wohl fühlen werden.

Dein freundliches Weihnachtsgeschenk für Johnny ist gestern richtig angekommen, aber es liegt noch auf dem Zollamte, von wo ich es erst nach dem Feste bekommen kann. Ich danke Dir in seinem, so wie in meiner Frau und meinem Namen herzlich für die Freundlichkeit, mit der Du Dich seiner erinnert hast, und bedaure nur die Unbequemlichkeit, die Du dadurch gehabt hast.

Deine Schrift über Faraday habe ich mit großem Interesse gelesen. Du bist in allem, was Faraday und seine Arbeiten anbetrifft, so vollkommen zu Hause, daß mein Urtheil dabei nur von sehr untergeordneter Bedeutung sein kann. Auch stimme ich überall mit Dir überein, und habe nur ein Paar unwesentliche Bemerkungen zu machen.

Was Faraday's Berechnung der Electricitätsmenge, die zur Zersetzung von 1 Gran Wasser nöthig ist, anbetrifft, so glaube ich daß sie im Wesentlichen richtig ist, da sie mit Berechnungen von Buff und von Wilhelm Weber und Kohlrausch im Grossen und Ganzen übereinstimmt. Buff berechnet die Electricitätsmenge, die nöthig ist um 1^{mgr} Wasserstoff aus 9^{mgr} Wasser auszuscheiden gleich 45480 Ladungen seiner Leidener Flaschen von 480^{mm} Höhe und 160^{mm} Durchmeßer bis zu 100^{mm} Schlagweite, und Weber und Kohlrausch berechnen, daß diese Electricitätsmenge (welche 1^{mgr} Wasserstoff aus 9^{mgr} Wasser ausscheiden kann) so groß ist, daß sie, wenn sie in einer Wolke von 1000^{m} Höhe befindlich wäre, auf eine gleich große in der Erde befindliche Menge entgegengesetzter Electricitaet eine Anziehungskraft von 2268000^{kil} ausüben würde. (Electrolytische Maaßbestimmungen von Kohlrausch und Weber v. J. 1856 S. 262)

Auf S. 13 sagst Du, bei der Zersetzung von schwefelsaurem Natron werde Schwefelsäure an der einen und Alkali an der anderen Polplatte frei. Obwohl dieses, wenn es richtig verstanden wird, ganz richtig ist, so würde ich es doch anders ausdrücken, weil es leicht missverstanden werden könnte. Man könnte nämlich hieraus schliessen, die electrolytische Zersetzung finde so statt:

$NaO|SO_3$

während sie doch so stattfindet:

$Na|OSO_3$

und erst durch secundäre chemische Wirkung das Na in HO zersetzt so daß gebildet wird NaO und H.

Auf S. 19 u. 22 steht mehrmals <u>dialectric</u>. Das scheint mir ein Druckfehler zu sein indem es, soviel ich weiß, <u>dielectric</u> heisst.

Auf S. 23 unten, wo Du von der Drehung der Polarisationsebene durch magnetische oder electrische Wirkung sprichst, sagst Du: Not so, however, when this same oil of turpentine is subjected to the action of an electric current. Hier würde ich lieber von dem schweren Glase sprechen wenn es der Wirkung des Magnets oder eines herumgehenden Stromes ausgesetzt wird, weil beim Terpentinoel das Verhalten weniger einfach ist, indem die eigene drehende Wirkung und diejenige, welche durch den Magnetismus oder den Strom hervorgerufen wird, gleichzeitig in Betracht kommen.

Im Uebrigen wüsste ich nichts zu bemerken. Wenn Du beim Erscheinen dieser Schrift ein Exemplar übrig hast, wirst Du mich sehr durch Zusendung desselben erfreuen, da ich sie sowohl in Bezug auf Faraday's Persönlichkeit als auch in wissentschaftlicher Beziehung sehr interessant finde.

Mit freundlichen Grüßen und herzlichen Glückwünschen zum neuen Jahre von meiner Frau und mir für Dich und Hirst.

Dein | Clausius

Meine Frau verlangt, daß ich noch hinzufügen soll, daß der kleine Johnny sagt, ich solle den großen Johnny, wie er Dich nennt, grüßen und ihm danken, und einen Kuß schicken.

Würzburg.[2] 26/12 67.

Dear Tyndall,

That I have let so much time pass without writing is only due to the fact that I am very busy here in the initial phase of my job,[3] in that the physical cabinet was in great disarray and I therefore have to first fix all the apparatuses that I want to use. Other than that, I like it here quite a lot, and I hope that once the inconveniences that are necessarily connected with a shift have been overcome, we shall feel quite well here.

Your kind Christmas present for Johnny[4] arrived safely yesterday, but it is still in the customs office, from where I shall only be able to get it after the holidays. I thank you sincerely in his, as well as my wife's,[5] and my name for the kindness with which you have remembered him, and regret only the inconvenience that you have had because of it.

I have read your paper on Faraday[6] with great interest. You are so completely at home in everything that concerns Faraday and his works that my judgment in this connection can only be of very subordinate importance. I also concur in all points with you, and only have a couple of unimportant remarks to make.

As far as Faraday's observations of the amount of electricity that is necessary to decompose 1 grain[7] of water is concerned, I think that it is essentially correct, as it concurs by and large with calculations by Buff,[8] and by Wilhelm Weber[9] and Kohlrausch.[10] Buff calculates the amount of electricity that is necessary to separate 1^{mgr} of hydrogen from 9^{mgr} water as equal to 45480 charges of his Leiden jar of 480^{mm} height and 160^{mm} diameter up to 100^{mm} spark distance, and Weber and Kohlrausch calculate this amount of electricity (which can separate 1^{mgr} hydrogen from 9^{mgr} water) is so large that if it were in a cloud of 1000^{m} height, it would exert an attractive force of 2268000^{kil} on an equally large amount of opposite electricty in the earth. (Electrolytische Maaßbestimmungen by Kohlrausch and Weber from 1856, p. 262[11])

On p. 13 you say, with the decomposition of sodium sulphate, sulphuric acid is set free on one polarisation plate and alkali on the other. Although this, if it is properly understood, is quite correct, I would express it differently all the same, because it could easily be misunderstood. One could deduce from it, for instance, that the electrolytic decomposition occurs thus:

$NaO|SO_3$

while it really occurs thus:

$Na|OSO_3$

and only by a secondary chemical action does the Na decompose into HO so that NaO and H is produced.

On p. 19 and 22 it says several times dialectric. This seems to me to be a printing error, in that, as far as I know, it means dielectric.

On p. 23 at the bottom, where you speak of the rotation of the polarisation plane by magnetic or electric action, you say "Not so, however, when this same oil of turpentine is subjected to the action of an electric current." Here, I would rather speak of the heavy glass, if it is subject to the action of the magnet or a circulating electric current, because with the oil of turpentine, its behavior is less simple, in that its own rotating action and that which is caused by the magnetism or the current come into consideration at the same time.

Other than that, I would not know of anything else on which to remark. If, when this paper appears, you have a copy left over, you would please me very much by sending it to me, as I find it very interesting both regarding Faraday's personality as well as in a scientific connection.

With kind regards and warm best wishes for the New Year from my wife and myself for you and Hirst.[12]

Your | Clausius

My wife insists I should add that little Johnny says I should give my regards to big Johnny—as he calls you—and thank him, and send a kiss.

RI MS JT/1/C/27

1. An English language summary of the German letter, probably written by LT, is included with the letter at the RI.
2. *Würzburg:* Clausius held a professorship at the University of Würzburg.
3. *initial phase of my job:* earlier in 1867 Clausius had accepted a professorship at the University of Würzburg.
4. *Johnny:* Rudolf John Clausius.
5. *my wife's:* Adelheid Clausius.
6. *your essay on Faraday:* see letter 2665, n. 3.

7. *grain:* in the German, Clausius used 'Gran', for which our German translator noted should have been 'Gram' (meaning 'grain').
8. *Buff:* Johann Heinrich Buff (1805–78), a German physicist and chemist. He was a student of Justus Liebig and had worked with Robert Bunsen at Kassel before obtaining a professorship at the University of Giessen in 1838 (*NDB*; J. R. Partington, *A History of Chemistry* (London: Macmillan, 1864), p. 125, n. 1).
9. *Wilhelm Weber:* Wilhelm Weber.
10. *Kohlrausch:* Rudolph Herrmann Kohlrausch (1809–58), a German physicist and mathematician best known for his work in electricity and electromagnetism (*CDSB*).
11. *Electrolytische Maaßbestimmungen by Kohlrausch and Weber from 1856, p. 262:* R. Kohlrausch and W. Weber, "Elektrodynamische Maassbestimmungen, insbesondere Zürückführung der Stromintensitäts-Messungen auf mechanisches Maass', *Abhandlungen der Mathematisch-Physischen Classe der Königlich-Sächsischen Gesellschaft der Wissenschaften,* 3 (1857), pp. 219–92.
12. *Hirst:* Thomas Hirst.

From Heinrich Gustav Magnus 27 December 1867 2713

Berlin 27 Dec. 67

Mein theurer Tyndall!

Sie haben mir eine große Freude gemacht durch die wenigen Worte die Sie aus Faraday's Notizen über meine Person am Ende Ihres Briefes vom 17 d. M. hinzugefügt haben. Ich weiß keinen Menschen den ich so geliebt und so hoch geachtet habe wie Faraday und von einem solchen nicht ungünstig beurtheilt worden zu sein, ist ein wohlthuendes Gefühl. Daß Sie mir das bereitet haben dafür bin ich Ihnen besonders dankbar. Ich freue mich daß Sie es übernommen haben ihn als einen Discoverer and Investigator darzustellen. Gewiß ist niemand geeigneter hierfür als Sie. Ich bin aber überzeugt Sie werden dabei auch ein Bild von ihm als Mensch, von seinem liebenswürdigen einfachen Character geben, mir scheint das gehört dazu wenn man ihn als Discoverer verstehen soll. Ich erinnere mich noch wie ich 1849 in Birmingham mit ihm zusammen war und eines Tages Arm in Arm mit ihm nach Hause ging und er mir sagte: What a pleasure, Magnus, to ask nature a question.

Ich sehe noch das heitere Gesicht mit dem er dies sagte, in dem sich sein ganzes Streben ausdrückte, nicht einzelne Zahlen festzustellen, sondern der Natur immer neue Geheimnisse abzufragen.

Bitte empfehlen Sie mich der Mrs Faraday, auf deren Photographie die sie mir ein mal schenkte, meine Blicke häufig mit einem wohlthuenden Gefühle fallen.

Für die Sendung des Reports der Aeronautical Society danke ich Ihnen noch besonders. Bitte lesen Sie die Einlage an Glaisher und wenn Sie sie für geeignet halten so senden Sie sie ab.

In alter Anhänglichkeit | Ihr | G Magnus.

Berlin 27 Dec. 67

My dear Tyndall!

You have given me great pleasure through the few words that you have taken from Faraday's notes about me and added at the end of your letter of the 17th inst.[1] I do not know of any person whom I have loved so much and respected so highly as Faraday, and to have been judged not unfavourably by such a person is an agreeable feeling. For having given me this, I am particularly grateful to you. I am pleased that you have taken it upon yourself to portray him as a discoverer and investigator.[2] Certainly no one is more suited for this than you. I am convinced, though, that in doing so you will also provide a picture of him as a human being, of his likeable, simple character; that seems to me to all be part of understanding him as a discoverer. I still remember how I was together with him in Birmingham in 1849[3] and was returning home, arm in arm with him one day, and he said to me: What a pleasure, Magnus, to ask nature a question. I still see the cheerful face with which he said this, in which were expressed all his efforts not to ascertain individual numbers, but rather to always question Nature on new mysteries.

Please give my compliments to Mrs Faraday, on whose photograph that she once gave me as a present my eyes frequently fall with an agreeable feeling.

For sending me the Report of the Aeronautical Society[4] I thank you especially. Please read the enclosure[5] addressed to Glaisher[6] and, if you think that it is suitable, then send it off.

In old devotion | Your | G Magnus.

RI MS JT/1/M/36

---

1. *your letter of the 17th inst.:* letter missing.
2. *depicting him as a discoverer and investigator:* see letter 2665, n. 3.
3. *Birmingham in 1849:* the BAAS held its nineteenth annual meeting in 1849 at Birmingham.
4. *Report of the Aeronautical Society:* probably the first volume of *Annual Report of the Aëronautical Society of Great Britain,* which was published in 1867. The society had its first meeting in April 1866, and James Glaisher was a founding member and treasurer.
5. *enclosure:* not identified.
6. *Glaisher:* see letter 2579, n. 14.

From Mary Egerton 28 December [1867][1] 2714

Mountfield Court | Hurst Green.[2] | Dec[embe]r 28th

Dear Mr Tyndall,

Thank you most cordially for your kind note.[3] Pray accept our warmest good wishes, (not compliments!) in return, for this season & the coming year. Your handwriting looked like an answer to my thoughts, for I have often wondered how you were getting on, & I am looking forward with intense interest to the lectures on Faraday[4] I see announced to begin the 28th. I hope we shall get to town just in time for them. I once thought I might have had a chance of your opening Friday's lecture,[5] (if you would admit me,) but I am afraid I shall not manage it. I hope it is not on some subject I shall be dreadfully sorry to miss.

I have been making acquaintance with Emerson.[6] It is very extraordinary poetry in the literal meaning of the word, & some bits very beautiful. It cannot always be said of his verses that they "march in time", & some I confess I don't understand; but he has such a wonderful insight into things & spirits, some passages cast a spell over and there is one I shall be ever grateful to him for—you have it in your book[7]—"Spring still makes spring in the mind" &c.[8] The poems made me turn to the Essays;[9] there again what a power he has of bringing out truths one had felt without knowing how to express them. And then that curious Pantheism,[10] so beautiful, so true, up to a certain point, but then suddenly leaving one afloat in a chilling sea of mist, instead of in the looming presence of the Living God! It is a book one longs to talk over but it is rather too harsh to inflict a "review" upon you in return for your good wishes, so I will conclude. I do hope you have been sleeping better, and and[11] are feeling more up to your work physically than you were. No chance of seeing you in the country now lectures have begun, I conclude! I wish I could take my boy[12] to hear the Christmas ones,[13] though I c[oul]d not wish my husband's[14] only holiday to be spent in town.

With our united kindest regards,

Believe me | Y[ou]rs very sinc[ere]ly | M F Egerton

RI MS JT/1/E/65

---

1. *[1867]:* the year is given based on the mention of Tyndall's upcoming lectures on Faraday.
2. *Mountfield Court | Hurst Green:* see letter 2665, n. 2.
3. *your kind note:* letter missing.
4. *lectures on Faraday:* Tyndall gave nine lectures 'On the Discoveries of Faraday' at the RI on Tuesdays and Thursdays from 30 January to 27 February 1868 (*Roy. Soc. Proc.,* 5 (1866–9),

p. 197). In his journal Tyndall noted that Faraday's niece Jane Barnard attended all of these lectures (RI MS JT/2/13c/1335).

5. *opening Friday's lecture:* the first of Tyndall's two Friday Evening Discourse on Faraday was on 17 January 1868 (see letter 2665, n. 3).
6. *Emerson:* Ralph Waldo Emerson.
7. *your book: Mountaineering in 1861.*
8. *"Spring still makes spring in the mind" &c:* Tyndall quoted lines 105–12 from Emerson's poem 'The World-Soul' in *Mountaineering in 1861* (on p. 9): 'Spring still makes spring in the mind | When sixty years are told, | Love wakes anew this throbbing heart, | And we are never old. | Over the winter glaciers | I see the summer glow, | And through the wild-piled snow drift | The warm rose buds below'.
9. *Essays:* R. W. Emerson, *Essays* (Boston: James Munroe and Company, 1841); and R. W. Emerson, *Essays: Second Series* (Boston: James Monroe and Company, 1844).
10. *Pantheism:* a belief or philosophical theory that God is immanent in or identical with the universe; the doctrine that God is everything and everything is God (*OED*).
11. *and:* Egerton repeated the word 'and'.
12. *my boy:* Hugh Edward Egerton (1855–1927), an English historian and son of Mary Frances and Edward Egerton (*ODNB*).
13. *Christmas ones:* see letter 2671, n. 5.
14. *husband's:* Edward Egerton.

To Peter Guthrie Tait 29 December 1867 2715

29th Dec. 1867

My dear Tait,

I have been lately endeavouring to give a sketch of Faraday as a scientific worker,[1] and by this post I send you a proof of a part of what I have been writing.

Dare I ask you to cast your eye over it? There is one allusion in it that I more particularly wish you to see, so that you may correct it if put either too strongly or incorrectly.

Thomson[2] called upon me on Christmas day. I am told he is now in Paris, otherwise I should write to him. That however will do when I send him the complete copy of my memoir on Faraday. I was staying with my friend Lubbock[3] & thus missed Thomson. It would have been a special pleasure to me to press his hand on Christmas day, and then to satisfy at a time sacred to both of us, though doubtless sacred in determined senses, the compact that you & he & I should go to war no more.[4]

I was sorry not to be able to rejoin you at St. Andrews, but the fact is that for some time previous to my visit to Dundee[5] I had been under an accursed spell of sleeplessness. I longed for the air of the hills & went to them[6] as soon as I could, taking hardly any part in the business of the meeting.

I would give a thousand pounds tomorrow for your capacity to smoke & drink beer. It is simply the want of that power which prevents me from being a real worker.

A happy new year to you & yours & believe me

Ever yours faithfully | John Tyndall

University of Edinburgh GN 237 Coll-234 Gen. 2169, Letter 229

---

1. *sketch of Faraday as a scientific worker:* see letter 2665, n. 3.
2. *Thomson:* William Thomson.
3. *Lubbock:* John Lubbock.
4. *go to war no more:* Tyndall, Thomson, and Tait reconciled their differences at the annual meeting of the BAAS at Dundee in September 1867 (see letter 2646; Journal, RI MS JT/2/13c/1334). On the differences between these physicists, see C. Smith, 'North Britain versus Metropolis: Territorial Controversy in the History of Energy', in *The Science of Energy: A Cultural History of Energy Physics in Victorian Britain* (Chicago: University of Chicago Press, 1998), pp. 170–91. This reconciliation would prove not to last, since, in 1873, Tait would join James Clerk Maxwell to critique Tyndall on his popular lectures and apparent mathematical incompetence (see *The Ascent of John Tyndall*, pp. 322, 324–5, 327, and 329).
5. *my visit to Dundee:* see letter 2627, n. 22 and n. 31.
6. *went to them:* see letter 2650, n. 2.

To Jane Barnard [1867][1] 2716

Tuesday

My dear Miss Barnard.

Have you that little Interlaken scrap & the Giessbach[2] scrap ready?

Can you tell me the date when Mr. Faraday was appointed Professor here?[3]

Also the date when he gave his first lecture in the theatre, & below stairs?

You will I hope excuse & help me; my brain is under great pressure.[4]

Yours ever | John Tyndall

RI MS JT/1/T/130
RI MS JT/1/TYP/2/4194

---

1. *[1867]:* the year is given by Tyndall's seeking detailed information about Faraday and his brain being 'under great pressure' (see n. 4).
2. *Giessbach:* Giessbach Falls, located near lake Brienz, Switzerland. Tyndall quoted Faraday's account of his 1841 visit to the Giessbach Falls in the RI discourses 'On Faraday as a

Discoverer' (see *Roy. Inst. Proc.*, 5 (1866–9), p. 232). Tyndall would himself visit the falls in July 1868, writing in *Hours of Exercise in the Alps:* 'I went to the Giessbach with a prejudice against the illumination of the fall. The crowd of spectators may suggest the theatre, but the lighting up of the water is fine. I like the colourless light best; it merely intensified the contrast revealed by ordinary daylight between the white foam of the cascades and the black surrounding pines' (pp. 273–4).

3. *here:* at the RI.
4. *my brain is under great pressure:* through November and December 1867, Tyndall had been preparing his Friday Evening Discourses on Faraday (see letter 2665, n. 3).

To Elizabeth Dawson Steuart [1867][1] 2717

Royal Institution. | Monday

My dear Mrs Steuart,

The family of Caleb Tyndall is a terrible nuisance to me. You know how willing I am to help the girls,[2] but this conduct sickens me. What can this woman[3] be thinking of? I know nothing of Capt. Beresford[4] and if I did I should as soon think of flying as of writing to him on any subject of the kind.

I am working terribly hard, and these pests augment my labours. It was only a few days ago that I received a begging letter[5] from the blockhead that I sent to America.[6]

Yours ever | John Tyndall.

RI MS JT/1/TYP/10/3366
LT Typescript Only

---

1. *[1867]:* the year is given based on relation to letters 2644 and 2718 of this volume, and that this letter likely comes after a letter in volume 9 but before a letter in volume 11 (see n. 6).
2. *the girls:* possibly Sarah Tyndall and Emma Young (née Tyndall). In letters 2644 (dated 11 September 1867) and 2718 (undated but after letter 2644), Tyndall shared with Steuart his intention to provide dowries for two of Caleb Tyndall's daughters.
3. *woman:* not identified.
4. *Capt. Beresford:* not identified.
5. *begging letter:* letter missing.
6. *blockhead that I sent to America:* John Tyndall, a son of Caleb Tyndall. In a letter to Steuart dated 23 March 1866, Tyndall wrote: 'It was only the day before yesterday that I lost 12 pounds in sending a son of Caleb Tyndall's to America' (letter 2386, *Tyndall Correspondence*, vol. 9). In letter 2634 of this volume, dated to mid-August 1867, Tyndall wrote to Steuart, 'Caleb Tyndall's son has again written to me, but I have thrown his letter in the fire. He wants the means of carrying him home to Ireland. If his sister thinks it right to devote a portion of the twenty pounds to this purpose she may do so. But I can have

no direct communication whatever with this very troublesome and very stupid person'. Finally, in a letter dated 20 November 1870, Steuart wrote to Tyndall concerning his offering financial assistance to relatives: 'Under any circumstances, John Tyndall is very unsteady: went twice to America, (I believe at your expense) but was too idle to remain' (RI MS JT/1/TYP/10/3389; forthcoming in the eleventh volume of *The Correspondence of John Tyndall*).

To Elizabeth Dawson Steuart [1867][1] 2718

Wednesday.

My dear Mrs Steuart,

I am very much obliged to you for your most kind letter.[2] At some future time I may avail myself of your offer of hospitality; for I suppose the longing to see the Barrow[3] will come back to me again: but how forward this will be in the future I cannot now say.

The work I am engaged upon at present is an attempt to respond to the desire of our managers to give our members and the world generally some picture of "Faraday as a Discoverer" I intend to occupy the two first Friday evenings of our coming Session with two discourses on this subject.[4] They will afterwards be published, in France and Germany as well as in England.[5]

You may rest assured of it that I have done with Mrs Doyle.[6] While in Scotland[7] I received a letter from Caleb Tyndall[8] desiring assistance in her behalf, and as he did not seem to be in any way moved towards her when I was in Leighlin[9] I took it for granted that her case was one of real distress.

Did you communicate to Caleb Tyndall my desire and intention, provided their behaviour warranted me in doing it, to make some provision for a dowry for his two daughters?[10] I was pleased with the appearance of his children when I was in Leighlin and at intervals since have thought of giving some substantial expression to my feelings. I dare say this proposal of doing something towards bestowing a marriage portion on the two girls is as good a way of proceeding as any other.

Goodbye and thanks—| Ever Yours | John Tyndall.

RI MS JT/1/TYP/10/3367
LT Typescript Only

---

1. *[1867]:* the year is given based on relation to letter 2644 dated 11 September 1867 (also to Steuart); the mention of having been in Scotland, from which he returned to London on 25 August 1867 (see n. 7); and that Tyndall was asked in September by Henry Bence Jones and the RI Managers to give a lecture on Faraday (see n. 4). Thus this letter would have been written sometime in the autumn of 1867 but a more precise date cannot be

determined.

2. *your most kind letter:* letter missing.
3. *Barrow:* the river Barrow, the second longest river in Ireland, which flows through Leighlinbridge.
4. *desire of our managers . . . two discourses on this subject:* see letter 2665, n. 3.
5. *published, in France and Germany as well as in England:* published in English as *Faraday as a Discoverer.* For the German translation, see letter 2686, n. 4; for the French translation, see letter 2750, n. 3.
6. *Mrs Doyle:* Abbey Doyle (see letters 2675 and 2681).
7. *While in Scotland:* see letter 2627, n. 22 and n. 31, and letter 2650, n. 2.
8. *letter from Caleb Tyndall:* letter missing.
9. *when I was in Leighlin:* Tyndall travelled to Ireland for his mother's funeral 2–5 May 1867 (Brock & MacLeod, *Hirst Journals,* 2–5 May 1867, pp. 1802–3).
10. *Did you communicate to Caleb Tyndall . . . his two daughters?:* in letter 2644, Tyndall had asked Steuart if she would 'manage this for me in your wise way'. The daughters were probably Sarah Tyndall and Emma Young (née Tyndall).

1868

To Heinrich Debus 1 January 1868 2719

New Year's day 1868

My dear Heins

Of course you come to us[1] when you come to London. This vapour[2] will instantly vanish when you begin work And remember your work is a noble one. You have to raise aloft the name & fame of science among those boys[3] I should not be surprised if in the end you had a class of 50, and instruments corresponding.

I am working like a Trojan with,—my boys here[4]—

Yours ever | John Tyndall

RI MS JT/1/T/263
RI MS JT/1/TYP/7/237

---

1. *come to us:* Tyndall was then living with Thomas Hirst at 14 Waverley Place, though he would move to the RI at the end of January 1868 (see letter 2707, n. 8). During his stay with Hirst and Tyndall, Debus had a dispute with Tyndall on 7 January 1868 regarding his English pronunciation (Brock & MacLeod, *Hirst Journals*, p. 1025).
2. *vapour:* the word *vapour* is used metaphorically in reference to Debus probably having a 'clouded' state of mind regarding his own work.
3. *those boys:* at the beginning of 1868 Debus left Queenwood College to become Science Master at Clifton College, near Bristol (Brock & MacLeod, *Hirst Journals*, 11 January 1868; E. M. Oakeley (ed), *Clifton College Register* (London: Rivingtons, 1890), p. 21).
4. *my boys here:* Tyndall gave the RI's 1867 juvenile Christmas Lectures on 'Heat and Cold' (see letter 2671, n. 5). There would have been girls present as well.

To Charles Kingsley[1] 1 January 1868 2720

10 Walton Place, | Hans Place, S. W.[2] | New Years Day 1868

My dear Kingsley

I am most glad to receive this note from you,[3] though you have not me to thank for that lecture[4]—In fact I sent it to nobody but to Carlyle[5] himself.

If I do not greatly err the coming times will more & more unfold the strength & wisdom of our friend, and it is only I imagine by the recognition & application of such strength & wisdom that this England can continue to be the noble England what she has hitherto been.

It has always given me pleasure to see spirits like your own unswervingly loyal to Carlyle. I suppose nobody among his recent acquaintances has wrangled more pertinaciously with him about science & some other matters than I have myself. Still these differences relate to matters of detail—to the mere wrap of the man—while the globe of the man remains to me one of the noblest, if not the very noblest, that the ages have produced.

Yours ever, | John Tyndall—| All kindly wishes to you and yours

RI MS JT/1/T/1032
Transcript Only[6]

1. *Charles Kingsley:* see letter 2645, n. 28.
2. *10 Walton Place, | Hans Place, S. W.:* embossed address. Located in London, Tyndall was likely writing from a friend's residence. A court case from 1869 shows that the residence belonged to a Harriet Spedding in at least 1863 (*English Reports Annotated. 1869,* 2 vols (London: Reports & Digest Syndicate, 1869), vol. 1, p. 1530). She may have been a member of the Spedding family, who through his friends William Frederick and Juliet Pollock, Tyndall was acquainted, most notably James Spedding (1808–81), an English literary editor and biographer known especially for editing the works of Francis Bacon (*ODNB*).
3. *this note from you:* letter missing.
4. *lecture:* probably Tyndall's lecture on 'Matter and Force', delivered on 5 September 1867 to working-class men at Kinnaird Hall, Dundee (see n. 5 and letter 2627, n. 31).
5. *Carlyle:* at the end of his lecture on 'Matter and Force', Tyndall mentioned and quoted Carlyle.
6. This handwritten transcript letter appears to be written in LT's hand. There is another letter in the same hand on the same embossed paper to Mrs Kingsley dated 20 May (year unknown), RI MS JT/1/T/1033.

To Thomas Frederick Elliot[1] 7 January 1868 2721

7th. January 1868.

My dear Sir,

I by no means think your little paper[2] "nonsense", but very sound sense, and probably calculated to do more good than mere pretentious writing. It is weighty & serious amid all its fun.

Yours very truly | John Tyndall

T. F. Elliot Esq[ui]re | &c. &c. &c.

NLS, Papers of Sir Thomas Frederick Elliot (1868), Ms. 19421

1. *Thomas Frederick Elliot:* Thomas Frederick Elliot (1808–80), a British civil servant. He served as an assistant Under-Secretary in the Colonial Office from 1847–68 (*ODNB*).
2. *your little paper:* probably 'A Discourse on Table Rapping', a privately published satirical paper dated 21 December 1867 (see *ODNB*; K. Bell, *The Magical Imagination: Magic and Modernity in Urban England, 1780–1914* (Cambridge: Cambridge University Press, 2012), p. 273).

From Jules Jamin 7 January 1868 2722

Paris 7 Janvier 1868.

Mon cher ami,

Nous avons, en France la bonne habitude de prendre un congé dans les environs du premier jour de l'année. J'étais absent quand votre lettre m'est arrivée et je vous demande pardon du retard de la mienne. Nous avons une autre habitude encore celle de souhaiter à nos amis tous les bonheurs possibles; permettez-moi de ne pas vous oublier et de vous envoyer ici tous mes vœux.

J'ai vu Mr Gauthier-Villars et il promet de se comporter avec vous comme un homme honnête et délicat. Il vend ordinairement beaucoup d'exemplaires de ses livres en Allemagne et quand il a appris que vous avez fait une édition allemande de votre livre sur le son, il a compris que son débit allait être restreint. C'est un négociant, vous ne pouvez vous étonner qu'il en ait été contrarié. La réflexion lui est venue ensuite, et vite; il est plus sensible à l'honneur de vous éditer qu'à la possibilité de perdre, et il ne vous demande absolument rien. Quand vous viendrez à Paris, il vous montera ses livres de vente. S'il a couvert ses frais il vous fera partager à son bénéfice. S'il a fait des pertes vous en jugerez; et au cas tout à fait improbable où ces pertes auraient une certaine importance, il vous laissera juge de faire ce que vous voudrez. Il a ajouté qu'il me prendrait pour arbitre auprès de vous. Vous voyez que vous pouvez vous confier à lui et que c'est un honnête homme.

Pour ce qui est de votre mémoire sur Faraday je crois qu'il faut le publier vite et j'ai la conviction qu'on en pourra vendre un nombre considérable d'exemplaires. Faraday était en très haute estime; votre réputation, d'un autre côté est bien établie et vos deux noms réunis enleveront le succès. Il ne faut donc pas mêler cette publication à celles que fait Babinet lesquelles ne sont pas biens appréciées; il faut un livre spécial, de vous seul.

Mr Gauthier offre de le publier dans les conditions suivantes.

Aussitôt qu'il aura soit votre manuscrit ou une copie imprimée anglaise, il vous dira ce que couterait en France le tirage et sa composition pour un nombre des exemplaires que vous ferez d'un commun accord—soit mille par exemple.

Vous lui devrez cette somme après quoi il se chargera de vendre le livre aux conditions suivantes.

1e un exemplaire sur 13 lui sera accordé

2e il touchera 33% du prix de vente

3e ce qui reste (67%) vous sera réservé au bout d'un certain temps vous compterez avec lui, si ce qui vous reviendra pour la vente dépasse la somme que vous lui devrez pour le tirage, il vous donnera la différence. Au cas contraire c'est vous qui lui redevrez quelque chose. Pour mon compte je suis certain que vous ne lui redevrez rien.

Voila mon cher ami les renseignements que je suis chargé de vous transmettre, disposez de moi, je vous prie, et croyez moi

Votre dévoué | J. Jamin

Voulez vous me donner le renseignement suivant. Je viens de constituer à la Sorbonne un grand laboratoire. Je voudrais y recevoir les manuscrits et le Philosophical Magazine—ne pourrais je point les recevoir sans les payer comme une gracieuseté faite de la Société Royale à la Sorbonne. Votre influence dans cette négociation me serait bien utile—il est bien entendu que cela ne m'est point personnel c'est un établissement public de France qui donnera à un établissement public d'Angleterre.

Paris, January 7 1868.

My dear friend,

In France we have the good habit of taking a vacation around the first day of the year. I was absent when your letter[1] arrived, and I apologize for the tardiness of mine. We have yet another habit to wish our friends as much good fortune as possible. Allow me not to forget you and send you all my best wishes.

I saw Mr Gauthier-Villars[2] and he promises that he will behave towards you as an honest and tactful man. Usually he sells a lot of copies of his books in Germany, and when he learned that you did a German edition of your book on sound,[3] he realized that his output will be reduced. He is a merchant; you should not be surprised that he was upset about it. Then he thought about it, and quickly realized that he is more touched by the honour of publishing your work,[4] rather than the possibility of a loss, and he does not ask anything from you. When you come to Paris,[5] he will show you his books for sale. If he has covered his expenses, he will share with you his earnings. If he has made some losses, you will decide. And in the entirely unlikely case that those losses would be of significant importance, he will let you decide what to do. He added that he would take me as a mediator with you. You see that you can trust him and that he is an honest man.

As for your essay on Faraday,[6] I think that it should be published soon and I am convinced that we will be able to sell a large number of copies. Faraday was highly esteemed; your reputation, on the other hand, is well established, and both of your names together will attract success. Thus, this publication should not be combined with those by Babinet,[7] which were not well-liked. It should be a unique book, by you alone.

Mr Gauthier offers to publish it under these conditions:

As soon as he has either your manuscript or a printed English copy, he will tell you how much it will cost to print it in France, and its typesetting for a number of copies that you will agree on—for a thousand, for example.

You will owe him this sum, after what he will sell the book in the following conditions:

1st Every one copy out of 13 will be granted to him.

2nd He will receive 33% of the sales made.

3rd What is left (67%) will be kept for you, after a certain period of time you will count with him, if your returns from the sales surpass the sum that you owe him for the printing, then he will give you the difference. Otherwise, it is you who would need to return something to him. On my behalf, I am certain that you will not have to give him anything.

Here is, my friend, the information that I was asked to forward to you. Please, I am at your service, trust me

Your devoted | J. Jamin

Would you give me the following information: I have just set up a large laboratory at the Sorbonne.[8] I would like to receive there the manuscripts and the Philosophical Magazine—could I not receive them without paying, as a courtesy from the Royal Society to Sorbonne? Your influence in this negotiation will be very useful—it is well understood that this is not for my personal use; it is a French public institution that will give to an English public institution.

RI MS JT/1/J/8

---

1. *your letter:* letter missing.
2. *Mr. Gauthier-Villars:* Jean-Albert Gauthier-Villars.
3. *German edition of your book on sound:* see letter 2677, n. 5.
4. *publishing your work:* see letter 2617, n. 2.
5. *come to Paris:* Tyndall did visit Paris during his trip to Naples, Italy in April 1868 (see letter 2801, n. 3, and letter 2802), and again while on his way to the Alps, departing London in early July arriving in Paris on 5 July (see letter 2859, n. 2). There is no indication, however, that he visited any publishers.
6. *your essay on Faraday:* see letter 2665, n. 3.
7. *those by Babinet:* see letter 2690, n. 8.
8. *large laboratory in Sorbonne:* in 1867 Jamin founded the first university physics research laboratory in France. He directed the laboratory at the Faculty of Sciences of Paris from 1867–86.

To Thomas Henry Huxley 8 January 1868 2723

8th. Jan. 1868.

Dear old Hal,[1]

You have given me a great pleasure—I might have missed it if you had not sent it to me, for I require to have things thrust under my nose to induce me to read them. This is the true voice of Jacob,[2] very different from either howling radicalism or hidebound stupid toryism. I ran half a dozen times to my inkstand to mark passages that delighted me, and at each move I muttered Bravo! Go on my dear fellow in this line, and keep yourself equally aloof from Henry Sandwith[3] and the Briton.[4]

Yours ever | John Tyndall.

IC HP 8.60
RI MS JT/1/TYP/9/2917
Typed Transcript Only

---

1. *Hal:* nickname for Huxley.
2. *true voice of Jacob:* a reference to Jacob of the Bible. Here in the typescript LT left the following note: 'prob[ab]ly "A Liberal Education & where to find it" delivered at the Working Men's Coll[ege]. on Jan 4. 68'. The essay was later printed in the third volume of Huxley's *Collected Essays,* 9 vols (London: Macmillan and Co, 1893–4).
3. *Henry Sandwith:* not identified. Possibly Humphry Sandwith (1822–81), an English army physician who served in the Crimean War (*ODNB*). It is possible Tyndall is referring to Sandwith's *Hints on the Philosophy of Education* (London: Hamilton, Adams, and Co., 1864) which discussed science and university education reform.
4. *Briton:* not identified.

From Emma Seiler[1] 12 January 1868 2724

Hochgeehrter Herr!

Gleichzeitig mit diesen Zeilen erlaube ich mir Ihnen mein eben erschienenes kleines Buch „the voice in singing" zu übersenden, und bitte Sie daßelbe als ein kleines Zeichen meiner tiefgefühltesten Verehrung anzunehmen. Gerade als Rev. Dr. Furniß das deutsch geschriebene Manuscript meines Buches ins Englische übersetzte, erhielten wir Ihr unschätzbares Werk über den Klang. Daßelbe war nicht nur eine große Hülfe für den Übersetzer dem es dadurch so viel leichter wurde die richtigen wißenschaftlichen Ausdrücke zu finden, sondern es feßelte und intereßirte uns wie noch selten ein anderes Buch. Wie viel Neues lehren Sie darin, und wie schön und klar verständlich ist es ausgedrückt; besonders was Sie über Licht und Wärme sagen. Ihr Name, hochgeehrter Herr, ist jetzt hier in Amerika ein hochgefeierter und man wüncht[2] und hofft sehnlichst, daß Sie dazu bestimmt werden könnten auch in unserem Lande Ihre Vorlesungen zu wiederholen. Die Anerkennung die Sie in Ihrem Buche Herrn Profeßor Helmholtz zollen und die ich in höchstem Grade theile, veranlaßte mich hauptsächlich Ihnen mein Buch zu senden. Vielleicht hat es Intereße für Sie zu sehen wie es mir gelungen die neuesten Entdeckungen des Profeßor Helmholtz praktisch zu verwerthen. Mein Wunsch, daß auch die Componisten mehr die Entdeckungen von Profeßor Helmholtz berücksichtigen möchten, wird hoffentlich sich auch erfüllen, denn sie sind für die Musik von unberechenbarem Werthe.

Mit der Versicherung der tiefgefühltesten Hochachtung empfiehlt sich Ihnen hochgeehrter Herr

Emma Seiler | care of Rev. W. H. Furniß | Philadelphia U.S. | 12/1 1868.

Esteemed Sir!

I am taking the liberty to send you, at the same time as these lines, my short book that has just appeared "the voice in singing",[3] and ask you to accept it as a small token of my most deeply felt admiration. Just as Rev. Dr. Furniss[4] was translating the manuscript of my book, which had been written in German, into English, we received your priceless work on sound.[5] It was not only a great help to the translator, for whom it thereby became so much easier to find the correct scientific expressions, but it captivated and interested us as any other book rarely has. How much that is new do you teach in it, and how beautifully and clearly understandably is it expressed; particularly what you say about light and heat. Your name, esteemed Sir, is now a highly celebrated one here in America, and one wishes and hopes most ardently that you could be induced to repeat your lectures in our country as well. The recognition that you pay to Professor Helmholtz in your book and that I share to the highest degree was the main reason for me to send you my book. Perhaps it will interest you to see how I managed to make practical use of Professor Helmholtz's most recent discoveries. My wish that composers also might take more consideration of Professor Helmholtz's discoveries will hopefully come true also, for they are of incalculable value for music.

With the assurance of the most deeply felt respect, I give my compliments to you, esteemed sir,

Emma Seiler | care of Rev. W. H. Furniss | Philadelphia U.S. | 12/1 1868.

RI MS JT/1/S/64

---

1. *Emma Seiler:* Emma Diruf-Seiler (1821–87), a Bavarian music teacher who founded a school in Philadelphia in 1873. She took part in Helmholtz's studies on sound and published a book on singing that was later translated into English (see n. 3) (K. K. Price, 'Emma Seiler: A Pioneering Woman in the Art and Science of Teaching Voice', *Journal of Singing,* 68:1 (2011), pp. 9–22).
2. *wüncht:* misspelled; 'wünscht' is the correct spelling.
3. *"the voice in singing":* E. Seiler, *The Voice in Singing* (Philadelphia: J. B. Lippincott & Co., 1868), translated from the original German, E. Seiler, *Altes und Neues über die Ausbildung des Gesangorganes* (Leipzig: Leopold Voss, 1861).
4. *Rev. Dr. Furniss:* William Henry Furness (1802–96), an American Unitarian minister and abolitionist (*ANB*).
5. *work on sound: Sound* (1867).

From Jane Barnard 13 January 1868 2725

13 Jan. 1868 | 12 Barnsbury Villa[1] | Islington N.

Dear Dr Tyndall

As my Aunt[2] answered, for me the other day so I now write for her. Mr Claudet of Regent S^t^.[3] (who died since Xmas) took the portraits of my Uncle[4] & Aunt to which you refer.

My Aunt is longing to see you, & did intend writing to ask you whether you would come some evening this week & take a supper with us at 6 o'clock—but now there is only tomorrow & Thursday remaining & we do not expect you can spare the time before the 17th.[5]

But you know that either before or after or when you will we shall be very glad to see you here.

I am | yours very sincerely | Jane Barnard

Dr Tyndall | &c &c &c

RI MS JT/1/B/33

1. *12 Barnsbury Villa:* see letter 2695, n. 2.
2. *Aunt:* Sarah Faraday.
3. *Mr Claudet of Regent St.:* Antoine François Jean Claudet (1797–1867), a French photographer and inventor who had resided in London since the 1830s. His Regent Street shop was known as the 'temple of photography' (*ODNB*).
4. *Uncle:* Michael Faraday.
5. *before the 17th:* on 17 January 1868, Tyndall delivered the first of two Friday Evening Discourses on Faraday (see letter 2665, n. 3).

To Hermann Helmholtz 13 January 1868 2726

13th January 1868

My dear Helmholtz,

Would you kindly give me your assistance and counsel in connexion with the following point?

During the meeting of the British Association at Dundee[1] Thomson[2] met me in a cordial friendly manner. I had also the great gratification of shaking hands with Tait.[3] So we have buried the hatchet, and I trust and believe will never go to war any more.[4]

I sent Tait the memoir on Faraday,[5] and he gave himself the trouble of reading it all through and of giving me his opinion upon it.[6] At pages 24, 29 & 39 he refers to Thomson's researches and thinks that they ought to be dwelt upon. Now you are Thomson's intimate friend, and I am anxious to do all just honour to Thomson: would you point out the places where you think his labours might be referred to? Tait refers to his views of the Contact theory,[7] also to his having proved that the "diamagnetic" in the case of the magnetization of light is necessarily in a state of molecular rotation.[8] He also refers to his papers in the Comptes Rendus on Magne-crystallic action[9] as describing some of the most extraordinary results of theory ever verified by experiment. Finally at the end of p. 39. he speaks of Maxwell[10] and Thomson having immensely simplified & extended Faradays reasoning.[11] I wish you would give me your notion on those points; believing that I am anxious not only to do justice to Thomson, but to express in the most liberal manner my admiration of his intellect.

Yours ever | John Tyndall

If you could write to me[12] in Roman letters I should be much obliged. Your Deutsche Buchstaben sind nicht immer leicht zu verstehen.[13]

I will send you the remainder of the memoir in a few days. I think Mrs. Helmholtz[14] will like the conclusion best.

BBAW, NL Helmholtz, Nr. 477, BI_10–11
RI MS JT/1/T/485

1. *Meeting of the British Association at Dundee:* see letter 2627, n. 22.
2. *Thomson:* William Thomson.
3. *Tait:* Peter Guthrie Tait.
4. *buried the hatchet, and I trust and believe will never go to war any more:* see letter 2646, and letter 2715, n. 4.
5. *Memoir on Faraday: Faraday as a Discoverer.*
6. *giving me his opinion upon it:* letter missing.
7. *his views of the Contact Theory:* see W. Thomson, 'New Proof of Contact Electricity', *Proceedings of the Literary and Philosophical Society of Manchester,* 2 (1860—2), pp. 176–8. For an overview of contact theory, see letter 2710, n. 6.
8. *"diamagnetic" in the case . . . state of molecular rotation:* W. Thomson, 'Dynamical Illustrations of the Magnetic and the Heliocoidal Rotatory Effects of Transparent Bodies on Polarized Light', *Roy. Soc. Proc.*, 8 (1856), pp. 150–8.
9. *his papers in the Comptes Rendus on Magne-crystallic action:* W. Thomson, 'Remarques sur les oscillations d'aiguilles non cristallisées de faible pouvoir inductif paramagnétique', *Paris, Comptes Rendus,* 38 (1854), pp. 632–40.
10. *Maxwell:* James Clerk Maxwell.
11. *Maxwell and Thomson . . . extended Faraday's reasoning:* referring to the manner in which they gave mathematical expression to Faraday's developing field theory (see P. M. Harman, *The Natural Philosophy of James Clerk Maxwell* (Cambridge: Cambridge University Press, 1998), pp. 71–80).
12. *write to me:* Helmholtz's reply, in German, is letter 2731.
13. *Your Deutsche Buchstaben . . . leicht zu verstehen:* your German letters are not always easy to understand (German). This sentence was handwritten by LT on the typescript.
14. *Mrs Helmholtz:* Anna Helmholtz.

To Robin Allen 14 January [1868][1] 2727

Royal Institution of Great Britain | 14th. Jan[uar]^y.

Dear M^r. Allen.

If you could do me the favour of coming here on Saturday, I would go with you over the Report[2] on M^r. Douglass's[3] Report,[4] and excise every thing which you might consider in the least degree personal—leaving the facts in their connected order to speak for themselves.

faithfully yours | John Tyndall.

I should also be disposed to add something regarding his remarks on "cuts".[5]

RI MS JT/1/T/95

1. *[1868]:* the year is given based on relation to Tyndall's report of 20 March 1868 (letter 2791)
2. *Report:* Tyndall's report is letter 2791.
3. *Douglass's:* James Nicholas Douglass (1826–98), an English civil engineer who served as chief engineer to Trinity House from 1862–92. He was apprenticed to the engineering firm Hunter and English at Bow, joined Trinity House as an engineer in 1839, and was appointed as assistant to his father, also Nicholas, in the erection of the Bishop's Rock lighthouse in the Isles of Scilly and subsequently became manager at R. J. and R. Laycock's railway carriage works on the Tyne. Following work as a resident engineer on a number of lighthouses he was appointed resident engineer on the Wolf Rock lighthouse in deep water between Land's End and the Isles of Scilly. This dangerous work was not completed until 1870. He also worked with Michael Faraday and Tyndall on experiments to test fog signals, Argand oil lamps, and electric arc lamps (*ODNB*).
4. *Report:* J. N. Douglass, 'First Order Lighthouse Lanterns', in *Correspondence Relative to the Comparative Advantages of the Lighthouse Lanterns adopted by the Corporation of the Trinity House and the Commissioners of Northern Lighthouses, 1867–8* (London: Eyre and Spottiswoode, 1868), pp. 21–6.
5. *"cuts":* in his report, Douglass wrote of the Trinity House lantern that 'a better defined line of demarcation' . . . 'can be produced from the larger lantern where coloured sections of light are required to be sent from the lighthouse, in other words, increasing the sharpness of the cut' (p. 21). In his report (letter 2791, on p. 27 of original), Tyndall responded on this point thus: 'The advantages claimed for it by Mr. Douglass as regards definition of cutting in the case of coloured arcs, and as regards the reduction of obstruction due to greater distance from the flame, are justly claimed'.

From Charles Frédéric Martins[1] 14 January 1868 2728

JARDIN DES PLANTES DE MONTPELLIER

Montpellier, le 14 janv. 68.

Mon cher Monsieur,

J'ai bien regretté de ne pas vous revoir à Dundee pour vous faire compliment sur votre Conférence aux ouvriers. Je n'ai jamais entendu de Professeur plus clair et plus chaleureux que vous. J'ai été heureux de le dire dans un article sur le meeting de la british association qui a paru dans la revue des Deux Mondes du 1er janvier. Vous recevrez un exemplaire tiré à part revu, corrigé et augmenté. Nous avons eu des nuits très sereines et par conséquent très froides au commencement de janvier. J'en ai profité pour remplir une bombe avec de l'eau et la tamponner elle a éclaté en se brisant suivant un méridien passant par le trou. On cite dans tous les livres de Physique une expérience analogue du Major Williams à Québec. Ne sauriez vous pas vous qui vous êtes tant occupé des propriétés de la glace où se trouve le mémoire original de cette expérimentation. J'ai besoin de le consulter. Les bouchons lorsque les bombes ne se tendent pas ne sont pas projetés comme on le lui fait dire mais seulement soulevés doucement. Il y avait une boule d'eau au centre de la sphère comme de glace qui s'en gelée quand la compression a cessé après l'éclatement cela est d'accord avec les expériences de Mousson, Thomson et les vôtres. M. Griffith m'apprend que le Comité de l'Association britannique se propose de me nommer correspondant. Je suis très flatté de cet honneur. Votre bien dévoué serviteur. | Ch. Martins

P.S.: Pardonnez moi d'avoir intitulé votre Conférence Matière et Mouvement au lieu de Force et Matière. Ce mot de force a en français un profil métaphysique (einen metaphysischen anhauch) qui me répugne. J'aime mieux rester dans les effets que de m'élever dans les nuages des causes. Au fond pour des savants cela revient au même, mais j'écrivais pour le public. Là est mon excuse si j'ai péché. Vous verrez que j'ai rétabli certains passages indiqués par des points dans la traduction de l'abbé Moigno sur lequel le soutane déteint toujours un peu.

JARDIN DES PLANTES DE MONTPELLIER[2]

Montpellier, 14 January 68

My dear Sir,

I regretted not seeing you again in Dundee[3] to congratulate you on your Lecture to working men.[4] I have never heard such a clear and warm professor as you. I was happy to say it in an article on the meeting of the British Association that appeared in the journal of Les Deux Mondes on January 1st.[5] You will receive a copy printed apart, reviewed, corrected and enhanced.

We have had very serene and therefore very cold nights at the beginning of January. I took the opportunity to fill up a bomb with water and dab it, it exploded and broke following a meridian by the hole. We cite in all the physics books a similar experiment of Major Williams[6] in Quebec. Would not you know, you who were so busy with ice properties, where the original memoir of this experiment[7] is.

I need to consult it. The stoppers when the bombs do not stretch out are not thrown like it is told but slightly kicked up. There was a ball of water at the centre of the sphere like ice that froze when the compression stopped after the bursting. This matches the experiments of Mousson,[8] Thomson[9] and yours.

Mr. Griffith[10] tells me that the Committee of the British Association offered to appoint me correspondent.[11] I am very flattered by this honour.

Your very devoted servant, | Ch. Martins

P.S.: Please forgive me for having entitled your lecture Matter and Movement instead of Force and Matter. This word force in French has a metaphysical profile (a metaphysical tinge) that bothers me. I like to stay with the effects rather than elevate myself to the clouds of causes. At the end, for scientists it comes to the same, but I wrote for the public. Here is my excuse if I have sinned. You will see that I have restored some passages indicated by dots in the translation by Father Moigno[12] on which the cassock always fades a little.[13]

RI MS JT/1/M/53

---

1. *Charles Frédéric Martins:* Charles Frédéric Martins (1806–89), a French physician, botanist, and geologist. He held several professorships at the Faculté de Médecine, Paris, the Sorbonne, and Montpellier ('Charles Frédéric Martins', *Annals of Botany,* 3 (1889–91), pp. 480–2).
2. *JARDIN DES PLANTES DE MONTPELLIER:* France's oldest botanical garden, founded in 1593.
3. *Dundee:* see letter 2627, n. 22.
4. *your Lecture to working men:* see letter 2627, n. 31.
5. *an article on the meeting of the British Association that appeared in the journal of Deux Mondes on January 1st:* C. Martins, 'La trente-septième réunion de l'Association britannique pour l'avancement des sciences', *Revue des Deux Mondes,* 73 (1868), pp. 212–38, on pp. 231–4.
6. *Major Williams:* Major Edward Williams of the Royal Artillery.
7. *original memoir of this experiment:* E. Williams, 'Experiments on the Expansive Force of Freezing Water, made by Major Edward Williams of the Royal Artillery, at Quebec in Canada, in the Years 1784 and 1785', *Transactions of the Royal Society of Edinburgh,* 2 (1790), pp. 23–8.
8. *Mousson:* Albert Mousson (1805–90), a Swiss physicist, glaciologist, and malacologist. After the founding of the University of Zurich in 1833 he became a lecturer, then professor of physics there. He held the professorial chair of physics at the École Polytechnique Fédérale and the University of Zürich (1855–78). He contributed to the study of glaciers and the role of pressure on the melting of ice. In 1854 he published *Die Gletscher der Jetztzeit* (Zürich: Fr. Schulthess) (*HLS*). For Mousson's experiment, see A. Mousson, 'Sur la fusion et la solidification de l'eau', *Bibliothèque Universelle de Genève,* 3 (1858), pp. 296–304.
9. *Thomson:* William Thomson. For Thomson's experiments, see W. Thomson, 'On the Stratification of Vesicular Ice by Pressure', *Roy. Soc. Proc.*, 9 (1859), pp. 209–13; and W. Thomson, 'Remarks on the interior melting of ice', *Roy. Soc. Proc.*, 9 (1859), pp. 141–3.
10. *Mr. Griffith:* George Griffith.
11. *appoint me correspondent:* see 'Corresponding Members', *Brit. Assoc. Rep. 1867,* p. xxxviii.
12. *Father Moigno:* François Moigno.

13. *on which the cassock always fades a little:* probably a reference to Moigno's possible religious bias in deciding to translate the final words of Tyndall's Dundee lecture in J. Tyndall, *La matière et la force. La Force: deux conférences,* trans. F. Moigno (Paris: Gauthier-Villars, Les Mondes, 1867). In his translation, Moigno placed ellipses after the following sentence, thus ending the lecture: 'Mais que votre conception de l'Architect de cet univers soit digne et noble, faites-en votre pensee la plus haute et la plus saint!' (p. 23; in Tyndall's words: 'But let your conception of it not be an unworthy one', with 'it' referring to the insoluble 'mystery of this universe'). Martins, on the other hand, continued the text of the lecture's conclusion following this sentence, which he translated as 'Que votre imagination ne lui prete que les attributs les plus nobles, les plus grands, les plus saints' (pp. 232–3). The whole of the text from Tyndall's lecture struck by Moigno but restored by Martins is: 'Invest that conception with your highest and holiest thought, but be careful of pretending to know more about it than is given to man to know. Be careful, above all things, of professing to see in the phenomena of the material world the evidences of Divine pleasure or displeasure. Doubt those who would deduce from the fall of the tower of Siloam the anger of the Lord against those who were crushed. Doubt equally those who pretend to see in cholera, cattle-plague, and bad harvests, evidences of Divine anger. Doubt those spiritual guides who in Scotland have lately propounded the monstrous theory that the depreciation of railway scrip is a consequence of railway travelling on Sundays. Let them not, as far as you are concerned, libel the system of nature with their ignorant hypotheses. Looking from the solitudes of thought into this highest of questions, and seeing the puerile attempts often made to solve it, well might the mightiest of living Scotchmen that strong and earnest soul, who has made every soul of like nature in these islands his debtor well, I say, might your noble old Carlyle scornfully retort on such interpreters of the ways of God to men: The Builder of this universe was wise, | He formed all souls, all systems, planets, particles; | The plan he formed his worlds and AEons by, | Was—Heavens!—was thy small nine-and-thirty articles!' (*Fragments of Science,* vol. 2, pp. 73–4).

From Thomas Carlyle 15 January 1868 2729

CHELSEA,[1] 15 January, 1868.

Dear Tyndall,

My sister[2] wishes to come up with me to hear you,[3] (has abundance of rugged intelligence for that!)—a seat one or two from the extreme front would suit us best.

We will contrive (unless you forbid) to be up 8 or 10 minutes before 9 p.m.—If there are not two seats attainable, let us know some time tomorrow (Thursday)

Yours ever | T. Carlyle.

RI MS JT/1/TYP/1/204
LT Typescript Only

1. *CHELSEA:* 5 Cheyne Row, Chelsea, Carlyle's London address since 1834.
2. *My sister:* Jean Aitken (née Carlyle, 1810–88), who lived in Dumfries. She visited Thomas Carlyle from 9 December 1867 to 14 March 1868 (Aileen Christianson of *CLO,* personal communication).
3. *to hear you:* on 17 January 1868, Tyndall delivered the first of two Friday Evening Discourses on Faraday (see letter 2665, n. 3). Carlyle wrote in letter 2740 that he was unable to make the first lecture, but planned to be at the second on 24 January. In his journal for 27 January 1868, Carlyle wrote: 'Attended Tyndall's lecture (on Faraday, his genius and merits), which Tyndall treated as quite heroic. A full and somewhat distinguished audience, respectful, noiseless, attentive, but not fully sympathetic, I should say; such, at least, was my own case, feeling rather that the eulogy was perhaps overdone. As to myself, "the grandeur of Faraday's discoveries," &c., excited in me no real enthusiasm, nor was either his faculty or his history a matter I could reckon heroic in that high degree. In sad fact, I cared but little for these discoveries—reckoned them uncertain—to my dark mind, and not by any means the kind of "discoveries" I wanted to be made at present. "Can you really turn a ray of light on its axis by magnetism? and if you could, what should I care?" This is my feeling towards most of the scientific triumphs and unheard of progresses and miracles so trumpeted abroad in these days, and I sadly keep it a secret, a sorrowful private possession of my own. Saw a good many people there, ancient friends of mine, to whom I wished right well, but found it painful to speak beyond mere salutations. Bishop Thirlwall, Sir Henry Holland, Dean Stanley and his wife. Lecture done. I hurried away, joined by Conway, American nigger friend, innocent and patient" (J. A. Froude, (ed), *Thomas Carlyle: a History of His Life in London,* 2 vols (New York: Charles Scribner's Sons, 1910), vol. 2 (1834–1881), p. 312). Carlyle refers to the English bishop and historian Connop Thirlwall (1797–1875), the President of the RI Henry Holland, the Dean of Westminster Arthur Stanley, and Moncure Conway (see letter 2552, n. 1).

From James Behan[1] 18 January 1868 2730

Leighlin Bridge.[2] | 18th Jan. 1868.

Sir, some time ago I wrote to you[3] relative to an account of Miss Tyndall[4] which account never entered in the pass book which was Closed and Settled at the time. She was always in the habit of having a little private account exclusive of what was entered in the pass book. My application I understand you sent to Mrs Steuart[5] which account she refused to pay. So she paid me the amount of the pass book for which I gave a receipt, being delicate to acquaint her of Miss Tyndall's private account thinking Miss Tyndall would have a recollection of it and pay it. She being removed from the locality I had no opportunity of reminding of it. I wrote to her since but got no answer. So I appeal to you as it was your money always paid the amount of the pass book and Miss Tyndall private account with me hoping that you will not leave the small sum of £1.13.8. due of your mother[6] and sister due to the orphan children of the late Michael Lynch[7] whose business is entrusted to my care. The whole of this account docs not belong to Miss Tyndall as there was a bottle of Brandy got the day of Mrs Tyndall's death[8] and £1.0.0 in Cash the day Mr Wynne[9] auction to Mrs Doyle[10] and the remainder is due solely by Miss Tyndall the pass book was not running at the time those things were got

I am Sir your ob[e]d[ien]t. Servant | James Behan | for the representatives of Michael Lynch

[the following text is presumably either a response to Behan that was copied onto the original letter, or simply notes about a forthcoming reply, which may be letter 2749]

Dr. Tyndall's affairs in Leighlin Bridge are entirely in the hands of Mrs Steuart, who will see that all just claims are fully liquidated. No further attention can be paid to applications of this kind.

RI MS JT/1/TYP/10/3369
LT Typescript Only

1. *James Behan:* not identified.
2. *Leighlin Bridge:* Leighlinbridge, County Carlow, Ireland, where Tyndall was born and his mother lived.
3. *I wrote to you:* letter missing.
4. *Miss Tyndall:* Emma Tyndall.
5. *Mrs Steuart:* Elizabeth Dawson Steuart.
6. *your mother:* Sarah Tyndall (née Macassey).
7. *Michael Lynch:* probably Michael Lynch who occupied a house opposite where Tyndall was born on Bridge Street, Leighlinbridge (*Griffith's Primary Valuation of Tenements* (1852), p. 179).
8. *day of Mrs Tyndall's death:* Tyndall's mother died on 30 April 1867 (see letter 2589).
9. *Mr Wynne:* possibly a Samuel Wynne, who according to *Griffith's Primary Valuation of Tenements* (1852), lived at the Fair-Green in Leighlinbridge (p. 181).
10. *Mrs Doyle:* Abbey Doyle.

From Hermann Helmholtz 18 January 1868 2731

Heidelberg. 18. Jan. 1868.

Lieber Tyndall

auf Ihre Fragen erlaube ich mir folgendes zu antworten:

p. 18. Was die Contact theorie für galvanische Ströme betrifft, so kann ich im Augenblick nichts darüber finden in den Aufsätzen von Thomson, die ich hier habe. Ich glaube aber, dass er darüber dieselbe Meinung hat, wie ich und Kirchhoff, und deren erste Formulirung vielleicht in meiner Erhaltung der Kraft zu finden ist.

Die Condensatorversuche, die nun jetzt mit Thomson's Divided Ring Electrometer sehr leicht und sicher auszuführen sind, zeigen unzweifelhaft, dass zwei sich berührende Metalle sich elektrisch laden unter Umständen, die sich auf eine wirklich nachweisbare Thatsache gründet, während, wie ich noch vor wenigen Tagen mit Kirchhoff besprach, kein einziger Fall vor liegt, dass chemische Processe ohne Bildung einer galvanischen <u>K</u>ette (eventualiter einer solchen von lauter flüssigen Leitern) elektricität unzweifelhaft entwickeln.

Aber allerdings ist Faraday's und Ihr Eigenes Verdammungsurtheil der älteren Form der Contact theorie gegenüber vollkommen gerechtfertigt. Ich glaube aber, Sie thäten gut zu erwähnen, dass in dieser neueren Form der Theorie jener Einwand beseitigt ist, wodurch denn Faradays wesentlichem Bestreben vollkommen Rechnung getragen ist.

P.24. Ich glaube allerdings, dass Sie Thomson's Theorie von rotirenden Theilchen gegenüber nicht sagen können, dass dieses Räthsel "has defeated the best intellects". Es ist wahr, dass wir uns noch keine vollständig klare Vorstellung von dieser Bewegung der Theilchen sagen können, und es scheint mir unnütz, schon jetzt Hypothesen darüber auszubilden. Aber ich halte Thomson's Beweis dafür, dass Körper mit ruhenden Molekeln die magnetische Rotation der Polarisationsebene nicht zeigen könnten, für allgemeingiltig und genügend.

P.29. Ich weiss nicht recht, was für Versuche Tait meint; ich kenne nur Thomson's Vorschlag einen magnekrystallischen Krystall in einer Eisenoxydlösung aufzuhängen, so dass er nach einer Axe stärker magnetisch als die Lösung wäre, nach der anderen schwächer. Wenn nichts weiter gemeint ist, finde ich nicht, dass Thomson wesentlich Neues hinzugebracht hat; er hat nur die Summe der Erfahrungen in eine für den Mathematiker leicht übersichtliche und klare Form gebracht. Diese Aufgabe war, denke ich, nach den Versuchen von Ihnen, Weber, Reich nicht sehr schwer; Sie war nur durch so wüste Köpfe, wie Plücker ist, in Verwirrung gebracht worden.

P.39. Maxwell's Theorie, welche die Lehre vom Magnetismus, der Elektricität und dem Lichte in eine Hypothese vereinigt, ist eine der genialsten mathematischen Conceptionen, welche je gemacht sind. Sie ist aber gleichzetitig so abweichend von allen bisherigen Vorstellungen über die Natur der Kräfte, dass ich es eigentlich noch nicht wagen möchte, ein Urtheil öffentlich darüber auszusprechen. Ich glaube aber wohl, dass man sie als eine in streng mathematische Form gebrachte Entwickclung von Faraday's Ideen bezeichnen kann. Und dass daraus die Möglichkeit folgt, die Geschwindigkeit des Lichts aus den elektrischen Constanten zu berechnen, ist ein Umstand, der mir ungeheuer imponirt hat. Ich selbst würde mich sehr freuen, das Meinige an dieser Arbeit zu thun, da ich selbst noch eine grosse Schuld von Dankbarkeit gegen Faraday, der immer äusserst liebenswürdig und zuvorkommend gegen mich war, abzutragen habe.

Ich glaube, dass dieses Buch hier in Deutschland auf guten Erfolg rechnen kann, und dass Vieweg allen Grund hat, es gern zu übernehmen. Sie werden wahrscheinlich selbst mit ihm verhandeln wollen. Wünschen Sie, dass ich es thun, so bin ich gern bereit.

Was die Übersetzung des On Sound betrifft, so kann erstens *[ja]* von einer Rivalität wegen der Tonempinddungen nicht die Rede sein. Im Gegentheil je mehr darüber von wissenschaftlichen Autoritäten im Sinne seiner Theorie gedruckt wird, desto mehr macht es allmälig Eindruck auf das *[grössen]* Publicum und die Musiker, die hier in Deutschland sehr träge herangehn. Im Ganzen scheint es in Frankreich besser zu *[werde];* jetzt soll dort auch eine französische Übersetzung erscheinen. Eine der Physiologischen Optik ist eben erschienen. Übrigens glaube ich nach einigen Aeusserungen von Wiedemann annehmen zu dürfen, dass er und seine Frau gern die Übersetzung des Buches übernehmen werden; sonst hätte ich vielleicht selbst Schritte in diesem Sinne gethan.

Was Faraday's Berechnung betrifft, so glaube ich, man darf hier nicht die Quantität der Arbeit vergleichen, sondern die Quantität Electricität. In der That, wenn man in zwei riesigen Conductoren so viel Electricität aufspeichern wollte, dass diese sich langsam durch Wasser entladend ein Gram zersetzte, so würden das riesige Mengen sein müssen. Nur ist zu bemerken dass die Schagweite dieser Electricitäten die möglichst geringste sein dürfte, wenn sie zum Beispiel in einer fast unendlich grossen und fast unendlich dünnen Leydener Flasche aufgehauft wären. Denn für die Wasserzersetzung kommt es nur auf die Quantität nicht auf die Spannung oder Schlagweite an. (die Differenz des Potentials an den beiden Leitern)*

*Das heisst die Differenz des Potentials braucht nur etwas grösser zu sein, als die an den beiden Platten eines Wasserzersetzungsapparats.

Die Arbeitsgrösse der Electricität in dem Gramm Wasser kann relativ klein sein, troz eines ungeheuern Quantums, weil positive und negative E. dicht zusammen gelagert sind.—Ich habe übrigens Faradays Abhandlung nicht hier und kann nicht direct nachsehen.

Wollte man die Electricitätsmenge, welche, in einer ungeheuer grossen und dünnen Leydener Flasche aufgehäuft, zur Wasserzersetzung würden, von einander trennen und auf entfernte Conductoren übertragen, so wurden dazu viel grössere Arbeitskräfte nöthig sein, als die Wasserzersetzung kostete<.> Faraday's Vergleich ist richtig, wenn man ihm auf das Quantum der Electricität allein bezieht; willkürlich und nicht ganz der Sache entsprechend, ist, wenn er sich die Electricitäten dabei als weit von einander getrennt denkt.

Der Goldschmuck hat seine Schuldigkeit vollkommen gethan; mit nochmaligen Versicherungen unseres besten Danks und vielen Grüssen

Ihr | H. Helmholtz.

Heidelberg.[1] 18. Jan. 1868.

Dear Tyndall

In response to your questions,[2] I take the liberty to comment:

p. 18. As far as the contact theory for galvanic currents[3] is concerned, I cannot, at the moment, find anything about it in the essays by Thomson[4] that I have here. I believe, though, that he has the same opinion about it as I and Kirchhoff[5] do, and the first formulation of which can perhaps be found in my conservation of force.[6]

The experiments with capacitors,[7] which can now be carried out very easily and safely with Thomson's divided ring electrometer, show, without doubt, that two metals that are in contact charge each other electrically under conditions that <are> based on a truly provable fact, while, as I was discussing with Kirchhoff a few days ago, not a single case exists where chemical processes clearly produce electricity[8] without forming a galvanic circuit (and possibly one such made of nothing but fluid conductors).

Certainly though, Faraday's and your *[own]* condemnatory judgement with regard to the older form of contact theory is completely justified. I believe, though, that you would do well to mention that in this newer form of the theory, that objection, through which Faraday's essential efforts have been completely accounted for, has been eliminated.

P. 24. I certainly believe that with regard to Thomson's theory of rotating particles,[9] you cannot say that this riddle "has defeated the best intellects". It is true that we still do note <have> a completely clear idea of this motion of particles and it seems useless to me to form hypotheses about it even now. But I consider Thomson's evidence that bodies with stationary molecules could not show the magnetic rotation of the plane of polarisation to be universally valid and sufficient.

P. 29. I am not quite sure what kind of experiments Tait[10] means; I know only Thomson's suggestion to suspend a magnecrystallic crystal in a solution of iron oxide, so that towards one axis it would be more strongly magnetic than the solution, towards the other, weaker. If nothing further is meant, I do not think that Thomson has added anything that is essentially new; he has just put the sum of experiences into a form that is easily viewable for a mathematician and clear. This task was, I think, after the experiments from you, Weber,[11] Reich[12] not very difficult; it had only got muddled up by such chaotic minds as Plücker[13] is.

P. 39. Maxwell's[14] theory,[15] which combines in one hypothesis the study of magnetism, electricity, and light, is one of the most brilliant mathematical concepts which have ever been devised. At the same time, though, it is so different from all previous ideas of the nature of energies that I would still really not like to venture to express a judgement on it in public. I certainly do believe, though, that one can describe it as a development of Faraday's ideas, brought into a strictly mathematical form. And, the fact that the possibility follows from it of calculating the speed of light from electrical constants is a circumstance that has impressed me enormously. I myself would be very happy to do my part on this work, as I still have to pay a great debt of gratitude myself to Faraday, who was always extremely kind and obliging to me.

I believe that this book[16] can count on a good success here in Germany, and that Vieweg[17] has every reason to be happy to take it on. You will probably want to negotiate with him yourself. If you wish that I do so, then I am happy and willing.

As far as the translation of On Sound[18] is concerned, there can, firstly, be no question [of course] of a rivalry due to *Tonempinddungen*.[19] On the contrary, the more printed about it by scientific authorities according to the spirit of his theory, the more it will gradually make an impression on the *[greater]* public and musicians, who here in Germany approach things very sluggishly. On the whole, it seems to be getting better in France; a French translation[20] is supposed to be appearing there now as well. One of the Physiologischen Optik[21] has just appeared. By the way, I believe I may assume from some remarks by Wiedemann[22] that he and his wife[23] will happily take on the translating of the book;[24] otherwise I would perhaps have taken steps in this regard myself.

As far as Faraday's calculation is concerned, I believe that here one may not compare the quantity of work, but rather the quantity of electricity. In fact, if one wanted to store in two gigantic conductors so much electricity that it would decompose one gram of water while slowly discharging through it, then that would have to be enormous amounts. The only thing that needs to be noted is that the spark distance[25] of these electricities might be the least possible if they had been accumulated,[26] for instance, in an almost infinitely large and almost infinitely narrow Leyden jar. This is because the decomposition of water depends only on the quantity, not on the voltage or the spark distance. (the difference of the potential on both conductors)*

*This is, the difference of the potential needs to be only somewhat larger than that on the two plates of an apparatus for decomposing water.

The amount of work done by electricity in the gram of water can be relatively small, despite an enormous quantum, because positive and negative e. are stored closely together.—I do not have Faraday's essay[27] here, by the way, and cannot check directly.

If one wanted to separate from each other the amounts of electricity which had been accumulated in an enormously large and narrow Leyden jar in order to decompose water, and transmit them to remote conductors, then[28] a much larger labour force would be necessary for this than the decomposition of water would cost<.> Faraday's comparison is correct if one applies it[29] to the quantum of electricity alone; it is gratuitous and not quite relevant to the matter, if, in doing so, he imagines the electricities as being widely separated from each other.

The gold jewellery[30] has done its duty completely; with assurances once again of our best thanks and many regards

Your | H. Helmholtz.

RI MS JT/1/H/46
RI MS JT/1/H/66[31]

1. *Heidelberg:* see letter 2677, n. 1.
2. *your questions:* in letter 2726, Tyndall had asked Helmholtz about some topics in his memoir on Faraday.
3. *contact theory for galvanic currents:* on contact theory, see letter 2710, n. 6.
4. *Thomson:* William Thomson.
5. *Kirchhoff:* Gustav Kirchoff (see letter 2710, n. 5).
6. *force:* our translator noted that this could also be translated as 'energy'.
7. *capacitors:* our translator noted that this could also be translated as 'condensers'.
8. *electricity:* in the German, Helmholtz used 'elektricität', for which our German translator noted should have been capitalized as 'Elektricität'.
9. *theory of rotating particles:* Thomson described this theory in 'On Vortex Atoms', *Phil. Mag.*, 34 (1867), pp. 15–24. See also C. Smith and M. N. Wise, *Energy and Empire: A Biographical Study of Lord Kelvin* (Cambridge: Cambridge University Press, 1989), pp. 396–444.
10. *Tait:* Peter Guthrie Tait.
11. *Weber:* Wilhelm Weber.
12. *Reich:* Ferdinand Reich (1799–1882), a German mineralogist who taught at the Freiburg Mining Academy and co-discovered indium in 1863. Reich published numerous investigations of magnetism (*NDB*; *ADB*).

13. *Plücker:* Julius Plücker.
14. *Maxwell's:* James Clerk Maxwell.
15. *theory:* see letter 2726, n. 11.
16. *this book: Faraday as a Discoverer.* For the German translation, see letter 2686, n. 4.
17. *Vieweg:* Heinrich Rudolf Vieweg.
18. *translation of On Sound:* see letter 2677, n. 5.
19. *Tonempinddungen:* H. Helmholtz, *Die Lehre von den Tonempfindungen als physiologische Grundlage für die Theorie der Musik* (Braunschweig: Vieweg und Sohn, 1863). In the German, Helmholtz used 'Tonempinddungen', for which he must have meant 'Tonempfindungen', which is in the title of his book.
20. *French translation:* H. Helmholtz, *Théorie physiologique de la musique, fondée sur l'étude des sensation auditives,* trans. M. G. Guéroult (Paris: Victor Masson, 1868).
21. *Physiologischen Optik:* H. Helmholtz, *Handbuch der Physiologischen Optik,* 3rd edn (Leipzig: Leopold Voss, 1867).
22. *Wiedemann:* Gustav Wiedemann.
23. *his wife:* Clara Wiedemann.
24. *translating of the book:* see letter 2677, n. 5.
25. *spark distance:* in the German, Helmholtz used 'Schagweite', for which our German translator noted should have been 'Schlagweite'.
26. *accumulated:* in the German, Helmholtz used 'aufgehauft', for which our German translator noted that Helmholtz left out an Umlaut over the second 'a' in this word.
27. *Faraday's essay:* probably M. Faraday, 'Experimental Researches in Electricity.—Seventh Series', *Phil. Trans.,* 124 (1834), pp. 77–122.
28. *then:* in the German, Helmholtz used 'wurden', for which our German translator noted that Helmholtz left out an Umlaut over the 'u' in this word
29. *it:* in the German, Helmholtz used 'ihm', for which our German translator noted should have been 'ihn'.
30. *gold jewellery:* see letter 2677, n. 3.
31. This letter was pieced together from fragments contained in two different folders at the RI.

From Jane Barnard 18 January 1868 2732

Sat. 18 Jan. 1868 | 12 Barnsbury Villas[1]

Dear Dr Tyndall

I am gradually recovering from the effects of the feast I enjoyed last night.[2] It was a very great pleasure to me tho' of course mixed with some pain & I thank you much for your part of it. My Aunt[3] received your packet of tickets & we availed ourselves of 3 of them—now we encroach on your kindness &

ask whether you can repeat this act of friendship for her—no one who was present last night likes to lose the next Friday evening, and we should much value the power of admitting these friends.[4]

The No of the last paragraph which you were asking about when here, is 16041 but these are only of papers for the Experimental Researches[5]—we met among the many manuscripts left, other papers such as Experiments with Mr. Gassiot[6]—Steinheils apparatus[7] &c which were numbered by themselves.

I am yours | very sincerely | Jane Barnard. | Dr Tyndall | &c &c &c

RI MS JT/1/B/34

---

1. *12 Barnsbury Villas:* see letter 2695, n. 2.
2. *the feast I enjoyed last night:* on 17 January, Tyndall delivered the first of two Friday Evening Discourses on Faraday (see letter 2665, n. 3).
3. *My Aunt:* Sarah Faraday.
4. *these friends:* not identified. Frank Barnard, Sarah Faraday's nephew, was possibly among the party (see letter 2741).
5. *papers for the Experimental Researches:* from 25 August 1832 to 6 March 1860, Faraday numbered the paragraph entries in his experimental research notebooks, the final entry being 16041. See R. Tweney, 'Faraday's notebooks: the active organization of creative science', *Physical Education,* 26 (1991), pp. 301–6, on p. 302.
6. *Experiments with Mr. Gassiot:* on 30 January 1858 Faraday had conducted experiments on electrical discharges with John Gassiot (see letter 2688, n. 1). For Faraday's experiments with Gassiot, see T. Martin (ed), *Faraday's Diary,* 8 vols (London: G. Bell and Sons, 1932–6), vol. 7, pp. 423–30; and J. P. Gassiot, 'On the Stratifications and Dark Band in Electrical Discharges, as observed in Torricellian Vacua', *Phil. Trans.,* 148 (1858), pp. 137–60, on p. 144.
7. *Steinheils apparatus:* the last experiments Faraday conducted (on 25 November 1861, and 13 January, 28 January, and 12 March 1862) were with the Steinheil apparatus. Described by Henry Bence Jones as an instrument 'for producing the spectrum of different substances' that 'gave a new method by which the action of magnetic poles upon light could be tried', the apparatus was invented by Carl August Steinheil (1801–70), a German astronomer, physicist, and instrument maker (*CDSB*). For Faraday's experiments with Steinheil's apparatus, see T. Martin (ed), *Faraday's Diary,* 8 vols (London: G. Bell and Sons, 1932–6), vol. 7, pp. 462–5; and H. Bence Jones, *The Life and Letters of Faraday,* 2 vols (London: Longmans, Green, and Co., 1870), vol. 1, pp. 449–50.

To Jane Barnard 18 January [1868][1] 2733

18th Jan.

My dear Miss Barnard.

Many thanks for your little note:[2] I hope you will like the feast next Thursday[3] still better. It was no easy task to *[omit]* & weld together; but the audience was very indulgent. I hope it will continue so.

Tomorrow when I go to the Institution I will send the tickets to your aunt:[4]—why did you not use all the five?

Thanks for the number.

I was awfully tired yesterday morning, but I am now all right.

Yours ever | John Tyndall

RI MS JT/1/T/114
RI MS JT/1/TYP/12/4195

1. *[1868]:* the year is given by relation to letter 2732, in which Barnard referred to the first Friday Evening Discourse on Faraday as a 'feast'.
2. *your little note:* letter 2732.
3. *feast next Thursday:* Tyndall probably confused Friday and Thursday when referring to the second Friday Evening Discourse (on 24 January). Barnard had described the first discourse on 17 January as a 'feast' (see letter 2732).
4. *your aunt:* Sarah Faraday.

From John Tyndall[1] 18 January 1868 2734

The Lodge | Gorey[2] | 18th Jan. 1868.

My dear Friend,

The enclosed[3] came here yesterday, and I thought better not to show it to Emma,[4] as all such are calculated to estrange her mind from the great quietness she has, and is latterly enjoying indeed, it could hardly be thought she could have been brought to the quiet settled and pretty contented state of mind she enjoys, but still at times she gets low which by tact she is herself again, she now dines at the table with us, takes her bit of mutton, her favourite, because as she says "it is yours," she also takes at dinner porter, all of which agrees with her admirably, you would hardly know her she is so much improved in her looks, and her mind is kept engaged by reading, knitting, sewing, &c.

I should not go into details so much, only I know how glad you would be to know all respecting her,—hoping you are enjoying good health, all unite in kindest regards

ever yours very sincerely | John Tyndall
Professor Tyndall

RI MS JT/1/TYP/10/3368
LT Typescript Only

---

1. *John Tyndall:* Dr. John Tyndall of Gorey, County Wexford, Ireland.
2. *The Lodge | Gorey:* Tyndall's residence in Gorey.
3. *enclosed:* not identified.
4. *Emma:* Emma Tyndall; see letter 2627 in which John Tyndall told Hector Tyndale that his sister had been placed in the care of Dr. John Tyndall of Gorey.

From Rudolf Clausius 19 January 1868[1] 2735

Würzburg, 19/1 68.

Lieber Tyndall,

Ich beeile mich, Dir Deinem Wunsche gemäß zu schreiben, was mir etwa bei Lesung Deiner Schrift aufgefallen ist. Ich habe den Schluß ebenso wie das Uebrige mit dem größten Vergnügen gelesen, und kann nur wiederholen, was ich früher geschrieben habe, daß ich das Ganze sehr schön finde.

Nur bei einigen wenigen Stellen würde mir eine andere Fassung zweckmäßig scheinen.

S. 38, Zeile 12 sagst Du: „heat is in fact converted into work." Stattdessen würde ich schreiben: „chemical action is in fact converted into mechanical work." Ich glaube nämlich dass in diesem Falle die chemische Wirkung nicht erst Wärme erzeugt und diese dann Werk, sondern dass die chemische Wirkung das Werk in anderer Weise hervorbringt, weil eine Verwandlung von Wärme in Werk unter Umständen, wie sie hier stattfinden, nicht möglich ist.

Ebenso würde ich auf Zeile 23 statt „muscular heat" sagen „muscular action".

S. 39 Zeile 16: „Here we have rudely represented" . . . bis zum Schluße des Absatzes. Hier würde ich doch auf den Unterschied aufmerksam machen, daß bei einer Bewegung in einem widerstehenden Medium die Wärme zum großen Theile in dem Medium erzeugt wird, während sie, wenn ein Leiter in der Nähe eines Magneten bewegt wird, nur im Leiter erzeugt wird, und zwar zum Theile in einem Theile des Leiters erzeugt werden kann, der gar nicht in Bewegung ist.—Auch bei dem Beispiele im Anfange der S. 39, wo eine magnetische Kugel bei jeder Umänderung der Bewegung gedreht werden soll, würde ich darauf hindeuten, daß die bei der Drehung stattfindende Arbeit grade die übrige Arbeit compensirt, so daß, wenn die Kugel wieder genau in ihre Anfangslage gebracht ist, die Gesammtarbeit Null ist.

Eine Stelle, wo Thomson zu erwähnen wäre, ist mir nicht aufgefallen.

Das Spiel, welches Du Johnny geschickt hast, ist allerliebst. Er ist freilich noch zu klein, um mehr dabei zu beachten, als den hübschen Anblick des sich drehenden Kreisels; aber das Spiel wird aufbewahrt, und später wird er auch verstehen lernen, worum es sich dabei handelt.

Mit herzlichen Grüßen von meiner Frau und mir an Dich und Hirst | Dein Clausius

NB. S 40 Zeile 21. | Das Wort: „Revulsive" verstehe ich nicht

Würzburg,[2] 19/1 68.

Dear Tyndall,

I hasten to write to you, in accordance with your wish, what I noticed while reading your work.[3] I read the end just as I did the rest with the greatest pleasure, and can only repeat what I have written earlier, namely that I find the whole thing very nice.

Only with a few places would a different version seem more appropriate to me.

On p. 38, line 12 you say: "heat is in fact converted into work." Instead of this I would write: "chemical action is in fact converted into mechanical work." I believe, you see, that in this case chemical action does not produce heat first and this then work, but rather that chemical action creates work in a different way because a conversion of heat into work is not possible under circumstances as they occur here.

Likewise, on line 23, I would say instead of "muscular heat" "muscular action".

P. 39 line 16: "Here we have rudely represented" . . . up to the end of the paragraph. Here I would draw attention to the difference that with a movement in a resistive medium, heat is produced to a part in the medium, whereas when a conductor is moved in the vicinity of a magnet, heat is only produced in the conductor, and indeed can be produced in part in a part of the conductor that is not moving at all.—With the example at the beginning of p. 39 also, where a magnetic sphere is supposed to be turned at every change in movement, I would point out that the work occurring with the turn just compensates for the remaining work, so that if the sphere is brought back exactly to its initial position, the total work is zero.

I did not notice a place where Thomson[4] could be mentioned.

The game which you sent Johnny[5] is absolutely delightful. He is admittedly still too little to notice anything more about it than the lovely sight of the spinning top, but the game will be kept in a safe place, and later he will also learn to understand what it is all about.

With warm regards from my wife[6] and me to you and Hirst[7] | Your Clausius

NB. P 40 line 21. | I do not understand the word "revulsive"[8]

RI MS JT/1/C/28

1. Tyndall wrote to Clausius on the same day; see letter 2736.
2. *Würzburg:* Clausius held a professorship at the University of Würzburg.
3. *your work:* the German translation of *Heat* (see letter 2607, n. 3).
4. *Thomson:* William Thomson.
5. *Johnny:* Rudolf John Clausius.
6. *my wife:* Adelheid Clausius.
7. *Hirst:* Thomas Hirst.
8. An English language summary of the German letter, probably written by LT, is included with the letter at the RI.

To Rudolf Clausius 19 January 1868[1] 2736

19th. January 1868.

My dear Clausius.

I want your wise head to help me under the following circumstances, though I am ashamed to pester you so much. At the meeting of the British Association at Dundee[2] Thomson[3] met me with great cordiality, and seemed to desire that he and I should be friends for the future; I had also the pleasure of shaking hands with Tait.[4]

Thomson being on the continent I asked Tait, in as much as I had introduced an allusion to him and Thomson, to look over the paper[5] which you have been so good as to read for me: He very kindly did so, and has sent me his remarks upon it. It is quite evident that Tait's admiration for Thomson is perfectly unbounded. Indeed it is interesting to see one scientific man so entirely devoted to the assertion & maintenance of the reputation of another.

He thinks Thomson's labours ought to be referred to at various points in the memoir on Faraday.[6] I did intend to refer to Maxwell,[7] but I did not, I confess, see my way towards any extensive reference to Thomson. It seemed to lead too far from my subject. With regard to the Contact Theory[8] Tait says that it is still held in a modified form by Thomson, and he yesterday sent me an extract from a letter written by Thomson, which I copy here:—"Do not let Tyndall make an opposition between Faraday and the Contact theory. It was really the want of reference to the source of energy (or rather the negation of it implied in Volta's[9] unhappy word "simple" conductor, as applied to water &c.) that Faraday was opposed to. "Simple conductor" was a particularly unfortunate expression for a body that we now know conducts in the manner implied by that expression, essentially because it is not simple, but compound. In all other respects Volta's theory and statements are literally true."

Tait refers to a paper in the Royal Society Proceedings last year,[10] and also to a paper in the Philosophical Magazine for the present month,[11] for an exposition of Thomson's views.

Now I was educated by Bunsen[12] in the belief in the Contact theory, and I had an opportunity of seeing all the experiments of both himself and Kohlrausch[13] in connexion with this subject. I never therefore doubted the existence of contact electricity; but I confess I cannot understand the position of Thomson in the foregoing extract; nor do I see how it is possible to represent Faraday otherwise than as opposed to the contact theory. Can you understand Thomson?

With reference to what I have said in page 24 about magnetic strain Tait says that "Thomson (R.S. Proceedings for 1856)[14] shows that the diamagnetic is necessarily in a state of molecular rotation". I confess I look with so much doubt upon conclusions of this kind that I should hardly feel justified in bringing them into the memoir.

At page 29 Tait writes thus with reference to Magne-crystallic action:—"Here again something is due to Thomson, who gave (after Poisson[15]) the whole theory (Phil. Mag. 1851[16]) and whose papers in the Comptes Rendus[17] describe some of the most extraordinary results of theory I have ever seen verified by experiment." I am anxious to be not only just but generous to Thomson & if you can advise me how I can be so you would render me a great service. I was however under the impression that Faraday's & Plucker's[18] results were first completely explained by myself in the Phil. Mag. for February 1856.[19]

Tait also refers to Maxwell & Thomson having immensely simplified & extended Faraday's reasoning from experiment in the case of the magnetization of Light.

Now Tait is evidently an enthusiast as regards Thomson: he speaks for example of his memoirs being so clear and simple that their profoundness is almost disguised by this simplicity. I must say that Thomson's writings are to me among the most difficult I ever sought to read. Hence I am anxious for your advice. If you think I can in any way bring Thomson's merits prominently forward nothing will give me greater pleasure than to do so.

I hope Johnny[20] has received his toy:—but I fear you have had some trouble with it. I gave particular directions to the person from whom it was bought to defray every expense connected with the carriage of the toy to Würzburg.[21] Give my kindest regards to your wife,[22] and excuse the immense trouble I give you.

Yours ever | John Tyndall

RI MS JT/1/T/196

1. Clausius wrote to Tyndall on the same day; see letter 2735.
2. *meeting of the British Association at Dundee:* see letter 2627, n. 22.
3. *Thomson:* William Thomson.
4. *Tait:* Peter Guthrie Tait.
5. *paper:* a draft of Tyndall's Friday Evening Discourses on Faraday for the RI (see letter 2665, n. 3).
6. *Faraday:* Michael Faraday.
7. *Maxwell:* James Maxwell.
8. *Contact Theory:* for an overview of contact theory, see letter 2710, n. 6.
9. *Volta's:* Alessandro Volta (1745–1827), an Italian scientist and electrician credited with inventing the pile (*CDSB*).
10. *paper in the Royal Society Proceedings last year:* probably W. Thomson, 'On a Self-Acting Apparatus for Multiplying and Maintaining Electric Charges, with Applications to Illustrate the Voltaic Theory', *Roy. Soc. Proc.,* 16 (1867), pp. 67–72.
11. *paper in the Philosophical Magazine for the present month:* W. Thomson, 'On Volta-Convection by Flame', *Phil. Mag.,* 35 (1868), pp. 64–6. The January 1868 issue included two other papers also written by Thomson.
12. *Bunsen:* Robert Bunsen.
13. *Kohlrausch:* Rudolph Kohlrausch (see letter 2712, n. 10).
14. *Thomson (R.S. Proceedings for 1856):* see letter 2726, n. 8.
15. *Poisson:* Siméon-Denis Poisson (1781–1840), a French mathematician and physicist (*CDSB*).
16. *Phil. Mag. 1851:* W. Thomson, 'On the Theory of Magnetic Induction in Crystalline and Non-crystalline Substances', *Phil. Mag.,* 1 (1851), pp. 177–86.
17. *papers in the Comptes Rendus:* W. Thomson, 'Remarques sur les oscillations d'aiguilles non cristallisées de faible pouvoir inductif paramagnétique', *Paris, Comptes Rendus,* 38 (1854), pp. 632–40.
18. *Plucker's:* Julius Plücker.
19. *Phil. Mag. for February 1856:* J. Tyndall, 'On the Relation of Diamagnetic Polarity to Magnecrystallic Action', *Phil. Mag.*, 9 (1856), pp. 125–37. See R. Barton, J. Rankin, and M. Reidy (eds), *The Correspondence of John Tyndall, Volume 3: The Correspondence, January 1850–December 1852* (Pittsburgh: University of Pittsburgh Press, 2017).
20. *Johnny:* Rudolf John Clausius.
21. *Würzburg:* Clausius held a professorship at the University of Würzburg.
22. *your wife:* Adelheid Clausius.

From Herbert Spencer 20 January 1868 2737

37 Queen's Gardens | Bayswater W.[1] | 20th Jan. 1868

My Dear Tyndall,

Thanks for your note[2] of Friday. I was not aware that I am so far privileged, and am much gratified by being so. I was very sorry not to be able to attend last Friday;[3] but I was not sufficiently well. Though considerably better when I returned from Glostershire,[4] I find myself so much out of condition that I am obliged to avoid evening excitements as much as possible—even the excitement of listening to a lecture. I am at present regretting that I have committed myself to an engagement for to-morrow night.[5] Do you know any lively pleasant fellow who would make a good travelling companion?[6]

Very truly yours, | Herbert Spencer

RI MS JT/1/S/96
RI MS JT/1/TYP/3/1114

1. *37 Queen's Gardens | Bayswater W.:* Spencer's London residence since September 1866.
2. *your note:* letter missing.
3. *last Friday:* the first of Tyndall's two Friday Evening Discourses on Faraday, on 17 January 1868 (see letter 2665, n. 3).
4. *returned from Glostershire:* or Gloucestershire, an English county to the west of London. Spencer had gone to Gloucestershire on 3 January where, as he wrote in a letter to the American scientific writer and editor Edward Youmans, 'I hope to get some skating; and I count upon this for doing a great deal towards setting me right again' (H. Spencer, *An Autobiography,* 2 vols (New York: D. Appleton and Co., 1904), vol. 2, p. 206).
5. *engagement for to-morrow night:* Thomas Hirst recorded in his journal for 21 January 1868 that he 'Dined at Dr. Carpenter's; there were present Mr and Mrs Huth, Mr and Mrs Frankland, Spencer, Gwyn Jeffreys &c.' (Brock & MacLeod, *Hirst Journals,* p. 1825).
6. *travelling companion:* if Spencer was inviting Tyndall to go with him, it appears that Tyndall declined, as Hirst would have certainly listed him in his journal entry (see n. 5).

From Charles Frédéric Martins 21 February 1868 2738

JARDIN DES PLANTES DE MONTPELLIER.

Montpellier, le 21 Fev. 1868.

Mon cher Monsieur,

Je suis enchanté que vous ayez été satisfait de mon article sur la british association, je n'ai fait que rendre justice aux savants anglais. Nous attendons avec impatience votre notice sur Faraday: la physique française a été frappée dans la personne de Foucault mort à 49 ans: il proposait de belles expériences sur la lumière solaire.

J'ai reçu le mémoire du Major Edward Williams c'est un Note très incomplète; impossible de deviner comment la bombe s'en tendue. Il n'a rien ajouté à ce que les Physiciens de l'Accadémíe del Cimento nous avaient appris. Avant de communiquer une note sur ma petite expérience à l'Institut j'aimerais avoir quelques détails sur la vôtre et pour épargner votre temps, qui est précieux, je vous envoie ci-joint un petit questionnaire que je vous prie de remplir et de me renvoyer. Je n'ai pas besoin d'ajouter que tous les détails que vous voudrez bien ajouter me seront très précieux. Mon ami Cazelles avance dans sa traduction de l'ouvrage de M. Herbert Spencer qui fera certainement une grande sensation en France. Nous n'avons plus froid mais pas une goutte de pluie depuis neuf mois.

Votre affectionné confrère. | C. Martins

JARDIN DES PLANTES DE MONTPELLIER[1]

Montpellier, 21 January 1868

My dear Sir,

I am glad that you were satisfied with my article on the British Association;[2] I only did justice to the British Scientists. We are looking forward to your notice on Faraday:[3] the French physics have been struck in the person of Foucault[4] dead at 49 years old. He proposed some great experiments on solar light.

I have received the memoir of Major Edward Williams[5] it is a very incomplete note; impossible to guess how the bomb gets stretched. He did not add anything to what the physicists of the Accademia del Cimento[6] taught us.

Before submitting a note on my little experiment to the Institute[7] I would like to have some details on yours and, to spare some of your time, which is precious, I have attached a little questionnaire[8] that I beg you to fill out and send back to me. I do not need to add that all the details you would like to add will be very valuable to me. My friend Cazelles[9] makes good progress in his translation of Mr. Herbert Spencer's book[10] which will probably make a big impression in France. We are not cold anymore but we have not had a single drop of rain for nine months.

Your affectionate colleague. | Ch. Martin

RI MS JT/1/M/54

---

1. *JARDIN DES PLANTES DE MONTPELLIER:* see letter 2728, n. 2.
2. *my article on the British Association:* see letter 2728, n. 5.
3. *your notice on Faraday:* probably just a reference to Tyndall's forthcoming *Faraday as a Discoverer* (see letter 2665, n. 3).

4. *Foucault:* Léon Foucault (1819–68), a French physicist best know for his pendulum experiments demonstrating the rotation of the earth and for his experiments to determine the velocity of light (*CDSB*).
5. *memoir of Major Edward Williams:* see letter 2728, n. 6 and n. 7.
6. *Accademia del Cimento:* a scientific society founded in Florence in 1657.
7. *note on my little experiment to the Institute:* not identified.
8. *I have attached a little questionnaire:* enclosure missing.
9. *my friend Cazelles:* Émile-Honoré Cazelles (1831–1907), a French politician, philosopher, and translator (*DBF*).
10. *translation of Mr Herbert Spencer's book:* Cazelles translated Spencer's *First Principles* (1862) into French as H. Spencer, *Les Premiers Principes,* trans. É. Cazelles (Paris: Germer Baillière, 1869).

To Arthur Penryn Stanley 22 January 1868 2739

The probable—almost certain—absence of Mr. Tyndall from London on Saturday next[1] prevents him from accepting the invitation with which he has been honoured by the Dean of Westminster & Lady Augusta Stanley.[2]

22nd. Jan. 1868.

Wellcome MS. 7777/10

---

1. *on Saturday next:* according to Thomas Hirst's journal entries of 25 and 26 January 1868, on 25 January Tyndall left London for High Elms (see letter 2635, n. 13), returning to London on 26 January (Brock & MacLeod, *Hirst Journals,* pp. 1825–6).
2. *Lady Augusta Stanley:* see letter 2573, n. 1.

From Thomas Carlyle 23 January 1868 2740

CHELSEA,[1] 23 Jan 1868.

Dear Tyndall,

Thanks for your manifold attentions to me! My misfortunes began last Friday at 5 a.m., when sleep fled prematurely, and they did not end till 8½ p.m.—when the fruit of my hard struggles thro' the day went suddenly to air. My sister[2] had surrendered, but for my own share, I was positive and had driven all things and arrangements vigorously on; determined on Albermarle Street[3] against odds; not even an intrusive "Friend" could detain me at the last moment, but got dismissed after a vapid cup of coffee [ice tea];—when, stepping triumphantly down, I discovered that it was raining torrents, the half-hour fairly come or more and nobody to run for a conveyance except a poor woman in her cottons. Pity me if you like. I am not very fortunate! At all events, don't bother yourself more with me in thick of many other cares and agitations.

My sister gives it up for good: I not at all; but am bent for this Friday, as I was for last,[4] tho' nobody can say whether the luck will be again too strong for me, or not strong enough! I have my ticket; don't mind me farther;—and right good speed be yours athletic man.

Yours sincerely | T. Carlyle.

RI MS JT/1/TYP/1/205
LT Typescript Only

---

1. *CHELSEA:* see letter 2729, n. 1.
2. *My sister:* see letter 2729, n. 2.
3. *Albermarle Street:* Albemarle Street, location of the RI.
4. *this Friday, as I was for last:* as Carlyle described, he was unable to attend the first of Tyndall's two Friday Evening Discourses on Faraday on 17 January, but planned to attend the second on 24 January, which he did (see letter 2665, n. 3, and letter 2729, n. 3).

From Frank Barnard[1] 25 January 1868 2741

43 Hemingford Road[2] | Jan 25. 68

My dear Sir

Although nearly a stranger to you, I cannot help expressing a few words of admiration for the beautiful discourse[3] you gave us last night on the character of our late loved and honoured uncle—a discourse so heartfelt, so worthy in tone and sentiment throughout. It is not too much to say that Faraday's dearest, & oldest friends might in many instances, derive a new and enlarged conception of his mind from your masterly and eloquent summary.

It would be out of place for me to give any opinion on its scientific bearings,—for I am a complete ignoramus in science—but you will have the most cordial sympathy of all Faraday's personal friends in your sentiments—if I understand you rightly—that after all, the character of the man was his greatest part, and must overtop in general estimation the eminence of the philosopher. In truth Faraday was to most of his personal, his non-scientific friends, in his scientific aspect, a sort of veiled prophet. We had a vague idea that his speculations, his labours and his name were linked with many of the chiefest scientific facts of the age; but we could not appreciate his steps, join link to link, track his knowledge, nor describe how his divinations grew into certainties, how these certainties took shape in practical developments, the hinges of the working day world of the present moment. It needed a mind like your own of similar calibre to his, to trace these connections and epitomise the intellectual history of the man, and the function he discharged in the story of the world's progress.

I speak merely for myself and so speaking, thank you from my heart for the eloquent summary of my illustrious uncle's character which you gave us last night. If these few words be impertinent you will excuse them, if they be erroneous you will smile at them, but if they be sympathetic you will approve them.

With best regards. Believe me | my dear Sir | yours very truly | and respectfully | Frank Barnard

John Tyndall Esq. FRS

RI MS JT/1/B/24

1. *Frank Barnard:* Frank Barnard (1828–95), son of John and Margaret Barnard (née Faraday), and brother of Jane Barnard ('Deaths', *Morning Post,* 21 August 1895, p. 1).
2. *43 Hessingford Road:* Barnard's residence in London.
3. *beautiful discourse:* on 24 January 1868, Tyndall delivered the second of two Friday Evening Discourses on Faraday (see letter 2665, n. 3).

From Rudolf Clausius 25 January 1868 2742

Würzburg 25/1 68.

Lieber Tyndall,

Auch ich habe schon die Erfahrung gemacht, daß Tait in Bezug auf Thomson sehr übertriebene Ansichten hat. Im vorigen Jahre schickte er mir eine Brochüre, „Historical Sketch of the Dynamical Theory of Heat" auf welcher stand „Not Published" zur Ansicht und Meinungsäußerung. Nachdem ich sie gelesen hatte, mußte ich ihm schreiben, die Schrift erscheine mir nicht als eine historische Darstellung, sondern als eine Parteischrift zu Gunsten einiger Personen (Joule und Thomson); ich selbst schätze diese Personen sehr hoch, aber ich glaube doch daß man nicht um ihretwillen ungerecht gegen Andere sein dürfe; ich könne nicht begreifen wie ein Mann, der selbst wißenschaftliche Verdienste habe, sich zu einer solchen Parteischrift hergeben könne.

Auf diesen Brief habe ich keine Antwort erhalten, und ich weiß nicht, ob und in welcher Form die Schrift publicirt ist. Wenn Du mir darüber gelegentlich einmal etwas schreiben willst, wird es mir lieb sein.

Was nun speciell die in Deinem Briefe erwähnten Puncte anbetrifft, so bin ich in Bezug auf die Contacttheorie ganz Deiner Ansicht. Auch ich gebe zu, daß durch den Contact electrische Spannungsdifferenzen entstehen, aber ich glaube nicht, daß sich die Entstehung des galvanischen Stromes hieraus erklären läßt, sondern daß die ihren Grund in der chemischen Wirkung des

Elementes hat. Auch glaube ich, daß allgemein Faraday als Gegner der Contacttheorie betrachtet ist. NB Das Heft der Phil. Mag. von diesem Monat ist hier noch nicht angekommen, ich weiß daher nicht, was Thomson darin sagt.

In Bezug auf Thomsons Erklärung der diamagnetischen Wirkungen (welche Tait bei Gelegenheit dessen, was Du auf S. 24 über magnetic strain gesagt hast, anführt,) kannst Du vielleicht die kleine Aenderung machen, daß Du statt der Worte (S. 24, Schluss des 3ten Absatzes): „has up to the present hour defeated the best intellects that have been brought to bear upon the question" sagst, daß Faraday die Erklärung nicht gegeben hat, daß freilich von Anderen (oder noch bestimmter von Thomson) interessante Arbeiten hierüber veröffentlicht sind, dass es aber an dieser Stelle zu weit führen würde, darauf einzugehen.

Ich selbst habe Thomson's Ansichten über Diamagetismus und die damit zusammenhängenden Erscheinungen zu wenig studiert, um ein sicheres Urtheil darüber abzugeben; aber ich glaube daß es in Deiner Schrift über Faraday auch gar nicht nöthig ist, darauf einzugehen. Wenn Du eine Geschichte der Physik schreiben wolltest, könnte Thomson vielleicht verlangen daß Du seine Theorien speciell anführtest; aber in einer Biographie von Faraday ist das nicht blos unnöthig, sondern es würde sogar für diesen Zweck schädlich sein, indem es die Schrift zu breit und schwerfällig machen würde.

Meinen Brief vom 19ten d. hast Du hoffentlich erhalten.

Der Zoll, welcher für Johnny's schönes Geschenk zu zahlen war, war so *[unbedeutend]*, daß er nicht der Rede werth ist. Er hatte nur zur Folge dass wir das Packet während der Festtage nicht bekommen konnten.

Mit herzlichen Grüssen | Dein Clausius.

Würzburg[1] 25/1 68.

Dear Tyndall,

I too have already had the experience of Tait[2] having very exaggerated views regarding Thomson.[3] Last year, he sent me a brochure called "Historical Sketch of the Dynamical Theory of Heat",[4] on which it said "Not Published", to view and comment on. After I had read it, I had to write to him that the paper did not appear to me to be an historical account, but rather a pamphlet for the benefit of some persons (Joule[5] and Thomson); I myself valued these persons very highly, but I still believed that one was not allowed to be unjust towards others for their sakes; I could not understand how a man who himself had scientific honours could lend his name to such a pamphlet.

I did not receive an answer to this letter, and I do not know whether and in what form the paper was published. I shall be glad if you will write and tell me something about it sometime or other.

Now as far as the points that you mention in your letter[6] in particular are concerned, I am entirely of your view regarding contact theory.[7] I also acknowledge that differences in electric voltage originate through contact, but I do not believe that the origin of the galvanic current can be explained from this, but rather that it has its basis in the chemical action of the element. I also believe that Faraday is generally regarded as an opponent of contact theory. NB This month's issue of the Phil. Mag has not arrived here yet and I therefore do not know what Thomson says in it.[8]

Regarding Thomson's explanation of diamagnetic actions (which Tait quotes, on the occasion of what you have said on p. 24 about magnetic strain), you can perhaps make the small alteration that, instead of the words (p. 24, end of the 3rd paragraph) "has up to the present hour defeated the best intellects that have been brought to bear upon the question", you say that Faraday has not given the explanation that interesting works on it have been published by others (or still more specifically by Thomson), but that, at this point, it would be leading too far to go into it.

I myself have studied Thomson's views on diamagnetism and the related phenomena too little to make a definite judgement about it; but I do believe that in your paper on Faraday,[9] it is not at all necessary to go into this. If you were wanting to write a history of physics, Thomson could perhaps insist that you quote his theories in particular, but in a biography of Faraday this is not only unnecessary, but rather it would even be harmful for this purpose, in that it would make the paper too broad and clumsy.

I hope that you have received my letter of the 19th inst.[10]

The customs duty which had to be paid for Johnny's[11] nice present was so /insignificant/ that it is not worth mentioning. The only consequence of it was that we were not able to get the package during the holidays.

With warm regards | Your Clausius.[12]

RI MS JT/1/C/29

---

1. *Würzburg:* Clausius held a professorship at the University of Würzburg.
2. *Tait:* Peter Guthrie Tait.
3. *Thomson:* William Thomson.
4. *"Historical Sketch of the Dynamical Theory of Heat":* probably a draft of the first chapter of Tait's *Sketch of Thermodynamics* published later in 1868. P. G. Tait, 'Historical Sketch of the Dynamical Theory of Heat', in *Sketch of Thermodynamics* (Edinburgh: Edmonston and Douglas, 1868), pp. 1–45.
5. *Joule:* James Prescott Joule (1818–89), an English physicist and brewer whose studies of heat led him to formulate the mechanical equivalent of heat for which he was awarded the RS's Royal Medal in 1852 (*ODNB*).
6. *your letter:* letter 2736.
7. *contact theory:* see letter 2710, n. 6.
8. *This month's issue of the Phil. Mag . . . Thomson says in it:* see letter 2736, n. 11.

9. *your paper on Faraday:* the manuscript for *Faraday as a Discoverer.*
10. *my letter of the 19th inst.:* letter 2735.
11. *Johnny's:* Rudolf John Clausius.
12. An English language summary of the German letter, probably written by LT, is included with the letter at the RI.

From Mary Egerton [25 January 1868][1] 2743

45 Eaton Place[2] | Saturday

One word of heartfelt thanks I must send, dear Mr Tyndall, as I could not express them last night.[3] I saw you for one moment, but you had vanished again, & I hardly regretted it, for it w[oul]d. have seemed almost a sacrilege to speak to you just then, & with a full heart, to utter some platitude of thanks, such as w[oul]d. come to one's lips in a crowd! The interest was almost pain. It is hard to analyze, but I am afraid along with the full tone of sympathy, there are "over-tones", (or rather under ones,) of every jealousy & "dépit",[4] at the thought of spirits so beautiful having been in the world, & oneself shut out from all contact with them, that make the "clang-tint"[5] of ones feelings less harmonious than it ought to be. (vide "Sound"![6]) Besides, the scientific push made one's brain reel, and left one but little hope of understanding the lectures on the discoveries.[7] Well can I understand now, the double excitement of thought & feeling that cost you so many sleepless nights! I trust now the work is done, you are more able to rest bodily, if not mentally.

Y[ou]rs very truly | M F Egerton | Excuse blots!

RI MS JT/1/E/83

---

1. *[25 January 1868]:* the date is given based on the reference to the second of Tyndall's Friday Evening Discourses about Faraday on 24 January (see n. 3), making this 'Saturday' letter 25 January.
2. *45 Eaton Place:* in London.
3. *last night:* the second of Tyndall's Friday Evening Discourses at the RI, 'On Faraday as a Discoverer' (see letter 2665, n. 3).
4. *"dépit":* spite (French).
5. *"clang-tint":* in *Sound* (1867), Tyndall wrote: '. . . an assemblage of tones, such as we obtain when the fundamental tone and the harmonics of a string sound together, is called by the Germans a *Klang.* May we not employ the English word *clang* to denote the same thing, and thus give the term a precise meaning akin to its popular one? And may we not, like Helmholtz, add the word *colour* or *tint,* to denote the character of the clang, using the term *clang-tint* as the equivalent of Klangfarbe?' (pp. 117–8).
6. *"Sound": Sound* (1867).
7. *lectures on the discoveries:* see letter 2714, n. 4.

To William Thomson 25 January 1868 2744

25th. Jan. 1868

My dear Thomson.

Do me the favour of accepting from me a copy of the last edition of my little book on Heat.[1]

I must give up this kind of writing for a long time to come.

Your labours are referred to in the preface. I trust the hope I have expressed[2] will be one day realized.

Yours faithfully | J. Tyndall

MS Kelvin T211, by permission of University of Glasgow Library, Archives & Special Collections

1. *last edition of my little book on Heat: Heat* (1868).
2. *the hope I have expressed:* Tyndall wrote in the preface to the third edition, 'It is hoped that the investigations of Joule, Helmholtz, Thomson, and Rankine, on this, the greatest scientific subject hitherto unfolded by the human mind, may ultimately be rendered equally accessible. The memoirs of Sir William Thomson, at once varied and profound, would be of especial interest and importance' (J. Tyndall, 'Preface to the third edition', in *Heat a Mode of Motion,* 4th edn (London: Longmans, Green, and Co., 1870), p. xii).

To Juliet Pollock [27 January 1868][1] 2745

Monday

My dear Mrs Pollock

I was very much obliged to you for your lenient & useful criticism. The delivery of my second lecture[2] did not quite satisfy myself, but this is as it ought to be—for without this I should have no ideal.

I am greatly pleased to find that you & Pollock[3] liked the lecture. I was also immensely gratified by a note sent to me by Mr Donne.[4]

I have changed my free Bedouin[5] kind of life with a mourning at heart; but I dare say I shall become reconciled to the change by & by.[6]

Thanks dear friend for all kindnesses past & to come.

Yours ever | John Tyndall

There is a certain antique grandeur in Donne that I admire exceedingly.

RI MS JT/1/T/1182
RI MS JT/1/TYP/6/1932

1. *[27 January 1868]:* the date is given based on the reference to the second of Tyndall's Friday Evening Discourses about Faraday on 24 January 1868 (see n. 2), making this 'Monday' letter 27 January.
2. *my second lecture:* the second of Tyndall's Friday Evening Discourses at the RI, 'On Faraday as a Discoverer' (see letter 2665, n. 3).
3. *Pollock:* William Frederick Pollock.
4. *note sent to me by Mr Donne:* letter missing. Possibly William Bodham Donne (1807–82), an English journalist and at the time the Examiner of Plays under the Lord Chamberlain (*ODNB*).
5. *Bedouin:* gypsy-like.
6. *I have changed . . . the change by & by:* probably a reference to his move into the apartment at the RI (see letter 2707, n. 8 and n. 9).

To Jane Barnard 28 January 1868 2746

28th Jan. 1868

My dear Miss Barnard.

I want you to instruct me—please tell me am I to use the Institution[1] stationery when I write upstairs? What has been the practice hitherto?

Yours ever | John Tyndall

I was a little amused at your wrath with me:[2] but really immensely relieved when I found it was the science you objected to

RI MS JT/1/T/115
RI MS JT/1/TYP/12/4195

1. *Institution:* on 25 January Tyndall had moved his residence from 14 Waverley Place to the apartment above the RI (see letter 2707, n. 8).
2. *your wrath with me:* in her reply (letter 2747), Barnard wrote that 'I was not wrath this morning'.

From Jane Barnard 28 January 1868 2747

12 Barnsbury Villas[1] | Islington N.

Dear Dr Tyndall

My Uncle's[2] feelings were that the R.I. should provide all his stationary while he paid all the postage of his letters for he considered he wrote much in the service of the Institution, & he thought that a fair arrangment.

I was not wrath this morning[3]—I had been, sometime before, but when you saw me I was repentent & humble!!

My Aunt[4] desires her thanks for the portraits.[5] She says it is very difficult to please her, but she does not object to either; indeed I think we like the layer one as well as the other, & I have no doubt they will please the generality of people.

Many thanks for the letters you have enclosed to her, & which I should have spoken of this morning

Mr Abbott,[6] I intend going to see; the others we do not think much of. Several letters have passed between him & my Aunt.

I should have told you this morning that I am, & have been for some time, suffering much from indigestion—this will account & partly excuse my captiousness

I am | yours very sincerely | Jane Barnard

John Tyndall Esq | &c &c &c | 28 Jan. 1868

RI JT/1/B/35

---

1. *12 Barnsbury Villas:* see letter 2695, n. 2.
2. *My Uncle's:* Michael Faraday.
3. *I was not wrath this morning:* in letter 2746, Tyndall had written: 'I was a little amused at your wrath with me: but really immensely relieved when I found it was the science you objected to'.
4. *My Aunt:* Sarah Faraday.
5. *portraits:* not identified.
6. *Mr Abbott:* Benjamin Abbott (1793–1870), a teacher and longtime friend and correspondent of Michael Faraday (Frank A. J. L. James, 'The tales of Benjamin Abbott: a source for the early life of Michael Faraday', *British Journal for the History of Science,* 25 (1992), pp. 229–40).

To unidentified 29 January 1868 2748

29th. Jan. 1868

Most hearty thanks, dear friend, for your beautiful & useful present. I have just put the first invitation in the dog's mouth.[1]

I bless the gods, not for dogs & inkstands, but for the friends that give them to me.

Yours ever | John Tyndall

Wellcome MS. 7777/11

---

1. *dog's mouth:* possibly a letter holder in the shape of a dog's head, the letter being inserted into the mouth.

To [James Behan][1] 1 February 1868 2749

Royal Institution. | 1st. Feb. 1868.

Sir,

Mrs Steuart[2] has kindly consented to be my liquidator at Leighlin Bridge.[3] In her hands I have left totally and entirely the winding up of the affairs of the late Mrs Tyndall.[4]

All just claims upon me will be promptly satisfied by Mrs Steuart. Have therefore the kindness to send your account to that lady.

Your ob[e]d[ien]t. Serv[an]t. | John Tyndall

RI MS JT/1/TYP/10/3371
LT Typescript Only

1. *[James Behan]:* this letter is possibly a formal but belated reply to Behan's letter of 18 January 1868 (see letter 2730).
2. *Mrs Steuart:* Elizabeth Dawson Steuart.
3. *Leighlin Bridge:* see letter 2612, n. 2.
4. *late Mrs Tyndall:* Sarah Tyndall (née Macassey).

From Jules Jamin 2 February 1868 2750

Paris 2 Février 1868

Mon cher ami,

J'ai attendu avec impatience la fin de votre article sur Faraday. J'ai lu la première partie avec le plus vif intérêt. Je l'ai même fait traduire à ma petite fille qui apprend l'anglais. Mr Gauthier Villars pense que cet ouvrage aura du succès en France, et il sera très heureux de le publier dans la forme et dans les conditions dont je vous ai parlé.

Quant à la traduction, elle ne sera pas faite certainement par l'abbé Moigno lui-même, pas plus que celle de votre ouvrage sur le son; mais il saurait la revoir, et il ne manque pas d'habileté pour résoudre les parties délicates. Au reste, s'il ne le pouvait pas je pourrais vous offrir un jeune homme, un de mes élèves qui possède parfaitement votre langue et qui se chargerait de ce travail. Je pourrais moi-même le revoir et ne permettre l'impression qu'après avoir lu une dernière fois les épreuves. Il en sera ce que vous voudrez.

Je vous suis bien obligé d'avoir pris si amicalement la demande indiscrète que je vous avais faite des transactions philosophiques; j'en suis très reconnaissant. Soyez assez bon toutefois pour les faire inscrire pour le Laboratoire des Recherches Physiques à la Sorbonne autrement on pourrait affecter cette publication à la bibliothèque générale de l'établissement.

Vous recevrez probablement dans quelques jours la visite de mon élève et collègue Mr Cornu pour lequel je vous demande la bienveillance que j'ai autrefois trouvée auprès de vous à Londres.

Ma femme et mes enfants se rappellent à votre bon souvenir. Croyez-moi sincèrement.

Votre ami dévoué | J. Jamin

Paris February 2 1868

My dear friend,

I eagerly awaited the conclusion of your article on Faraday.[1] I read the first part with much delight. I also had it translated to my daughter,[2] who is learning English. Mr Gauthier-Villars thinks that this book will be a success in France, and he will be very glad to publish it[3] in the manner and under the conditions I told you.

As for the translation, it will not be done certainly by Father Moigno[4] himself, no more than the translation of your book on sound;[5] but he could review it, and he does not lack in skill to amend delicate sections. As for the rest, if he could not do it, I could offer you a young man, one of my students who has a perfect command of your language, and who could take care of this work. I could review it myself and not allow it to go into print until after I have read the samples one last time. It will be as you wish.

I am much obliged to you for having taken my indiscreet request about the philosophical transactions[6] so kindly; I am very thankful for it. Be kind enough to register it for the Laboratory of Physics Research in Sorbonne otherwise this publication could be assigned to the general library of the establishment.

You probably will get, in a few days, a visit from my student and colleague Mr Cornu[7] for which I ask your benevolence, just as I once received from you in London.

My wife and children[8] send you their best wishes. Believe me sincerely.

Your devoted friend | J. Jamin

RI MS JT/1/J/9

---

1. *your article on Faraday:* probably the manuscript for Tyndall's *Faraday as a Discoverer* (see letter 2665, n. 3).
2. *my daughter:* Lucie Zoé Marie Jamin (*c.* 1858–78), daughter of Jules and Thérèsa Jamin (see n. 8). In 1878 she married the physicist Henri Becquerel.
3. *publish it:* J. Tyndall, *Faraday Inventeur,* trans. F. Moigno (Paris: Gauthier-Villars, 1868).

4. *Father Moigno:* François Moigno.
5. *translation of your book on sound:* see letter 2617, n. 2.
6. *indiscreet request about the philosophical transactions:* see letter 2722.
7. *Mr Cornu:* Marie Alfred Cornu (1841–1902), a French physicist who had been named professor of physics at the École Polytechnique in 1867 (*CDSB*).
8. *My wife and children:* Thérèsa Joséphine Jamin (née Lebrun, 1832–80), who married Jules Jamin in 1851, and their two children, Paul Joseph (1853–1903) and Lucie (see n. 2).

From François-Napoléon-Marie Moigno — 3 February 1868 — 2751

Paris, le 3 février 1868

LES MONDES | Revue Hebdomadaire | DES SCIENCES | ET DE LEURS APPLICATIONS | aux arts et à l'industrie | par | M. l'Abbé MOIGNO | RÉDACTION | 2, RUE D'ERFURTH, 2

Mon cher M. Tyndall

Je me suis bien aperçu que vos leçons de Noël étaient improvisées et prêtes par la sténographie, mais elles ont une forme nouvelle et sont très intéressantes. Je les reverrai avec le plus grand soin, et elles auront l'avantage de faire attendre la seconde édition de la chaleur.

Cette seconde édition je serais bien tenté de la faire moi-même, car je vois de plus en plus que les éditions prennent tout pour eux. Depuis que j'ai repris mes mondes ils ont grandi à vue d'œil. Les réabonnements et les abonnements nouveaux dépassent toutes mes espérances. Dans un mois ou six semaines nous aurons en caisse de 25 à 30 mille francs, et le meilleur emploi que l'on puisse faire avec cet argent est d'imprimer des livres dont la vente est certaine comme La Chaleur; Le Son; Les travaux de Faraday, etc.

Dites moi votre avis à cet égard. Avez-vous envoyé à M^r^. Gauthier-Villars comme je vous le demandais votre procuration à l'effet de demander à M^r^. Giraud le compte de la chaleur.

J'ai adressé il y a quelque temps à M. Stokes pour le présenter à la Société Royale, un exemplaire de ma Statique Analytique. Je n'ai reçu aucune réponse. Oserais-je vous prier quand vous verrez M. Stokes de lui rappeler mon envoi.

Je suis désolé de voir que l'on ne commence pas l'impression du beau livre sur le son. Il paraît que les figures se font beaucoup attendre.

Je suis dans les sentiments respectueux d'une affection dévouée.

Votre humble serviteur | L'Abbé F. Moigno

Paris, February 3 1868

LES MONDES[1] | Revue Hebdomadaire | DES SCIENCES | ET DE LEURS APPLICATIONS | aux arts et à l'industrie | by the | Father MOIGNO | EDITORIAL BOARD | 2 RUE D'ERFURTH, 2

My dear Mr Tyndall,

I noticed that your Christmas lessons[2] were improvised and made with the use of stenography, but they have a new format and are very interesting. I will reread them with the utmost care, and they will have the advantage of delaying the second edition of la chaleur.[3]

I would be very tempted to do this second edition myself, because I see more and more that editors take everything for them. Since I resumed my Mondes, they grew visibly. The amount of renewals and new subscriptions exceed all my expectations. In a month or six weeks, we will have on hand 25 to 30 thousand francs, and the best thing that we can do with this money is to print books whose sales would be certain, like the La Chaleur, Le Son;[4] Faraday's works,[5] etc.

Tell me what you think about this. Have you sent Mr. Gauthier-Villars your power of attorney as I asked you in order to ask Mr. Giraud[6] for the account of the Heat?

A while ago I sent Mr. Stokes[7] a copy of my Analytical Statics[8] to be presented at the Royal Society. I have not received any response. Would you be so kind as to remind Mr. Stokes about my mail when you see him?

I am sorry to see that we have not started to print the beautiful book on sound yet. It seems that the figures are being much delayed.

I am sending you my most respectful and affectionate wishes of devotion.

Your humble servant | Father F. Moigno

RI MS JT/1/M/112

1. *LES MONDES:* see letter 2706, n. 9.
2. *Christmas lessons:* Tyndall gave the 1867 RI Christmas lectures on 'Heat and Cold' between 26 December 1867 and 7 January 1868. The lectures were published in French as J. Tyndall, *Chaleur et froid: six leçons faites devant un jeune auditoire pendant les vacances de Noël 1867,* trans. Moigno (Paris: Les Mondes, 1868).
3. *second edition of Heat:* J. Tyndall, *La Chaleur, mode de mouvement,* 2nd edn, trans. F. Moigno (Paris: Gauthier-Villars, 1874)
4. *Sound:* see letter 2617, n. 2.
5. *Faraday's works:* see letter 2750, n. 3.
6. *Mr. Giraud:* Étienne Giraud.
7. *Mr. Stokes:* George Stokes.
8. *Analytical Statics:* F. Moigno, *Leçons de mécanique analytique* (Paris: Gauthier-Villars, 1868).

To Sarah Faraday 6 February 1868 2752

Royal Institution of Great Britain | 6th Feb. 1868.

My dear Mrs. Faraday.

I send you the first three sheets of my memoir on "Faraday as a Discoverer"[1] I expect the remainder of the memoir before the conclusion of the week.

The proofs are sent to you before publication to enable you to judge, and to inform me of your judgement, whether the memoir contains any thing inconsistent with the love and reverence of a free man for your husband's memory.

I hope the state of your health will permit you to read the memoir. Your wishes to me in the matter are so far paramount that if I thought you desired it I would suppress the memoir altogether.

Yours ever truly | John Tyndall

Mss.Ms.C011.200. American Philosophical Society

1. *my memoir on "Faraday as a Discoverer"*: see letter 2665, n. 3.

From Edward Sabine 7 February [1868][1] 2753

13. Ashley Place[2] | Feb. 7.

Dear Tyndall.

Will you read, if you have not done so already, Mr. Baxendell's[3] paper.[4]

The Stonyhurst Observatory[5] have undertaken at my request to make a series of observations for the same objects in the next half sun spot period—(probable maximum about the end of 1868, or beginning of 1869: and minimum 1873–1874)—they write to me to direct them in the procurement of the radiation Therm[omet]er. They have a self recording Thermer: for the shade temperature. It is on this point that I wish to consult you. The Thermer. should have its bulb covered, I suppose, with a thin coating of lamp-black.[6] It should be placed horiz[ont]ally over a surface of short grass, as near the ground as is consistent with being well above the grass: it must be so placed that the sun may shine directly on the bulb for the greater part (the middle hours) of the day. The bulb should project beyond the support & scale. (Should it, or should it not, be enclosed in a thin glass tube, hermetically sealed, the air being withdrawn).

The Observations to be as follows—in summer the hours of 9, 10, 11, 12, 1, 2 and 3: in winter, 10, 11, 12, 1 and 2—The observer to have his watch carefully compared with the clock of the photographic cylinder of the shade Therm[er]. To record the time of observation, hour & minute, and the degree of the solar Radiation Therm[er]—the record to be compared with the same minute in the shade Therm[er]. Photogram.

It needs one precaution that of wiping dry the solar radiation Therm[er], if there has been rain previous to the observation. Pray correct or improve in any way what I have written, and believe me,

always faithfully y[ou][rs]. | Edward Sabine

I regret that D[r]. Miller[7] will not permit us to rebuild your laboratory.[8]

The observation may of course be made before or after the exact hour, within 5 minutes or so. The comparison with the photogram should be exact.

RI MS JT/1/S/31
RI MS JT/1/TYP/4/1322

---

1. *[1868]:* the year is given by reference to Baxendell's paper (see n. 3 and n. 4) and Sabine referring to the year 1868.
2. *13. Ashley Place:* Sabine's residence in London.
3. *M[r]. Baxendell's:* Joseph Baxendell.
4. *paper:* probably J. Baxendell, 'On Solar Radiation', *Proceedings of the Literary and Philosophical Society of Manchester,* 7 (1867–8), pp. 36–46, 97–106.
5. *Stonyhurst Observatory:* constructed in 1866, an astronomical and meteorological observatory located at Stoneyhurst College in Lancashire, England.
6. *lamp-black:* a colour pigment consisting of a very high concentration of carbon, historically obtained by collecting soot from oil lamps (*OED*).
7. *Dr. Miller:* probably William Allen Miller (see letter 2599, n. 5), who was treasurer of the RS from 1861–70.
8. *rebuild your laboratory:* presumably Tyndall sought funds from the RS to rebuild the RI laboratory, but we have not found anything specific on this either at the RI or RS. A new laboratory was built at the RI a few years later, while Tyndall went on his American lecture tour from October 1872 to February 1873 (see F. A. J. L. James and A. Peers, 'Constructing Space for Science at the Royal Institution of Great Britain', *Physics in Perspective,* 9 (2007), pp. 130–85, on pp. 162–3).

From Jane Barnard 8 February 1868 2754

12 Barnsbury Villas[1] | 8 Feb. 1868

Dear D[r] Tyndall

We feel your kind attention in sending us the memoir[2] to read & you encouraged me so much, that I have ventured to note down a few points. Into the scientific portion we cannot enter & are unable to judge at all (or ought not) of the way in which the differences between his[3] views & others is referred to—but many of your allusions to his personal character are very gratifying to us, and such as we might be inclined to think might have been omitted, will we hope, prove as interesting to general readers as you expect—on this point of course you are the best & right judge.

With warm thanks for giving us the perusal of it I am

yours very sincerely | Jane Barnard.

I found that what I had heard at the 3 lectures[4] enabled me to read much of this with far more understanding.

D[r] Tyndall | &c &c &c

RI MS JT/1/B/36

1. *12 Barnsbury Villas:* see letter 2695, n. 2.
2. *memoir:* a draft of *Faraday as a Discoverer* (see letter 2665, n. 3). Here Jane was writing for herself and Sarah Faraday, in reply to Tyndall's letter to Sarah Faraday (2752).
3. *his:* Michael Faraday.
4. *3 lectures:* the first three of Tyndall's course of nine lectures 'On the Discoveries of Faraday' at the RI on Tuesdays and Thursdays from 30 January to 27 February 1868 (*Roy. Soc. Proc.*, 5 (1866–9), p. 197). In his journal Tyndall noted that Barnard attended all of these lectures (RI MS JT/2/13c/1335).

From Edward Sabine 9 February [1868][1] 2755

13. Ashley Place[2] | February 9th

Dear Tyndall.

You are interested as I expected you to be with Baxendells[3] statements;[4] but you have given me very few replies on the points on which I consulted you.[5]

I cannot say that I am quite satisfied with the local of the Oxford Solar Radiation Therm[omet][er] "In a niche in a stone wall".—This led me to specify what seemed to me a preferable arrangement "Stationed over but not in contact with a surface of closely shorn grass" But you say nothing in reply on a point on which I should be glad of your opinion.

Several years ago, (nearer 50 than 40) I tried various modes of cutting off the earth's influences, & preferred the surface of shorn grass.

The Oxford Solar Radiation Therm[omet][er] is a simple therm[omet][er] with a blackened bulb, (how blackened is not said). It has no surrounding case of thin glass, having the Air excluded. Mine, used nearly 50 years since had that protection. Is it better with or without?

You wish the bulb to be darkened not with Lampblack,[6] but with some special mixture. Can you refer me to a maker who understands this recommendation, and to whom I can trust for supplying the Therm[omet]er.

The bulb of mine made so long since (by Newman[7]) was certainly not darkened by Lamp black; but I do not remember what was substituted for it.

Ought there to be an instruction to wipe the Therm[ometer] dry before observing, if rain had recently fallen?

I hope to get a series of five years observations made at Stonyhurst,[8] at one of our Scottish Observatories,[9] at Bombay[10] and in the Himalaia[11] by Mr. Hennessey.[12]

I confess that I was much disappointed at the little interest shown on this subject of Solar Radiation by the Hennessey Committee;[13] From the great difference in the amount of Aqueous vapour at Heights in the Himalaia & on the sea coast, I expected that you at least would have had something to say. I hope that I shall still have sufficient influence to set the observations on foot, though I cannot expect to live to see the result. As soon as I can be satisfied in regard to the most suitable instruments I will order & pay for them, and set the Observatories to work.

In my observations so long since I employed, in addition to the blackened bulb in vacuo, a Leslie's differential Therm[omet]er[14] also in a glass case from which the air was excluded. One of the bulbs was silvered, & the other blackened (by Newman) The apparatus was kept in a wooden case, and when taken out was instantly exposed to the sun till the liquid in the stem reached a stationary point, which was the degree registered. I had, (and have still) the record of Observation with this instrument in the Tropics, on the sea-shore, and on mountains in fair exposure.

Sincerely yours | Edward Sabine

As so old an observer on Solar Radiation I have some claim on you for advice, besides our mutual regard.

RI MS JT/1/S/32
RI MS JT/1/TYP/4/1323

---

1. *[1868]:* the year is given by relation to letter 2753.
2. *13. Ashley Place:* Sabine's residence in London.
3. *Baxendells:* Joseph Baxendell.
4. *statements:* see letter 2753, n. 4.
5. *very few replies on the points on which I consulted you:* letter missing.
6. *Lampblack:* see letter 2753, n. 6.
7. *Newman:* probably the firm of John Newman (*c.* 1783–1860), a scientific instrument maker who supplied instruments to the RI and had been a laboratory assistant there (*ODNB*).
8. *Stonyhurst:* see letter 2753, n. 5.

9. *one of our Scottish Observatories:* in a paper presented to the Royal Society about observations at Kew Observatory, Sabine referred to other British observatories or meteorological stations that he organized observations to be made: Aberdeen (Scotland), Armagh (Ireland), Falmouth (England), Glasgow (Scotland), Stonyhurst (England), and Valencia (Ireland), noting that he sought but was unable to secure the establishment of an eighth station in the north of Scotland (see E. Sabine, 'Results of the first year's performance of the Photographically Self-recording Meteorological Instruments at the Central Observatory of the British System of Meteorological Observations', *Roy. Soc. Proc.,* 18 (1869), pp. 3–12, on p. 7).
10. *at Bombay:* probably the Colaba Observatory in Bombay, India.
11. *Himalaia:* Mussoorie, a hill station in the Himalayas near Dehradun, India (see n. 13).
12. *Mr. Hennessey:* John Baboneau Nickterlien Hennessey (1829–1910), a surveyor and astronomer of Indian and English descent, who was employed in the Great Trigonometrical Survey of India and made important observations of the 1874 Transit of Venus (*ODNB*).
13. *Hennessey Committee:* in 1866 Hennessey had written a letter to Sabine, then president of the RS, offering to conduct experiments of Sabine's choosing at the Mussoorie hill station in the Himalayas. A committee was struck by the RS to provide instructions for the observations (see 'Proceedings of the Council of the Royal Society with reference to the undertaking of certain Physical Observations in India', *Roy. Soc. Proc.,* 16 (1868), pp. 283–94).
14. *Leslie's differential Therm[omet]^er:* a thermometer for displaying and measuring differences in temperature invented by John Leslie (1766–1832), a Scottish mathematician and natural philosopher (*ODNB*).

Fom George Biddell Airy 10 February 1868 2756

Royal Observatory, Greenwich, | London, S. E. | 1868 Feb[ruar]^y. 10

My dear Sir

Stimulated in some measure by your reference, I have been labouring a little on Helmholz.[1] But really he is the heaviest of heavy German writers. I wish his book[2] had been only $^1/_4$ or $^1/_6$ as long as it is.

I am somewhat surprised that writers on the organs of speech have not given greater attention to whistling. The notes produced are very pure, the range is much greater than in singing (taking 2 octaves for a singer and nearly 3 octaves for a whistler), and all the organs are easily under observation.

Will you allow me to remark on the sense in which you have used one word which necessarily occurs often in your excellent book,[3] namely interference. You have confined it absolutely to the meaning destructive interference. But this limitation is not recognised in Physical Optics, from which science, I apprehend, the application to Acoustics is derived. Thus in Herschel's "Light", Encyclopaedia Metropolitana, 1827,[4] you will find in Articles 611, 618, &c,

that interference is either corroborative, modificative, or destructive. In my "Tracts", 1831,[5] you will find the same breadth of sense, passion. I refer to Herschel and myself as the persons who, principally, introduced the Undulatory Theory[6] into England.—Pardon my adverting to this: I can only hope to be excused on account of the importance of precision in scientific phraseology.

I am, my dear Sir, | Yours very truly | G B Airy

The printing of my book on Sound[7] is advancing. I think you will be surprised to see how different it is from yours:—to which however in several different places it is much indebted.

Professor Tyndall

RI MS JT/1/A/44
RI MS JT/1/TYP/1/34

1. *Helmholz:* Hermann Helmholtz.
2. *his book:* H. Helmholtz, *Die Lehre von den Tonempfindungen als physiologische Grundlage für die Theorie der Musik* (Braunsweg: Freidrich Vieweg und Sohn, 1863). The first English translation appeared in 1875.
3. *your excellent book: Sound* (1867).
4. *Herschel's "Light", Encyclopaedia Metropolitana, 1827:* see J. Herschel, 'Light', in E. Smedley, et al. (eds), *Encylopaedia Metropolitana*, 30 vols (London: B. Fellowes, 1845), vol. 4, pp. 341–586 (originally written in 1827)..
5. *"Tracts", 1831:* G. B. Airy, *Mathematical Tracts*, 2nd edn (Cambridge: J. & J. J. Deighton, 1831). To this second edition Airy added 'On the Undulatory Theory of Optics' (pp. 249–410); this was later republished in separate form as: G. B. Airy, *On the Undulatory Theory of Optics, designed for the use of students in the university* (London and Cambridge: Macmillan and Co., 1866).
6. *Undulatory Theory:* the theory that light consists in an undulatory movement of an elastic medium pervading space (*OED*); light as a wave.
7. *my book on Sound:* G. B. Airy, *On Sound and Atmospheric Vibrations: with the mathematical elements of music* (London: Macmillan and Co., 1868).

To Jane Barnard — 10 February 1868 — 2757

10th. Feb. 1868.

My dear Miss Barnard.

You have shown in connexion with my memoir of your uncle[1] an intellectual insight and force of character which have excited my admiration, and which now induce me to place some of my thoughts before you with as much distinctness as I can. The conversations which I have had with you have caused me to consult some wise friends as to the desirability of suppressing the memoir altogether, but they consider such a step, not only not desirable but unjustifiable. Thus I find myself in a strait between two. If I consulted my own feelings solely I should probably quench the memoir, but such an act would only provoke public comment which would be agreeable to neither you nor me.

Could I do so I should place with the utmost fearlessness and confidence that memoir in your uncle's own hands for he would be the last to desire his name and fame to be raised to that of an infallible pope of science. But in his absence your feelings, and still more so the feelings of your aunt[2] are very sacred to me, hence my desire, which I think you must have seen to meet your wishes in every possible truthful & practicable way.

Your last note to me[3] contains these words. "Many of your allusions to his personal character are very gratifying to us, and such as we might be inclined to think might have been omitted, will, we hope, prove as interesting to general readers as you expect". You will excuse me my dear Miss Barnard if I speak to you with great distinctness here, and ask you with very great earnestness whether there are <u>any</u> allusions to your uncle's personal character that are <u>not</u> gratifying to you. You will excuse me, moreover, when I say that regarding my personal references to Michael Faraday, I must, after the publication of this memoir, have no whisper of discontent from you. It must never be said that I introduced into the memoir anything of a personal character that you felt inclined to see omitted. As regards science I write in the presence of the world, and there are numbers there—keen eyes and wise heads—to discover my errors and to chastise me, if I do anything approaching to injustice to your uncle's memory; but in personal matters I submit wholly to the guidance of your aunt and yourself. I therefore again send you the memoir, and ask you to point out to me such personal allusions as you would wish to see omitted. Until you can certify to me your approval, or rather your aunt's approval of every allusion to personal character which the memoir contains, the Memoir must remain unpublished.

I hoped to get it off my thoughts this week, but you are at liberty to keep it as long as you like—a month if that be necessary—you are also at liberty to consult such friends of yours as your aunt may deem it desirable to consult. You mistake my character very much if you imagine that I could allow any allusions to your uncle's personal character to go forth to the world under the stamp of your aunt's implied disapprobation.

I say "your aunt's" because though I believe the feeling of dissent from me to be exclusively <u>yours</u>—you use the pronoun <u>we</u> to include your aunt also.

You sent me a few notes[4] on points which you wished to have modified. These notes are of such a trivial character that I cannot imagine that the points to which they refer could have roused the "wrath"[5] of any human-being. There surely must be something else, and I therefore ask you to point out to me, if you can, where I have erred in my appreciation of his scientific work, where I have done him injustice, where I have expressed myself otherwise than lovingly and reverently regarding him. I ask for distinct information on these points, assuring you of my willingness to do everything that truth and reason will permit me to do to meet your wishes.

Let me again say in conclusion that is on points of personal interest that your opinion will carry weight with me. Of the science—I say it with respect—you know nothing. But if the tone or style of even the scientific portion of the memoir are in your opinion capable of improvement, they shall be improved to the best of my ability.

yours truly | John Tyndall

RI MS JT/1/T/109[6]
RI MS JT/1/TYP/12/4196

---

1. *my memoir of your uncle:* see letter 2665, n. 3.
2. *your aunt:* Sarah Faraday.
3. *Your last note to me:* letter 2754.
4. *few notes:* letter 2754 was probably accompanied by Jane Barnard's notes on the manuscript, as she wrote in that letter 'I have ventured to note down a few points'.
5. *"wrath":* this is a reference to letter 2746, in which Tyndall wrote 'I was a little amused at your wrath with me: but really immensely relieved when I found it was the science you objected to', and letter 2747, in which Barnard replied, 'I was not wrath this morning—I had been, sometime before, but when you saw me I was repentent & humble!!'.
6. The RI holds a handwritten draft of this letter that differs in some ways from this final version. The text of the draft, written on 9 February, is as follows: '9th. Feb | My dear Miss Barnard. | You have shown in connexion with my memoir of your uncle an intellectual insight and force of character which have excited my admiration and I therefore think that you will not be unwilling to see me showing something of the same kind. In scientific matters I can hardly hope to satisfy you, for I feel it would involve the annihilation of my intellectual self.—not that annihilation would be of much consequence—but any thing approaching to insincerity or untruth would be of great consequence. Still I can ask you whether the memoir has produced on you such an impression as to cause you to prefer its having been unwritten? I here state my perfect readiness to suppress the memoir if it should have made any such impression.

    But supposing you allow it to pass on the whole, then I have another statement to make. It must be evident to you that I write in the presence of the whole scientific world, and that your uncle is not without friends to chastise me if I do him anything approaching to injustice. But there is one domain in which your feelings would have a paramount influence on me. I never dreamt that a fault would be found with my treatment of the scientific questions, but I felt from the first that it might be otherwise with the private & personal questions. Your last note to me contains these words:—"many of your allusions to his personal character are very gratifying to us, and such as we might be inclined to think might have been omitted, will we hope, prove as interesting to general readers as you expect." You say "many" of my allusions to his personal character are gratifying to you. Might I ask whether there are any that are not gratifying? And now my dear Miss Barnard you will pardon me if I speak with distinctness. Regarding my personal references to Michael Faraday I must have no whisper of discontent from you. No intimation must ever be made that I inserted in the memoir any thing of a personal character that you felt inclined to see omitted. I shall therefore suspend the publication of the memoir until you are able to certify to me that it contains nothing of a personal character that does not meet your entire approval. I am perfectly willing to omit every thing that you wish to see omitted. And if in praise of your uncle's character you can say anything better than what I have said, then I do not apprehend any difficulty in the way of my subscribing to it, and inserting it.

Among all those who have read this memoir, and heard the lecture, I have met but one dissentient voice and that is yours. It is natural that it should be yours, and no effort shall be wanting on my part to remove all cause of dissent on your part. But now is the time to make alterations, and I would repeat that until your approval has been expressed, not regarding "many" but regarding all personal matters the memoir shall remain unprinted: | I am | yours very truly | John Tyndall'.

To Jane Barnard 10 February 1868 2758

10th Feb 1868

My dear Miss Barnard

The enclosed note[1] was written before your note of Saturday the 8th.[2] reached me this morning. I am half inclined not to send it to you, for your last note is so pleasant to me, and shows so clearly that your mind is softened towards me as to render the sending of the note unnecessary. On the whole it is perhaps best to send it so that we may understand each other better.

Yours ever | John Tyndall

RI MS JT/1/T/116
RI MS JT/1/TYP/12/4195

1. *enclosed note:* letter 2757.
2. *your note of Saturday the 8th:* letter missing.

From Jane Barnard 11 February 1868 2759

12 Barnsbury Villas[1] | Tuesday. 11 Feb. 1868

Dear Dr Tyndall

I thank you for your notes received yesterday,[2] and now endeavour to reply to them. As I said before, neither my Aunt[3] nor I, can enter into, or judge of the scientific portion of the memoir.[4] I never pretended to know anything of Science & therefore could form no opinion of the way in which you have treated my Uncle's[5] discoveries & researches, but I do feel very grateful for the alterations you have made in some of the impressions & which have removed all my objections.

I have glanced through the book again, & I find as a whole it reads very pleasantly, & the general result is most favourable to him.

I think I can now freely say, that all your personal allusions to my dear Uncle are very gratifying to us & we shall rejoice with you, when it is fairly out of your hands.

Let me remind you in conclusion that I never sought to criticise this memoir, I felt all the way through that it was a difficult subject for us to interfere in, but you asked it each time.

I am yours | very sincerely | Jane Barnard

P.S. I send the book by the same post.

Dr Tyndall. | &c &c &c

RI MS JT/1/B/37

---

1. *12 Barnsbury Villas:* see letter 2695, n. 2.
2. *your notes received yesterday:* letters 2757 and 2758.
3. *my Aunt:* Sarah Faraday.
4. *memoir:* the manuscript for Tyndall's *Faraday as a Discoverer* (see letter 2665, n. 3).
5. *my Uncle's:* Michael Faraday.

From Jane Barnard [15 February 1868][1] 2760

12 Barnsbury Villas[2] | Saturday Night

Dear Dr Tyndall

I have read this sheet of the memoir[3] through to my aunt[4] & we both like it much. Some parts are beautiful—you <u>do</u> paint his character with a loving hand; & this makes me think, & say, that if in these parts which I can enter into a judge, I see so much to admire, why should I worry you & myself with other parts, & which I may well believe myself presumptuous in speaking of.

Your last words[5] grieved me; I did not think I was wronging you—I thought only of my Uncle.[6] As you say I may expect some sharp lessons & may well fear that I shall not find another to listen & bear with me, so patiently as you have done.

My Aunt does not wish any part omitted of this concluding sheet, nor I either; we think the part of the letter already crossed out was not of general interest, & therefore much better left out. I do not find it so very sad now, I think it was your mode of reading it[7] which affected me so much.

How glad you will be when you have finished the subject.

I am | yours very sincerely | Jane Barnard.

Dr Tyndall &c &c &c

RI MS JT/1/B/77

1. *[15 February 1868]:* the date is given by relation to letter 2757, dated 10 February 1868 (a Monday), the following Saturday being 15 February.
2. *12 Barnsbury Villas:* see letter 2695, n. 2.
3. *memoir:* see letter 2754, n. 2.
4. *my aunt:* Sarah Faraday.
5. *Your last words:* in letter 2757, Tyndall wrote to Barnard of his reluctance to publish his memoir of Faraday and asked her to inform him of 'where I have erred in my appreciation of his scientific work, where I have done him injustice, where I have expressed myself otherwise than lovingly and reverently regarding him'.
6. *my Uncle:* Michael Faraday.
7. *reading it:* see letter 2665, n. 3.

From George Biddell Airy 17 February 1868 2761

[It is requested that all correspondence on the subject of this letter may be addressed to the Astronomer Royal.]
Royal Observatory, Greenwich, | London, S.E. | 1868 Feb[ruar]^{y}. 17

My dear Sir

I am almost afraid that my awkward expressions[1] may have conveyed an idea quite different from what I intended.—Sir J. Herschel[2] and myself have no right to fix the meaning of the term <u>interference</u>: we are followers of Young[3] and Fresnel:[4] and I cited Sir J. H. and myself only as authors of the books now most commonly read,[5] these books having already attained a respectable antiquity.

In Optics we <u>must</u> have a word to include all the degrees of difference of phase. Almost every problem in Optics is of this kind.

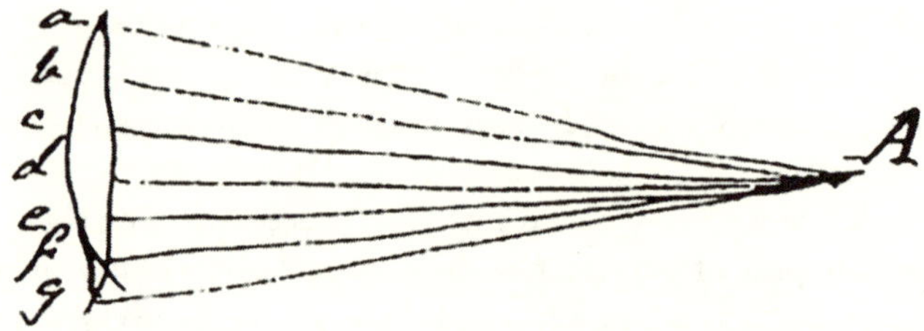

Waves of Light from every point of a lens or of an aperture, as from a, b, c, d, e, f, g, &c, unite at A. They unite in every possible difference of phase. This is what, in Optics, is always called interference. Now if you deprive us of this word as available in this sense, and so recognized for many years, we must ask you to give us another.

I am, my dear Sir, | Yours very truly, | G B Airy

Professor Tyndall

RI MS JT/1/A/45
RI MS JT/1/TYP/1/35

---

1. *my awkward expressions:* see letter 2756.
2. *Sir J. Herschel:* John Herschel.
3. *Young:* see letter 2667, n. 7.
4. *Fresnel:* Augustin-Jean Fresnel (see letter 2706, n. 8).
5. *books now most commonly read:* see letter 2756, n. 4 and n. 5.

From Jules Jamin 20 February 1868 2762

Paris 20 février 1868

Mon cher ami,

Mr Gauthier-Villars un peu malade est dans ce moment à Cannes et j'ai dû attendre sa réponse. La dépense à faire pour éditer votre livre à 1000 exemplaires se monterait à 1200 francs environ sans compter les frais de traduction et de correction qui seront de 500 francs à peu près. C'est la la dépense totale que vous seriez obligé de faire tout d'avance.

Après cela Mr Gauthier vendra votre livre moyennant un droit de librairie de 33% et ce qui restera vous reviendra intégralement soit pour solder les frais d'impression soit pour vos droits d'auteur définitifs.

Le traducteur, un de mes élèves, a déjà commencé son travail qui sera bientôt terminé; mais il est souvenu deux incidents dont je dois vous faire part. Je viens de recevoir la visite de Mr Alglave directeur de sa Revue des cours scientifique qui a déjà publié plusieurs de vos leçons et qui demande à publier votre éloge de Faraday. Il offre de faire la traduction à ses frais et après l'impression dans son journal de faire un tirage à part. Ce tirage à 300 exemplaires serait vendu par Germer Baillière, un libraire très honorable; il serait exécuté à ses frais et vous recevrez la moitié des bénéfices. Cette combinaison ne mettrait à votre charge aucune dépense et vous n'auriez à toucher que le revenu qui vous serait afférent. Il y a dans cette combinaison un avantage évident celui de vous exonérer de toute chance; mais en publiant dans la revue à 6000 exemplaires vous détruisez à peu près toute chance de vente ultérieure puisque les lecteurs de cette revue sont ceux qui achèteraient votre livre. En second lieu, la composition de cette revue très entassée et en petits caractères ne fera pas un livre de forme agréable et je me rappelle vos propres paroles que vous voulez paraître en une forme honorable en France. Je vais attendre vos ordres pour répondre à cette proposition.

Le second indicent est très différent. Mr Dumas récemment nommé secrétaire perpétuel de l'académie veut inaugurer sa fonction en faisant l'éloge de Faraday. Il a dû vous écrire pour vous demander des renseignements et il m'a demandé communication des épreuves que vous m'avez envoyées. Il ne me parait pas convenable de les refuser d'autant plus que un pour le but que vous-même poursuivez. Vous pourriez même envoyer à Mr Dumas un exemplaire, ce qui serait tout à fait gracieux.

Ma fille et moi avons été bien sensibles à votre cadeau. Elle travaille à vous faire en anglais une lettre de remerciements, mais comme elle est très inhabile et que je veux qu'elle la fasse seule vous l'attendrez peut-être encore un peu. Soyez indulgent pour elle autant que vous avez été bon.

à vous sincèrement | J. Jamin

Paris February 20 1868

My dear friend,

Mr Gauthier-Villars, who is a little sick is in Cannes[1] at the moment, and I had to wait for his reply. The cost of publishing your book[2] at 1000 copies would come to about 1200 francs, without counting the costs of the translation and the correction that will be around 500 francs. This is the total cost that you will have to provide in advance.

Afterward, Mr Gauthier will sell your book with a 33% fee for the rights of the bookstore, and he will fully return the rest to you either to pay off the cost of the printing or your final copyright.

The translator, one of my students,[3] has already begun his work and it will be completed soon. But he remembered two incidents that I have to share with you. I have just received a visit of Mr Alglave,[4] the director of his Revue des cours scientifiques who already published many of your lessons, and who asks to publish your commendation of Faraday.[5] He offers to do a translation at his expense, and after printing it in his journal he would make a separate printed edition.[6] This printing of 300 copies would be sold by Germer Baillière,[7] a very honourable bookseller. He would run it at his own expense and you will receive half of the profits. This combination would not cost you anything, and you would have to deal with nothing but the income that will be allotted to you. There is an obvious advantage in this arrangement that exempts you from any chance; but by being published in the Review at 6000 copies, you would destroy almost all chances of future sales because the readers of this Review are those who would buy your book. On the other hand, the composition of this Review, which is very crowded and in small characters, will not make a well-styled book, and I remember your own words, that you want to appear in an honourable form in France. I will wait for your orders before responding to this proposition.

The second incident is very different. Mr Dumas,[8] recently appointed Permanent Secretary to the Academy,[9] wants to inaugurate his office by making a commendation for Faraday. He probably wrote to you to ask for information, and he asked me to send him the samples that you sent me. It does not seem appropriate for me to refuse, especially for the goal that you pursue as well. You could also send a copy to Mr Dumas, which would be quite gracious.

My daughter[10] and I were very touched by your gift.[11] She is working on a thank you letter to you in English.[12] But since she is unskilled and I want her to do it on her own, you might have to wait for a little while longer. Please be as kind to her as you have been.

Yours truly | J. Jamin

RI MS JT/1/J/10

---

1. *Cannes:* a seaside city in France, on the Mediterranean.
2. *your book: Faraday as a Discoverer* (see letter 2750).
3. *translator, one of my students:* not identified, but he was referred to in letter 2750.
4. *Mr Alglave:* Émile Alglave (1842–1929), a French jurist and popularizer of science. In 1863 he and Eugène Yung (1827–87) founded the *Revue des cours scientifiques,* a widely read scientific periodical (*DBF*).
5. *your commendation of Faraday:* see n. 2.
6. *after printing it in his journal he would make a separate printed edition:* Tyndall's memoir of Faraday does not appear to have been printed in *Revue des cours scientifiques de la France et de l'Étranger* (a scientific periodical published between 1863–70, edited by Eugène Yung and Émile Alglave), nor as a separate printed French edition in addition to that published by Gauthier-Villars (see letter 2750, n. 3).
7. *Germer Baillière:* a French publisher specializing in medical and scientific works founded by brothers Germer Ballière (1807–59) and Jean-Baptiste Baillière (1797–1885).
8. *Mr Dumas:* Jean Baptiste André Dumas (1800–84), a French chemist and politician who worked with the French administrator Georges-Eugène Haussmann (1809–91) on the modernization of Paris (*CDSB*).

9. *Academy:* Dumas delivered a paper, 'Éloge historique de Faraday' at the Académie des Sciences on 18 May 1868. See J. Dumas, 'Éloge historique de Faraday', *Revue des cours scientifiques,* 5 (1868), pp. 393–407.
10. *My daughter:* Lucie Jamin (see letter 2750, n. 2).
11. *your gift:* not identified.
12. *thank you letter to you in English:* if this was sent to Tyndall, it is missing.

To Rudolf Clausius 20 February 1868 2763

20th Feb. 1868.

Lieber[1] Clausius

Deinen Brief werde ich Morgen antworten. Jetzt habe ich nur Zeit zu sagen, daß Heute die Royal Society sich die Ehre gethan hat Dich als Foreign Member zu wählen.[2]

I am rejoiced my dear Clausius at this result: and Huxley[3] desires me to send his best congratulations. Hirst[4] will also be rejoiced.

Yours ever | John Tyndall

RI MS JT/1/T/198

1. *Lieber:* Dear (German).
2. *Deinen Brief werde . . . Foreign Member zu wählen:* I shall reply to your letter tomorrow. For now, I only have time to say that today the Royal Society has done itself the honour of electing you a Foreign Member (German). Clausius's letter is missing, but Tyndall's reply is letter 2764 (see 'Anniversary Meeting', *Roy. Soc. Proc.,* 17 (1868–9), p. 134). The RS would officially elect Clausius as a Foreign Member on 26 March 1868; strangely, Tyndall's signature is not on the certificate (RS, Certificates of Election and Candidature, EC/1868/02).
3. *Huxley:* Thomas Henry Huxley.
4. *Hirst:* Thomas Hirst.

To Rudolf Clausius 22 February 1868 2764

22nd. Feb. 1868.

My dear Clausius,

The first part of your Memoirs on the Dynamical Theory of Heat[1] reached me duly. I have no clear memory regarding the Memoir for the Royal Society. But of one thing I am certain, that my first act on receiving the memoir would be to put it into the hands of one of our porters & to send it to the Royal Society.

The three copies of the Second part[2] have also reached me. Hirst[3] has received his copy, I have mine, and you inform me[4] that the Royal Society has received its copy.

In my brief note on Thursday[5] I had not time to inform you that your reception by the Council of the Royal Society[6] was unprecedented in my experience. Members of the council have always their favourite candidates which always more or less destroy unanimity. But every member of the council present voted for you[7]

I congratulate Johnny[8] on the possession of a brother.[9] It will add greatly to the happiness of his life. I congratulate you also my dear Clausius on this new transmission of your own good blood.

Yours ever | John Tyndall

RI MS JT/1/T/199

1. *first part of your Memoirs on the Dynamical Theory of Heat:* probably the first volume (1864) of R. Clausius, *Abhandlungen über die mechanische Wärmetheorie,* 2 vols (Braunschweig: Friedrich Vieweg und Sohn, 1864–7), a collection Clausius's original papers on thermodynamics. Published in English (with Thomas Hirst as editor and an introduction by Tyndall) as R. Clausius, *The mechanical theory of heat, with its applications to the steam-engine and to the physical properties of bodies* (London: John van Voorst, 1867).
2. *Second part:* see n. 1.
3. *Hirst:* Thomas Hirst.
4. *inform me:* letter missing.
5. *my brief note on Thursday:* letter 2763.
6. *Council of the Royal Society:* for the members of the RS Council elected on 30 November 1867, see 'Anniversary Meeting', *Roy. Soc. Proc.,* 16 (1867–8), p. 180.
7. *voted for you:* as Tyndall informed him in letter 2763, Clausius was elected a Foreign Member of the RS in 1868.
8. *Johnny:* Rudolf John Clausius.
9. *brother:* Alfred Clausius (1868–1950).

From Rudolf Clausius 25 February 1868 2765

WURZBURG, 25/2/68.

Lieber Tyndall,

Du hast mich durch Deine in so freundlicher Weise gleich an demselben Tage abgesandte Benachrichtigung von der Wahl der Royal Society ausserordentlich überrascht und erfreut. Ich weiss die grosse Ehre, welche mir durch die Aufnahme in eine der ersten wissenschaftlichen Gesellschaften der Welt erwiesen ist, vollkommen zu schätzen, und ich weiss auch, dass ich diese

Ehre zum grossen Theile Deinen freundlichen Bemühungen zu verdanken habe, und ich sage Dir meinen herzlichsten Dank dafür. Auch meine Frau welche sich für meine wissenschaftlichen Beziehungen lebhaft interessiert, hat sich sehr gefreut.

Was die Angelegenheit wegen des ersten Theiles meiner Abhandlungensammlung anbetrifft, so bin ich von Anfang an davon überzeugt gewesen, dass das Versehen welches gemacht dabei zu sein scheint, nur von der Buchhandlung gemacht ist, welche bei der Versendung des ersten Theiles auch andere mir zur Kentniss gekommene Versehen gemacht hat.

Ich muss Dich gleich wieder mit der Bitte ähnlicher Art belästigen. Ich habe von dem kleinen Vortrage, von dem ich Dir vor einiger Zeit ein deutsches Exemplar geschickt habe, eine Anzahl Exemplare einer französischen Uebersetzung enthalten, und möchte diese einigen Gelehrten schicken, denen ich keine deutschen Exemplare geschickt habe, unter anderen Grove und Maxwell. Da ich aber deren Adressen nicht weiss, so erlaube ich mir sie an Dich zu adressiren, mit der Bitte sie ihnen gelegentlich zukommen <zu> lassen.

Besten Dank für Deine Gratulation zur Geburt des kleinen neu angekommenen Sprösslings. Sei so gut auch Huxley für seine freundliche Theilnahme an meiner Wahl meinen Dank zu sagen, und ihn von mir zu grüssen, und nimm selbst für Dich und Hirst die herzlichsten Grüsse von meiner Frau und mir.

Dein | Clausius.

Würzburg,[1] 25/2/68.

Dear Tyndall,

The kindness with which you notified me[2] on the same day of the Royal Society's election[3] has surprised and delighted me extraordinarily. I do appreciate the great honour of being admitted to one of the leading scientific societies in the world, and I also know that I owe this great honour to a great degree to your kind efforts, for which I want to say my most heartfelt thanks. My wife,[4] who takes a lively interest in my scientific relationships, was also delighted.

Regarding the matter with the first part of my collection of articles,[5] I was convinced from the beginning that the mistake that seems to have been made was made by the bookstore alone, of which I have learned that it had also made other mistakes with the shipping of the first part.

I have to importune you immediately again with a request of a similar nature. I have received a number of copies of a French translation of a small lecture[6] I had sent you a German version[7] of some time ago. I would like to send these to some scholars to whom I had not sent a German version, Grove[8] and Maxwell,[9] among others. Since I do not know their addresses, I take the liberty of sending these to you with the request to convey them to them when you have occasion to do so.

I thank you for your kind congratulations on the birth of our newly arrived offspring.[10] Will you be so kind as to also thank Huxley[11] for his kind participation in my election and to greet him from me? Please accept my and my wife's best regards to you and Hirst.

Your | Clausius

RI MS JT/1/TYP/7/2304–5
LT Typescript Only

1. *Würzburg:* Clausius held a professorship at the University of Würzburg.
2. *You notified me:* letter 2763.
3. *Royal Society's election:* in 1868, Clausius was elected a Foreign Member of the RS (see letter 2764, n. 6).
4. *My wife:* Adelheid Clausius.
5. *first part of my collection of articles:* see letter 2764, n. 1.
6. *French translation of a small lecture:* this refers to Clausius's 23 September 1867 lecture before the German Scientific Association in Frankfurt, 'Über den zweiten Hauptsatz der Mechanischen Wärmetheorie', in which he distinguished between the first and second principles of thermodynamics and introduced the term 'irreversibility' into physics. Later in 1867 he published this—not in a journal but on its own—as *Über den zweiten Hauptsatz der Mechanischen Wärmetheorie* (Braunschweig: Vieweg und Sohn, 1867). This was translated and published in French in February 1868 (R. Clausius, 'Le second principe fondamental de la théorie mécanique de la chaleur', *Revue des cours scientifiques,* 5 (1868), pp. 153–9), and into English in June 1868 (R. Clausius, 'On the Second Fundamental Theorem of the Mechanical Theory of Heat', *Phil. Mag.*, 35 (1868), pp. 405–19).
7. *German version:* see n. 6.
8. *Grove:* William Grove.
9. *Maxwell:* James Clerk Maxwell.
10. *birth of our newly arrived offspring:* see letter 2764, n. 9.
11. *Huxley:* Thomas Huxley.

To Hallam Tennyson 26 February 1868 2766

26th. Feb. 1868.

My dear Hallam

I have been thinking of sending you & Lionel[1] two books[2] that I have written in the last six weeks, but I have been so awfully busy that the sending of the books has been postponed until now. You, I suppose, are at Marlborough[3] & I therefore send your book there, Lionel is a wanderer I know not where, so I send his to Freshwater.[4]

Give my kind regards to your father[5] & mother[6] when you write to them.

I may see them next week, as I am weary & am thinking of resting myself by walking round the Isle of Wight.[7]

I was terribly frightened by your illness,[8] and terribly glad that you got over it.

Give my kind remembrances to M[r]. & M[rs]. Bradly.[9]

Yours affectionately | John Tyndall

TRC Letters 6244

---

1. *Lionel:* Lionel Tennyson.
2. *two books:* probably *Faraday as a Discoverer* and *Heat* (1868).
3. *Marlborough:* Hallam was a pupil at Marlborough College (see letter 2575, n. 2).
4. *Freshwater:* Freshwater Bay, a village on the Isle of Wight, where the Tennysons had a home—Farringford House—since 1853 (and kept by the family after Tennyson's death in 1892). From 1869, the Tennyson's main residence was Aldworth House near Haslemere.
5. *father:* Alfred Tennyson.
6. *mother:* Emily Tennyson.
7. *Isle of Wight:* the Isle of Wight, where Tyndall visited the Tennyson family on 4–5 March 1868. In letter 2769 (28 February), he wrote of 'running away from London for a few days next week', while in letter 2770 (28 February), he wrote 'next week I intend to rush down to the Isle of Wight'. Letter 2773 shows he arrived on 4 March, and letter 2777 implies that he visited for only two days. During this trip, Tyndall missed an X Club meeting (with Darwin as a guest) on 5 March (see letter 2774).
8. *Your illness:* see letter 2575, n. 4.
9. *M[r]. & M[rs]. Bradly:* George Bradley (see letter 2575, n. 1) and his wife, Marian Bradley (née Philpot, d. 1910).

From Jane Barnard — 27 February 1868 — 2767

12 Barnsbury Villas[1] | N. | Thursday Night

Dear Dr Tyndall

I was so overpowered by the feelings raised by the conclusion of your lecture today,[2] that I felt unable to say more than a passing word, & so must have appeared cold & ungrateful: I cannot now lose this opportunity of expressing my warm thanks, & trying in some degree to efface the unpleasant impression I have given you, of my captious fault finding & the difficulty of my being pleased.

It would be presumptuous of me to offer any opinion of the way, you have treated my dear uncle's[3] work; but I may tell you that every word you have said of the fine, & lovely qualities, of his mind, has touched & delighted me.

I have had the privilege of attending the whole course,[4] & I have listened to all that has fallen from your lips, with unmixed pleasure—todays lecture seemed to me, a fine conclusion to a beautiful course.

My Aunt[5] intended writing to you herself, but has now commissioned me to say she is looking for your promised visit. Any evening this week except Wednesday would suit us, if you can name one, but perhaps you may be going away for a rest.

With our grateful thanks for all you have done

I am | yours very sincerely | Jane Barnard | J Tyndall Esq | &c &c &c | 27 Feb. 1868

RI MS JT/1/B/38

1. *12 Barnsbury Villas:* see letter 2695, n. 2.
2. *your lecture today:* the last of a course of nine lectures Tyndall gave at the RI on 'The Discoveries of Faraday' which began on 30 January 1868 (see letter 2714, n. 4).
3. *my dear uncle's:* Michael Faraday.
4. *whole course:* see n. 2.
5. *My Aunt:* Sarah Faraday.

From John Tyndall[1] 27 February 1868 2768

The Lodge | Gorey[2] | 27th. Feb. 1868.

My dear Friend,

Your consideration and kindness excel, in sending me <u>two</u> cheques for which accept my best thanks.

With regard to Emma[3] I am happy to say, I have nothing new to add, only that she still enjoys excellent health out every day in the fresh air, during this unprecedented beautiful winter. She is going on well except at intervals she experiences despondences which soon pass off

Mrs Tyndall[4] and my family cordially unite with me in kindest regards

Believe me | ever yours very sincerely | John Tyndall.

RI MS JT/1/TYP/10/3372
LT Typescript Only

1. *John Tyndall:* John Tyndall of Gorey.
2. *The Lodge | Gorey:* Tyndall's home in Gorey, County Wexford, Ireland.
3. *Emma:* Emma Tyndall.
4. *Mrs Tyndall:* the wife of Dr. John Tyndall is not identified.

To Jane Barnard 28 February 1868 2769

28th Feb 1868

My dear Miss Barnard.

It is I who am your debtor for the extreme pleasure which your note[1] has given me.

As you surmise I am thinking of running away from London[2] for a few days next week, but when I return I shall be very much delighted to pay your aunt[3] & yourself a visit.

Yours ever | John Tyndall

RI MS JT/1/T/110
RI MS JT/1/TYP/12/4198

1. *your note:* letter 2767.
2. *running away from London:* to the Isle of Wight (see letter 2766, n. 7).
3. *your aunt:* Sarah Faraday.

To Elizabeth Dawson Steuart 28 February [1868][1] 2770

Royal Institution, | 28th. Feb.

My dear Mrs Steuart,

Yesterday I finished my course of lectures at the Royal Institution.[2] Tomorrow I commence another, but not by any means so arduous, at the School of Mines.[3]

The completion of this piece of work gives me a little breathing time; which I need. I have been at times low, but next week I intend to rush down to the Isle of Wight,[4] and enjoy there the healthy breezes of the downs for a couple of days!

This morning I received a note from Gorey[5] which I send you—it gives a very fair account of Emma's[6] health. She had always a tendency to despond so that her present occasional lowness is not surprising.

What charming weather we have had lately—as I write the sun is shining on the walls of St. George's Chapel,[7] which is just in front of me.

I lived for several years in St. John's Wood, but some weeks ago I shifted to Albermarle Street to occupy the rooms vacated by Faraday.[8] This I suppose will be my residence to the end of the Chapter.

I hope your own health is better.

Believe me | ever yours | John Tyndall.

RI MS JT/1/TYP/10/3373
LT Typescript Only

1. *[1868]:* the year is given by relation to letter 2768 from John Tyndall of Gorey, in which he related to Tyndall the health of his sister Emma. Also, it was in 1868 that Tyndall moved into Faraday's old rooms at the RI.
2. *course of lectures at the Royal Institution:* see letter 2767, n. 2.
3. *another . . . at the School of Mines:* this is likely to refer to Tyndall's lectures to working men, which were given in the evenings (see letter 2776). He also may have been referring to the commencement of his work preparing the course of thirty-two lectures he would deliver there starting on 27 April 1868 (see letter 2804, n. 3). On Tyndall and the School of Mines, see letter 2601, n. 2.
4. *Isle of Wight:* see letter 2766, n. 7.
5. *note from Gorey:* letter 2768.
6. *Emma's:* Emma Tyndall.
7. *St. George's Chapel:* the fourteenth-century Gothic chapel at Windsor Castle in London.
8. *I lived for several years . . . to occupy the rooms vacated by Faraday:* see letter 2707, n. 8.

To Jane Barnard [February 1868][1] 2771

Sunday

My dear Miss Barnard,

Your corrections[2] are all excellent & your wishes shall be all attended to. One point however though I will modify it to suit your wishes I must make clear; and that is your uncle's's[3] care to keep himself in a position which could at any moment be justified before the face of men. His care in this particular was extraordinary: I could give you half a dozen different illustrations of it:—it was not that he feared the world, or was in any way anxious about it, save through the wish to avoid all complication. The distinction between him & Scipio[4] has been in my mind for many years; and remember he was Christian whereas Scipio was a heathen. And if I lean towards Scipio it is because I am a heathen too.

Yours ever | John Tyndall

John or Paul[5] would have acted the part of your uncle:—but the devil in Scipio takes my fancy.

With this I leave a quantity of proofs[6] at the Institution:[7] will you kindly read them and tell me whether you have any quotation or reference that you think might be introduced with advantage in this portion of the Memoir.

RI MS JT/1/T/135
RI MS JT/1/TYP/3/4194

1. *[February 1868]:* the month and year are given by Tyndall's seeking Jane Barnard's assistance in looking over the manuscript of Tyndall's *Faraday as a Discoverer,* published in March 1868 (working on the manuscript would have come after Tyndall's Friday Evening Discourses on Faraday in January; see letter 2665, n. 3).
2. *your corrections:* letter missing. Jane Barnard appears to have objected to Tyndall's comparison of Michael Faraday to Scipio. Scipio was not mentioned in *Faraday as a Discoverer* (1868) nor in the published discourses from January.
3. *your uncle's:* Michael Faraday.
4. *Scipio:* probably a reference to the Roman general and politician Scipio Africanus (*c.* 236–183 bc) or his grandson Scipio Aemilianus (*c.* 185–129 bc), also a Roman general and politician.
5. *John or Paul:* John the Apostle (*c.* AD 6–100) and Paul the Apostle (*c.* AD 5–64), two of the apostles of Jesus.
6. *proofs:* the draft of *Faraday as a Discoverer.*
7. *Institution:* the RI.

To Juliet Pollock [2 March 1868][1] 2772

Monday

My dear M[rs]. Pollock.

There were a great number of people elected today at the Royal Institution: but I noticed when the Chairman[2] read out <u>your</u> name he paused & said "I have the pleasure to announce &c." Perhaps nobody noticed it but myself—It gave me pleasure too.

Yours ever | John Tyndall

RI MS JT/1/T/1179
RI MS JT/1/TYP/6/2220

1. *[2 March 1868]:* the date is given based on the date Juliet Pollock was elected a member of the RI, 2 March 1868 ('General Monthly Meeting', *Roy. Inst. Proc.,* 5 (1866–9), p. 308).
2. *Chairman:* William Spottiswoode.

To Hallam Tennyson 4 March [1868][1] 2773

Plumbly's Freshwater Bay Hotel S.W.[2] | 4th March

My Dear Hallam.

Thus far am I come:—I will call to see you tomorrow morning & shall deem myself lucky if I can persuade you & Lionel[3] to throw your books away for couple of hours & come with me to the end of the Island.[4] I will not tempt you down the cliffs. Be sure to make this promise known to your mother.[5]

Yours affect[ionate]ly | John Tyndall

TRC Letters 6245

1. *[1868]:* the year is given by reference to Tyndall visiting the Tennyson's on the Isle of Wight (see letter 2766, n. 7).
2. *Plumbly's Freshwater Bay Hotel S.W.:* a hotel in the village of Freshwater Bay, Isle of Wight.
3. *Lionel:* Lionel Tennyson.
4. *Island:* Isle of Wight.
5. *your mother:* Emily Tennyson.

From Joseph Dalton Hooker 7 March 1868 2774

Royal Gardens Kew[1] | March 7th. 1868.

Dear Tyndall,

What is doing about Faraday and the Abbey?[2] We had a talk on the subject at X[3] on Thursday. (where, by the way, I got Darwin to dine with us[4]) Suppose that the Club resolves itself (or rather creates itself) into a Committee upon this matter? Or had we better be fewer? and confine the first steps to the joint action of ourselves Huxley,[5] Lubbock[6] and Spottiswoode.[7]

The idea is abroad that the Royal Institution should take a leading part in the matter, in which case we should memorialize that body.

I would come to town and meet you on the subject on any day you chose and could get the other fellows together will you see about this?

I came here to see my mother,[8] whom I find very stout and well for her.

Ever yours | J. Hooker

Miss Hawthorn[9] is at Kew,[10] if you care to see her you will find her on Monday. She leaves on Tuesday. My wife[11] is laid on the sofa with a sprained ankle.

I wish you would go to Woolner[12] and would bring him out on Monday evening, I have told him he will find food and no ceremony at 6½.

We might talk over the monument with him to advantage.

RI MS JT/1/TYP/12/4208
LT Typescript Only

1. *Royal Gardens Kew:* see letter 2554, n. 2.
2. *Faraday and the Abbey:* Hooker, Thomas Huxley, and Tyndall led efforts to honor Faraday with a memorial at Westminster Abbey (see letter 2778).
3. *X:* in 1864 Tyndall joined Edward Frankland, Thomas Hirst, Joseph Hooker, Thomas Huxley, and four others as members of the X Club, a monthly dining club in which they discussed the promotion of science in British society. This refers to their meeting of 5 March 1868, which Tyndall did not attend. On the X Club, see R. Barton, *The X Club: Power and Authority in Victorian Science* (Chicago: University of Chicago Press, 2018).
4. *Darwin to dine with us:* in a letter from 3 March, Hooker invited Darwin to join him at the X Club meeting, to which Darwin replied on 4 March, despite not being sure of his ability to actually attend, 'I sh[oul]d like *excessively* to dine at your Club' (Joseph Hooker to Charles Darwin, [3 March 1868]; and Charles Darwin to Joseph Hooker, [4 March 1868], *Darwin Correspondence,* vol. 16, part 1, pp. 229–30, 231; these are letters 5917 and 13816 in the online edition of the Darwin correspondence).
5. *Huxley:* Thomas Huxley.
6. *Lubbock:* John Lubbock.
7. *Spottiswoode:* William Spottiswoode.
8. *my mother:* Maria Sarah Hooker (née Turner, 1797–1872), mother of Joseph Dalton Hooker. She married William Jackson Hooker in 1815 (*ODNB*).
9. *Miss Hawthorn:* Grace Agnew Hawthorn (1824–1914), a friend of Frances Hooker, and daughter of the Clergyman Robert Hawthorn. She married Thomas Boycott in 1871 (see Leonard Jenyns to Charles Darwin, 9 May 1868, *Darwin Correspondence,* vol. 16, part 1, p. 480; this is letter 6168 in the online edition of the Darwin correspondence).
10. *Kew:* the home of Joseph and Frances Hooker at Kew Green.
11. *My wife:* Frances Hooker.
12. *Woolner:* Thomas Woolner (1825–92), an English poet and sculptor who was among the founders of the Pre-Raphaelite Brotherhood. He was known for sculpting portrait medallions and busts, and among his subjects were Tennyson, Tyndall, and Bence-Jones (*ODNB*). It does not appear that Woolner produced any commission for Faraday.

From Emily Tennyson 8 March 1868 2775

Farringford, Freshwater, Isle of Wight.[1] | March 8th 1868

My dear Professor,

Accept my best thanks for your welcome gift. The book[2] will be dear to us all both as your gift and as what we know we shall find it, the able and loving memorial of a good and great man[3] by one worthy of the task if I may be allowed to say so—

We hope that it may not be long before we have the pleasure of another visit from you[4]

With the affectionate remembrances of all | Ever yours | Emily Tennyson

RI MS JT/2/10/330–1

---

1. *Farringford, Freshwater, Isle of Wight:* see letter 2766, n. 4.
2. *book:* probably *Faraday as a Discoverer,* which Tyndall may have gifted to the Tennysons during his visit.
3. *good and great man:* Michael Faraday.
4. *another visit from you:* see letter 2766, n. 7.

To Joseph Dalton Hooker 9 March [1868][1] 2776

9th. March.

My dear Hooker

I lecture to the working men tonight[2] otherwise I would join you.[3]

I am willing to fall in with your suggestion; but I think it would be a peaceful act to introduce the subject to our managers.[4] They are all so devoted to Faraday. What do you say if I bring the question before them on Monday next when they meet?

Still I do not see any objection to the formation of this preliminary committee if you like it.

I hope to see you soon: I am used up & will run out to get a little air tomorrow or next day.

Yours ever | John Tyndall

On reflection I think you have hit upon the best possible mode of proceeding. Let a few of us memorialise the managers. It could be done by Monday next; and I would present & explain the memorial.

Unfortunately I am fixed to the Trinity House[5] for some days otherwise I w[oul]d. come out to you.

Kew JDH, vol. 4, 1267–67a

---

1. *[1868]:* the year is given based on Faraday having died in 1867 and the beginning of the effort to honor him with a memorial statue (see *The Ascent of John Tyndall,* p. 209).
2. *lecture to the working men tonight:* Tyndall gave evening lectures to working men in his position with the Royal School of Mines (see letter 2601, n. 2).
3. *join you:* this letter is a reply to letter 2774 from Hooker.

4. *introduce the subject to our managers:* Tyndall, Hooker, and Thomas Huxley proposed to the Managers of the RI (see letter 2592, n. 6) that there should be a national monument to Faraday (see letter 2778). Also see letters 2774, 2781, 2784, and 2785.
5. *fixed to the Trinity House:* see letter 2590, n. 5.

To Elizabeth Dawson Steuart 10 March 1868 2777

10th. March 1868.

My dear Mrs Steuart,

Thank you for your letter,[1] and thank you also for those sweet violets. I am still very tired but I hope to get away for a week or so very soon. I was only two full days in the Isle of Wight,[2] where I stayed with the Poet Laureate,[3] and had a charming walk with him over the breezy downs.

He has a charming wife[4] and two sons,[5] and a most beautiful place at Farringford near Freshwater in the Island.[6]

And now to business—Sarah Tyndall[7] in my time of sorrow made a very pleasing impression upon me; and the thought then occurred to me of reserving for her a little marriage portion—I am quite ready to carry out this notion; but let me confess to a point of sentiment. I do not like to hear of a man[8] about to marry a girl he ought to love "expecting" £100:—However on this score Sarah herself is the best judge, and if you deem it desirable I am willing upon her marriage to place a cheque for £50 at her disposal. It is the sum I intended to set apart for her. If you advise it I am willing, if no other source of funds exist to make up the required sum, to add another £50 at the end of six months from her marriage.

His offer of a valueless security simply arouses in me indignation and contempt. But I like the girl and although the draft is a heavy one I will do my best to meet it.

Yours ever | John Tyndall.

RI MS JT/1/TYP/10/3374
LT Typescript Only

---

1. *your letter:* letter missing.
2. *Isle of Wight:* see letter 2766, n. 7.
3. *Poet Laureate:* Alfred Tennyson.
4. *wife:* Emily Tennyson.
5. *two sons:* Hallam and Lionel Tennyson.
6. *at Farringford near Freshwater in the Island:* see letter 2766, n. 4.

7. *Sarah Tyndall:* Sarah Tyndall (not Tyndall's mother but a cousin; see Biographical Register).
8. *man:* not identified. Probably the fiancé of John Tyndall's cousin, Sarah Tyndall.

To the Managers of the RI[1] 11 March 1868 2778

The Royal Society | Burlington House, London. W. | March 11th. 1868.

Gentlemen[2]

There is an opinion prevalent that no time should be lost in raising a national Monument to the late Prof. Faraday.[3] Having obtained from the Dean of Westminster[4] the assurance that every facility will be afforded for placing such a Memorial in Westminster Abbey,[5] we take this opportunity of calling your attention to this subject in the hopes that the members of that Institution with which Prof. Faraday was most conspicuously and intimately connected will take steps to promote this object.

We have the honor to be | Y[ou]r. most obedient | J. Hooker | T. H. Huxley | John Tyndall

To the | Managers of | The Royal Institution.

RI MS JT/1/TYP/12/4209
LT Typescript Only

1. *To the Managers of the RI:* this letter is from Joseph Dalton Hooker, Thomas Henry Huxley, and John Tyndall.
2. *Gentlemen:* for the Managers of the RI at this time, see letter 2592, n. 6.
3. *late Prof. Faraday:* Michael Faraday died on 25 August 1867.
4. *Dean of Westminster:* Arthur Stanley.
5. *Westminster Abbey:* a church in Westminster, London founded in 960, notable for being the burial site of famous scientists among monarchs, statesmen, and national figures, including Charles Darwin (1882), John Herschel (1871), Charles Lyell (1871), Isaac Newton (1727), and William Thomson (1907).

From Jules Jamin 11 March 1868 2779

Paris 11 mars 1868.

Mon cher ami,

Je me reproche de vous avoir beaucoup négligé depuis quelque temps. J'ai eu une grande préoccupation scientifique et vous me pardonnerez j'en suis sûr d'y avoir un peu trop pensé. J'ai d'ailleurs été obligé d'attendre la réponse de Mr Gauthier-Villars qui était toujours en voyage.

Mr Gauthier regrette beaucoup que vous ayez cru devoir envoyer un cheque pour couvrir les frais de la publication, il ne veut en aucun jours le recevoir et je me vois obligé de vous le renvoyer. Votre livre est bientôt traduit. On va l'imprimer—Mr Gauthier est sûr qu'il ne vous coutera rien et que loin de là, il vous rapportera quelque chose. Je partage sa confiance et je comprends sa délicatesse. Vous pouvez vous confier à lui et je me déclare garant de son honnêteté.

Je vous enverrai bientôt comme spécimen la traduction de l'un des vos chapitres et si vous en êtes content on commencera l'impression. J'ai vu hier Mr Dumas qui a beaucoup apprécié votre lettre. Il doit faire l'éloge académique de Faraday dans un mois. Je pense que votre traduction sera finie et imprimée pour cette époque et il sera convenable de la mettre en vente juste quand l'attention sera appelée sur Faraday.

Je reçois aujourd'hui même votre lettre donc je vous remercie.—et vous renvoie le cheque

a vous sincèrement | J. Jamin

£ 21–14–6

14–16–6

————————

£ 36–11–0

Paris, March 11 1868

My dear friend,

I am reproaching myself for having neglected you for some time. I had a great scientific concern and you will forgive me, I am sure, for having thought about it a bit too much. I also had to wait for a reply from Mr Gauthier-Villars, who was still travelling.

Mr Gauthier regrets very much that you thought you had to send a cheque in order to cover the expenses of the publication.[1] He does not want to receive it, in any day, and I am obliged to send it back to you. Your book is overall translated. It will be printed—Mr Gauthier is sure that it will not cost you anything, and that; far from it, it will provide a return for you. I share in his confidence and I understand his delicacy. You can trust him and I will guarantee his honesty.

Soon I will send you the example of the translation of one of your chapters, and if you are happy with it, we will start the printing. Yesterday I saw Mr Dumas,[2] who greatly appreciated your letter.[3] He has to put forth an academic commendation for Faraday[4] in the next month. I think that your translation will be finished and printed by then, and it will be convenient to start selling it just when the attention will be drawn to Faraday.

Today I also received your letter,[5] so thank you.—and I am returning you the cheque

Sincerely yours | J. Jamin

£ 21–14–6

14–16–6

————————-

£ 36–11–0[6]

RI MS JT/1/J/11

---

1. *publication:* probably J. Tyndall, *Faraday Inventeur,* trans. F. Moigno (Paris: Gauthier-Villars, 1868).
2. *Mr Dumas:* Jean Baptiste Dumas (see letter 2762, n. 8).
3. *your letter:* letter missing.
4. *academic commendation for Faraday:* Dumas gave a eulogy for Faraday at the Académie des Sciences in Paris on 10 May 1868 (see J.-B. Dumas, *Éloge historique de Michel Faraday* (Paris: Firmin-Didot, 1868)).
5. *your letter:* letter missing. Probably Tyndall's reply to letter 2762.
6. *£ 21–14–6 | 14–16–6 |————————- | £ 36–11–0:* written on the back of the letter; presumably the amount of money being returned to Tyndall with a cheque enclosed with this letter.

From Emily Peel 11 March 1868 2780

Drayton Manor | Tamworth[1] | March 11/68

Dear Professor Tyndall,

I have just received the copy of your work on the late Michael Faraday[2] which you have been good enough to send to me & I need scarcely say how much I appreciate your gift, or with what interest I shall read your notice of that truly great mind, & all that it achieved. I hope when I am in town after Easter to have the pleasure of seeing you & thanking you in person for your remembrance of me.

I remain yours | very truly | Emily Peel

RI MS JT/1/P/46

---

1. *Drayton Manor | Tamworth:* see letter 2693, n. 1.
2. *your work on the late Michael Faraday: Faraday as a Discoverer* (see letter 2665, n. 3).

From Henry Bence Jones 12 March 1868 2781

Dear Tyndall

After you spoke to me some days since regarding a monument to Mr Faraday[1] I saw the Dean of Westminster[2] and I find from him that his fees amount to about 200 guineas. He said that there was a place for a tablet and bust near Sir Humphry Davy's tablet[3] which I find is a poor thing; put up by Lady Davy,[4] I should think at a cost of £50.

A tablet and bust of Mr Faraday by Woolner[5] would cost £250 or £300.

Now I feel by no means inclined to a subscription to pay the Dean 200 guineas and upwards to put up such a monument.

If any thing is done it seems to me that a monument like that of Sir Isaac Newton[6] should be raised to Mr Faraday. For such a monument it would be far more worth paying the Dean what he requires. Such a memorial would cost some thousands and it seems to me that it should be started by a public meeting at which the President of the Royal Society[7] should preside and probably such a meeting had best be held by advertisement in May at the Royal Societies rooms.

The subscription should be unlimited and the Presidents of the different scientific societies should be asked to start the subscription and form the nucleus of the committee which should include every one. Whilst the working committee should be limited to those who will work.

In my opinion the Managers of the Royal Institution[8] would not start the subscription as widely as even the Council of the Royal Society and I do not think the Council of the R.S. would be followed as well as the presidents of the different societies

I am | yours truly | H. Bence Jones. | March 12th 1868

RI MS JT/1/TYP/2/693
LT Typescript Only

---

1. *monument to Mr Faraday:* see letters 2774 and 2778.
2. *Dean of Westminster:* Arthur Stanley.
3. *tablet:* a white marble memorial to Davy is in the north wall of the chapel of St Andrew in Westminster Abbey.
4. *Lady Davy:* Jane Davy (see letter 2654, n. 5).
5. *Woolner:* Thomas Woolner (see letter 2774, n. 12).
6. *Sir Isaac Newton:* Isaac Newton (1642–1727), an English natural philosopher, mathematician, and Lucasian Professor at University of Cambridge noted for many important discoveries in the physical sciences. He was buried in Westminster Abbey (*ODNB, CDSB*).

7. *President of the Royal Society:* Edward Sabine.
8. *Managers of the Royal Institution:* see letter 2592, n. 6.

From Henrietta Huxley 12 March [1868][1] 2782

26 Abbey Place | St. John's Wood[2] | 12th March (1868)

Dear Brother John

Thank you exceedingly for sending me your book about "Faraday".[3] I shall always associate it with your lectures at the Royal Institution in which you so feelingly and to us so unexpectedly alluded to your comradeship with dear Hal.[4] It stirred me deeply and so I could not speak to you about it then.

I have had a book[5] sent to me by a friend[6] to read—by the Rev. T. Cooke[7]—about his brother's[8] claim to be the first who applied electricity to the working of the telegraph.[9] Any one who reads this book must believe that Mr Wheatstone[10] did not behave well in the matter—but of course it represents one side of the case only. May I ask you whether Mr Cooke or Mr Wheatstone should have the credit of having first put electricity to telegraphic use—my friend being anxious to have a competent opinion. If faces are an index of character I should think all true that is alleged of Wheatstone in the book.

Hal is very very tired. For three years I have been urging him to give up the College of Surgeons lectures[11]—which coming as they do just after the course at Jermyn Street[12] are too much for him—and now I think he will not take them again. So much lecturing too has this disadvantage that he never has time to publish the work he has done and so it is in a manner lost. I long for Easter to come that he may get away for his walk. Are you going to be of the party?[13]

I am reading a delightful German book The Life of Fanny Lewald[14] which Dr Dohrn[15] recommended us very strongly—I don't know if you met him at the British Association last year.[16] He was there—and afterwards came on to visit us at Swanage[17]—A clever man and an agreeable companion.

How do you like living in town? My heart has been very sore at the thought of strangers going into the little house at Waverley Place.[18] I wanted to get up the courage to go inside the other day but could not.

If the children[19] knew that I was writing to you they would all send their love by acclamation. When Hal gets a little more leisure I hope that you will come and dine with us.

With very true regards | I am | Yours sincerely | Nettie Huxley.

RI MS JT/1/TYP/9/2918
LT Typescript Only

1. *[1868]:* the year is given based on the publication of Tyndall's book about Michael Faraday (see n. 3).
2. *26 Abbey Place | St. John's Wood:* Huxley's residence in London.
3. *your book about "Faraday": Faraday as a Discoverer* was published on 5 March 1868.
4. *Hal:* a nickname for Huxley.
5. *book:* T. F. Cooke, *Authorship of the Practical Electric Telegraph of Great Britain: Or, The Brunel Award Vindicated, in VII Letters Containing Extracts from the Arbitration Evidence of 1841, Edited in Assertion of His Brother's Rights* (Bath: R. E. Peach, 1868).
6. *friend:* not identified.
7. *Rev. T. Cooke:* Thomas Fothergill Cooke (1814–84), an Anglican clergyman and brother of William Fothergill Cooke (see n. 8). He wrote a number of books supporting his brother's claim to have invented the electric telegraph (*Alum. Cantab.*).
8. *brother's:* William Fothergill Cooke (1806–79), an English inventor who began his career in medicine and served in the British Army in India. He shares the invention of the electric telegraph with Charles Wheatstone (see n. 9) which they patented in 1837. Cooke also co-founded the world's first public telegraph company (Electric Telegraph Company) in 1846 (*ODNB*).
9. *claim to be the first who applied electricity to the working of the telegraph:* Cooke and Wheatstone become embroiled in a bitter dispute over the patents for the electric telegraph, and Wheatstone was not one of the founders of the Electric Telegraph Company. See J. Liffen, 'The Introduction of the Electric Telegraph in Britain, a Reappraisal of the Work of Cooke and Wheatstone', *International Journal for the History of Engineering and Technology,* 80 (2010), pp. 268–99.
10. *Mr Wheatstone:* Charles Wheatsone.
11. *College of Surgeons lectures:* Huxley held the Hunterian Professorship in Comparative Anatomy and Physiology at the Royal College of Surgeons in London from 1863–9.
12. *course at Jermyn Street:* Huxley was Professor of Natural History at the Royal School of Mines from 1854–85.
13. *Are you going to be of the party?:* in letter 2801, Tyndall wrote to Huxley: 'I have been looking forward to a scamper with you over the Welsh Hills', but informed him that Thomas Hirst persuaded him instead to join John Lubbock and others on a trip to Naples, Italy (see letter 2801, n. 3).
14. *German book The Life of Fanny Lewald:* F. Lewald, *Meine Lebensgeschichte,* 6 vols (Berlin: Otto Janke, 1861–62). Fanny Lewald (1811–89), a German novelist and author of a number of books supporting women's emancipation (*NDB*).
15. *Dr Dohrn:* Felix Anton Dohrn (1840–1909), a German zoologist and supporter of Charles Darwin's theory of evolution. In 1872 he founded the Stazione Zoologica in Naples, the world's first zoological research station (*CDSB*).
16. *British Association last year:* the annual meeting of the BAAS at Dundee (see letter 2627, n. 22).
17. *Swanage:* following the annual meeting of the BAAS in 1867, Huxley and his family

holidayed at the seaside in Swanage on the southern England coast. In his biography of Huxley, Adrian Desmond writes, '. . . Huxley and Dohrn fished in cooler Swanage seas. They all remembered it as a time of laughter. Up early for lobster breakfasts, then out to dredge, while at night the herd of seven children was taught "bovine vocalisations" by the affable Dohrn, leaving Harry [Henry] toddling round the house shouting "Mroo"' (A. Desmond, *Huxley: From Devil's Disciple to Evolution's High Priest* (Reading, MA: Perseus Books, 1994), p. 355).

18. *How do you like . . . house at Waverley Place:* see letter 2707, n. 8.
19. *children:* Jessie, Marian, Leonard, Rachel, Henrietta, Henry, and Ethel Huxley.

To Thomas Archer Hirst 13 March 1868[1] 2783

My dear Tom,

It is hard to catch you now so I send you the enclosed letter[2] from the Head Master of King Edward VI School at Norwich.[3] You are going to High Elms[4] I am informed on Sunday & may then be able to arrange about our lodging at Norwich.[5] Mr Jessopp was good enough to ask me in a very hearty fashion to stay with him—but as in the case of Buxton[6] I pleaded a prior engagement.

Yours aff[ectionate]^ly^. | John Tyndall | 13th March 1868

RI MS JT/1/T/666
RI MS JT/1/HTYP/542

1. The date 'March 13. 1868.' Is written on the letter, at top right, in a different hand.
2. *enclosed letter:* letter missing.
3. *Head Master of King Edward VI School at Norwich:* Augustus Jessopp (1823–1914), an English clergyman and educator who was headmaster of King Edward VI School, Norwich from 1859–79 (*ODNB*).
4. *High Elms:* the home of John Lubbock (see letter 2635, n. 13). Tyndall joined Hirst on Sunday 15 March from Victoria Station to Bromley, but Hirst proceeded to High Elms alone (Brock & MacLeod, *Hirst Journals,* 15 March 1868, p.1828).
5. *our lodging at Norwich:* Tyndall and Hirst traveled to Norwich in August 1868 for the thirty-eighth annual meeting of the BAAS.
6. *Buxton:* probably Charles Buxton (1822–71), an English brewer, Member of Parliament, and social reformer. In 1850 he married Emily Mary (d. 1871), the eldest daughter of Henry Holland (*ODNB*).

To Joseph Dalton Hooker 13 March [1868][1] 2784

13th March

My dear Hooker,

We must take the thing[2] in our own hands and work it as we see fit. Bence Jones objects to its being brought before the Managers.[3] His notion is that the presidents of the various societies ought to be the origin of it. Let us form a committee and if necessary have a meeting at Burlington House.[4] The country will respond to our appeal without a doubt.

Yours ever | John Tyndall

RI MS JT/1/TYP/8/2596
LT Typescript Only

---

1. *[1868]:* the year is given by relation to letter 2781.
2. *the thing:* Joseph Hooker, Thomas Huxley, and Tyndall had written to the managers of the RI (letter 2778) about erecting a monument in Faraday's honour, possibly in Westminster Abbey.
3. *Bence Jones objects to its being brought before the Managers:* Tyndall enclosed with this letter one he received from Henry Bence Jones (letter 2781). For the Managers of the RI at this time, see letter 2592, n. 6.
4. *Burlington House:* a private mansion purchased by the British government and made home to the Royal Society, Linnean Society, and Chemical Society in 1857. Three other learned societies and the Royal Academy of Arts joined them later.

From Henry Bence Jones 15 March [1868][1] 2785

Royal Institution

Dear Friend

I received a letter from Miss Barnard[2] last night in which she says

> "I repeated to my aunt[3] what you told me about the proposed Memorial in Westminster Abbey.[4] She does not feel that she could interfere in the matter at all; but at the same time would prefer that nothing should be done. She does feel a very strong objection to subscriptions for she thinks (and it was my dear uncle's[5] opinion too) that it is a wrong way of doing such things, for many are obliged to give because they do not like to refuse and thus it becomes a tax."

I think it would be well if <u>you</u> mentioned to the Managers[6] to morrow that Dr Hooker[7] was going to write a formal letter to them[8] on the subject but that it was afterwards thought better to begin with the R.S.

Yours ever | H. Bence Jones

March 15th. | The Prince of Wales[9] must be got to take the chair at the meeting at Burlington House[10] if it is held in May.

RI MS JT/1/TYP/12/4208
LT Typescript Only

---

1. *[1868]:* the year is given in relation to the efforts to erect a memorial for Faraday at Westminster Abbey.
2. *Miss Barnard:* Jane Barnard.
3. *my aunt:* Sarah Faraday.
4. *proposed Memorial in Westminster Abbey:* see letters 2774, 2778, and 2781.
5. *my dear uncle's:* Michael Faraday.
6. *Managers:* see letter 2592, n. 6.
7. *Dr Hooker:* Joseph Hooker.
8. *formal letter to them:* Hooker, Thomas Huxley, and Tyndall had written to the Managers a few days before (letter 2778), and it appears this 'formal letter' did not happen.
9. *Prince of Wales:* see letter 2788, n. 1.
10. *Burlington House:* a private mansion purchased by the British government and made home to the RS, Linnean Society, and Chemical Society in 1857. Three other learned societies and the Royal Academy of Arts joined them later.

From William Thomson 15 March 1868 2786

Glasgow College[1] | March 15/68

My dear Tyndall

I have been excessively busy ever since my return or I would long ago have written to thank you for the book[2] I found, from you, waiting my arrival, and which I value very highly. I have now also to thank you for your paper on Faraday[3] which came a few days ago, and of which I have been only able today, to read a considerable part. As soon as I finish reading it I shall send it to Mentone[4] as I know that my wife[5] will be greatly interested in reading all you have said about Faraday himself. I have been greatly interested in it today.

Tait[6] and Joule[7] are with me and both send you their kind regards.

I am yours always truly | William Thomson

P.S. I am to be in London from Thursday the 9th April till the following Monday and hope to bring you a quadrant electrometer.

RI MS JT/1/T/15
RI MS JT/1/TYP/5/1553

1. *Glasgow College:* Glasgow College is in the University of Glasgow, where Thomson held the Chair of Natural Philosophy from 1846–99.
2. *book: Heat* (1868). Tyndall had sent Thomson a copy on 25 January (see letter 2744).
3. *your paper on Faraday: Faraday as a Discoverer.*
4. *Mentone:* see letter 2535, n. 5. Margaret Thomson spent the winter in Mentone because of ill health (S. P. Thompson, *The Life of Lord Kelvin,* 2 vols (London: Macmillan, 1910), vol. 1, p. 526). She died in 1870.
5. *my wife:* Margaret Thomson (née Crum, d. 1870), a daughter of a wealthy cotton manufacturer. In 1852 she married her second cousin William Thomson (*ODNB*).
6. *Tait:* Peter Guthrie Tait.
7. *Joule:* James Joule (see letter 2742, n. 5).

From Henri Étienne Sainte-Claire Deville[1] 16 March 1868 2787

Laboratoire de Chimie | Ecole Normale Supérieure.
Paris, le 16 Mars 1868

Mon cher Tyndall, Je vous remercie de vos deux livres. J'en connais un & je l'admire beaucoup: car il a fait beaucoup d'effet en France, c'est votre livre sur la chaleur. Je vais lire votre Faraday avec le plus grand plaisir. Je l'ai beaucoup aimé votre illustre maître & il m'a écrit quelquefois qu'il daignait avoir quel qu'amitié pour moi. J'ai publié ici une notice sur lui qui d'après ce qu'il m'a fait écrire par *[2 words illeg]* ne lui a pas déplu—Je suis donc bien attaché à sa mémoire & je suis heureux de penser que c'est un homme aussi savant & aussi bon que vous qui *[écrivez]* sa vie. Merci mille fois

Votre ami bien dévoué | H Sainteclair Deville

Chemistry Laboratory | Ecole Normale Supérieure[2]
Paris, 16 March 1868

My dear Tyndall, I thank you for your two books.[3] I knew about one of them & I greatly admire it: because it made a big impression in France, it is your book on heat. I will read your Faraday with much pleasure. I very much liked your illustrious teacher & he sometimes wrote to me that he did have some kind of friendship towards me. I published here a note about him,[4] which according to what he wrote to me by *[2 words illeg]* did not displease him.—I am thus much attached to the memory of him & I am glad to think that it is such a scientific & good man as you who writes his life. Many thanks.

Your devoted friend | H Sainteclair Deville

RI MS JT/1/D/130

1. *Henri Étienne Sainte-Claire Deville:* see letter 2660, n. 1.
2. *Ecole Normale Supérieure:* the École Normale Supérieure in Paris, founded in 1794, served as a teacher-training school.
3. *your two books: Heat a Mode of Motion* (1868) and *Faraday as a Discoverer.*
4. *note about him:* probably the biographical note on Faraday written by Deville which preceded the French translation of Faraday's *Chemical History of a Candle* (1861) (see M. Faraday, *Histoire d'une chandelle,* trans. W. Hughes (Paris, J. Hetzel, 1865), pp. 1–26).

From Albert Edward, Prince of Wales[1] 16 March 1868 2788

Marlborough House[2] | March 16th 1868

Dear Professor Tyndall,

Accept my very best thanks for your letter,[3] and for sending me your Book[4] on the great and lamented Professor Faraday.

You may be sure that I shall read your book with the very greatest interest, which will remind me not only of the great man Science has lost, but of the good and upright man who England has lost.

Believe me, | Dear Professor Tyndall | Very sincerely yours | Albert Edward.

RI MS JT/2/13c/1335
LT Typescript Only

1. *Albert Edward:* Albert Edward (1841–1910), 'Prince of Wales', eldest son of Queen Victoria and Prince Albert, who succeeded to the throne of the United Kingdom as King Edward VII when Queen Victoria died in 1901 (*ODNB*).
2. *Marlborough House:* a mansion in Westminster London used as a residence for members of the Royal Family.
3. *your letter:* letter missing (the Royal Archives do not hold a letter from Tyndall to the Prince of Wales).
4. *your Book: Faraday as a Discoverer.*

From [Thomas Henry Farrer][1] 12 March [1868][2] 2789

Professor Tyndall

Sir,

With reference to your letter of the 4th August last[3] undertaking the duty of Scientific Adviser in respect of the inspector of Lighthouse Apparatus on the terms (£200 p an[4]) stated in the letter from this Department of the 10 July last[5] I am *[1 word illeg]* to transmit to you the enclosed two orders for the sum of £40.4.4 & £59.15.8 amounting together to £100 less income tax for H. M. Paymaster General[6] for the half year ending 31 December last.

I am to add that the sum due to the late Professor Faraday at the time of his death[7] has been paid to his widow.[8]

T H F | 12–3

[initials illeg] 13/3

[at top of letter] Done 16th March 1868 | *[initials illeg]*

National Archives, MT 10—128, H2727/3

---

1. *Thomas Henry Farrer:* Thomas Henry Farrer (1819–99), an English civil servant and statistician who was permanent secretary of the Board of Trade from 1867 (*ODNB*). This letter appears to be a draft, initialled 'T H F | 12–3', and perhaps then sent by a secretary. A further initialled note in contrasting red ink reads 'Done 16th March 1868', but it is not clear whose initials these are.
2. *[1868]:* the year is given based on reference to a letter from the previous year and the red date note (see n. 1 and n. 3).
3. *your letter of the 4th August last:* letter 2624.
4. *p an:* per annum.
5. *letter from this Department of the 10 July last:* letter 2614.
6. *H. M. Paymaster General:* Stephen Cave (1820–80), an English lawyer and politician who served as paymaster-general from 10 July 1866 until 1 December 1868 (*ODNB*).
7. *at the time of his death:* Michael Faraday died on 25 August 1867.
8. *widow:* Sarah Faraday.

From Louisa Baring 19 March 1868 2790

Mentone[1]— | March 19—68.

My dear Professor Tyndall—

Please accept my sincere & affectionate thanks for the precious book[2] you so very kindly have sent me—& of which I had already had sorts of foretastes through the papers—I am sure you must know how much I value you as having your writing, & your gift, & how much also as having the record of so great & good a man—I am hungry to read it—& nothing but absence of a few days, c[oul]d would have kept me so long. I found it on my return. This my last winter at Mentone is nearly at an end—but I have had no midnight arrivals—no beloved friends—it has been very quiet—& except for the companionship of my dear Miss Coutts Trotter[3] who has spent it under my roof, I don't know how I w[oul]d have come through its lonely hours—We hope to be home the end of May—& I hope you will look on my house in England, as one of your houses. I hope to be back in June at the latest—& there are few friends of my dear Bingham's[4] who will be so welcome as yourself. My Mary[5] is quite well—but there is scarlet fever all round us—Hoping soon to meet, I am as ever,

Ever yours truly | L. Ashburton. | Again aff[ectiona]te thanks—

RI MS JT/2/10/331
RI MS JT/2/13c/1336

1. *Mentone:* see letter 2535, n. 5.
2. *precious book: Faraday as a Discoverer.* Tyndall wrote in his journal in March 1868: 'Received from Lady Ashburton an exceedingly warm acknowledgement of my little book: she has a fine affectionate nature. Her eye is still magnificent in its fire' (Journal, RI MS JT/2/13c/1335).
3. *Miss Coutts Trotter:* Margaret Coutts Trotter (d. 1882), a friend who shared a passion for collecting art and in the 1860s shared romantic affection for each other that would waver in the early 1870s; she also sometimes helped Baring care for her daughter Mary (Virginia Surtees, *The Ludovisi Goddess: The Life of Louisa Lady Ashburton* (Michael Russell, 1984), pp. 129–30, 155–6; and M. Vicinus, *Intimate Friends: Women Who Loved Women, 1778–1928* (Chicago: University of Chicago Press, 2004), pp. 87–8).
4. *my dear Bingham's:* Bingham Baring (1799–1864), 2nd Baron Ashburton, an English politician born in Philadelphia, USA. He served as MP from 1826–48 when he succeeded to the peerage in 1848 after the death of his father (*ODNB*).
5. *Mary:* Mary Florence Baring.

To Robin Allen 20 March 1868 2791

Royal Institution, 20 March 1868.

Sir,

I HAVE carefully considered the report of Messrs. D. & T. Stevenson,[1] engineers, relative to the First Order Lighthouse lantern[2] used by the Commissioners of Northern Lighthouses,[3] and that adopted by the Trinity House.[4] I have also considered the report addressed by Mr. Douglass to yourself, entitled, "First Order Lighthouse Lanterns," and dated 28th February 1868.[5]

With the view of making myself intimately acquainted with the lantern adopted by the Trinity House, I inspected the lantern erected at Blackwall.[6] I also visited the works of Messrs. Simpson,[7] engineers, and saw the manner in which the various parts of the lantern are adjusted to each other and united together. Finally, as no dioptric[8] apparatus as yet surrounds the lamp at Blackwall, I thought it desirable to visit the Portland lighthouses,[9] and from this inspection I have just returned. The panes of the Scotch lantern[10] are triangular in shape, with plane surfaces, and ¼ of an inch thick. Those of the Trinity lanterns are lozenge-shaped, with cylindrical surfaces, of a radius equal to that of the lantern, and ½ an inch thick.

The light from the dioptric apparatus falls upon the plane surfaces of the Scotch lantern at a mean angle of 5¼ degrees. In the Trinity lantern, if the adjustment be perfect, the light falls normally, or very nearly so. Mr. Douglass urges, in his report,[11] that in the case of the Scotch lantern the light reflected from the surfaces of the panes is not only lost to the purposes of the mariner, but that it acts injuriously where coloured arcs are employed, by sending white light into the coloured areas, and thus rendering their distinction from the adjacent white areas more difficult. The remarks of Mr. Douglass upon this head I consider to be sound.

In the Trinity lantern, on the contrary, the light striking normally, retraces its path through the dioptric apparatus to the flame.

If the incidence be perfectly normal, the coincidence of the direct and reflected rays is perfect, and all the rays so reflected reunite to form an image of the flame co-incident with the flame itself.

Thus the light intercepted by reflection in one direction may be utilised in another direction, and not rendered useless or injurious, as in the case of reflection from the plane lantern glass.

Mr. Douglass considers that 88 per cent. of the light reflected from the Trinity lantern is again utilised, the remaining 12 per cent. being dissipated in the dioptric apparatus, through which the reflected rays have to return to their source.

While believing the bending of the glass to be decidedly advantageous, I do not think the gain likely to reach the amount stated by Mr. Douglass. The gain could only be so high if the apparatus were optically perfect (which, considering the mode of its manufacture, it not to be expected). Actual examination of the Portland lanterns shows that the rays reflected from the lantern glass do not in all cases return accurately to their source, so that, though a gain accrues, it does not reach the amount claimed for it by Mr. Douglass. Stricter attention and increased experience in the process of bending the glass will, in all probability, render the approximation to perfection closer than it now is.

I have examined Mr. Douglass's mode of testing the strength of lantern panes, both plane and cylindrical, and consider it to be skillful and correct. His experiments show, what indeed might be expected on theoretic grounds, that the bending of the glass materially augments its strength.

Taking this increase of strength into account, as well as the improvements in the reflection, I do not hesitate to recommend the employment of bent glass in the Trinity lanterns to be continued.

Glass ¼ of an inch in thickness would doubtless be able to resist the pressure of a very strong wind, but possible injury from accidental blows, as well as from such causes as those assigned by the Messrs. Stevenson, must also be taken into account. The gain in point of strength to resist such occurrences obtained by raising the thickness of the glass from ¼ of an inch to ½ an inch, in my opinion, far more than compensates for the minute increase of absorption due to the increased thickness of the glass.

To Mr. Alan Stevenson[12] is due the merit of introducing the important improvement of helical joints in the dioptric apparatus. To him also belongs the credit of the diagonal lantern. By those helical and diagonal arrangements the flame is never obscured from top to bottom of the apparatus at the same time, which would be the case if vertical framing were employed.

The helical principle Mr. Douglass has extended to the lantern which surrounds the dioptric apparatus, and I am bound to say that the design of it has been formed, and the execution of the design carried out in a manner which excites my admiration.

The framing of the Douglass lantern is helical, at every point of it its edge is presented to the flame, and the power of deepening the framing which this construction confers, permits great strength to be arrived at with a comparatively small amount of obstruction.

By the inspection of their respective shadows I have compared the amount of obstruction presented by the framing of both lanterns.

In the revolving apparatus there is little apparent difference, but in the fixed apparatus the quantity of light intercepted by the Scotch framing is sensibly greater than that intercepted by the helical framing.

The framing thus examined was formed in wood, of the dimensions employed respectively by Mr. Douglass and the Messrs. Stevenson.

Taking into account its apparent lightness and real strength, I am strongly inclined to recommend the Elder Brethren[13] to adhere to the helical construction of lantern.

The arrangements for the ventilation of the new Trinity lantern I consider to be very satisfactory.

The cleaning of the lantern externally is justly considered by the Messrs. Stevenson to be a point of great importance, and they recommend two cleaning paths to be attached to the framing outside, and so placed that the shadows of the horizontal brass setting of the dioptric apparatus shall fall upon them.

In this position the cleaning paths are stated by the Messrs. Stevenson to offer no additional obstruction to the light. Mr. Douglass, on the contrary, contends that the simple expedient of a ladder supported on the parapet below, and hooked above to a bar surrounding the lantern, meets the necessities of the case.

The cleaning of the outside of the lantern is of especial importance during a snowstorm at night. While the storm lasts the removal of the snow ought to be almost incessant. The ladder in a storm appears to me an unwieldy and objectionable instrument. If the wind be high it is very unmanageable, and may readily come into ruinous collision with the glass of the lantern.

I also doubt whether a man, probably numbed by cold, is likely to mount the cleaning paths sufficiently often to effectually remove the snow. A ladder, I think, ought always to be at hand; but, in addition to it, I would earnestly recommend a strong hair brush, shaped like a mop, with a handle so long that a man standing on the parapet might readily reach to the top of the lantern. With such a brush the keeper can readily move from pane to pane, and I think it probable that by means of it and the sponge now supplied to the keepers, the work of cleaning will in reality be more effectually done in a snowstorm than with either ladder or cleaning paths.

I need scarcely say that the cleaning paths would furnish resting-places for the snow, and would materially interfere with the application of the brush. In a snowstorm the cleaning can only be approximate, but in fairer weather the use of the ladder and the perfect cleansing of the glass ought to be insisted on.

With regard to the material of the lantern framing, the Messrs. Stevenson use gun metal where Mr. Douglass employs steel. All that Messrs. Stevenson urge as to the incorrodibility of their framing is correct. Nevertheless, what Mr. Douglass urges on behalf of painted steel, showing by the examples of the old Eddystone and the St. Bees lanterns[14] its permanence, when duly cared for, is very strong.

I think, however, that this is a point to which the attention of the light-keepers ought to be directed. When a spot of oxidation appears it ought instantly to be obliterated. The appearance of such spots is detected with extreme ease on the white pain of the lanterns. They really force themselves upon the attention of the keeper. If proper care be employed, I have no doubt that what Mr. Douglass says regarding both the strength and durability of the steel framing may be relied on.

A point of importance as regards the comparative cost of the two lanterns is their difference of size. The Scotch lantern measures 12 feet 6 1/4 inches, while that of the Trinity House measures 14 feet in diameter. The Trinity lantern is both roomy and airy.

The advantages claimed for it by Mr. Douglass as regards definition of cutting in the case of coloured arcs, and as regards the reduction of obstruction due to greater distance from the flame, are justly claimed. It may be added that the greater diameter permits of greater obliquity on the part of the astragals. This is an advantage now, but with a smaller source of light-the electric light, for example-its advantage would be very great. I am not, however, prepared to say that, if economy be made the point of chief importance, the lantern might not be somewhat contracted. But this is a subject on which the Elder Brethren are more competent than I am to form an opinion.

I would say, in conclusion, that I agree with Mr. Douglass in thinking that a considerable excess of strength over and above that required under ordinary circumstances should exist in the lanterns of our lighthouses.

I do not doubt that the claim made by the Messrs. Stevenson for their lantern is a perfectly just claim, namely, that it is "sufficiently strong to resist all ordinary trials."

In my opinion, the strength of the lantern ought to go beyond this, that strength being regulated, not by reference to "ordinary trials," but extraordinary and violent ones.

I am not prepared to recommend any reduction of materials which would sensibly reduce the strength of the present helical lantern.

I have, &c. | [(signed)] *John Tyndall.*

Robin Allen, Esq., | Secretary to the Trinity House.

'Correspondence relative to the Comparative Advantages of Lighthouse Lanterns adopted by Trinity House and Commissioners of Northern Lighthouses', HC 5, Parliamentary Papers (1872), pp. 26–8

---

1. *report of Messrs. D. & T. Stevenson:* David (1815–86) and Thomas Stevenson (1818–87), Scottish civil engineers whose works included many lighthouses (*ODNB*). See D. & T. Stevenson, 'Report relative to the First-order Lighthouse Lantern used by the Commissioners of Northern Lighthouses ad that proposed to be adopted by the Trinity House', in *Correspondence Relative to the Comparative Advantages of the Lighthouse Lanterns adopted by the Corporation of the Trinity House and the Commissioners of Northern Lighthouses, 1867–8* (London: Eyre and Spottiswoode, 1868), pp. 17–19.
2. *First Order Lighthouse lantern:* a lantern with a focal length of 920 mm. (*Glossary of Lighthouse Optics Terminology* (London: World Lighthouse Society, 2005), p. 55). A lantern of the first order is one made from eight panels, each of 45° (a second order is six at 60°, a third order is five at 72°).
3. *Commissioners of Northern Lighthouses:* the lighthouse authority for Scotland.
4. *Trinity House:* see letter 2590, n. 5.
5. *report addressed by Mr. Douglass . . . dated 28th February 1868:* see letter 2727, n. 4.
6. *Blackwall:* a lighthouse at the Trinity Buoy Wharf (where the rivers Thames and Lea meet), designed by James Douglass and established in 1846, and primarily used for experimental and training purposes.
7. *works of Messrs. Simpson:* James Simpson and Co., a steam engine and pump manufacturing firm founded by the engineer James Simpson (1799–1869) (*ODNB*).
8. *dioptric:* in lighthouses, directing rays from their light source into a beam via refraction; relating to the refraction of light (*OED*).
9. *Portland lighthouses:* probably two lighthouses, the Old Lower and the Old Higher, on the Isle of Portland in Dorset, England.

10. *Scotch lantern:* a lantern using kerosene (paraffin) rather than oil developed by Henry Harrison Doty, an American ship captain from Norfolk, Virginia who patented improvements to lighthouse lamps in both the United States and Britain. See *The Ascent of John Tyndall,* p. 284.
11. *his report:* see n. 5.
12. *Alan Stevenson:* Alan Stevenson (1807–65), a Scottish civil engineer and eldest brother of David and Thomas Stevenson. From 1830 he served as the clerk of works for the Northern Lighthouse Board. After studying Fresnel's lenses in France Stevenson introduced dioptric lights in English lighthouses (*ODNB*).
13. *Elder Brethren:* see letter 2614, n. 5.
14. *old Eddystone and the St. Bees lanterns:* the Eddystone Lighthouse on the Eddystone Rocks in English Channel, and the St Bees Lighthouse near the village of St Bees, Cumbria.

From Jules Jamin 20 March 1868 2792

Paris 20 mars 1868.

Mon cher ami

Je vous remercie de l'envoi que vous me faites aujourd'hui. J'ai déjà parcouru les premières pages et je les ai trouvées très intéressantes. Vous avez un talent original et vous dites les choses d'une manière qui réveille la curiosité des personnes, lors même qu'elles connaissent bien le sujet. Je crois votre livre destiné à un succès réel aussi bien en France qu'en Angleterre.

Je viens précisément de commencer mon cours d'acoustique à la Sorbonne et vous me fournissez des idées nouvelles. J'ai terminé aussi la 2e édition de cette partie de la science dans mon livre. J'ai eu l'occasion de faire exécuter des figures avec beaucoup de soin; elles pourront être employées pour la traduction française, particulièrement celle de la sirène de Helmoltz que vous placez à la page 77 et qui me paraît dans votre livre très inférieure au dessin que j'en ai suis faire. Je le tiens à votre disposition et je vous en enverrai une épreuve aussitôt qu'il sera terminé.

J'ai fait dernièrement une leçon sur les vents et la pluie. Je vous l'envoie, il m'a paru que l'on pouvait expliquer les phénomènes en admettant ce que j'ai nommé les <u>dérivations descendantes.</u> Je désire, moi aussi, vous imiter et faire un petit livre sur ce sujet. Vous seriez vrais bon de me dire votre impression générale et de me faire les objections que me suggérera votre amitié pour moi.

Continuez, je vous prie, de m'envoyer vos leçons, elles ont pour moi un grand intérêt d'actualité. Je vais les porter chez Mr Gauthier et *[1 word illeg]* la traduction française.

Ma femme se joint à moi pour vous offrir les plus sincères amitiés. N'oubliez pas que si vous venez à Paris vous avez chez nous un logement qui ne sera pas grand, mais que nous sommes heureux de vous offrir avec sincérité—ma fille est bien heureuse du souvenir que vous lui envoyez.

Votre ami dévoué | J. Jamin

Paris 20 March 1868.

My dear friend,

I thank you for what you have sent me today.[1] I already skimmed through the first pages and I found them to be very interesting. You have a unique talent, and you say things in such a manner that it arouses people's curiosity, even when they know the subject matter well. I think that your book will be a real success in both France and England.

I just started my course on Acoustics at the Sorbonne[2] and you are giving me new ideas. I also finished the 2nd edition of this part of the science in my book.[3] I had the opportunity to complete the figures very carefully. They could be used for the French translation, particularly that of Helmholtz's siren that you placed on page 77 in your book,[4] and which seemed to me a worse drawing than what I was able to do. It will be at your disposal and I will send you a sample as soon as it will be finished.

Lastly, I did a lesson on wind and rain.[5] I am sending it to you, and it appeared to me that we could explain the phenomena by suggesting that which I have termed descending derivations.[6] I wish to imitate you as well and to write a small book on this subject.[7] You will be honest enough to let me know of your general impression and to tell me of any objections because of your friendship towards me.

Please continue sending me your lessons;[8] their topic is of great interest to me. I will bring them to Mr Gauthier[9] and *[1 word illeg]* the French translation.

My wife[10] and I send you our warmest wishes. Do not forget that if you come to Paris, you have a place to stay at our home, which is not large, but which we would be sincerely happy to offer you—my daughter[11] is very happy about the souvenir that you sent her.

Your devoted friend | J. Jamin

RI MS JT/1/J/12

1. *what you have sent me today:* probably a copy of *Sound* (1867).
2. *my course on Acoustics in Sorbonne:* Jamin taught at the Sorbonne from 1863–85.
3. *my book:* J. Jamin, *Cours de physique de l'École polytechnique,* 2nd edn, 3 vols (Paris: Gauthier-Villars, 1868–9).
4. *Helmholtz's siren . . . in your book: Sound* (1867) contained an illustration (fig. 27) of Helmholtz's double siren (p. 77).
5. *lesson on wind and rain:* J. Jamin, 'Les vents et la pluie', *Revue des Deux Mondes,* 67 (1867), pp. 915–43.

6. *descending derivations:* Jamin introduced 'dérivations descendantes' on p. 921 of 'Les vents et la pluie'.
7. *small book on this subject:* Jamin did publish 'Les vents el la pluie' separately with the Parisian publisher Jules Claye (1806–86) in 1867 (*Bibliographie de la France* (Paris: Cercle de l'Imprimerie, de la Librairie et de la Papeterie, 1867), p. 167), but it does not appear he extended this work into a more substantial book. However, he did incorporate much on wind and rain in his *Quelques phénomènes atmosphériques* some years later (Paris: Hachette, 1880).
8. *your lessons:* see n. 1.
9. *Mr Gauthier:* Jean-Albert Gauthier-Villars.
10. *My wife:* Thérèsa Jamin (see letter 2750, n. 8).
11. *my daughter:* Lucie Jamin (see letter 2750, n. 2)

From Jane Barnard 31 March 1868 2793

12 Barnsbury Villas[1] | N. | 21 March 1868

Dear Dr Tyndall

I received from my Father,[2] last evening the book[3] you gave him for me; many thanks for it—I value it very much. I have been cutting it open & have glanced through it with great interest.

We feel that such attentions from my dear Uncles'[4] friends, are marks of the warm feelings of love & respect felt for him & they are very precious to us.

Believe me | always yours very sincerely | Jane Barnard

Dr Tyndall | &c &c &c

RI MS JT/1/B/39

1. *12 Barnsbury Villas:* see letter 2695, n. 2.
2. *my Father:* John Barnard (1797–1880), brother of Sarah Faraday.
3. *book: Faraday as a Discoverer.*
4. *my dear Uncles':* Michael Faraday.

To Jane Barnard 22 March 1868 2794

22nd. March 1868

My dear Miss Barnard

You will not I think interpret my words as denoting any enfeeblement of my feelings towards your uncle[1] when I say that the offer of the book[2] which you refer to was prompted by respect for yourself.

Yours very since[re]ly | John Tyndall

I shall know tonight whether M^r. Carlyle[3] will go to the Isle of Wight.[4] I should like, if I could, to heighten that proud old fellow's life by improving his health, and hence I gladly devote three days or four to the promotion of this object. If he declines the expedition, then I am ready any even[in]g named by your aunt[5] or yourself to come up to Barnsbury.[6] Though I should shrink from intruding upon her, as she knows right well, it is always a delight to me to see your aunt.

RI MS JT/1/T/117
RI MS JT/1/TYP/12/4198

---

1. *your uncle:* Michael Faraday.
2. *book: Faraday as a Discoverer.*
3. *M^r. Carlyle:* Thomas Carlyle.
4. *Isle of Wight:* Tyndall had planned to leave for the Isle of Wight with Carlyle the following day (Monday, 23 March 1868), but according to his journal Carlyle refused to join him and it is unclear whether Tyndall went to the Isle of Wight alone (Journal, RI MS JT/2/13c/1336–7).
5. *your aunt:* Sarah Faraday.
6. *Barnsbury:* see letter 2695, n. 2.

From François-Napoléon-Marie Moigno — 25 March 1868 — 2795

Paris, le 25 mars 1868
LES MONDES | Revue Hebdomadaire | DES SCIENCES | ET DE LEURS APPLICATIONS | aux arts et à l'industrie | par | M. l'Abbé MOIGNO | RÉDACTION | 2, RUE D'ERFURTH, 2

Mon cher Monsieur Tyndall

J'attends de vous par l'intermédiaire de M. Jamin une réponse à la quelle j'attache une très grande importance; permettez moi de vous la demander directement.

Convaincu que vous m'avez réservé la traduction de votre notice sur Faraday, d'après ce que vous avez écrit à M. Gauthier-Villars, et recevant deux exemplaires de cet opuscule, l'un de l'institution royale, l'autre de Longmann, je me mis aussitôt à l'œuvre avec mes habiles collaborateurs, et ma traduction fut prête en huit jours. J'allais annoncer que j'étais prêt quand M. Mallet-Bachelier m'appris en l'absence de M. Gauthier-Villars, que vous avez confié cette traduction à M. Jamin, qui l'avait donnée à son tour à un ancien élève de l'école normale.

Cette nouvelle me fut d'autant plus pénible que je n'avais pas pu la prévoir; et il m'en couterait je vous assure de renoncer à ma traduction. Elle a été très difficile, et je ne suis arrivé au but qu'après une lutte très opiniâtre. Il fallait tout ce que je sais et ma grande habitude pour réussir à rendre le texte anglais parfaitement intelligible et agréable.

J'aurais voulu faire de cette notice une brochure tout à fait semblable à celle que je vous envoie sur l'éclairage. Il y aurait le même nombre de pages, et l'impression ne vous couterait rien, M. Gauthier-Villars et moi, nous pourrions vous donner 50 centimes par exemplaire, et je suis sur d'en vendre *[plus]* de mon *[plus]* de quinze cent exemplaires.

Autorisez moi je vous en prie à continuer mon entreprise.

Rien n'empêche qu'il y ait en France comme en Angleterre deux éditions; l'une populaire qui ne vous coutera rien et vous aidera à payer les frais de la seconde édition de luxe.

Si je me suis trop empressé, vous ne pouvez pas m'en vouloir, puisque c'est pour vous honorer Faraday et vous.

J'ai complètement achevé les huit leçons sur le son.

La seconde édition de la chaleur complétée par la troisième édition anglaise que vous m'avez envoyée est toute prête à être imprimée quand M. Gauthier-Villars sera prêt.

Les six leçons sur la chaleur et le froid sont charmantes et toutes traduites; elles feront une jolie brochure; j'attends les dernières figures.

J'espère vous donner bientôt des nombres de la première édition de la chaleur je poursuis M. Giraud en mon *[1 word illeg]* nom pour le forcer à rendre ses comptes en à me dire la part qui vous revient afin que je vous l'envoie.

Vous voyez que je ne perds pas mon temps et que vos intérêts me sont grandement chers. Ne me condamnez pas à jeter au feu la traduction de la notice sur Faraday; d'autant plus que pour cette traduction je vous suis encore plus utile, sinon plus nécessaire que pour le reste, à cause de mon *[1 word illeg]* au travail.

Je compte sur un mot de réponse | Votre humble et dévouée serviteur | L'abbé F. Moigno

Paris March 25 1868
LES MONDES[1] | Revue Hebdomadaire | DES SCIENCES |
ET DE LEURS APPLICATIONS | aux arts et à l'industrie | by the |
Father MOIGNO | EDITORIAL BOARD | 2 RUE D'ERFURTH, 2

My dear Mr Tyndall,

I am waiting for an answer from you through Mr. Jamin.[2] It is an answer which I consider very important; allow me to ask you directly.

Having been convinced that you saved for me the translation of your memoir about Faraday,[3] from what you have written to Mr. Gauthier-Villars, and receiving two copies of that booklet, one from the Royal Institution and one from Longmann,[4] I immediately began to work with my skilled colleagues, and my translations were ready in eight days. I was about to announce my completion, when Mr. Mallet-Bachelier[5] told me in the absence of Mr. Gauthier-Villars, that you entrusted Mr. Jamin with this translation, who himself had given this task to a former student of the training college.

This news was all the more painful as I could not foresee it; and it would cost me, I assure you, to give up my translation. It was very difficult, and I only reached my goal after a stubborn fight. It took all that I know and my extensive experience to succeed in making the English text perfectly understandable and easy to read.

I would have liked to do a brochure out of this memoir, very similar to that which I sent you on the lighting. There would be the same number of pages, and the printing would not cost you anything, Mr. Gauthier-Villars and I could give you 50 cents per copy, and I am sure that it will sell more than my more than fifteen hundred copies.

Please, allow me to continue my undertaking.

Nothing prevents there being two editions in France like in England; one that is popular, which will cost you nothing, and which will help you pay the cost of the second deluxe edition.

If I am too eager, you cannot blame me, since it is to honour Faraday and yourself.

I completed all the eight lessons on sound.[6]

The second edition of la chaleur,[7] completed by the third English edition[8] that you sent me is ready to go into printing when Mr. Gauthier-Villars is ready.

The six lessons on Heat and Cold are delightful and entirely translated;[9] they will make a nice brochure; I am waiting for the final figures.

I hope to give you soon numbers about the first edition of la chaleur.[10] I insist with Mr. Giraud[11] *[1 word illeg]* to make him do the reports and tell me the part of your returns so that I can send it to you.

You see that I do not waste my time and that your interests are very dear to me. Do not condemn me to throw away the translation of the memoir on Faraday; especially that for this translation I can be even more useful for you, even more *[1 word illeg]* than for the rest, because of my *[1 word illeg]* at work.

I am counting on your answer | Your humble and devoted servant | Father F. Moigno

RI MS JT/1/M/113

---

1. *LES MONDES:* see letter 2706, n. 9.
2. *Mr. Jamin:* Jules Jamin.

3. *translation of your memoir about Faraday: Faraday as a Discoverer.*
4. *Longmann:* see letter 2617, n. 5.
5. *Mr. Mallet-Bachelier:* a Parisian publisher.
6. *eight lessons on sound:* see letter 2617, n. 2.
7. *second edition of Heat:* J. Tyndall, *La chaleur, mode de mouvement,* 2nd edn, trans. F. Moigno (Paris: Gauthier-Villars, 1874). This was translated from the fourth English edition (1870).
8. *third English edition: Heat* (1868).
9. *six lessons on Heat and Cold are delightful and entirely translated:* see letter 2751, n. 2.
10. *first edition of Heat:* J. Tyndall, *La Chaleur considérée come un mode de mouvement,* trans. F. Moigno (Paris: Étienne Giraud, 1864).
11. *Mr. Giraud:* Étienne Giraud.

From Clémentine Soret[1] 28 March 1868 2796

Genève le 28 Mars 1868—

Cher Monsieur Tyndall

Je ne sais comment vous témoigner, au point où je le sens, la reconnaissance que j'éprouve, pour la pensée qui vous a dirigé, en comptant <u>pauvre moi</u> au nombre des cours, que vous *[*sentiez*]* assez sympathique à votre *[<u>1 word illeg</u>]*, pour lui faire partager les souvenirs, et <u>péné</u>trer dans l'affection qui vous trait, avec ce noble et touchant Faraday!

Aussi j'ai tout lu, quand même bien des choses étaient au dessus de ma portée, comme <u>science</u>! Mais je craignais trop de perdre, dans l'exposé de l'un ou l'autre de ces profonds travaux, quelque trait lumineux et caractéristique, de cette nature si extraordinaire, dans son élévation, sa pureté, et cette ardentée <u>seconde vue</u>, aussi bien morale que scientifique.—La contemplation de cette vie, consacrée à l'étude de la vérité de ce caractère à la fois si fier et indépendant, vis à vis des puissants de la terre, et si humble et tendre vis à vis de Dieu et de ses amis, quelque chose qui possède un charme d'attraction unique! Enfin, après l'avoir suivi, travers vos intéressants récits, et ceux de M. de la Rive, dans cette longue vie de travaux, dont il sentait et décrivait si bien les enchantements, il me semble le voir maintenant, et je suis sûre que votre coeur le cherche aussi, de l'autre côté du voile, avec la confiance qui s'impose à vous du bonheur avec lequel cette nature d'élite, possédée de la soif de connaître, plonge maintenant dans tous les mystères, qui préoccupent sa pensée ici-bas; de la joie avec laquelle le cœur ardent, cette âme bien organisée, est rassasiée du bonheur d'aimer; et de celui d'entrer dans cette grande harmonie, où toutes les notes fausses de la terre se taisent, tous les sombres voiles se lèvent; et où, comme dans les laboratoires de l'Institution Royale "He is still looking for you"!—Cette pensée est douce n'est ce pas? O que ferions nous sur cette pauvre froide terre, avec tant d'aspirations qui ne trouvent pas à se réaliser, même et peut-être surtout, pour les plus nobles coeurs, si nous ne sentions que ce n'est qu'un commencement! Et à l'heure des adieux, si une voix, que nous ne pouvons repousser, ne murmurait à notre oreille: Non pas adieu mais au revoir!

Au revoir donc! Cher Monsieur Tyndall, au revoir, ce doux mot, si doux à dire sur la terre, et au-delà. Mais non pour tout sans vous avoir remercié encore chaudement, et demandé d'excuser ma réponse tardive, mais ma santé me prive souvent du plaisir d'écrire; des crises au cœur pénibles, quoiqu'on les dise sans gravité, mais (à condition de renoncer souvent à ce que j'aime le mieux, à ce qui fait le prix de la vie—son *[1–2 words illeg]* et ses vives émotions—) limitent tristement mes possibilités; m'obligent à réclamer et à espérer beaucoup d'indulgence—

Mon mari vous serre affectueusement la main, il travaille avec ardeur, et ses travaux l'intéressent et l'absorbent de plus en plus, j'en jouis vivement et aimerais pouvoir l'aider davantage. Il me charge de vous dire, qu'il vous écrira sous peu, au sujet de ses expériences sur le rayonnement solaire.

Vous avez su que l'été passé il a été étudier les rayons du grand astre, jusqu'au somment du Mt Blanc, il a pu passer une heure et demi au sommet, par un temps admirable, j'étais à Chamonix avec mon fils, et nous l'avons trois fois vu, lui, et ses instruments qui reluisaient au soleil—Je savais d'avance son projet, et ne l'avais pas découragé . . . Vous voyez que moi aussi je peux être un peu heroïque. Mais j'ai besoin de *[1 word illeg]* en *[1 word illeg]* vite . . . ce qui est mauvais signe *[1 word illeg]* à l'avenir.

Encore adieu et pardon de cette longue causerie—J'ai eu tant de plaisir à vous lire—que je n'ai pu résister à vous le dire en trop de mots.—

Votre bien dévoué et reconnaissante | Clémentine Soret—

Geneva 28 March 1868—

Dear Mr Tyndall,

I do not know how to express how grateful I feel for the thought that drove you, thinking about poor me in a number of courses, to *[feel]* sympathetic enough to your *[1 word illeg]*, to share memories with him and to enter into the affection with which you have for the noble and touching Faraday!

I read everything,[2] even though many things were beyond me, like science! But I was too afraid to lose, in the presentation of one or another of these profound works, some bright and characteristic feature of that extraordinary nature, in his nobility, his purity, and this passionate second view, as much morally as scientifically.—The contemplation of this life, devoted to the study of truth—of this character, both so proud and independent, in relation to the powers of the earth, and so humble and gentle in respect to his God and his friends, something that has a uniquely magnetic charm! Finally, after having followed it through your interesting stories, and through those of Mr. de la Rive,[3] in that long life devoted to work in which he felt and described the delights so well, it seems that I see it now, and I am sure that your heart searches for it too. And on the other side of the veil, with confidence which requires happiness, with which this elite nature, possessing the thirst for knowledge, now dives into all the mysteries that concern his mind down here. Of the joy with which the heart burns, that well organized soul is satisfied with the happiness of love; and that of entering this grand harmony, where all the wrong notes of the earth are silent and where all the dark veils are lifted, and where, like in the laboratories of the Royal Institution "He is still looking for you"!—That thought is sweet, is it not? What would we be doing here on this poor cold earth, with so many aspirations that end up not being accomplished, even or perhaps especially for those with the noblest of hearts, if we did not feel like it is only the beginning! And at the time of farewell, if a voice that we cannot dismiss would not whisper in our ear: Not goodbye, but farewell, until we meet again!

Farewell then! Dear Mr Tyndall, farewell, this sweet word, so sweet to say on earth and beyond. But not without thanking you deeply once more, and asking you to forgive my late reply, but my health often deprives me from the pleasure of writing. Painful heart seizures, although they say that they are minor, but (under the condition of often renouncing that which I like best, that which is the price of life—his *[1–2 words illeg]* and his strong emotions—) sadly limited my abilities; forcing me to demand and expect a lot of indulgences.

My husband[4] shakes your hand with affection. He works hard and his work interests him and draws him in more and more. I strongly enjoy it and I wish that I could have helped him more. He asks me to tell you that he will soon write to you, on the subject of his experiments on solar radiation.[5]

Did you know that last summer he was studying the rays of the great heavenly body, up until the summit of Mt. Blanc. He could spend an hour and a half at the summit by an admirable weather. I was in Chamonix with my son,[6] and we saw him three times, him, and his instruments that shine in the sun.—I knew about his project ahead of time, and I did not discourage him . . . You see that I can also be a little heroic. But I need *[1 word illeg]* to *[1 word illeg]* quickly . . . which is a bad sign *[1 word illeg]* in the future.

Farewell again, and forgive me for this long conversation—I enjoyed reading yours[7] so much—that I could not resist telling you that through too many words.—

Your devoted and grateful | Clémentine Soret—

RI MS JT/1/S/82

---

1. *Clémentine Soret:* Clémentine Soret (née Odier, 1831–99), who married Jacques-Louis Soret (see n. 4) in 1853. They had a son, Charles Soret (see n. 6).
2. *I read everything:* Tyndall appears to have sent Clémentine Soret a copy of *Faraday as a Discoverer* (1868).
3. *those of Mr. de la Rive:* see letter 2670, n. 3.
4. *My husband:* Jacques-Louis Soret (1827–90), a Swiss physicist and chemist who taught at the University of Geneva from 1866 (*HLS*).
5. *experiments on solar radiation:* J.-L. Soret, 'Sur la radiation solaire,' *Paris, Comptes Rendus,* 66 (1868), pp. 810–13.
6. *my son:* Charles Soret (1854–1904), a Swiss physicist and chemist best known for his work on thermodiffusion, also known as the Soret effect (*HLS*).
7. *yours:* letter missing.

From Cecilia Anne Barlow[1] [March 1868][2] 2797

5 Berkeley St.[3] | Sunday night

Dear M^r^. Tyndall,

I ought sooner to have written to you to thank you for the book[4] you so kindly sent me, but I wished to read it first. In that however I was forestalled by my husband,[5] to whose claims I could not but yield.—but now I can tell you with what intense delight I have read it, and how much I regret that I could not hear these most interesting lectures[6] delivered. It was a source of much annoyance to me that I could not be present, but these pleasures of the past are not for me now, and I must be content with the privilege of reading your words.

You have done full justice to the character of the dear old man, whose loss we all deplore, and very often your words brought the tears into my eyes.

My husband has been equally gratified with myself, and begs me to tell you so. He is very very feeble, and quite unequal to any conversation, but still finds pleasure in his books—.

With reiterated thanks, Believe me Very sincerely Y[ou]^rs^. | C. A. Barlow.

RI MS JT/1/B/16
RI MS JT/1/TYP/1/147

---

1. *Cecilia Anne Barlow:* Cecilia Anne Barlow (née Law, d. 1868), the daughter of an MP who made his fortune in India. She and John Barlow (see n. 5) married in 1824.
2. *[March 1868]:* the month and year are given based on the publication of Tyndall's *Faraday as a Discoverer* in March 1868 and that Tyndall received similar letters during this month from friends whom he sent a copy (2780, 2782, 2786, 2787, 2788, 2790, 2793, 2796, and 2798).

3. *5 Berkeley St.:* the Barlow residence in London.
4. *book: Faraday as a Discoverer* was published on 5 March 1868.
5. *Barlow:* John Barlow (1799–1869), an Anglican clergyman and secretary of the RI. Barlow studied at Trinity College, Cambridge, and was ordained in 1823. From 1842–60 he was the secretary of the RI, and together with Michael Faraday he undertook the day-to-day running of the institution. Barlow lived near the RI and entertained many of its wealthy and aristocratic supporters as well as fellow scientists, including Tyndall, who was a frequent guest (*Alum. Cantab.*).
6. *most interesting lectures:* Tyndall's RI lectures on Faraday in January that were the basis for the book (see letter 2665, n. 3).

From Ellen Busk [March 1868][1] 2798

Dear Mr. Tyndall,

How very kind you are to give me your beautiful book[2]—It is most intensely interesting & doubly valuable as your gift in remembrance of the great & good man[3] you had the great privilege to call almost Father. Many many thanks.—

Believe me | ever y[ou]rs & truly | Ellen Busk

RI MS JT/1/B/156
RI MS JT/1/TYP/1/172

1. *[March 1868]:* the month and year are given based on the publication of Tyndall's *Faraday as a Discoverer* in March 1868 and that Tyndall received similar letters during this month from friends whom he sent a copy (2780, 2782, 2786, 2787, 2788, 2790, 2793, 2796, and 2797).
2. *your beautiful book: Faraday as a Discoverer* was published on 5 March 1868.
3. *the great & good man:* Michael Faraday.

From Mary Egerton [March 1868][1] 2799

45 Eaton Place[2] | <u>Monday</u>

Dear M^r^. Tyndall,

We have been looking out for you at the Tuesday lectures,[3] thinking you might be perhaps learning about your own dominions;[4] but I have just discovered that you are again giving a daily course in Jermyn Street.[5] I wonder whether you would give us, as you did last year, an order to go to a few of them? I am almost ashamed to ask such a thing, it seems so dreadfully superficial! but then you see, we poor women have got to be superficial (all owing, as

M^r^. Mill[6] tells us,[7] to not having a vote for members of Parliament!) and gold is gold even if picked up in stray grains on the surface instead of being worked for with pickaxe & shovel, though it will never make one's fortune! I have not got the programme so I do not know where you are & must leave house you see. Excuse my troubling you & believe me

Y[ou]^rs^ truly | M F Egerton

RI MS JT/1/E/99b

---

1. *[March 1868]:* the date is given by reference to a lecture course Tyndall gave at the Royal School of Mines (see n. 5), and appears to relate to letter 2804 the following month, in which Egerton again asked Tyndall for tickets to the lecture course.
2. *45 Eaton Place:* in London.
3. *Tuesday lectures:* Egerton wrote that she 'was looking out for you', implying that she did not know if Tyndall would be at the RI or not when she visited, and thus the lectures must have been some delivered by another person. The two lecture courses delivered in 1868 *before* the School of Mines course referred to (see n. 5) were a course of six lectures 'On Historical Portraiture of Various Times and Countries' by George Scharf (1820–95) on Tuesdays and Thursdays from 3–19 March, and four lectures 'On the various Modes of the Development of Animals' by Michael Foster (1836–1907) on Tuesdays and Thursdays from 24 March to 2 April (*Roy. Inst. Proc.,* 5 (1866–9), pp. 197).
4. *learning about your own dominions:* Tyndall had in January 1868 moved in the Faraday's rooms at the RI. See letter 2707, n. 8 and n. 9.
5. *daily course in Jermyn Street:* see letter 2804, n. 3.
6. *M^r^. Mill:* John Stuart Mill.
7. *tells us:* Mill supported women's suffrage, and as a Member of Parliament he put forth on 20 May 1867 an amendment to the 1867 reform bill for the House of Commons to vote on whether or not women should be allowed to vote for members. While it was not passed (75 in favor and 196 in opposition), Mill would later write in his autobiography that 'by far the most important, perhaps the only really important, public service I performed in the capacity of a Member of Parliament: a motion to strike out the words which were understood to limit the electoral franchise to males, and thereby to admit to the suffrage all women who, as householders or otherwise, possessed the qualification required of male electors' (J. S. Mill, *Autobiography* (London: Longmans, Green, Reader, and Dyer, 1873), pp. 303–4). Mill would later lose his parliamentary seat in the 1868 general election, and the following year he published *The Subjection of Women* (London: Longmans, Green, Reader, and Dyer, 1869). Women would not gain the parliamentary vote until 1918, and equal voting rights in 1928. See also E. L. Pugh, 'John Stuart Mill and the Women's Question in Parliament, 1865–1868', *The Historian,* 42 (1980), pp. 399–418.

From Charles Lennox Bruce Cumming[1] — 7 April 1868 — 2800

Coconada | Godavery District | Madras Presidency[2] | India. April 7th 1868.

My dear Dr Tyndall,

I hope you will forgive me for not having written before to thank you for your kind note of the 22nd Oct. last[3]—and the friendly wishes expressed by you in it. I would have done so, but have scarcely yet got over the excitement of a first arrival in India, or settled myself down to work in my new career.

I am sorry to say that since you wrote I have learnt to my disappointment that my small vol[ume]. of Rhymes[4] has proved a failure—and met with the unsatisfactory fate of so many others: viz—the privilege of slumbering undisturbed on the publisher's shelves.

I have to thank myself in great measure for the result, for if I had not been so pugnacious—so uncommonly self-confident and independent in my tone. I do not think the Reviewers would have 'sat upon me'[5] so severely. But 'experientia docet',[6] and I shall be wiser next time i e if I ever do publish again.—But after this first reception, I have little encouragement to do so. However some friends have advised me to remould the vol[ume]: omitting much, and adding to it—besides altering and emending &c. and if I am spared to return from India in 4 or 5 years time on furlough I may publish again—but it shall be anonymously—and only provided I can between this and then write something (as Keats[7] says) <u>with judgment</u>,[8] and more worthy of exhibition.

But I do not think I have been altogether treated fairly or justly—as the higher papers such as the 'Saturday Review',[9] and the 'Spectator'[10] did not deign to notice my wee book—and those which did, do not seem to have read more than the first poem—before they quietly proceeded to condemn it <u>in toto</u>.[11] I must confess it was rather a trial after Rev. Gilfillan[12] and other competent judges had professed to discover the dawnings of merit in my verses—to be told by 2nd rate Reviews that my verses were but 'nursery twaddle'—and that I was audacious to publish them. It was a rude shock to one who from boyhood's time had imagined himself a 'born poet' and possessed of the 'Art Divine'. But I suppose it is useless repining—Since then I have got into a healthier frame of mind—and the idea of never leaving my mark upon the world, or winning 'fame' is not so torturing as it used once to be: I am thankful to say I have enough serious work at present to distract my thoughts from such dreams,—as I am engaged nearly all day in trying cases, and sentencing different natives to various terms of imprisonment. It is rather fun, as the work is quite new. Besides I ought to be impressed with a sense of my own dignity—since I rejoice in the Titles of Assistant Collector and Magistrate of the Godavery District. And Secretary to the (with a salary of £500 per annum) Municipal commission of the Township of Coconada. An Honourary Post—chief duty of which is to make the Town Roads &c. &c. But these I fear will have more weight with the good folks at home who do not quite realize their meaning—than with persons out here. I like India far better than I expected. The climate is not oppressively hot (altho' 96 now in shade) since we are on the sea, and get a cooling breeze from it all day and night. But the really hot weather is only just commencing. The country here is very flat and uninteresting—inland it is better—by well wooded—while here we have nothing but arid plains, void of trees—save a few palms. There is good shooting in the interior—lots of big game, bears, <*1 word missing*>, tigers, and panthers: The skins are often brought in by natives to get the Government Reward for killing these animals. This is a quaint little town. Most of the European Inhabitants are French and German Merchants. There are very few English. The harbour is bad, but there is a good deal of shipping and commerce. We boast a club,[13] and a Chamber of Commerce, and are within 2 days sail of Madras,[14] I was so sorry to hear of poor Faraday's death[15]—A great loss to the scientific world. You must have felt his death much too since you knew him so well. Sir D. Brewster's death[16] is also a great loss: I think aunt Coxe[17] told me you had met him at Kinellan.[18] I see by the papers that you have been writing a sketch of Faraday's 'Life and Labours'[19] I should much like to read the work—but such luxuries are not to be got here, although we have an excellent library of old works—Among which is your 'Glaciers of the Alps'.[20] Have you seen Aunt Coxe lately? She has written to me once only since I left England—but has sent me no end of "Scotsmen"[21]—and I know she is too busy to write often. I much regret having always missed meeting you at Kinellan, as I have not forgotten your kindness to us all in the old Queenwood days[22]—and—should have muched liked to renew our acquaintanceship but I never liked to call in London, as I feared you might have forgotten me. However I think now I need not have been afraid. My last acc[oun]t. of all at Kinellan were good, both uncles W.[23] and Coxe[24] flourishing—and Aunt Coxe looking younger than ever. I must now conclude as I have come to the end of my paper. With best thanks your kind note and wishes, and kindest regards from myself, believe me,

My dear Dr Tyndall | Very truly Yours | Chas. L. B. Cumming.

N.B. | I shall address this to the Royal Institution, and trust it will reach | C. L. B. C.

RI MS JT/1/TYP/1/294
LT Typescript Only

---

1. *Charles Lennox Bruce Cumming:* Charles Lennox Bruce Cumming (1844–1912), a nephew and ward of Mary Anne Coxe, after the death of his father George Vanbrugh Cumming in 1856. Cumming attended Queenwood College along with his brothers where Tyndall taught them. He joined the Indian Civil Service in 1867 retiring in 1888 (*The India List and India Office List for 1905* (London: Harrison and Sons, 1905), p. 474).
2. *Coconada | Godavery District | Madras Presidency:* now Kakinada, a seaport city in eastern India during British rule.
3. *your kind note of the 22nd Oct. last:* letter missing.
4. *my small vol[ume]. of Rhymes:* C. L. B. Cumming, *The Grecian Maid, And other Poems* (London: Griffith and Farran, 1867).
5. *'sat upon me':* not identified.
6. *'experientia docet':* experience teaches (Latin).
7. *Keats:* John Keats (1795–1821), an English Romantic poet (*ODNB*).
8. *with judgment:* in an 1818 letter to his publisher James Augustus Hessey (1785–1870), Keats rose above the critics of his poetry and wrote: 'I will write independently. I have written independently *without judgment.* I may write independently and *with judgment,* hereafter' (R. M. Milnes (ed), *Life, Letters, and Literary Remains of John Keats,* 2 vols (London: Edward Moxon, 1848), vol. 1, p. 214).
9. *'Saturday Review':* a weekly British newspaper published from 1855–1938.
10. *'Spectator':* a weekly British magazine on current affairs, first published in 1828.
11. *in toto:* as a whole (Latin).
12. *Rev. Gilfillan:* George Gilfillan (1813–78), a Scottish author and Presbyterian minister. Cumming had dedicated his book of poetry (see n. 4) to Gilfillan (*ODNB*).
13. *club:* not identified.
14. *Madras:* the former name of both the Indian state of Tamil Nadu and Chennai, the state's capital city.
15. *poor Faraday's death:* Michael Faraday died on 25 August 1867.
16. *Sir D. Brewster's death:* David Brewster died on 10 February 1868.
17. *aunt Coxe:* Mary Anne Coxe (née Cumming, 1811–c. 1875), a daughter of Lesley Baillie and Robert Cumming of Logie, County Moray. Her mother was the subject of Robert Burns's 'Bonnie Lesley', as noted in the novelist Charles Gibbon's dedication of *In Honour Bound,* vol. 1 (London: Richard Bentley and Son, 1874) to Lady Coxe. In 1841 she married the Scottish physician James Coxe (1811–78), and became Lady Coxe after he was knighted in 1863. Tyndall claimed to have toasted her on the summit of Mont Blanc in 1857.
18. *met him at Kinellan:* the Coxe family estate in the Scottish county of Midlothian, which borders Edinburgh to the southeast. In his 'Personal Recollections of Thomas Carlyle (1890)', originally published in the *Fortnightly Review* in January 1890, Tyndall wrote of a visit to Kinellan on 1 April 1866, during a trip to see Carlyle deliver an Address as Lord Rector of Edinburgh University: 'The evening before the eventful day I dined at Kinellan with my well-beloved friends, Sir James and Lady Coxe, whose permanent guest I was at

the time. Sir David and Lady Brewster were there, and Russell of the *Scotsman.* The good sir David looked forward with fear and trembling to what he was persuaded must prove a fiasco. "Why," he said to me, "Carlyle has not written a word of his Address; and not Rector of the University ever appeared before his audience without this needful preparation"' (J. Tyndall, *New Fragments* (London: Longmans, Green, and Co., 1892), pp. 347–97, on p. 362).

19. *sketch of Faraday's 'Life and Labours': Faraday as a Discoverer.*
20. *'Glaciers of the Alps': Glaciers of the Alps.*
21. *"Scotsmen":* a daily newspaper established in Edinburgh, Scotland in 1817.
22. *old Queenwood days:* Charles and his brothers William, Alexander, and George all attended Queenwood College where they came to know Tyndall (see letter 0794, n. 7, *Tyndall Correspondence,* vol. 4); M. E. Cumming Bruce, *Family Records of the Bruces and the Cumyns* (Edinburgh: William Blackwood and Sons, 1870), p. 660).
23. *W.:* William Fullerton Cumming (1804–92), a physician and brother of Mary Anne Coxe and George Cumming, Charles Cumming's father. He was employed by the East India Company after having obtain his MD from Edinburgh University (1827) but returned to Scotland in 1836 due to poor health ('Obituary', *Lancet,* 140 (1892), pp. 349–50).
24. *Coxe:* James Coxe (1811–78), a Scottish physician who married Mary Anne Coxe in 1841. From 1857 he served as one of the commissioners of the General Board of Lunacy for Scotland, and was knighted in 1863 for his services ('Obituary of Eminent Persons', *Lancet,* 111 (18 May 1878), p. 471).

To Thomas Henry Huxley [8 April 1868][1] 2801

Royal Institution. | Wednesday.

My dear Huxley,

I have been looking forward to a scamper with you over the Welsh Hills.

But at this season of the year I deliver myself up like a physician, who fears to prescribe for himself, to the guidance of Hirst,[2] and he last night finally abolished Wales and established Naples.[3]

I start tomorrow morning; Lubbock[4] starts tomorrow evening and I halt for him at Amiens.[5]

I suppose your turning up at Amiens is quite out of the question?

Well if you do not turn up and thus come into my keeping, the next best thing to be done is to commit you to the keeping of the higher powers.

Yours ever | John Tyndall.

IC HP 8.61
RI MS JT/1/TYP/9/2919
Typed Transcript Only

1. *[8 April 1868]:* the date is given by relation to letter 2802, dated 13 April 1868, in which Tyndall wrote to Juliet Pollock, 'On Thursday morning last I quitted London'. This letter, written on a Wednesday and in which Tyndall wrote, 'I start tomorrow morning', would then be 8 April 1868. Hirst also wrote in his journal on 9 April 1868 that 'Tyndall started for Naples' (Brock & MacLeod, *Hirst Journals,* p. 1829).
2. *Hirst:* Thomas Hirst.
3. *established Naples:* according to his journal, Tyndall departed London on 9 April 1868, and he and his companions John Lubbock, Robert Birkbeck (husband of Lubbock's sister Mary), and Archie Hamilton arrived in Naples on 16 April (a Thursday). They went to Pompeii on 17 April, climbed Vesuvius on either 17 or 18 April, and climbed it again on 19 April. They left the following Tuesday (21 April), visiting Rome during their return (Journal, RI MS JT/2/13c/1338–46). Tyndall was back in London no later than 27 April (the date of the earliest surviving letter written from London after his return; see letter 2805). Thomas Hirst also noted in his journal on 9 April 1868 that 'Tyndall started for Naples. Lubbock and others to follow' (Brock & MacLeod, *Hirst Journals,* p. 1829). On 1 May, following Francis Turner Palgrave's Friday Evening Discourse at the RI on 'How to Form a Good Taste in Art', Tyndall spoke briefly about his and Lubbock's trip to Vesuvius ('Professor Tyndall at Vesuvius', *Glasgow Evening Citizen,* 5 May 1868, p. 4; 'A Visit to Vesuvius', *Dublin Builder,* 15 June 1868, p. 12).
4. *Lubbock:* John Lubbock.
5. *Amiens:* a city in northern France.

To Juliet Pollock 13 April [1868][1] 2802

Genoa[2] 13th. April

My dear Mrs Pollock

I wrote you a scrap[3] from Amiens,[4] but failed to throw it into the letter box: since then I have lived a great deal. On Thursday morning last I quitted London & reached Amiens that night. There I found myself for nearly two hours in the thick of a crowd in the noble Cathedral[5] listening to a priest discoursing on the life and death of Christ. At first I thought he was going to be earnest & inspired; but his inspiration became art, sound & gesticulation, towards the end: so I was not comforted. I halted at Amiens for the purpose of seeing in company with Lubbock[6] the pits where the famous flint implements which have caused so much sensation are found in some abundance.[7] There on the Friday I was joined by Lubbock, and having done our duty to the flints we went on to Paris. We dined at Vefours,[8] and went on to Marsailles[9] at 7.15; travelling all night. It was desperately cold; and the wind in the streets of Marsailles after our arrival searched us through and through. I could hardly have thought that so sunny a sky could be so bitter. Boats would not venture out, so we had to pass the night there in the hope that the morrow would allow us to escape. The hope was not disappointed: The wind sank, and Sunday morning a perfectly cloudless heaven over arched the world. I have rarely passed so delightful a day. The blue of the sea; the blue of the air, the noble line of coast, and the perfect absence of all internal disturbance rendered the voyage exceedingly enjoyable. And at night we had the stars; Jupiter casting a pillar of light upon the sea, and the sea itself sparkling and gleaming with phosphorescent creatures. We travelled all night and this morning reached Genoa. We saw a whale this morning rolling and spouting & cutting all manner of elephantine antics in the water.

We have had a scramble among the streets and palaces here: it is a beautiful place. This evening at 9 we loose from the harbour & dart into Leghorn.[10] There we halt for several hours and visit Pisa.[11] It is hard to get on in consequence of the numerous halts of the steamers: on Wednesday night we reach Naples[12] & on Thursday if the Gods favour us I hope we shall scramble up Vesuvius:[13]—good bye. Love to all. I have written to nobody else since I came away—ought you not to be proud.

I intended to call to see you before I came away, but I was most horribly low & very occupied. I hope the journey will diminish the lowness, for really one feels at times as if the boon of life were scarcely worth acceptance.

Yours ever & ever | John Tyndall

I shall have only two or three days at Naples; otherwise I would ask you to write to me there. | I wish you patience to read this letter.

RI MS JT/1/T/1164
RI MS JT/1/TYP/6/2094–5

---

1. *[1868]:* the year is given by reference to Tyndall's trip to Naples, Italy (see letter 2801, n. 3).
2. *Genoa:* a coastal city in northwestern Italy.
3. *wrote you a scrap:* letter missing.
4. *Amiens:* see letter 2801, n. 5.
5. *noble Cathedral:* the Cathedral Basilica of Our Lady of Amiens, built in the thirteenth century.
6. *Lubbock:* John Lubbock.
7. *famous flint implements . . . in some abundance:* see J. Prestwich, 'On the Occurrence of Flint-Implements, Associated with the Remains of Animals of Extinct Species in Beds of a Late Geological Period, in France at Amiens and Abbeville, and in England at Hoxne', *Phil. Trans.*, 150 (1860), pp. 277–317.
8. *Vefours:* Le Grand Véfour, a famous Parisian restaurant opened in 1784.
9. *Marsailles:* Marseilles, France (now Marseille), a port city on the Mediterranean Sea.
10. *Leghorn:* Livorno, a coastal city in western Italy.
11. *Pisa:* a coastal city in central Italy to the west of Florence.

12. *Naples:* a city in southern Italy, the destination for Tyndall's trip (see letter 2801, n. 3).
13. *Vesuvius:* Mount Vesuvius, a volcano in western Italy south of Naples that decimated the Roman towns of Pompeii and Herculaneum in AD 79.

To Mary Somerville[1] [19 April 1868][2] 2803

Sunday

My dear Mrs. Somerville

It will give my friend Lubbock[3] and myself the greatest pleasure to call to see you; today we go up Vesuvius[4] for the second time, but if possible we shall reach home sufficiently early to call upon you this evening. If this cannot be managed then tomorrow evening we shall be sure to come to you.

I leave Naples I am sorry to say on Tuesday morning to return to London.[5]

It was only a few days ago that I met your daughter in law[6] in London, and I was delighted to hear from her that your health was good & your book[7] in progress.

most truly yours | John Tyndall

Miss Somerville | Poste Restante[8] | Florence | Italy.[9]

Oxford, Bodleian Library, Dep c. 372 (MST-1)

---

1. *Mary Somerville:* Mary Somerville (1780–1872), a Scottish science writer who studied mathematics and astronomy. Primarily educated at home, her works included *On the Connexion of the Physical Sciences* (1834), *Physical Geography* (1848), and *On Molecular and Microscopic Science* (1869). In 1835, Somerville and astronomer Caroline Herschel (1750–1848) became the first women elected members of the Royal Astronomical Society (*ODNB*).
2. *[19 April 1868]:* the date is given by reference to Tyndall visiting Naples, Italy (see letter 2801, n. 3). Having arrived in Naples on 17 April 1868 (a Friday) and left on the following Tuesday, this letter, written on a Sunday, would then be 19 April 1868.
3. *Lubbock:* John Lubbock.
4. *Vesuvius:* see letter 2802, n. 13.
5. *to return to London:* Tyndall and his companions left Naples on Tuesday, 21 April, and Tyndall returned to London no later than 27 April (the date of the earliest surviving letter written from London after his return; see letter 2805).
6. *daughter in law:* Agnes Greig (née Graham, *c.* 1808–74), widow of Somerville's son Woronzow Greig (1805–1865) from her first marriage to Samuil Samuilovich Greig (1778–1807) (T. McLean (ed), *Further Letters of Joanna Baillie* (Madison, NJ: Fairleigh Dickinson University Press, 2010), p. 236, n. 531).
7. *your book:* probably M. Somerville, *On Molecular and Microscopic Science* (London: John Murray, 1869).

8. *Poste Restante:* general delivery (French); when a post office holds mail until the recipient requests it.
9. *Miss Somerville | Poste Restante | Florence | Italy:* written as the address on the envelope.

From Mary Egerton 21 April [1868][1] 2804

45, Eaton Place, S. W.[2] | April 21st.

Dear Mr. Tyndall,

I see your Jermyn St. lectures announcement to begin on Monday next.[3] I could have wished for your sake; that there were to be no more this year but if they must take place, may we[4] profit by a few of them? Last year, you generously gave me "fifty welcomes", of which I was only able to exhaust a very small proportion, may I consider some of the remainder as still valid for this year's use? But do not accede to this unless you are quite sure we are not in the way, for last year there seemed to be only students; and are we not cheating somebody by going without paying? We are going out of town on Friday but hope to be back on Monday before 3 o'clock, which I see is the hour this year, as we should particularly like to hear those on Magnetism, they w[oul]d. help us to understand Faraday.

I dare say you have forgotten long ago the question you asked me on your last note,[5] but something has reproached me for not answering it. I find however that I cannot do so without some explanation so I am going to venture to send you what I wrote down at the time, originally without any intention of your seeing it, but merely to give shape to my own thoughts. It is solely for my own satisfaction that I recur to the subject at all, (for I am not quite so vain as to suppose my *<1 or more pages missing>* I had better not say much about debts, for I owe you, for stores of interest, of thoughts, & associations, both directly & indirectly, for which I have simply "nothing to pay", so that the balance is all the wrong way!

Believe me y[ou]rs. very sinc[ere]ly | M F Egerton

What an interesting article there is in Macmillan */about/* Faraday.[6] What is the latest time there w[oul]d. be a chance of meeting you at Zermatt, not that I think we shall go this year unless some political earthquake takes place soon. I am sure you ought to have three months holiday this year.

RI MS JT/1/E/31

---

1. *[1868]:* the year is given based on mention of the Macmillan article about Faraday (see n. 6). This letter appears to relate to letter 2799 the previous month, in which Egerton first asked Tyndall for tickets to the lecture course.

2. *45, Eaton Place, S. W.:* in London.
3. *Jermyn S[t]. lectures announcement to begin on Monday next:* Tyndall's course of thirty-two lectures at the Royal School of Mines on 'Magnetism, Electricity, Sound, Light, and Heat' began on 27 April 1868 (see *PMG,* 20 April 1868, p. 14). This would be his last lecture course for the School of Mines, where he had held the Chair of Natural Philosophy since 1859. On Tyndall and the School of Mines, see letter 2601, n. 2.
4. *we:* possibly Olivia Bennet (Lady Tankerville) (see letters 2807, 2808, and 2809).
5. *your last note:* letter missing.
6. *Macmillan [about] Faraday:* J. Scott Russell, 'Faraday, a Discoverer', *Macmillan's Magazine,* 18 (1868), pp. 184–91 (this is a review of Tyndall's *Faraday as a Discover*).

To Leonard Lyell[1] 27 April 1868 2805

Royal Institution, 27th. April, 1868.

Dear Leonard

It is always difficult to arrange matters so as to meet the wishes of the students at the School of Mines.[2] The two o'clock lectures were objected to by many.

I fear there is no hope of our being able to alter the published hour.

Yours very sincerely | John Tyndall

RI MS JT/1/TYP/3/840
LT Typescript Only

---

1. *Leonard Lyell:* Leonard Lyell (1850–1926), a Scottish politician who was created a baronet in 1894 and raised to the peerage in 1914. He was the nephew of Charles Lyell, earned a B.Sc. at University College London, and was sometimes a lecturer on geology at the School of Mines ('Lord Lyell', *Times,* 20 September 1926, p. 16).
2. *School of Mines:* see letter 2601, n. 2.

To Dr. Henry[1] 27 April [1868][2] 2806

27th. April

My dear D[r]. Henry

When your note[3] reached this place I was on my way to Naples.[4] I have just returned.

Had your invitation reached me in time it would have given me great pleasure to accept it.

Yours very truly | John Tyndall

Lehigh University Digital Library Archive

---

1. *Dr. Henry:* not identified, and certainly not Joseph Henry (1797–1878), the American scientist and first secretary of the Smithsonian Institution, as he was not a 'Dr.'.
2. *[1868]:* the year is given by reference to Tyndall's traveling to Naples, Italy.
3. *your note:* letter missing.
4. *my way to Naples:* see letter 2801, n. 3.

From Emily Peel 27 April 1868 2807

4 Whitehall Gardens[1] | April 27./68

Dear Professor Tyndall,

Lady Tankerville[2] tells me that you are lecturing every day, Saturdays excepted, at the School of Mines[3] & that you have given her a ticket of admission to your lectures. As she is under the impression that this is not a purchasable article, I venture to ask you whether you will confer the same favour upon me, for in such case Lady Tankerville & I propose to attend your lectures together.

I remain yours | very truly | Emily Peel

RI MS JT/1/P/47

---

1. *4 Whitehall Gardens:* in London.
2. *Lady Tankerville:* Olivia Bennet.
3. *lecturing every day . . . at the School of Mines:* see letter 2804, n. 3.

To Emily Peel 29 April [1868][1] 2808

Royal Institution of Great Britain | 29th April

Dear Lady Emily

I shall be so proud of two such pupils![2]

The lectures are from 3 to 4 in the afternoon at the School of Mines[3] in Jermyn Street.

Tomorrow (Thursday) I break ground upon a new subject—Electricity—and I would recommend you to begin your studies then.

Most truly yours, | John Tyndall

RI MS JT/1/T/1116
RI MS JT/1/TYP/3/957

1. *[1868]:* the year is given based on relation to letter 2807.
2. *two such pupils:* in letter 2807, Peel asked Tyndall for a ticket so that she could join her friend Olivia Bennet.
3. *The lectures . . . School of Mines:* see letter 2804, n. 3.

From Emily Peel 30 April 1868 2809

Thursday 30th | April/68

Dear Professor Tyndall,

Many thanks for your card of admission to the School of Mines.[1] I regret extremely that neither Lady Tankerville[2] or I are able to attend the Lecture today—but we propose going tomorrow.

Yours very truly | Emily Peel

RI MS JT/1/P/48

1. *card of admission to the School of Mines:* see letters 2807 and 2808.
2. *Lady Tankerville:* Olivia Bennet.

To Elizabeth Dawson Steuart [30 April 1868][1] 2810

Royal Institution. | Thursday.

My dear Mrs Steuart,

I wrote to you some time ago regarding a dowry for Sarah Tyndall.[2] I expressed my willingness to provide such for her. I have heard nothing since about the matter.

I have been at Naples, Vesuvius and Rome[3] since I wrote to you. My health has been very poorly, but I have now a prospect of seeing it revive.

Yours ever | John Tyndall.

RI MS JT/1/TYP/10/3375
LT Typescript Only

1. *[30 April 1868]:* the date is given in relation to Tyndall's having returned from his trip to Italy in 1868 (LT marked this letter as '[April 1868]'). He appears to have returned no later than 27 April 1868 (see letter 2801, n. 3), making this Thursday letter 30 April.
2. *I wrote to you . . . dowry for Sarah Tyndall:* letter 2777.
3. *I have been at Naples, Vesuvius and Rome:* see letter 2801, n. 3.

To the Editor of *PMG*[1] [4][2] May 1868 2811

CORRESPONDENCE. | MR. HOME. |
To the EDITOR of the PALL MALL GAZETTE.

SIR,—A few years ago Mr. Faraday received various pressing communications from a gentleman[3] who had been deeply impressed with the performances of Mr. Home,[4] and who was exceedingly anxious that our great experimenter should analyze them. The invitation to do so amounted almost to entreaty, and at length Faraday expressed to me his willingness to gratify his correspondent if I would accompany him. I gladly agreed to this, for I had long wished to make the acquaintance of the spirits, and had tried to do so in Germany, France, and England. The invitation was accordingly accepted, Faraday claiming only the right to test the "manifestations" by his own methods, and expressing his determination to publish in such form as he might deem fit the result of his examination of Mr. Home.

The proposed investigation never took place.[5]

Since that time it has been my fortune to meet the spirits under a variety of circumstances.[6] They have kindly dubbed me "medium," and given me "a name in heaven." Still, notwithstanding their civility, I cannot say that their acquaintance has been edifying to me. JOHN TYNDALL.[7,8]

PMG, 5 May 1868, p. 3.

---

1. *Editor of PMG:* Frederick Greenwood.
2. *[4]:* the day is given based on this letter being published in the *PMG* on 5 May 1868.
3. *Mr. Faraday received various pressing communications from a gentleman:* letters from James Tennent. One such letter is James Emerson Tennent to Faraday, 17 June 1861, *Faraday Correspondence,* 6:4022, which is a reply to Faraday's letter of 14 June 1868, *Faraday Correspondence,* 6:4019; reproduced in Tyndall's letter to the *PMG* of 8 May 1868 (see letter 2815).
4. *Mr. Home:* Daniel Dunglas Home.
5. *never took place:* the investigation had been proposed in 1861. For an account of Tyndall's spiritualism debate in the *PMG,* see E. Wadge, 'Spiritualism', pp. 95–106.
6. *my fortune to meet the spirits under a variety of circumstances:* one such circumstance was a visit, with Latimer Clark (1822–98), an English civil engineer, to the spiritualist Camilla Dufour Crosland (née Toulmin, 1812–95) on 8 May 1857 (see letter 1390, *Tyndall Correspondence,* vol. 6).
7. Tyndall was prompted to write this letter because his name and that of the RI were invoked in a 2 May 1868 editorial in the *PMG* discussing the case of spiritualist Daniel Dunglas

Home, that he be 'carefully examined by the first experimental philosophers of the day, until the laws of these anomalous phenomena have been satisfactorily determined. Surely Professor Tyndall would find lodgings for him in the Royal Institution, and every man of science would be glad to subscribe a moderate sum to keep him as the most singular of recorded curiosities' (quoted in E. Wadge, 'Spiritualism', p. 97). Legal action had been brought against Home in the spring of 1868, as he was accused of using undue influence to cheat a widow of money through séances with her dead husband.

8. Home's letter to the editor responding to Tyndall's was published in the following issue (*PMG,* 6 May 1868, p. 3): 'SIR,—If Professor Tyndall will kindly refer to the correspondence that passed between Professor Faraday and the gentleman he alludes to in his letter, he will ascertain that Professor Faraday's last letter was to this effect, that his time was so valuable that he could not attend a séance unless he could have a programme of what would take place. As my "performances" are beyond my own control, this gentleman had the good sense without even consulting me to decline. My wife was at the time very ill, and we were lodging with a friend who kept a diary, and this fact was noted in her diary. For my own part I never saw one of the letters. It would have given me pleasure to meet Professor Faraday. In 1855 he refused to meet me, alleging that his religious scruples prevented him so doing. Professor Faraday would have been quite at liberty "to publish in such form as he might deem fit" the result of his investigation. It will give me the same pleasure to meet Professor Tyndall and any two gentlemen he shall designate. On my side I will have at least two gentlemen whose names and position place them above the suspicion of aiding or abetting a fraud. I will meet Professor Tyndall and these gentlemen when and where they please, and under such circumstances as they may decide on. I must only crave their patience if nothing should occur at the first or even second séance. A patient and candid investigation is all I ask. Having been much harassed, and not being in good health, I would postpone this meeting till sometime in June.—I am, Sir, your obediently, D.D. HOME. *Cox's Hotel, Jermyn-street, May* 5, 1868.'.

To the Editor of *PMG*[1] 6 May 1868 2812

CORRESPONDENCE. | PROFESSOR FARADAY AND MR. HOME. | *To the* EDITOR *of the* PALL MALL GAZETTE.

SIR,—It would make me Mr. Home's[2] debtor if he would favour me with a sight of the original letter[3] in which Mr. Faraday refuses to meet Mr. Home, and alleges his religious scruples as a reason.

It would also clear up all apparent differences between Mr. Home and myself if the gentleman[4] who wrote to Mr. Faraday, and with whom Mr. Home was in communication, would publish Faraday's reply. A letter so remarkable and so characteristic must surely have been preserved.

I have no reason to doubt what Mr. Home alleges regarding the "religious scruples" of Mr. Faraday. Holding as Faraday did that modern "spiritualism" was a compound of knavery and delusion, he may well have shrunk from seeing its "manifestations" clothed in the garb of religion. Still, I should like to see the original letter, and I rely on the courtesy of Mr. Home to produce it.

JOHN TYNDALL. | *Athenaeum Club,*[5] *May* 6, 1868.

PMG, 7 May 1868, p. 3

---

1. *Editor of Pall Mall Gazette:* Frederick Greenwood.
2. *Mr. Home's:* Daniel Dunglas Home.
3. *original letter:* the 14 June 1861 letter from Faraday is reproduced in Tyndall's letter to the *PMG* of 8 May 1868 (see letter 2815).
4. *gentleman:* Robert Bell (see E. Wadge, 'Spiritualism', p. 98).
5. *Athenaeum Club:* see letter 2537, n. 8.

From James Tennent 7 May 1868 2813

66, Warwick-square, Pimlico,[1] May 7, 1868.

MY DEAR SIR,—Having seen in the *Pall Mall Gazette*[2] the allusion made by you to the willingness of Professor Faraday to investigate personally the phenomena reported to him to have occurred in the presence of Mr. Home[3] during the spiritual séances, I enclose you the letter which I received from Faraday,[4] indicating the conditions under which he would be prepared to conduct the inquiry.

On communicating them to the gentleman[5] with whom I corresponded on the subject [an eminent spiritualist, and a friend of Mr. Home], but who is since dead, I received from him a letter dated June 17, 1861, declining the conditions, and the matter fell to the ground.—Faithfully yours, | J. EMERSON TENNENT.

PMG, 9 May 1868, p. 6

---

1. *66, Warwick-square, Pimlico:* in London.
2. *in the Pall Mall Gazette:* see letter 2812.
3. *Mr. Home:* Daniel Dunglas Home.
4. *letter which I received from Faraday:* the 14 June 1861 letter from Faraday is reproduced in Tyndall's letter to the *PMG* of 8 May 1868 (see letter 2815).
5. *the gentleman:* Robert Bell.

From William Wilkinson 7 May 1868 2814

44. Lincoln's-inn-fields,[1] May 7, 1868.

DEAR SIR,—I have noticed your letter[2] to the *Pall Mall Gazette* about Mr. Home[3] and the late Mr. Faraday, and also Mr. Home's answer.[4] I can fully corroborate what Mr. Home there says about Mr. Faraday having refused to come, at the last moment almost, unless he had a programme of the performance previously put into his hands. Sir Emerson Tennent it was who conducted the correspondence with Mr. Faraday,[5] and as soon as he saw his foolish demand he stopped the proposed meeting. Indeed, no one with any sense could have done otherwise.

If you knew of this I think you should have mentioned it in your letters[6] to the *Pall Mall,* because the omission gave the public a wrong impression of the reason why you did not meet Mr. Home.

I believe the question of whether the alleged manifestations are facts has gone much too far to render any further inquiry necessary, and that even if you should authoritatively pronounce that they do not happen, they would be found to occur notwithstanding. In fact, it would be seen that your saying a thing was impossible would be no bar to its happening. I have known of such cases before in the history of science.—Yours faithfully, | W. M. WILKINSON.

PMG, 9 May 1868, p. 6

1. *44. Lincoln's-inn-fields:* in London.
2. *your letter:* letter 2811.
3. *Mr. Home:* Daniel Dunglas Home.
4. *Mr. Home's answer:* see letter 2811, n. 8.
5. *Sir Emerson Tennent it was who conducted the correspondence with Mr. Faraday:* see letter 2813.
6. *your letters:* letters 2811 and 2812.

To the Editor of *PMG*[1] 8 May 1868 2815

CORRESPONDENCE. | FARADAY AND THE SPIRITUALISTS. | *To the* EDITOR *of the* PALL MALL GAZETTE.

SIR,—At a critical point in the history of English law and of English common sense I thought it my duty to publish the letter which appeared in your journal on the 5th of May.[2]

In a letter published in the following number of the *Pall Mall Gazette*[3] Mr. Home[4] impugns the accuracy of my statements. According to him it was Mr. Faraday, and not Mr. Home, who declined the combat.

In your number of yesterday you were good enough to publish a note from me,[5] suggesting the desirableness of having the letter of Mr. Faraday, to which I referred, made known to the English public.

Thus the matter stands as regards published communications. I have also received others, to one of which, inasmuch as it very emphatically supports Mr. Home, it is now my duty to refer.

Mr. W. M. Wilkinson,[6] who is, I am informed, one of the most distinguished "spiritualists" in England, writes to me thus:—

[. . .][7]

Thus Mr. Home asserts, and Mr. Wilkinson "can fully corroborate" his assertion, that Mr. Faraday "refused to come at the last moment almost, unless he had a programme of the performance previously placed in his hands." Sir Emerson Tennent, according to this authority, like a sensible man, "stopped the proposed meeting."

"To the law and to the testimony," then. Of his own free will, and unsolicited by me, Sir Emerson Tennent (whose correspondence with Faraday arose, not out of his conversion to spiritualism, but from the desire to see a single experiment which perplexed him properly examined) has placed Faraday's letter, accompanied by the following explanatory note,[8] in my hands:—

[. . .][9]

I think most of your readers will join me in thanking Sir Emerson Tennent for his prompt action in reference to the letter referred to. Here is the document:—

Folkestone,[10] June 14, 1861.

MY DEAR SIR EMERSON,—I cannot help feeling that you are indiscreet in your desire to bring me into contact with the occult phenomena which is said are made manifest in Mr. Home's presence. I have investigated such in former times, during some years, and as much as I thought consistent with the self-respect that an experimental philosopher owes to himself. It would be a condescension on my part to pay any more attention to them now; and I can only do so under the persuasion that all concerned wish to have the phenomena unravelled and understood, and will do all they can to aid in such a result. To settle whether I can go or not I wish to put to you the following points:—

Who wishes me to go?—to whose house?—for what purpose?

Does Mr. Home Wish me to go?

Is he willing to investigate as a philosopher, and as such to have no concealments, no darkness, to be open in communication, and to aid inquiry all that he can?

Does he make himself responsible for the effects, and identify himself more or less with their cause?

Would he be glad if their delusive character were established and exposed, and would he gladly help to expose it, or would he be annoyed and personally offended?

Does he consider the effects natural or supernatural? If natural, what are the laws which govern them? or does he think they are not subject to laws? If supernatural, does be suppose them to be miracles or the work of spirits? If the work of spirits, would an insult to the spirits be considered as an insult to himself?

If the effects are miracles, or the work of spirits, does he admit the utterly contemptible character, both of them and their results, up to the present time, in respect either of yielding information or instruction, or supplying any force or action of the least value to mankind?

If they be natural effects without natural law, can they be of any use or value to mankind?

If they be the glimpses of natural action not yet reduced to law, ought it not to be the duty of every one who has the least influence in such actions personally to develop them, and aid others in their development by the utmost openness and assistance, and by application of every critical method, either mental or experimental, which the mind of man can devise?

I do not wish to give offence to any one, or to meddle with this subject again. I lost much time about it formerly, in hopes of developing some new force or power; he found nothing worthy of attention. I can only look at it now as a natural philosopher; and, because of the respect due to myself, will not enter upon any further attention or investigation unless those who profess to have a hold upon the effects agree to aid to the uttermost. To this purpose they must consent (and desire) to be as critical upon the matter and full of test investigation in regard to the subject as any natural philosopher is in respect of the germs of his discoveries. How could electricity, that universal spirit of matter, ever have been developed in its relation to chemical action, to magnetic action, to its application in the explosion of mines, the weaving of silk, the extension of printing, the electro-telegraph, the illumination of lighthouses, &c., except by rigid investigation, grounded on the strictest critical reasoning and the most exact and open experiment? and if these so-called occult manifestations are not utterly worthless they must and will pass through a like ordeal.

As I do not want to debate this matter with those who have already made up their minds in a direction contrary to my own, but (if I see sufficient reason) only to work it out with such a desire to find incontrovertible proofs independent of opinion or assertion, so I wish you would show this letter to Mr. Home, and those who want me to meet him and them on his ground; after which you will know whether you should persevere in asking me. You will understand that I decline to meet any whose minds are not at liberty to investigate according to the general principles I have here expressed.

Further, I claim the right of publishing the whole or any part of this letter, or any future written communication that may arise out of it, in any manner that I may think fit.—Ever, my dear Sir Emerson, your very faithful servant,

M. FARADAY.

You will see that I consent to all this with much reserve and only for your sake.—M. F.

This is the letter which Mr. Wilkinson, and such as he, pronounce "foolish." This is the fashion in which Faraday "refuses" to meet Mr. Home.

Leaving these facts to settle in the public mind, and more especially in the legal mind.—I have the honour to subscribe myself, your obedient servant,

JOHN TYNDALL | *Athenaeum Club,*[11] *May* 8, 1868.

P.S.—I hold myself in readiness to witness and investigate, in the spirit of the foregoing letter, such phenomena as Mr. Home may wish to reveal to me during the month of June.

PMG, 9 May 1868, p. 6.
Faraday Correspondence, 6:4019

---

1. *Editor of PMG:* Frederick Greenwood.
2. *letter which appeared in your journal on the 5th of May:* letter 2811.
3. *letter published in the following number of the Pall Mall Gazette:* letter 2811, n. 8.
4. *Mr. Home:* Daniel Dunglas Home.
5. *a note by me:* letter 2812.
6. *Mr. W. M. Wilkinson:* William M. Wilkinson.
7. *[. . .]:* here Tyndall inserted Wilkinson's letter (2814).
8. *explanatory note:* letter 2813.
9. *[. . .]:* here Tyndall inserted Tennant's letter (2813).
10. *Folkestone:* a town in Kent, England, on the English Channel, where Henry Bence Jones had a home.
11. *Athenaeum Club:* see letter 2537, n. 8.

From J. J. Doherty[1] 8 May 1868 2816

Model Schools, Bailieboro', County Cavan, Ireland,[2] 8th May, 1868
| John Tyndall, Esquire, F.R.S., &c. &c.

—Sir, I beg leave to remind you of your promise contained in your letter of 22nd January last,[3] in reply to mine of a previous date.[4] I had stated that Major Dalton,[5] the chairman of our local Committee, had offered seven medals; one gold, three silver, and three bronze, for competition to the members of our class especially (who have been engaged this session in 'acoustics, light, and heat'); and that it was the wish of all parties that the decision should be left in your hands. In your reply you very kindly consented to undertake the additional trouble of deciding the order of the first seven of our class. The members of the class are most anxious that it should be understood that Major Dalton has offered the prizes *in addition* to those given by the department, and that the special prizes are not to exclude them from their chance of obtaining medals and prizes of the department. The latter, of course, carry with them a greater distinction; they would be more highly valued, and the candidates therefore hope their prospects in regard to them may not be interfered with by the local rewards. These latter have contributed to a great extent in keeping the members of our class hard at work. I venture to say, that for the last six months, physical science has received as much attention here in the centre of a rural district as has been given to it by any class in the metropolis itself. I have the honour to be, sir, respectfully, your obedient servant, J. J. Doherty, Head Master, Model Schools.

Select Committee on Scientific Instruction, *Report, Proceedings, Minutes of Evidence*, 15 July 1868, HC 432, 1867–8, Ev. 121

---

1. *J. J. Doherty:* J. J. Doherty (dates unknown), an Irish teacher and Head Master of the National Model School, Bailieborough, County Cavan, Ireland.
2. *Model Schools, Bailieboro', County Cavan, Ireland:* the National Model School in Bailieborough, County Cavan was founded in 1849 by the Commissioners of National Education in Ireland (Cavan County Council, *National Schools: Archive Service,* 2006, p. 22, Cavan County Archive, NE/1/00001–ED/12/00001). Twenty-eight model schools were established throughout Ireland in the second half of the nineteenth century (J. Coolahan, *Irish Education: Its History and Structure* (Dublin: Institute of Public Administration, 1981), p. 23).
3. *your letter of 22nd January last:* letter missing.
4. *mine of a previous date:* letter missing.
5. *Major Dalton:* Edward Tuite Dalton (1815–80), an Irish military officer and anthropologist, of County Cavan, Ireland ('Obituary', *Times,* 6 January 1881, p. 8).

From Alfred Russel Wallace 8 May 1868 2817

May 8, 1868.[1]

Dear Mr. Tyndall,

I do not know Mr. Varley,[2] but I will forward him your note,[3] and he can reply if he thinks proper. I rather doubt if any single case would be conclusive to you. Hume's[4] argument is overwhelming against any single case, considered alone, however well authenticated. He himself admits that no facts could possibly be better authenticated than the (so-called) miracles which occurred at the tomb of the Abbé Paris.[5] But when you look at a series of such cases, amounting to thousands in our own day, and a corresponding series extending back through all history, Hume's argument entirely fails, because his major proposition—that such facts are contrary to the universal experience of mankind—ceases to be true.

During the last two years I have witnessed a great variety of phenomena, under such varied conditions[6] that each objection as it arose was answered by other phenomena. The further I enquire, and the more I see, the more impossible becomes the theory[7] of imposture or delusion. I know that the facts are real natural phenomena, just as certainly as I know any other curious facts in nature.

Allow me to narrate one of the scores of equally remarkable[8] things I have witnessed,[9] and this one, though it certainly happened in the dark, is thereby only rendered more difficult to explain as a trick.[10]

The place was the drawing-room of a friend of mine,[11] a brother of one of our best artists.[12] The witnesses were his own and his brother's family, one or two of their friends, myself, and Mr. John Smith, banker, of Malton, Yorkshire,[13] introduced by me. The medium was Miss Nichol.[14] We sat round a pillar-table in the middle of the room, exactly under a glass chandelier. Miss Nichol sat opposite me, and my friend, Mr. Smith, sat next her.[15] We all held our neighbour's hands, and Miss Nichol's hands were both held by Mr. Smith, a stranger to all but myself, and who had never met Miss N. before. When comfortably arranged in this manner the lights were put out, one of the party holding a box of matches ready to strike a light when asked.

After a few minutes' conversation, during a period of silence, I heard the following sounds in rapid succession: a slight rustle, as of a lady's dress, a little tap, such as might be made by setting down a wineglass on the table; and a very slight jingling of the drops of the glass chandelier. An instant after Mr. Smith said, 'Miss Nichol is gone.' The match-holder struck a light, and on the table (which had no cloth) was Miss Nichol seated in her chair, her head just touching the chandelier.

I had witnessed this phenomenon before, and was able to observe coolly; and the facts were noted down soon afterwards. Mr. Smith assured me that Miss Nichol simply glided out of his hands. No one else moved or quitted hold of their neighbour's hands. There was not more noise than I have described, and no motion or even tremor of the table, although our hands were upon it.

You know Miss. N.'s size and probable weight, and can judge of the force and exertion required to lift her and her chair on to the exact centre of a large pillar-table, as well as the great surplus of force required to do it almost instantaneously and noiselessly, in the dark, and without pressure on the side of the table which would have tilted it up. Will any of the known laws of nature account for this?

Yours very faithfully, | ALFRED R. WALLACE.

A. R. Wallace, *My Life: A Record of Events and Opinions*, 2 vols (London: Chapman & Hall, 1905), vol. 2, pp. 291–3
NHM WP1/8/287[16]

1. *1868:* in the draft of this letter, Hurstpierpoint, Sussex was included as the location. This is where Wallace and his wife Annie Wallace (née Mitten, 1846–1914) lived in her family home, from July 1867 and into early July 1868.
2. *Mr. Varley:* Cromwell Fleetwood Varley (1828–83), an English telegraph engineer who served as chief engineer on the laying of the 1865 Atlantic telegraph cable. Varley was a believer in spiritualism and conducted investigations of his own (*ODNB*).
3. *your note:* letter missing. Varley's reply to Tyndall is letter 2822.
4. *Hume's:* David Hume (1711–76), a Scottish philosopher who made important contributions to epistemology through his philosophy of empiricism (*ODNB*).
5. *He himself admits . . . the tomb of the Abbé Paris:* referring to Hume's essay 'Of Miracles' which appeared in his *Philosophical Essays Concerning Human Understanding* (London: A. Millar, 1748). Hume had discussed the miracles reported to have occurred at the tomb of François de Pâris (1690–1727), a Catholic priest and Jansenist who had sided with the Jansenist sect against the papacy and French monarchy. He was buried in the Parisian cemetery of Saint-Medard and from 1727–32, claims of miracles occurring at his tomb were reported.
6. *conditions:* in the draft letter, Wallace had used the words '& of such a varied nature' here.
7. *theory:* in the draft letter, Wallace had used 'explanation' in place of 'theory'.
8. *remarkable:* in the draft letter, Wallace had used 'extraordinary' in place of 'remarkable'.
9. *Allow me to . . . I have witnessed:* this séance took place on 15 May 1867 (see A. R. Wallace, 'Notes of a Seance with Miss Nicholl at the House of Mr. A. S___, 15th May', *Spiritual Magazine,* 2 (June 1867), pp. 254–5). Wallace gave an account of this séance in his *On Miracles and Modern Spiritualism: Three Essays* (London: James Burns, 1875), but the names of the host and witnesses were not given (p. 162–3). Wallace had also published about another séance with Agnes Elisabeth Nicholl on 14 December 1866 (see 'Postscript by Alfred R. Wallace', *Spiritual Magazine,* 2 (February 1867), pp. 51–2). On Wallace and spiritualism, see letter 2558, n. 2.
10. *as a trick:* ARW did not use the words 'as a trick' in the draft letter.
11. *friend of mine:* not identified. In the draft letter, Wallace had shared that this person was 'a late Chairman of the Stock Exchange Committee'.
12. *best artists:* not identified.
13. *Mr. John Smith, banker, of Malton, Yorkshire:* not identified.
14. *Miss Nichol:* Agnes Elisabeth Nicholl [Guppy; Guppy-Volckman] (née White, 1838–1917), an English medium. After having visions as a child Nicholl was convinced of her powers by Fanny Sims, sister of Alfred Russel Wallace. Nicholl began holding séances in 1866 at first in the home of Wallace (*ODNB*).
15. *sat next her:* in the draft letter, Wallace had used the words 'One of the party held a box of lucifer matches ready to strike a light when asked' here, but in the published version they are moved to the end of the paragraph.
16. This is possibly a handwritten draft of the letter; if so, the final letter sent to Tyndall is missing. The draft includes many words/fragments that were crossed out, and there are differences in punctuation between the draft and the published letter, from which this transcription is based. Italicized words in the published letter are given as underlined words to match convention of handwritten letters at the time.

From François-Napoléon-Marie Moigno 10 May 1868 2818

LES MONDES | Revue Hebdomadaire | DES SCIENCES | ET DE LEURS APPLICATIONS | aux arts et à l'industrie | par | M. l'Abbé MOIGNO | RÉDACTION | 2, RUE D'ERFURTH, 2

Paris, le 10 Mai 1868

Mon cher M[r]. Tyndall

J'ai été très peiné des notes que vous avez mises au bas des pages de l'épreuve. Ce n'est certes pas le soin qui a manqué à ma traduction; si vous pouviez voir le manuscrit vous seriez effrayé du travail que j'ai fait. La traduction suit la rédaction, qui a elle aussi ses originalités. Les leçons sur le son au point de crée de la perfection de la rédaction ne ressemblent nullement aux leçons sur la chaleur qui étaient un chef-d'œuvre.

Il y a aussi des fautes d'impression qui ne laissent pas bien voir le sens littéral du texte. Par exemple p. 138, lignes 11 et 12.

Au reste indiquez avec soin tout ce qui vous semble défectueux, c'est pour cela que j'ai voulu qu'on vous envoie les épreuves avant le tirage. Je ne me reposerai pas tant que je n'aurai pas fait tout ce que je peut pour atteindre la perfection relative: Je ne me lasse jamais de bien faire.

Ni vous, ni Jamin n'avez répondu relativement à Faraday inventeur.

Votre humble serviteur | F. Moigno

LES MONDES[1] | Revue Hebdomadaire | DES SCIENCES | ET DE LEURS APPLICATIONS | aux arts et à l'industrie | by the | Father MOIGNO | EDITORIAL BOARD | 2 RUE D'ERFURTH, 2

Paris, May 10 1868

My dear M[r]. Tyndall,

I was very distressed by the notes that you wrote at the bottom of the pages of the proof.[2] It is certainly not care that was lacking in my translation; if you could look at the manuscript, you would be frightened of the work that I have done. The translation follows the writing, which also has its own peculiarities. The lessons on sound,[3] with regard to perfect writing, are nothing like the lessons on heat,[4] which were a masterpiece.

There are also misprints, which do not allow for a proper understanding of the literal meaning of the text. For example, page 138, lines 11 and 12.

For the rest, carefully indicate everything that appears to you as faulty, that is why I wanted the proofs to be sent to you before printing. I will not rest until I have done everything that I can to reach relative perfection: I never tire of doing things well.

Neither you, nor Jamin[5] have answered about Faraday Inventeur.[6]

Your humble servant | F. Moigno

RI MS JT/1/M/114

1. *LES MONDES:* see letter 2706, n. 9.
2. *proof:* see letter 2617, n. 2.
3. *lessons on sound: Sound* (1867).
4. *lessons on heat: Heat* (1863), published in French as J. Tyndall, *La Chaleur considérée come un mode de mouvement,* trans. F. Moigno (Paris: Étienne Giraud, 1864).
5. *Jamin:* Jules Jamin.
6. *Faraday Inventeur:* see letter 2750, n. 3.

To Alfred Russel Wallace [11 May 1868][1] 2819

Royal Institution of Great Britain | Monday

My dear Wallace

Your sincerity & desire for the pure truth are perfectly manifest. If I know myself I am in the same vein. I would ask one question.

Supposing I join you[2] will you undertake to make the effects evident to my senses; will you allow me to reject all testimony, no matter how solemn or respectable? Will you allow me to touch the effects with my own hands, see them with my own eyes and hear them with my own ears. Will you in short permit me to act towards your phenomena as I act and successfully act in other departments of nature?

I really wish to see the things able to produce this conviction in a mind like yours which I have always considered to be of so superior a quality.

Yours very faithfully | John Tyndall

NHM WP1/8/224

A. R. Wallace, *My Life: A Record of Events and Opinions*, 2 vols (London: Chapman & Hall, 1905), vol. 2, pp. 278–9

1. *[11 May 1868]:* the date is given by relation to letter 2817 in which Wallace described to Tyndall the circumstances of a séance. That letter, written on Friday, 8 May 1868, establishes this reply, on a Monday, as 11 May 1868.
2. *Supposing I join you:* presumably at a séance. On the top of the first page of this letter is a note from LT: 'about coming to sit with Miss Nicholl' (see letter 2817, n. 14).

To the Editor of *PMG*[1] 17 May 1868 2820

CORRESPONDENCE. | SCIENCE AND SPIRITUALISM. | *To the* EDITOR *of the* PALL MALL GAZETTE

Sir,—From time to time it has been my privilege to be instructed and edified by the notions of men and things with which Mr. F. T. Palgrave[2] has favoured the English public, and I now find myself promoted to the distinction of being the object of his criticism.[3]

Mr. F. T. Palgrave knows the world, and the world's ways, and when it is desirable for the world to know, and how that knowledge should be presented to the world, far better than I do. He is also intimately acquainted with the frame of mind in which a philosopher like Faraday ought to approach the study of reputed spiritualistic phenomena. It would, therefore, be presumptuous in me to contest the opinion of this eminent person that Faraday was wrong in writing that letter[4] to Sir Emerson Tennent, and that I was wrong in publishing it. I would only ask permission to state, in a few words, the motives which induced me to act in a manner to which Mr. Palgrave applies the adverb "unadvisedly," a word which derives its power less from its intrinsic weight than from the elevation of the pen from which it falls.

Recent events had revived the subject of spiritualism. In the public courts of England men with heavy scientific appendages to their names had testified, on oath, their conviction that the phenomena reputed to manifest themselves in the presence, and through the agency of Mr. Home[5] are "not due to the operation of any of the known laws of nature." This solemn testimony had been circulated through the length and breadth of the land, and along with it a great deal of flying gossip regarding the relationship of Faraday to these spiritual phenomena.

Though it involved a kind of jar to the sanctity in which I hold his memory, I thought it right that the world should learn from Faraday himself what his latest opinions were in connection with this subject. As a matter of mere authority, by which so many are guided, I thought his published opinion would be a wholesome corrective of the evidence to which I have referred. I also thought it desirable to sweep away the gossip which surrounded his name in connection with this subject. These were my motives for publishing his letter.

Mr. F. T. Palgrave speaks of the letter as the "hasty and transient error of a great man." Now, as I before stated, I am willing to defer to the opinion of a gentleman so experienced, so well-informed, and so self-assured as Mr. Palgrave that the letter was an error. Here he is on his own high ground, where his supremacy is indisputable. But how did he find out that the "error" was "hasty and transient"? I would ask him, with all distinctness, whence come these adjectives? From the region of fact or from that of imagination? If the latter—and this is the case—I would in all submission say that the tendency of the imagination to assume a definite structural form, irrespective of fact, is a very perilous one, and that it will require a master equal to Mr. F. T. Palgrave to keep its possessor out of danger, if the imaginative faculty be allowed this licence.

The simple truth is that Faraday was not moved by either heat or haste in writing that letter. He thought over it, wrote it, and read it to me after it had been written, with the utmost coolness and deliberation. Here, as in all other things, he endeavoured to "walk circumspectly." It cannot be otherwise than a source of regret to all of us that so eminent a judge as Mr. F. T. Palgrave should regard his endeavour as a failure. I was not aware, until I had seen the last number of your journal, that the letter of Faraday could be rightly described as a challenge to Mr. Home. Faraday responded, reluctantly, to the entreaty of a friend. He had spent time upon this subject; he had examined it in his way, and reached his own conclusions regarding it. He had not the slightest hope of adding to his own knowledge by further investigation. He regarded the necessity even of discussing such phenomena as are ascribed to Mr. Home as a discredit, to use no stronger term, to the education of this age. Still, when pressed to do so, he overcame his repugnance, while avowing it in moderate language, and for the sake of others he agreed to undertake an investigation which he knew would be without profit to himself.

If, after having approached the subject in a spirit which might have secured the approval of Mr. F. T. Palgrave; if, after having in this spirit satisfied his own mind that these reputed spiritualistic phenomena were only worthy of the scorn or the pity of all intelligent persons, he permitted a modicum of that strong contempt with which he regarded the subject to ooze out in undertaking a repetition of his task; surely the *honesty* of making his feelings indirectly known by the series of questions he proposed, and to which he required no affirmative subscription, will, in the minds of simple people, be a set-off to the lack of that "judicial" tact which Mr. F. T. Palgrave would have preferred.

In all reverence I would now say, let the glorious shade rest. I endorse the opinions and accept the responsibilities of Faraday in this matter. In the spirit of his letter, which, cavil apart, was simply this, that he should be permitted to apply to the examination of this subject the methods which have made his name immortal, I will at any practicable time undertake to meet Mr. Home. I ask him to subscribe to no condition, to furnish me with no "programme." I only ask him to permit me to deal with him and his phenomena as I now deal with Nature. If he fully and frankly grants this condition, the investigation can have but one of two results: either his phenomena will be proved delusive, or I shall be converted to the ranks of spiritualism.—I am, Sir, your obedient servant,

John Tyndall | *Athenaeum Club,*[6] *May* 17, 1868.

PMG, 18 May 1868, pp. 2–3

1. *Editor of PMG:* Frederick Greenwood.
2. *Mr. F. T. Palgrave:* Francis Turner Palgrave (1824–97), an English anthologist, critic, and poet (*ODNB*).
3. *the object of his criticism:* Palgrave had written a letter to the editor of the *PMG* on 13 May. Although he referred to spiritualism as a 'mischievous delusion' he also characterized Tyndall's suggestions (see letters 2811 and 2815) as 'eminently unjudicial' ('Correspondence', *PMG,* 16 May 1868, p. 6). Palgrave responded to Tyndall's reply on 20 May ('Science *Versus* Spiritualism', *PMG,* 21 May 1868, p. 3–4). Palgrave wrote that Tyndall was 'angry', and criticized him for having published Faraday's letter to James Tennant (quoted in 2815) after his death, especially considering that the contents of the letter make claims about someone who is the 'subject of legal inquiry'.
4. *that letter:* Tyndall included Faraday's letter to Tennent in his 8 May 1868 letter to the *PMG* (see letter 2815).
5. *Mr. Home:* Daniel Dunglas Home.
6. *Athenaeum Club:* see letter 2537, n. 8.

From Heinrich Debus 17 May 1868 2821

Clifton Coll[ege][1] | Clifton | Bristol | May 17. 1868.

My dear John,

Please excuse me for not replying to your note[2] before this day for I had much business to transact. I will answer your questions in the same order in which you have placed them.

The boys at Clifton are divided into two classes, the Junior School and the College. The former contains the young boys up to 13 years of age and the latter consists of boys between the ages of 14 and 20.

The College boy's receive instruction in Nat[ural] Phil[osophy] and Chemistry, two hours a week in both subjects. As far as the effect upon the boys is concerned I can only say that the boys like it very much and make good progress. They are very attentive, write copious notes and will very soon get correct notions about the operations of nature into their heads, which latter I consider of inestimable value. As a discipline of the mind the exact scientific method is quite as good as Euclid and a good deal more interesting to the boy's.

I myself consider Physics and Chemistry better means of Education than Natural History subjects because there is much more science in the former. Botany may do very well for young Ladies.

In order to give you an idea I will mention that last term I had 16 boys in the Lab[orator]y. The class is a voluntary one and the boys must take the working time out of their play time. Yet, they come and work very hard. Out of these 16 boys three have left at the commencement of this term, but the rest have all come back to me with some new ones. There are now 17 in the Lab[orator]y. Besides I have two other voluntary classes. The fifth and sixth forms have the option to do or not to do Nat[ural] Sciences. These two forms contain about 40 boys. Out of these I have got 14 for Chemistry and 12 for mechanics. These classes have also to take the time out of playtime.

Class books, of course, are necessary. I use Roscoe's[3] book[4] with some of the boy's. In Nat[ural] Phil[osophy] I am in a dilemma so the older boys who wish to read for themselves we have recommended Atkinson's translation[5] of Ganots.[6] But for the younger boys I have not yet adopted a book and they accordingly attend only lectures and write notes on the lectures.

I wish you would come for a few days. I like to have often a talk with you. There is nobody about here with scientific knowledge to make a companion in science. Give my kind regards to Tom[7] and believe me.

Always your | H. Debus

I think I must go to London about June 1st for a day or two.

RI MS JT/1/D/37

---

1. *Clifton Coll[ege]:* a school in Bristol, which Debus joined from Queenwood in 1868.
2. *your note:* letter missing.
3. *Roscoe's:* Henry Roscoe.
4. *book:* H. E. Roscoe, *Lessons in Elementary Chemistry* (London: Macmillan and Co., 1866).
5. *Atkinson's translation:* E. Atkinson (ed), *Natural Philosophy for General Readers and Young People translated and edited from Ganot's Cours Élémentaire de Physique,* 2nd edn (Longmans, Green and Co., 1866). For more on this, see J. Simon, *Communicating Physics: The Production, Circulation and Appropriation of Ganot's Textbooks in France and England 1851–1887* (London: Pickering & Chatto, 2011).
6. *Ganots:* Adolphe Ganot (1804–87), a French author and publisher of science textbooks which were often translated (J. Simon, *Communicating Physics: The Production, Circulation and Appropriation of Ganot's Textbooks in France and England 1851–1887* (London: Pickering & Chatto, 2011).
7. *Tom:* Thomas Hirst.

From Cromwell Varley[1] 19 May 1868 2822

Fleetwood House, Beckenham.[2] | 19th May, 1868.

My dear Sir.—Mr. Wallace[3] has forwarded me your letter to him, dated 7th inst.[4]

In compliance with your desire, I will endeavor to describe briefly some of the 'physical phenomena' I saw on two occasions, in the presence of Mr. Home,[5] together with the precautions I adopted to guard against trickery. To facilitate the explanation, permit me to premise, that the object of the *séance* was to demonstrate to me that the physical manifestations'[6] were not the result of trickery, and that there was some other intelligence at work besides Mr. Home and the observers. I had repeatedly heard from well-informed people of extraordinary manifestations occurring in the presence of this much-abused gentleman, and became very anxious to ascertain for myself the nature of them.

Having no introduction, I called one Tuesday morning in the Spring of 1860, at his residence, 134, Sloane Street, and told him that I was Electrician of the Electric and International and the Atlantic Telegraph Companies, that I was well acquainted with electricity, magnetism and other physical forces—that I had heard of extraordinary phenomena occurring in his presence, and was desirous of seeing them and of investigating their cause. I asked him if he would permit me to witness them. His answer was 'He would be delighted to do so.' At the same time he warned me that he could not promise that any manifestations would take place, that the phenomena were of a delicate character, and that it generally required several sittings before the necessary rapport was established to permit of anything decisive.

The next day I received a formal invitation for myself and Mrs. Varley[7] for Thursday evening, between seven and eight o'clock. Mr. Home residing in furnished apartments, and the drawing room into which I was introduced on the Tuesday, and in which the phenomena occurred on the Thursday, was over a chemist's shop.

On each of those occasions I examined the staircase to see if the floor was of unusual thickness or not, so as to permit of the introduction of machinery, and I also went into the shop underneath and examined the ceiling under the drawing room, but there was nothing unusual. The room was rather scantily furnished, there was a couch, about a dozen chairs, and nothing else capable of concealing any quantity of apparatus.

Eight of us sat down to a large, heavy, circular table. I had previously arranged with Mrs. Varley to observe very closely, to look out keenly for anything resembling trickery, to observe the room, the furniture and those present for anything that might appear suspicious and if anything unusual happened to note it carefully and fix it in her memory, so that we might collate our observations. On a slip accompanying this I give you in confidence the names of those, who were present, all excepting one whose name I do not recollect, and their positions around the table. Finding so many ladies I felt some disappointment, fearing it would prevent a rigid scrutiny.

The lady, who sat on Mr. Home's left, *i.e.* between me and him (and whom I will call Mrs. A.,[8]) and who had devoted considerable attention to the subject, entreated me, as did Mr. Home, to use my utmost endeavours to satisfy myself with regard to the phenomena, and he again urged me on no account of etiquette or otherwise to hesitate to make a complete scrutiny.

I availed myself of this. After sitting at the table for 20 minutes the first phenomenon occurred. We heard a number of sounds or raps as they are frequently termed. I examined the table underneath while Mrs. Varley observed above. There were four full-sized gas burners in the room burning brightly, all hands were on the table, all feet, as previously desired by Mr. Home, were turned back under the chairs.

The following simple telegraphic alphabet was then explained to me:—

1 sound or movement, or action of any kind			*No.*
3 sounds	"	"	*Yes.*
2 sounds	"	"	*Doubtful.*

i.e. that neither *No* nor *Yes* would answer the question.

5 sounds or movements, &c. of any kind, the *Alphabet.*

i.e. that the *alphabet* was to be run over or repeated aloud, and on the right letter being reached the fact was indicated by three sounds. In this manner words could be telegraphed easily, though slowly, by any one capable of producing the signals in question. Mrs. A—audibly expressed a wish that I should be touched—immediately five sounds were heard and telegraphing commenced. We were informed by this means that the communicator was afraid to approach me. I made no remark, but simply kept a sharp look-out, and endeavoured to find whence the sounds proceeded. Shortly after, the table tipped up, first on one side and then on the other, remaining sometimes for many seconds in the inclined position, having described an arc of about 30 degrees. I examined under and over the table—all hands were resting gently on it, and I could see no indication whatever, of any exertion of muscular force on the part of any of us. About this time the sounds became louder, and there were two distinct kinds of raps—the one being a lower, but louder note than the other. The lighter and higher sound communicated that it had now lost its dread of me, and would move my coat. It was pulled or jerked three times—about half a second of time elapsing between each pull. As these jerks were given to my coat, low down on the right hand side, between my chair and that of Mrs. A., the thought flashed through my mind, that this could not be accepted as evidence; but that, if my coat were pulled above the table, so that I could see it, it would be far more satisfactory. The instant this thought had passed through my mind, the lap of my coat, on the right-hand side, within a foot of my face, was turned up three times. Immediately, and in order to test it carefully, I *mentally* wished the left hand collar to be moved in a similar manner. Before the thought had been put into words in my own mind, the left-hand flap moved three times, as requested.

Very shortly afterwards information was signalled that I should be touched on the knee; without moving I again inaudibly desired that my right knee should be touched three times, and instantly I felt three distinct pressures. I mentally said, 'Left knee,' and it was touched three times without waiting a second. I mentally said, 'Right shoulder,' and it was instantly touched, but I could see nothing. I then mentally said, 'Left shoulder,' and finally, 'Top of my head,' and was touched accordingly three times at each place. The whole occupying not more than 10 seconds. As I did not speak or move, or make any gesture, no one was aware of what had taken place until I described it to them. The table was tilted about a great many times, and by means of the signals we were instructed to stand up, which we did, resting our hands lightly on the table, the latter after rocking about for a little while suddenly rose up bodily off the floor 14 or 15 inches, moved about, and then descended. I examined underneath during this phenomenon, but there was nothing visible. The hands were well over the table and it certainly was not supported by them. It then occurred to me, as my *unuttered* thoughts had been responded to in the early phenomena, that if the table rose off the floor again I would wish it to be moved in various directions; after waiting three or four minutes the table rose again into the air about 12 or 15 inches and remained up fully half a minute, I think more nearly a minute. I mentally wished it to move in various directions, and it did so, before those thoughts had assumed the shape of words in my own mind.

During a portion of the *séance,* as many as four distinct kinds of raps were head at the same time, and as many people were putting questions and receiving answers at the same moment. When requested, the raps were produced on the walls, on the ceiling, on our chairs, which latter were distinctly shaken by them; and on several occasions, the whole vibrated sufficiently to simultaneously attract the attention of every one.

Several of us, while seated in our chairs, were suddenly twisted around; and subsequent experiment showed that it required a good vigorous effort to perform a like operation by the hands. These twistings occurred to those who were far from, as well as those who were near to Mr. Home. The *séance* continued till about 11 o'clock; Mr. and Mrs. Home, as well as Mrs. A., pronounced it unusually successful, and expressed surprise at such a display of force at my *first* investigation.

On returning home Mrs. Varley and I compared notes, I found that she had put a great many questions mentally and received answers as rapidly as I had.

It was between 12 and 1 o'clock when we reached our dwelling at the other end of London, five or six miles from Sloane Street. Before retiring I requested Mrs. Varley to sit down with me in the drawing room, and once more think over carefully what we had seen together with the tests we had employed to guard against delusions; while so engaged, although five miles away from the medium, the sounds or raps commenced upon the walls of our own house.

The next evening I received a letter from Mr. Home, in which he informed us of the fact that we had heard these sounds in our own house.

I saw him shortly afterwards and asked him how he knew this, and he said that the same power which had produced the phenomena at his house, had produced the sounds in my own room, and had informed him of the fact and instructed him to write to me as an additional evidence.

Second séance at my own house, at Beckenham, Kent.

Present:—D. D. Home, a lady, a city accountant, a merchant, a gentleman, the first officer of a ship,[9] Mrs. Varley, and myself.

I purchased this house in the latter part of 1863 while it was in course of erection, and before any of the flooring was laid; I am therefore well acquainted with its internal structure.

In the summer of 1864, I requested Mr. Home to keep his promise of giving me a *séance* at my own house, and I invited the people mentioned to meet him. He had never been in the house before.

We had a great many phenomena similar to those already described; there were some, however, different from those I had seen in his house. During the course of the evening he became apparently nervous, and requested me to hold his hands, and said: 'Oh! look behind you'—and became somewhat alarmed.

He then put both his legs over my left knee, and at his request I held both his legs between my legs, and grasped both his hands in my own. We all of us looked in the direction, which he indicated—there was a small side table close to the conservatory window, seven feet behind Mr. Home's back, Mr. Home and I being the nearest to it. We were seated thus:—

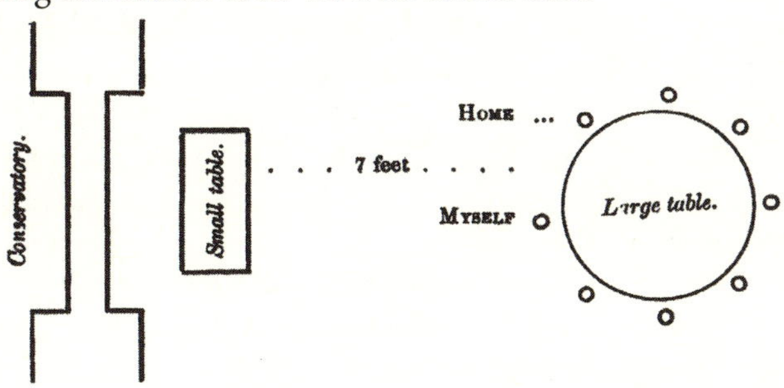

Shortly afterwards the side-table began to move. This table is mounted on casters, and it was driven up to me by some invisible means, no one being near it, and while I had hold of both Mr. Home's hands and legs. A large ottoman, capable of seating eight persons, was moved all over the room, and we were all driven by it up to the pianoforte.

Imposture was impossible. Phenomena of this kind were abundant this evening. As many of them occurred in the dusk of the evening they hardly come within the scope of your demand, *viz.* under a "bright light." I have witnessed the "physical phenomena" more than twenty times; but the higher psychical phenomena, which convey better proof to those who actually witness them than do the physical, have been witnessed by me more than a hundred times both in England and America.

You may ask why I have not published this before. The answer is simple—you yourself know how all new things are received in this world of contention.

I have endeavoured, whenever opportunity, health, and business would permit, to ascertain the nature of the force by which these phenomena are produced, but I have not progressed much farther at present than to find out the source whence the physical power is abstracted, *viz.*, from the vital systems of those who are present, and especially from the medium. The part of the subject under discussion, therefore, is not yet ripe for publication.

As to the phenomena themselves, there are numerous accounts on record—some excellently authenticated, both in this and the previous century. We are only now re-studying what has been investigated by the philosophers who lived 2,000 years ago; and when some good Greek or Latin scholar, who has made himself acquainted with the character of the phenomena which have been presenting themselves so numerously since the year 1848, will properly translate the writings of those great men, the world will find out that what is happening at the present time is only a new edition of an old page of history, studied by keen intelligences to an extent that will redound greatly to the credit of those good and clear-headed old sages, who seem to have risen far above the narrow-minded prejudices of their age, and to have investigated the matter under discussion to an extent in some respects transcending our present knowledge.

I am, my dear Sir, very truly yours, | [(Signed)] CROMWELL F. VARLEY. | J. Tyndall, Esq., F.R.S., | &c., &c., &c.

Spiritual Magazine, 3 (June 1868), pp. 273–8

1. *Cromwell Varley:* see letter 2817, n. 2.
2. *Fleetwood House, Beckenham:* in Kent, southeast of London.
3. *Mr. Wallace:* Alfred Russel Wallace. On Wallace and spiritualism, see letter 2558, n. 2.
4. *your letter to him, dated 7th inst:* letter missing.

5. *Mr. Home:* Daniel Dunglas Home.
6. *physical manifestations:* the opening quotation mark before 'physical' was not included in this published letter.
7. *Mrs. Varley:* Ellen Varley (née Rouse), married Cromwell Varley in 1855. Ellen later left Cromwell for Ion Perdicaris and settled in Morocco with her children (*ODNB*).
8. *Mrs. A.:* not identified.
9. *a lady, a city accountant, a merchant, a gentleman, the first officer of a ship:* not identified.

From Mary Egerton [21–22 May 1868][1] 2823

45, Eaton Place, | S. W.[2] | Thursday Morn[in]g

Dear Mr. Tyndall,

I feel as if I must tell you how intensely interesting it has been reading "Faraday"[3] again since the lecture.[4] Of course there are still many points "dark through excess of light",[5] to semi-blind scientific vision like mine; but one has a much more definite idea of the whole thing, & of the wonderful unity that every such discovery seems more & more to bring us in the realm of nature! How I hope it is reserved for you to complete the next link in the glorious chain, by proving another "mode of motion"![6] The greatest puzzle in that, seems to me to be, that one must divest oneself (I suppose?) of the analogy of propagated wave notions, if electricity begins from opposite ends at once.—And this motive "force"—whence is it?—Do you know Sir J. Herschel's paper on that subject?[7] If I was to tell you the odd thoughts that come into my mind on these subjects, I am afraid you might hold the B[isho]p. of Oxford intercession[8] over me again! But he is by no means my "Father Confessor"—my sympathies are with a broader school! You have looked so wearied lately, I have felt quite sorry for you to have to repeat all these elementary experiments over & over again To us they are pleasant as the meeting of old friends, after studying them in your book.[9]—

I have followed the "Spiritualist" Controversy[10] first with amusement, but afterwards with deeper feeling of which the first was, sympathy with your pain in having the—so to speak—sacred name of Faraday bandied about in Connection with such a subject! As to the merits of the Case, it would of course be absurd for me to pretend to an opinion; but, was it not Lord Palmerstone[11] who said every great man ought to have a Foolometer? Let me act as yours for a moment, & say that the closing sentence of your last letter[12] will do more good to the average general mind, than all the contempt which you are, very naturally, disposed to pour down upon such opponents (and of which your Conceited Mr. Palgrave[13] also comes in I must say for a very handsome show!) I feel very much disposed to agree with Mr. Stuart Glennie's[14] letter:[15] but what sort of an authority he may be, I know not. I only hope the investigation will really come off! Is there no performer upon whom it might be carried on without his being prepared beforehand? instead of that utter rogue M^r^. Home.[16] Of course this is <u>not</u> to be answered, it is only a bit of conversation on my side, to adjust the balance a little, as we have listened to <u>you</u> for many days!

<u>Friday.</u> I see that impossible M^r^. Palgrave again![17] Well, he could not perhaps be expected to sit down quietly under your <u>biting</u> sarcasms! but I confess he makes my blood boil! However, may I venture to say it? Now is the time to prove yourself like him whom you delight to honour, and by silence to him a greater triumph than they who stormed the cliffs of Magdala.[18]

Believe me | Y[ou]^rs^ very truly | <u>M. F. Egerton</u>

We are in all the miseries of preparing for a ball next week!

RI MS JT/1/E/78

1. *[21–22 May 1868]:* dated given in relation to Egerton's mention of Francis Palgrave's letter to the editor of the *PMG* (see n. 17).
2. *45, Eaton Place, | S. W.:* in London.
3. *"Faraday": Faraday as a Discoverer.*
4. *lecture:* see letter 2804, n. 3.
5. *"dark through excess of light":* a reference to line 380 in Book III of John Milton's (1608–74) poem 'Paradise Lost' (1667), referring to God: 'Drawn round about thee like a radiant shrine, | Dark with excessive bright thy skirts appear, | yet dazzle heaven, that brightest Seraphim | Approach not, but with both wings veil their eyes' (379–82) (J. Prendeville (ed), *Milton's Paradise Lost* (London: Samuel Holdsworth, 1840), p. 69).
6. *"mode of motion":* referring to Tyndall's experimental researches on heat, sound, and light as nothing more than molecular physics. Tyndall's book *Heat* (1863) had 'mode of motion' as part of the full title.
7. *Sir J. Herschel's paper on that subject:* possibly J. F. W. Herschel, 'On Light: Part II', in *Familiar Lectures on Scientific Subjects* (London: Alexander Strahan, 1867), pp. 268–339.
8. *B[isho]p. of Oxford intercession:* Samuel Wilberforce (1805–73), an Anglican clergyman known for his opposition to Charles Darwin's theory of evolution and his conflict with Thomas Henry Huxley at the annual meeting of the BAAS at Oxford in 1860. He was Bishop of Oxford (1845–69) before becoming Bishop of Winchester (1870–3) (*ODNB*).
9. *in your book:* probably *Faraday as a Discoverer* (1868). Egerton had attended Tyndall's lectures at the School of Mines on 'Magnetism, Electricity, Sound, Light, and Heat' (see letter 2804, n. 2), the syllabus of which remained largely unchanged over the years—hence Egerton's comment about Tyndall having to repeat the experiments 'over & over again'. The experiments which Egerton refers to were probably discussed in *Faraday as a Discoverer* and Tyndall may have also performed them in his nine lectures 'On the Discoveries of Faraday' at the RI on Tuesdays and Thursdays from 30 January to 27 February 1868 (*Roy. Soc. Proc.,* 5 (1866–9), p. 197).

10. *"Spiritualist" Controversy:* since 4 May 1858 Tyndall had been engaged in a public debate over the claims of the spiritualist Daniel Dunglas Home (see letters 2811, 2812, 2813, 2814, 2815, 2820, and following this letter, 2824 and 2825; also see Wadge, 'Spiritualism').
11. *Lord Palmerstone:* Henry John Temple (1784–1865), third Viscount Palmerston, an English politician who served twice as Prime Minister (1855–9 and 1859–65) (*ODNB*).
12. *your last letter:* in his most recent letter to the editor of the *PMG* (letter 2820), Tyndall took aim at Francis Palgrave.
13. *Mr. Palgrave:* Francis Palgrave (see letter 2820, n. 2).
14. *Mr. Stuart Glennie's:* John Stuart Stuart-Glennie (1841–1910), a Scottish lawyer and folklorist (P. Geddes, 'The Late Mr. J. S. Stuart Glennie', *Sociological Review,* a3 (1910), pp. 317–23).
15. *letter:* see 'Correspondence', *PMG,* 20 May 1868, p. 3.
16. *Mr. Home:* Daniel Dunglas Home.
17. *Mr. Palgrave again!:* Palgrave wrote a reply to Tyndall's letter (see n. 12) on 20 May ('Correspondence', *PMG,* 21 May 1868, pp. 3–4).
18. *they who stormed the cliffs of Magdala:* a reference to the Battle of Magdala (9–13 April 1868) in which British and Indian forces attacked Abyssinian forces to secure the release of missionaries who were being held captive.

From James Tennent 22 May 1868 2824

66, *Warwick-square, Pimlico,*[1] *May* 22, 1868.

MY DEAR SIR,—After reading the various letters in the *Pall Mall Gazette,* in which strictures have been passed on the letter which Mr. Faraday wrote to me,[2] on the subject of my request that he should be present at some of the "manifestations" displayed in the presence of Mr. Home;[3] and, above all, after seeing the *résumé* and comment on these communications in the *Pall Mall Gazette* of to day,[4] it strikes me that some injustice has been done to Faraday by insulating his letter from what had previously passed between him and me, and thus pronouncing upon its tone and temper, apart from the considerations which led to the expression of either.

I think, therefore, it might be well to set Faraday right so far as regards the previous circumstances, and they are simply these: You may remember a remarkable article, written by Mr. Bell,[5] which appeared in the *Cornhill Magazine,* under the title, if I recollect rightly, of "Stranger than Fiction."[6] This paper embraced in one condensed account the several phenomena occurring on several occasions in the drawing-rooms of a lady of rank in London,[7] and in, I think, nearly every instance I was present and saw what he described. You have already stated correctly that on these occasions I was a spectator merely, and not a convert or a spiritualist; but I regarded *some* of the things I saw as so extraordinary that, having puzzled me to imagine how they could be done (without a supposition altogether inadmissible, namely, the connivance and concert of the lady of the house with the express intention to deceive and delude the guests), I bethought me of asking Mr. Home and Mr. Bell whether there would be any objection to inviting a man like Faraday to be present, and, should they again occur, to encourage him to examine and if possible to discover by what means such unusual results were obtained.

With their full approval I called upon Faraday and in a friendly and cordial interview I narrated to him such of the occurrences as I thought were worthy of investigation. By this selection I mean that I promptly excluded everything that occurred when the room was darkened, because, being insusceptible of proof, they were not entitled to attention. I excluded equally everything for the production of which I could imagine some device common to the performances of the professors of legerdemain,[8] and I confined my enumeration as entitled to further scrutiny solely to two or three items so curious that, except on the conjecture alluded to of unworthy connivance on the part of the lady of the house, they seemed to me inexplicable.

Faraday received my request in the same gentle and genial spirit which endeared him to us all, but his first impulse was eminently characterized by a cautious regard for the interests of true science, of which he was one of the recognized trustees. He reminded me that he had already spent some time in an unsatisfactory search for some physical principle to which such phenomena might be referable, and, unable to discern any, he naturally was thrown back on the supposition that they might perhaps be more or less the result of deliberate prearrangement. If so, he felt naturally apprehensive that he might fail in detecting the process, because, as he said, although familiar with the observation of natural analysis, he had no sufficient experience with mere mechanical agency, or rather he might say, in such case, "mechanical contrivance," as to give him sufficient confidence that he might not fail. And failure in such a case would be regarded not merely as a discomfiture of truth, but as a corroboration of error.

However, he added, as he was desirous, personally, to gratify me in the matter, he would take a few days for consideration, and the result of this reflection was the letter I have already sent you, embodying the conditions under which alone he felt that he would be safe in acceding to my request without imperilling the interests which he felt to be in his own keeping.—Faithfully yours, | J. EMERSON TENNENT.

To Professor Tyndall, &c.

PMG, 25 May 1868, pp. 3–4

1. *66, Warwick-square, Pimlico:* in London.
2. *letter which Mr. Faraday wrote to me:* see letter 2815.

3. *Mr. Home:* Daniel Dunglas Home.
4. *résumé and comment on these communications in the Pall Mall Gazette of to-day:* see 'Science and Spiritualism', *PMG,* 22 May 1868, p. 1.
5. *Mr. Bell:* Robert Bell.
6. *"Stranger than Fiction.":* R. Bell, 'Stranger than Fiction', *Cornhill Magazine,* 2 (1860), pp. 211–24.
7. *lady of rank in London:* not identified.
8. *legerdemain:* skilful use of one's hands when performing conjuring tricks; sleight of hand (*OED*).

To the Editor of *PMG*[1] 23 May 1868 2825

FARADAY AND SPIRITUALISM. | *To the* EDITOR *of the* PALL MALL GAZETTE.

Sir,—The following letter has been sent to me by Sir Emerson Tennent. You will decide whether it is desirable that the letter should appear in the *Pall Mall Gazette.*[2]

JOHN TYNDALL. | *Athenaeum,*[3] *Saturday.*

PMG, 25 May 1868, pp. 3.

1. *Editor of PMG:* Frederick Greenwood.
2. *desirable that the letter should appear in the Pall Mall Gazette:* the letter, dated 22 May, was indeed published by the *PMG* with this one (see letter 2824).
3. *Athenaeum:* see letter 2537, n. 8.

From George Frederic Daniell[1] 25 May 1868 2826

Vicarage | Nr. Chichester. | Monday, May 25: 1868.

Dear Sir,

I have been reading with great interest your Essay on Faraday as a Discoverer.[2] Many, I daresay, will consider the anecdote about the Dean[3] as too good to be true, but you will find it confirmed in a letter by my Father Professor Daniell.[4] (page 5) of which I have great pleasure in sending you a copy[5] by this day's post

I remain | Dear Sir | Yours faithfully | G. F. Daniell

RI MS JT/1/TYP/12/4210
LT Typescript Only

1. *George Frederic Daniell:* George Frederic Daniell (*c.* 1818–88), an Anglican clergyman who was Vicar of Aldingbourne, West Sussex, near Chichester, from 1852 until his death (*Alum. Cantab.*).
2. *Essay on Faraday as a Discoverer: Faraday as a Discoverer.*
3. *anecdote about the Dean:* not identified. The Dean was not named in the anecdote (*Faraday as a Discoverer,* pp. 32–3).
4. *my Father Professor Daniell:* John Frederic Daniell (1790–1845), an English natural philosopher who served as manager and visitor at the RI and helped to publish twenty volumes of the *Quarterly Journal of Science.* His many achievements include having won all three RS medals (*ODNB*).
5. *copy:* letter missing.

From Fanny Tyndall 25 May 1868 2827

Bagnalstown[1] | 25th May 1868.

Professor Tyndall

Sir,

I hope you will forgive me for taking the liberty of writing to you this time. I never will forget your kindness to me when in Dublin six years since[2] I beg to leave before you our grievances we are after loosing a cow with mourine[3] that we were offered £17 for a few days since and had £4 paid for her grass and will be at the loss it when not able to buy one A cow is able to make from ten to twelve shillings a week besides giving milk to the children and I do not know what to do if not able to buy one this is a slack time of earning with William[4] he is going on well and drinks none and is in respect with all the gentlemen Mr Grogan[5] Mr Newton[6] Doctor Trayer[7] or Mrs Steuart[8] can tell you Sir if you would assist in getting a cow this time to help to support my family I hope I never will have to trouble you again and will ever pray.

Fanny Tyndall.

RI MS JT/1/TYP/10/3376
LT Typescript Only

1. *Bagnalstown:* see letter 2587, n. 1.
2. *when in Dublin six years since:* possibly referring to Tyndall's trip to Ireland in 1860 (and thus not six but eight years since).
3. *mourine:* alternative spelling of 'murrain', any virulent infectious disease of cattle or other livestock, such as anthrax, rinderpest, or babesiosis (redwater fever) (*OED*).
4. *William:* William Tyndall.
5. *Mr Grogan:* probably Rev. Charles J. Grogan, a rural dean of Dunleckney about two miles

from Leighlinbridge (*Thom's Irish Almanac and Official Directory, 1870* (Dublin: Alexander Thom, 1870), p. 1048).

6. *Mr Newton:* probably Philip Jocelyn Newton, former High Sheriff of Carlow who also served as Justice of the Peace and Deputy Lieutenant. He was the chairman of the Bagnalstown town commission and lived at Dunleckney, Bagnalstown (*Thom's Irish Almanac and Official Directory, 1870* (Dublin: Alexander Thom, 1870), pp. 1184–5, 1326).
7. *Doctor Trayer:* James John Trayer.
8. *Mrs Steuart:* Elizabeth Dawson Steuart.

To Elizabeth Dawson Steuart 28 May [1868][1] 2828

Royal Institution. | 28th. May.

My dear Mrs Steuart,

Mrs Fanny Tyndall, who I believe is the wife of one of the sons[2] of Caleb Tyndall has written to me[3] asking assistance towards purchasing a cow in the place of one which she and her husband have recently lost. I heard a very good character of Mrs Fanny Tyndall some years ago, but I also remember that her husband was only too ready to make application to me for assistance.

Nevertheless if you think that they really are likely to be distressed by the loss of their cow; and if you can corroborate the good character which Mrs Tyndall gives of her husband I shall be willing to assist them.

Perhaps the best plan will be to send you a cheque for ten pounds leaving it to you to decide what part of it, or whether the whole of it, should be devoted to the uses of William and Fanny Tyndall.

I shall soon be unable to pay any attention to applications of this character, for the state of my health causes me to think seriously of limiting my bread and butter work and thus considerably reducing my income.[4]

Of course what I have expressed my willingness to do for the girls[5] shall be duly carried out.

Yours ever | John Tyndall.

RI MS JT/1/TYP/10/3377
LT Typescript Only

1. *[1868]:* the year is given by relation to letters 2827 and 2831.
2. *one of the sons:* William Tyndall.
3. *written to me:* letter 2827.
4. *for the state of my health . . . reducing my income:* Tyndall resigned from his lecturing duties at the School of Mines shortly after this, but in a letter to Thomas Huxley (2833) he gives as the reason the need to create time for original research, rather than the state of his health. He did make two major research discoveries that autumn: that 'floating matter' in the air

could be burnt by a flame and which he concluded was thus organic, and that solar light could decompose the vapour of amyl nitrite (see *The Ascent of John Tyndall*, pp. 242–3).

5. *willingness to do for the girls:* see letter 2644, n. 5, and letter 2718, n. 10.

To Robin Allen 30 May 1868 2829

30 May 1868.

Sir,

I HAVE this day examined Captain Doty's[1] paraffin lamp,[2] and compared its performance with that of the oil lamp usually employed in first-class dioptric[3] apparatus.

The quantity of light emitted was in all cases in favour of the oil flame. At no time during the experiments did the paraffin flame emit the same amount of light as the oil flame.

The ratio of the one light to the other was in the first experiment 5:7, the larger number representing the emission from the oil flame. By nursing his flame, Captain Doty raised it so as to make the ratio 5:6; I even rendered it a little nearer to equality.

I do not doubt that improvements in his lamp might enable Captain Doty to render his flame fully equal to the oil flame. The question of light does not, however, excite my concern so much as the question of manageability.

In oil we have a substance comparatively sluggish to variations of draft; in paraffin we have a substance much more sensitive to such variations.

A very slight change in the draft is sufficient to convert a bright paraffin flame into a smoky one. Once or twice during the experiments, when attempts were made to vary the draft, the paraffin flame became very smoky—so much so, indeed, as to deposit soot upon the inferior surface of the lamp glass.

The paraffin lamp, therefore, in its present condition, would demand a far greater amount of attention than the oil lamp; there is, in its case, a far small margin for those minor acts of inattention to which all things under human control are more or less liable.

I will not say that the difficulties to which I refer cannot be overcome, but I will say that as yet they have not been overcome to anything like the extent which would justify me in recommending to the Elder Brethren[4] the displacement of oil by paraffin.

I have, &c. | [(signed)] *John Tyndall.*

The Secretary, Trinity House, London.

'Correspondence between Lighthouse Authorities and Board of Trade, relative to Proposals to substitute Mineral Oils for Colza Oil in Lighthouses', HC 2 264, Parliamentary Papers (1872), pp. 8–9

1. *Captain Doty's:* see letter 2791, n. 10.
2. *paraffin lamp:* a lamp which uses kerosene (paraffin) rather than oil.
3. *dioptric:* see letter 2791, n. 8.
4. *Elder Brethren:* see letter 2614, n. 5.

From Emily Peel [May 1868][1] 2830

Dear Professor Tyndall,

I have been prevented from attending any of y[ou][r] Lectures[2] during the past week.—Pray tell me what day you begin a fresh subject in order that I may then resume my attendance of those agreeable & instructive Lectures.—

Your very faithfully | Emily Peel

RI MS JT/1/P/73

1. *[May 1868]:* the month and year are given based on relation to letters 2807, 2808, and 2809.
2. *y[ou][r] Lectures:* see letter 2804, n. 3.

From Elizabeth Dawson Steuart 1 June 1868 2831

June 1st. 68

My dear John,

I delayed replying to your letter,[1] until I had made enquiries respecting the case of Mr Tyndall.[2] They[3] did lose cows, but not of such value as stated to you. He has the situation of Court-House keeper in Bagenalstown,[4] and is also Process-server—both of which must keep them very comfortably, and I really do not think you are called upon to give assistance beyond the extent of £5—for which they ought to be very thankful. He has given up drinking, and is now well conducted. I certainly would urge you to lay by all you can, and be prepared for any emergency which may arise. You have spent too much already on worthless people, and it is time to give it up. Please let me know if I shall give the £5—and return the rest to you. It grieves me to hear of your health being so broken, you must try to preserve it.

Ever and always | most sincerely yours | E. D. Steuart.

RI MS JT/1/TYP/10/3378
LT Typescript Only

1. *your letter:* letter 2828.

2. *Mr. Tyndall:* William Tyndall.
3. *They:* William Tyndall and his wife Fanny Tyndall.
4. *Bagenalstown:* see letter 2587, n. 1.

From Henry Clifton Sorby[1] 6 June 1868 2832

10 Cavendish Terrace | Buxton.[2] | June 6/68

My dear Tyndall,

I take it for granted that you fully sympathise with everything connected with heat. I have in the last few days made some experiments which at first seemed so remarkable that I felt scarcely able to believe my own eyes. If you have not seen such a thing, or have only read what I have, I think you will be equally unwilling to believe me when I say that I have been studying the law of the expansion of a liquid which actually increases in bulk 1/4 for <u>only one degree centigrade</u> viz expands from 100 to 125 parts when heated from 31° to 32° C. That is the liquid inclosed in the fluid-cavities in sapphire and no doubt is the same as that in topaz described by Brewster.[3]

In carefully examining a lot of cavities for a paper I am writing,[4] I saw how important it would be to ascertain the rate of expansion. I found a cavity moderately well filled for this experiment, and it gave an expansion of from 100° to 148° when heated from 0° to 30° C. Thilorier[5] says that liquid carbonic acid expands from 100° to 145° when heated from 0° to 30° C, but in another statement gives 100° to 138°. I was thus led to believe that the liquid in the cavities is really liquid carbonic acid. Fortunately I since found a cavity, which might have been made for the experiment. It is a tube of nearly equal diameter throughout; as may be proved by making the fluid pass to one end or the other. I therefore resolved to study the expansion at various temperatures, and find the results most remarkable. I copy down the volume at some different temperatures.

At 0° C	100.
17 1/2° (Cavity then half full)	109.
20° ------	113.
25° ------	122.
28° ------	130.
29° ------	139.
30° ------	150.
31° ------	174.
32° ------	217

The cavity is then quite full. Hence from 31° to 32° the expansion is equal to 1/4 of the bulk at 31° i:e: about 70 times as much as a gas. Now the only thing which makes me hesitate in concluding that it is a liquid carbonic acid, is that it seems possible that it may be liquid sulphurous acid. The only experiments that I know of are those of Kopp,[6] quoted in Gmelin's[7] Handbook of Chemistry[8] but they are all below the boiling point—below 8° C. I have discussed his statements by the method of differences, and find that they do not agree at all well, but by taking the volumes given for 5° and 25° and assuming that the law may be expressed by the form.

$v = 1. + A\,t + B\,t^2$ we obtain as an approximation to the volume at 20° C above the boiling point 8° C i:e: at 28° C only 102.16 instead of 130, which was the case in the liquid in sapphire. At 18° C the vol., would be 100.83 instead of 109. Hence you see that as far as one can judge by mere theory the expansion of liquid sulphurous acid is far too small. Still, since you see how irregularly such a liquid expands, it may really expand more than the above calculation.

My object in writing is to ask if you have ever made any experiments on the expansion of liquid sulphurous acid above its boiling point, or know of any that have been made. I have no chance of making them here and shall not return home for a number of weeks, and wish to write my paper here. Will you therefore kindly tell me if you know anything about it; or, if not, would you mind trying a rough experiment, if you have any liquid sulphurous acid at hand. It is easy to make such an experiment in a thermometer tube, of course sealed up, without any bulk. All that I wish to know is to what approximate bulk it expands when heated up to 30° or 32° C.

If the results are interesting, I will work it out more carefully in cold weather, when I could more easily prepare the acid. I expect the expansion will turn out very different to that of the liquid in the cavities; but it is very important to know it for certain; for, if it is so different, we must I think certainly look on the fluid in the cavities as liquid carbonic acid, and for the future call it a mineral substance occurring in nature class it with water &c.

Pardon my thus troubling you with so long a note, but trust that the great interest of the facts will be sufficient excuse.

Yours very truly | H. C. Sorby.

Notes on some polariscope objects | by H. C. Sorby F.R.S.; F.G.S.[9]

For some years past I have employed with my polariscope[10] several methods of illustrating certain well known facts, which are so very simple that I can scarce believe that they have not also been adopted by others. Nevertheless, I have never been able to find any account of them; and, since they are so very superior to those usually described in the various well known treatises of optics and polarised light, I venture to publish a few short notes on them.

In illustrating the fact of the production of white light by the overlapping of the complementary colours, when a double refracting prism is used for the analyzer it is usual to employ as object a thin plate of selenite. This however of course can only show one pair of colours. Instead of selenite I make use of a plate of quartz, about 1/5 of an inch thick, cut perpendicular to the principal axis, so as to exhibit the phenomena of circular polarization. Placing this over a black card, in which is a circular hole of about 1/5th of an inch in diameter, passing the polarized beam through it, having for an analyzer such a double image prism as is used for Leeson's[11] microscope goniometer,[12] and a magnifying power of two or three linear, the most admirable results are obtained. On account of the circular polarization of the quartz, when the analyzer is rotated, the two images of the hole in the card, having complementary colours, pass through every gradation in the series of the colours of the spectrum, and the part where they overlap constantly remains white. There is thus seen, with one object, every variety in tint, its true complement, and the white light due to their combination; whereas, when a simply double refracting substance like selenite is used, no such pair of colours is seen and no gradual passage from one to another through the whole series. By having a smaller hole in a thin black plate, these facts may be observed, but not so well, with an ordinary polarizing microscope.

If between two circular pieces of strong plate glass about a couple of inches in diameter, there be placed a double convex lens with a focal length of a few inches, and there be a framework to hold all together and three clamping screws, so that the two pieces of plate glass can be very strongly pressed on the lens or entirely released, a most excellent illustration is obtained of the production of double refraction by pressure. Where the whole is placed on the stage of the polariscope, with the ordinary analyzer so adjusted that no light passes through it, the tightening of the screws gives rise to the formation of a most perfect black cross, like that seen along the axis of uniaxial crystals, which disappears at once when the screws are relaxed. This simple arrangement, then, serves extremely well to show that symmetrical pressure gives rise to a symmetrical double refraction, not existing however in the exact line of the pressure; as proved by the production of a cross having equal branches, and the centre black in all positions, when the object is turned round on its axis.

RI MS JT/1/TYP/4/1354–7
LT Typescript Only

---

1. *Henry Clifton Sorby:* see letter 2632, n. 3.
2. *10 Cavendish Terrace | Buxton:* Buxton, in the Peak District southeast of Manchester, England, is a resort town.
3. *described by Brewster:* D. Brewster, 'On the Modification of the Doubly Refracting and Physical Structure of Topaz, by Elastic Forces emanating from Minute Cavities', *Phil. Mag.,* 31 (1847), pp. 101–4 (originally read at the Royal Society of Edinburgh in 1845).
4. *paper I am writing:* H. C. Sorby and P. J. Butler, 'On the structure of Rubies, Sapphires, Diamonds, and some other Minerals', *Roy. Soc. Proc.,* 17 (1868–9), pp. 291–302.
5. *Thilorier:* Adrien-Jean-Pierre Thilorier (1790–1844), a French lawyer and natural philosopher who invented a method for producing liquid carbon dioxide and was the first to produce solid carbon dioxide (dry ice) (L. G. Michaud (ed), *Biographie Universelle Ancienne et Moderne,* 2nd edn, 45 vols (Paris: Madame C. Desplaces, 1843–65), vol. 41, pp. 376–7).
6. *Kopp:* Hermann Franz Moritz Kopp (1817–92), a German chemist who studied among

other subjects the relationship between specific heat and the chemical composition of compounds (*CDSB*; *ADB*).

7. *Gmelin's:* Leopold Gmelin (1788–1853), a German chemist and professor at the University of Heidelberg, whose investigations included the chemical study of digestion. His *Handbuch der theoretischen Chemie* (Frankfurt am Main, 1817) was in its fifth edition at the time of his death (*CDSB*).
8. *Handbook of Chemistry:* see L. Gmelin, *Hand-book of Chemistry,* trans. H. Watts, 18 vols (London: Cavendish Society, 1848–72), vol. 1 (2nd edn, 1861), p. 230. Kopp's paper had appeared as H. Kopp, 'Untersuchungen über das specifische Gewicht, die Ausdehnung durch die Wärme und den Siedpunkt einiger Flüssigkeiten', *Poggend. Annal.,* 72 (1847), pp. 1–62, 223–93.
9. *F.G.S.:* Fellow of the Royal Geological Society of London.
10. *polariscope:* an instrument for showing the polarization of light, or for viewing objects in polarized light (*OED*).
11. *Leeson's:* Henry Beaumont Leeson (1803–72), an English physician and chemist who invented the Leeson Prism (*Alum. Cantab.*).
12. *microscope goniometer:* an optical apparatus for measuring angles in crystals. Also known as the Leeson Prism after its inventor Henry Beaumont Leeson.

To Thomas Henry Huxley 10 June [1868][1] 2833

Royal Institution. | 10th. June

My dear Huxley,

I wrote a letter[2] to Sir Roderick Murchison a few days ago which was preceded by an up and down struggle on my part, one, or the other of two opposing influences getting at intervals the upper hand. One of these influences was the desire to try my hand upon a heavy piece of original work, now before me,[3] the other was the strong reluctance that I feel to separate myself from the School of Mines.[4]

My mind is singularly inflexible as regards original work, I cannot quit it and return to it at pleasure. And thus it is that during my time of lecturing,[5] I can do nothing considerable in the way of research. The Managers of the Royal Institution[6] leave it optional to me to give 6 or 8 or 10 or 12 lectures just as I please. So that were it not for the four or five and thirty lectures that I am called upon to give at Jermyn Street I should be able to devote the autumn and the spring to one continuous stretch of original investigation. Something under these circumstances might be done. At all event I resolved to give myself the chance for the forth-coming two or three years.

Under these circumstances and not seeing any proper way of lessening my work at Jermyn Street, I have asked Sir Roderick to accept my resignation.

I have thought that in young Herschel[7] (he is 28) who has now had considerable practice at the Andersonian University in Glasgow you might find a suitable successor to the Chair of physics.[8] But in any and every way in my power I shall be happy to aid in the discovery of a suitable man.

Need I say my dear fellow that the severance which I most regret—though in reality it is no severance—is that from yourself.

Ever yours | John Tyndall.

IC HP 8.63
RI MS JT/1/TYP/9/2920

---

1. *[1868]:* the year is given by relation to letter 2834.
2. *letter:* letter missing.
3. *a heavy piece of original work, now before me:* it appears that Tyndall was thinking about experiments he would conduct a few months later in the autumn, resulting in: J. Tyndall, 'On a New Series of Chemical Reactions Produced by Light', *Roy. Soc. Proc.,* 17 (1868–9), pp. 92–102.
4. *School of Mines:* see letter 2601, n. 2.
5. *my time of lecturing:* for Tyndall's School of Mines lecturing at the time of this letter, see letter 2804, n. 3.
6. *Managers of the Royal Institution:* the following were elected Managers of the RI on 1 May 1868: Henry Wollaston Blake (1815–99), Charles Brooke (1804–79), Henry John Codrington (1808–77), Douglas Galton (1822–99), John Peter Gassiot (1797–1877), William Robert Grove (1811–96), Caesar H. Hawkins (1798–1884), John Lubbock (1834–1913), Roderick Murchison (1792–1871), William Pole (1814–1900), William Frederick Pollock (1815–88), Robert P. Roupell (1798–1886), Edward Sabine (1788–1883), Charles Wheatstone (1802–75), and Philip James Yorke (1799–1874). Henry Holland (1788–1873) was elected President, William Spottiswoode (1823–83) was elected Treasurer, and Henry Bence Jones (1813–73) was elected Secretary ('Annual Meeting', *Roy. Inst. Proc.*, 5 (1866–9), p. 375; *ODNB*).
7. *young Herschel:* Alexander Herschel.
8. *suitable successor to the Chair of physics:* see letter 2834, n. 7.

From Thomas Henry Huxley — 10 June 1868 — 2834

Jermyn St.[1] | June 10th. 1868.

My dear Tyndall,

All I can say is I am heartily sorry.[2]

If you feel that your lectures here[3] interfere with your original work,[4] I should not be a true friend either to science or yourself if I said a word against your leaving us.

But for all that I am and shall remain very sorry.

Ever yours very sincerely | T. H. Huxley.

If you recommend Herschel,[5] of course I shall be very glad to support him in any way I can. But at present I am rather disposed to d-n[6] anybody who occupies your place.[7]

IC HP 8.64
RI MS JT/1/TYP/9/2923
Typed Transcript Only

---

1. *Jermyn St.:* location of the Royal School of Mines (see letter 2601, n. 2).
2. *I am heartily sorry:* this letter replies to 2833.
3. *your lectures here:* see letter 2804, n. 3.
4. *interfere with your original work:* in letter 2828, Tyndall wrote to Elizabeth Dawson Steuart that 'the state of my health causes me to think seriously of limiting my bread and butter work and thus considerably reducing my income', offering a different reason for why he resigned from the School of Mines.
5. *Herschel:* Alexander Herschel.
6. *d-n:* damn.
7. *anybody who occupies your place:* in the event, Frederick Guthrie (1833–86), an English chemist and physicist, succeeded Tyndall at the School of Mines (*ODNB*), despite Tyndall's advocacy for Herschel.

To Thomas Henry Huxley 10 June [1868][1] 2835

Royal Institution. | 10th. June

Dear Hal,[2]

That d—n[3] was generated by the purest christian feeling. I have told Sir Roderick[4] that if it be deemed desirable I shall be happy to lecture to the working men.[5] I will add to you that if a short course on a special subject such as Light or Heat be deemed desirable I shall be glad to throw myself into it; but the present course[6] fetters me too much.

Yours ever | John Tyndall.

IC HP 8.65
RI MS JT/1/TYP/9/2921
Typed Transcript Only

---

1. [1868]: the year is given by relation to letter 2834.
2. *Hal:* nickname for Huxley.

3. *d—n:* damn. This letter replies to 2834.
4. *Sir Roderick:* Roderick Murchison.
5. *working men:* the series of evening lectures for working men at the Royal School of Mines that Tyndall took turns with others to give. It does not appear he gave them on this occasion. On Tyndall and the Royal School of Mines, see letter 2601, n. 2.
6. *present course:* see letter 2804, n. 3.

From John Herschel 14 June 1868 2836

Collingwood[1] June 14/68

My dear Sir

Let me thank you most heartily both in my son's[2] name and my own for your very kind and friendly suggestion of him to Sir. R. Murchison[3] as your successor in the R[oyal]. School of Mines Lecturership[4]—a suggestion the more gratifying as it would not have been made in the absence of a very well matured opinion of his fitness to succeed you. Sir Roderick has expressed himself as exceedingly well disposed to coincide with your suggestion—but his action in the matter must to a certain extent be dependent on or at least influenced by the approval of the other Professors. The position would be every way preferable to that which he holds at present at Glasgow,[5] and I need hardly say that I have advised him by all means to let nothing be wanting on his part to secure this appointment—should further testimonials, references &c. in addition to your recommendation (the most powerful testimonial of all) be needed—on this point you would perhaps give us your advice

Meanwhile believe me | My dear Sir | Yours very faithfully | J F W Herschel

RI MS JT/1/H/109
RI MS JT/1/TYP/2/546

---

1. *Collingwood:* see letter 2669, n. 1.
2. *son's:* Alexander Herschel.
3. *Sir. R. Murchison:* Roderick Murchison.
4. *your successor in the R[oyal]. School of Mines Lecturership:* Tyndall was succeeded by Frederick Guthrie (see letter 2834, n. 7).
5. *that which he holds at present at Glasgow:* Alexander Herschel was a professor of natural philosophy at Anderson's College, Glasgow (1866–71).

From George Campbell 15 June 1868 2837

Argyll Lodge,[1] | Kensington. | June 15/68

My dear Professor Tyndall

Could you come to breakfast with us on Wednesday next at 10 a.m.? It w[oul]^d give us great pleasure to see you

Yours v[ery tru]ly | Argyll

RI MS JT/1/A/99
RI MS JT/1/TYP/1/70

---

1. *Argyll Lodge:* see letter 2602, n. 1.

From Jean-Albert Gauthier-Villars 16 June 1868 2838

IMPRIMERIE ET LIBRAIRIE

Pour les Mathématiques, les Sciences et les Arts,

QUAI DES GRANDS-AUGUSTINS, 55.

Gauthier-Villars, Successeur de *Mallet-Bachelier,*

IMPRIMEUR-LIBRAIRE

Du Bureau des Longitudes; de l'École impériale-Polytechnique; de l'Observatoire impérial de Paris; de l'École impériale Centrale des Arts et Manufactures. — Éditeur des COMPTES RENDUS HEBDOMADAIRES DES SÉANCES DE L'ACADÉMIE DES SCIENCES; du JOURNAL DE MATHÉMATIQUES, par M. LIOUVILLE; du JOURNAL DE L'ÉCOLE IMPÉRIALE POLYTECHNIQUE; des ANNALES SCIENTIFIQUES DE L'ÉCOLE NORMALE SUPÉRIEURE, par M. PASTEUR; des NOUVELLES ANNALES DE MATHÉMATIQUES, par MM. GERONO et PROUHET; des ANNALES DE L'OBSERVATOIRE IMPÉRIAL DE PARIS; des ŒUVRES DE LAGRANGE; du BULLETIN DE LA SOCIÉTÉ FRANÇAISE DE PHOTOGRAPHIE.

Imprimerie et Librairie | *Pour les* Mathématiques, *les* Sciences *et les* Arts | Quai des Grands-Augustins, 55. | *Gauthier-Villars*, successeur de *Mallet-Bachelier* | Imprimeur-libraire | Du Bureau des Longitudes; de l'École impériale-Polytechnique; de l'Observatoire Impérial de Paris; de l'École impériale Centrale des Arts et Manufactures.—Éditeur des Comptes Rendus Hebdomadaires des Séances de l'Académie des Sciences; du Journal de Mathématiques, par M. Liouville; du Journal de l'École Impériale Polytechnique; des Annales scientifiques de l'École Normale Supérieure, par M. Pasteur; des Nouvelles Annales de Mathématiques, par MM. Gerono et Prouhet; des Annales de l'Observatoire Impérial de Paris; des Oeuvres de Lagrange; du Bulletin de la Société Française de Photographie.

N° 4903 | Paris, le 16 Juin, 1868

Monsieur

Vos observations sont extrêmement justes, et je regrette beaucoup que des épreuves ne vous aient pas été envoyées. Mais le mal est facile à réparer, et je ferai relire les pages dont vous n'êtes pas satisfait. J'attache, permettez-moi de le dire, un prix extrême à cet ouvrage dont la lecture m'a ravi. Rien ne sera épargné pour que l'édition française soit bonne.

J'accepte avec plaisir la proposition que vous voulez bien me faire de m'envoyer les clichés destinés à remplacer les figures, dont je me suis servi après avoir consulté l'abbé Moigno, et dont vous n'êtes pas satisfait. Je vous serai bien obligé de me faire faire cet envoi le plus tôt qu'il vous sera possible.

Veuillez agréer, Monsieur, l'assurance de ma considération la plus distinguée.

Gauthier Villars

Mr J. Tyndall, à Londres.

Printing and Bookstore | *For* Mathematics, Science *and* Arts | Quai des Grands-Augustins, 55. | *Gauthier-Villars*, successor of *Mallet-Bachelier*, | Printer-Seller | Of the Bureau of Longitude; of the Imperial Polytechnic School; the Imperial Observatory of Paris; the Central Imperial School of the Arts and Manufacturing.—Editor of the Comptes Rendus Hebdomadaires des Séances de l'Académie des Sciences; the Journal de Mathématiques by Mr. Liouville; the Journal de l'École Impériale Polytechnique; the Annales scientifiques de l'École Normale Supérieure by Mr. Pasteur; the Nouvelles Annales de Mathématiques by Mr. Gerono and Mr. Prouhet; the Annales de l'Observatoire Impérial de Paris; the Bulletin de la Société Française de Photographie.

N° 4903 | Paris, 16 June, 1868

Sir,

Your comments[1] are very much correct, and I deeply regret that the samples[2] were not sent to you. But the damage is easy to repair, and I will reread the pages with which you are not satisfied. I attach, allow me to say so, a great value to this book of which the reading greatly delighted me. Every effort will be deployed so that the French edition will be good.

I gladly accept the proposition that you are kind enough to offer me, that is, to send me the pictures meant to replace the figures, which I used after consulting with Father Moigno,[3] and with which you are not satisfied. I would be much obliged to you if you would have them sent to me as soon as possible.

Please accept, Sir, the assurance of my warmest regards.

Gauthier-Villars

Mr J. Tyndall, in London.

RI MS JT/1/V/19

1. *your comments:* letter missing.
2. *samples:* either the French translation of Tyndall's *Sound* (1867), which appeared as J. Tyndall, *Le Son,* trans. F. Moigno (Paris: Gauthier-Villars, 1869), or *Faraday as a Discoverer* (1868), which appeared as J. Tyndall, *Faraday Inventeur,* trans. F. Moigno (Paris: Gauthier-Villars, 1868).
3. *Father Moigno:* François Moigno.

From Rudolf Clausius 18 June 1868 2839

WÜRZBURG, 18 6/68.

Lieber Tyndall,

Wir haben so lange nichts von einander gehört dass ich mich gedrungen fühle, Dir einmal wieder Nachricht von uns zu geben, damit ich auch einmal wieder etwas von Dir und Hirst höre.

Wie Du weisst, sind wir seit Herbst vorigen Jahres hier in Würzburg. Der Ort ist recht hübsch, und die Leute sind uns sehr freundlich entgegen gekommen, so dass wir uns hier bald heimisch gefühlt haben. An der Universität fand ich ein physicalisches Cabinet vor, was ganz veraltet und in schrecklicher Unordnung war. Der alte Osann, mein Vorgänger, hatte in der letzten Zeit nichts mehr zur Vervollständigung gethan, sondern hatte im Gegentheil viel von dem Vorhandenen verdorben, und hatte alles im verdorbenen Zustande stehen lassen. Man hat mir nun zwar Geld genug bewilligt, um das Verdorbene *[wieder]* herzustellen, und Neues anzuschaffen, aber ich habe dadurch für das erste Jahr sehr viel Arbeit, indem ich fast jeden Apparat, den ich für die Vorlesungen gebrauche, erst in Stand setzen oder neu anschaffen muss. Es ist mir daher, seit ich in Würzburg bin, noch nicht möglich geworden, irgend eine eigene wissenschaftliche Arbeit zu machen. Hoffentlich werde ich jetzt bald dazu kommen.

Als ich Dir im vorigen Jahre mittheilte, dass ich nach Würzburg gehen würde, schriebst Du, dass Du vielleicht einmal für Deine Sommerreise das bayerische Oberland statt der Schweiz wählen und dann nach Würzburg kommen würdest. Wie steht es damit? Kannst Du diesen Plan vielleicht in diesem Jahre ausführen? Sowohl meine Frau als ich würden uns sehr freuen wenn wir Dich und Hirst hier in Würzburg begrüssen könnten.

Ich habe vor einiger Zeit auf die gütige Veranlassung von Hirst eine Einladung zur Versammlung der British Association in Norwich bekommen; ich fürchte aber, dass es mir in diesem Jahre nicht möglich sein wird, nach England zu kommen.

Du theiltest mir vor längerer Zeit einmal mit, Tait habe Dir angekündigt, er werde Dir eine Schrift schicken, mit der Du nicht ganz einverstanden sein werdest. Ich bin gespannt darauf, wie das ist, und es würde mir lieb sein, wenn Du mich gelegentlich einmal wissen liessest, ob Du schon etwas von ihm bekommen hast, und was es ist.

Bei uns geht jetzt alles gut. Eine Zeitlang waren die Kinder eins nach dem anderen unwohl, und eins wurde sogar gefährlich krank, wobei natürlich auch meine Frau viel zu leiden gehabt hat. Aber jetzt sind sie alle wieder hergestellt.

Nun lieber Tyndall, lebe wohl, und nimm für Dich und Hirst die herzlichsten Grüsse von meiner Frau und mir,

Dein | Clausius.

P.S. Als ich vom Secretär der Royal Society, Miller, die Anzeige meiner Ernennung zum auswärtigen Mitgliede erhielt, schrieb ich ihm ein Dankschreiben für die Royal Society. Jetzt habe ich nun das Diplom erhalten welches von Sabine unterschrieben ist. Ist es Sitte, dafür noch ein zweites Dankschreiben, vielleicht an den Präsidenten Sabine, zu schicken, oder hältst Du das für unnöthig?

WURZBURG,[1] 18 6/68.

Dear Tyndall,

We have not heard anything from one another for so long[2] that I feel compelled to give you some news from us once again, so that I shall hear something from you and Hirst[3] once again too.

As you know, we have been here in Würzburg since autumn of last year. The place is really nice, and the people received us in a very friendly manner, with the result that we soon felt at home here. At the university,[4] I found a physics laboratory which was quite antiquated and in a terrible disarray. Old Osann,[5] my predecessor, had not done anything more to improve it in recent years, but rather, on the contrary, had ruined much of what already existed, and had left everything in a ruined state. I have, admittedly, now been granted enough money to do up what was ruined and to buy new equipment, but because of this I am having a very great deal of work for the first year, in that I first have to repair almost every apparatus that I need for my lectures or buy it new. It has therefore still not become possible for me, since I have been in Würzburg, to do any sort of scientific work of my own at all. Hopefully I shall now come to this soon.

When I informed you last year that I would be going to Würzburg,[6] you wrote[7] that you would possibly choose the Bavarian Oberland[8] instead of Switzerland sometime for your summer trip and then come to Würzburg. How are things now with that? Can you possibly carry out this plan this year? Both my wife[9] and I would be very pleased if we could welcome you and Hirst here in Würzburg.

Some time ago, at the instigation of Hirst, I got an invitation to the meeting of the British Association in Norwich;[10] I fear, though, that it is not going to be possible for me this year to come to England.

You told me once a rather long time ago[11] Tait[12] had announced to you he was going to send you a paper[13] with which you were not going to entirely agree. I am curious to know how that is, and I would be glad if you would let me know sometime or other whether you have already got something from him, and what it is.

Everything is fine now at our place. For a while, the children[14] were unwell one after the other, and one of them even became dangerously ill, with which of course my wife has had a great deal to endure too. But now they have all recovered again.

Now, dear Tyndall, fare well, and accept the warmest regards from my wife and me to you and Hirst.

Your | Clausius.

P.S. When I received notice of my appointment as foreign member,[15] from the secretary of the Royal Society, Miller,[16] I wrote him a letter of thanks for the Royal Society. that I had been appointed as foreign member, I sent a letter of thanks for the Royal Society. I have now received the certificate, which is signed by Sabine.[17] Is it customary to send a second letter of thanks for this, perhaps to the president Sabine, or do you think this is unnecessary?

RI MS JT/1/TYP/7/2306–7
LT Typescript Only

---

1. *WURZBURG:* Clausius held a professorship at the University of Würzburg.
2. *We have not heard anything from one another for so long:* Tyndall and Clausius last exchanged letters in February 1868 (see letters 2763, 2764, and 2765).
3. *Hirst:* Thomas Hirst.
4. *university:* Julius Maximilian University of Würzburg (or University of Würzburg), where Clausius took a professorship in 1867.
5. *Old Osann:* Gottfried Wilhelm Osann (1796–1866), a German chemist and physicist who was professor of physics and chemistry at the University of Würzburg from 1828 until his death in 1866 (*ADB*).
6. *When I informed you last year that I would be going to Würzburg:* letter missing.
7. *you wrote:* in letter 2581, Tyndall wrote: 'I am exceedingly glad to hear that you have found a position better than the one you occupy in Zürich. In fact you ought to be in one of the first positions in Germany. No doubt dear Zürich will be very sorry for your departure. It will tempt me to make the Bavarian Highlands occasionally my home when I think that you are at Wurzburgh'.
8. *Bavarian Oberland:* see letter 2581, n. 5.
9. *my wife:* Adelheid Clausius.

10. *meeting of the British Association in Norwich:* the thirty-eighth annual meeting of the BAAS was held at Norwich from 19–26 August 1868 (*Brit. Assoc. Rep. 1868*).
11. *You told me once a rather long time ago:* if this was in a letter, it is missing.
12. *Tait:* Peter Guthrie Tait.
13. *paper:* possibly Tait's 'Historical Sketch of the Dynamical Theory of Heat' (see letter 2742, n. 4).
14. *Our children:* at this time, Rudolf and Adelheid Clausius had five children: Mathilde (1861–1907), Helene (1863–1919), Rudolf John (1864–1915?), Hedwig (1866–1946), and Alfred (1868–1950). A sixth child was born a few years later, Else (1872–1954).
15. *appointed as foreign member:* see letters 2763, 2764, and 2765.
16. *Miller:* William Hallowes Miller (1801–80), a Welsh mineralogist who held the professorship of mineralogy at the University of Cambridge (1832–70) and foreign secretary of the RS (1856–73) (*ODNB*).
17. *Sabine:* Edward Sabine.

To Rudolf Clausius 20 June 1868 2840

20th June 1868.

Lieber[1] Clausius

Es hat mich sehr erfreut Deinen Brief zu erhalten.[2] I have often thought of writing to you, but the multitude of matters which besets one's life in London, and the additional cares which the death of Faraday imposed upon me[3] prevented me from carrying out my good intention. I expected that you would have some heavy work in putting matters in order in Würzburg.[4] To teach aright always involves labour. For my own part I have been lessening the time I devoted to lecturing by giving up my chair at the School of Mines.[5] This will lessen my income considerably, but it will leave me a large amount of time for original work, which with me is the Hauptsache.[6] It does not look like an intention on my part to imitate your example, and take next myself a wife—and still I firmly believe that a good wife is the greatest blessing that a man can possess in this world. You must not forget to read this portion of my letter to Mrs. Clausius.[7] I have finished my lectures and have now only a few examination papers to look over, so that by the end of next week I hope to be ready to start for Switzerland. Hirst[8] is coming with me, and if we can combine a visit to Würzburg with our mountaineering I am sure it will be a delight to both of us to do so. I am not however certain whether this will be possible or not.

Since I wrote to you[9] I have not had any communication whatever from Tait;[10] and I am therefore in entire ignorance as to the nature of the memoir[11] to which he referred. He has very much changed with reference to Mayer,[12] and he may also change, as soon as he makes himself acquainted with the literature, with reference to you.

They are printing a translation[13] of my book on Sound[14] in Paris. Gauthier-Villars[15] is the publisher. Helmholtz[16] and Wiedemann[17] have undertaken a German translation.[18] I do not intend to publish anything of the kind for a long time to come—perhaps never again, because such work takes my thoughts so much away from original investigation. Still I think those two books on Heat & Sound[19] have done great good, by giving numbers of people an interest in scientific matters who had previously no interest in them.

I do not think there is the least necessity for you to write to Sabine.[20] You have already written to the Foreign Secretary[21] and that is enough. I was present when you were elected by the Society,[22] and your election was unanimous. You know the rule is to be chosen first by the council, and afterward by the Society—there has been scarcely a single case in which the unanimity was like that in yours.

And now my dear Clausius good bye. As usual at this season of the year I do not feel quite strong: but I shall soon recover my vigour in Switzerland. Give my kindest regards to your wife, and love to Johnny[23] & the rest of the children[24]

Yours ever | John Tyndall

RI MS JT/1/T/200

1. *Lieber:* Dear (German).
2. *Es hat mich sehr erfreut Deinen Brief zu erhalten:* It very much pleased me to receive your letter (German).
3. *additional cares which the death of Faraday imposed upon me:* see letter 2661, n. 4, and letter 2665, n. 3.
4. *Würzburg:* Clausius held a professorship at the University of Würzburg.
5. *giving up my chair at the School of Mines:* see letters 2833, 2834, and 2835.
6. *Hauptsache:* main thing (German).
7. *M^rs^. Clausius:* Adelheid Clausius.
8. *Hirst:* Thomas Hirst.
9. *Since I wrote to you:* the previous letter from Tyndall to Clausius in this volume is 2764, dated 22 February 1868.
10. *Tait:* Peter Guthrie Tait.
11. *memoir:* see letter 2742, n. 4.
12. *Mayer:* Julius Robert Mayer (1814–78), a German physicist known for his early articulation of the first law of thermodynamics, the conservation of energy (*CDSB*).
13. *translation:* see letter 2617, n. 2.
14. *my book on Sound: Sound* (1867).
15. *Gauthier-Villars:* Jean-Albert Gauthier-Villars.
16. *Helmholtz:* Hermann Helmholtz.
17. *Wiedemann:* Gustav Wiedemann.

18. *German translation:* see letter 2677, n. 5.
19. *those two books on Heat & Sound: Heat* (1863), of which a third edition was published in 1868, and *Sound* (1867).
20. *Sabine:* Edward Sabine.
21. *Foreign Secretary:* William Hallowes Miller (see letter 2839, n. 16).
22. *elected by the Society:* see letter 2763, n. 2.
23. *Johnny:* Rudolf John Clausius.
24. *rest of the children:* see letter 2839, n. 14.

From Heinrich Gustav Magnus 20 June 1868 2841

Berlin 20 Juni 1868

Mein theurer Freund!

Da Sie sich für die strahlende Wärme ganz besonders interessiren, so sende ich Ihnen einen Crystall von Sylvin (Chlorkalium). Diese Substanz ist zwar schon vor mehreren Jahren in Staßfurt gefunden worden, aber nur in derben nicht ganz klaren Massen. Jetzt hat man eine kleine Anzahl klarer Crystalle gefunden. Ich habe sie auf ihr Vermögen die Wärme durchzulassen untersucht und gefunden daß sie sich ganz wie Steinsalz verhalten. Eine kleine Notiz wird jetzt darüber gedruckt, und sende ich Ihnen dieselbe sobald ich sie aus der Druckerei erhalte. Aber da es mir möglich gewesen ist nur noch ein Paar Crystalle zu verschaffen, so erhalten Sie einen, der zwar nicht ganz so groß wie der von mir benutzte ist, aber doch klar und groß genug um eine Platte zur Untersuchung zu schneiden.

Mit vielen Grüßen von meiner Frau und meinen Kindern

Ihr | G Magnus.

Berlin 20 June 1868

My dear friend!

As you have a quite special interest in radiant heat, I am sending you a crystal of sylvite (potassium chloride). This substance was found several years ago in Stassfurt[1] but only in coarse and not entirely clear masses. A small number of clear crystals has now been found. I have examined them for their capacity to transmit heat and found that they behave quite like rock salt. A short note[2] is now being printed about this, and I shall send it to you as soon as I receive it from the printer. But as it has been possible for me to obtain just a couple of crystals, you are receiving one that is admittedly not quite as large as the one used by me, but is still clear and large enough to cut a plate for examination.

With many regards from my wife and my children[3]

Your | G Magnus.

RI MS JT/1/M/37

1. *Stassfurt:* a town in Germany known for its salt deposits and their extraction in the nineteenth century.
2. *short note:* G. Magnus, 'Ueber die Diathermansie des Sylvins', *Poggendorff's Annal.,* 134 (1868), pp. 302–4 (published in English as G. Magnus, 'On the Diathermancy of Chloride of Potassium', *Phil. Mag.*, 36 (1868), p. 320).
3. *my wife and children:* Bertha Magnus and their children, Anna, Christine, and Paul (see letter 2618, n. 3 and n. 4).

To Jane Barnard [22 June 1868][1] 2842

Monday

Dear Miss Barnard

Cottrell[2] is in the city & Brown[3] brought your note to me saying it referred to apparatus which you wished to have. I therefore thought I could not do wrong in opening it.

I send you all that I could find of the little steam engine[4] which I thought your uncle[5] told me had been made & given to him by Perkins.[6] It seems to be in a bad condition

Your uncle I believe never used it.

Yours very truly | John Tyndall

RI MS JT/1/T/128
RI MS JT/1/TYP/3/4198

1. *[22 June 1868]:* this letter precedes 2846 (written on Friday, 26 June), making this the Monday before, 22 June.
2. *Cottrell:* John Cottrell.
3. *Brown:* William Brown.
4. *little steam engine:* not identified.
5. *your uncle:* Michael Faraday.
6. *Perkins:* either Angier March Perkins (1799–1881), or his father Jacob Perkins (1766–1849), American-born heating and steam engineers who moved to England in 1821 (*ODNB*).

From Rudolf Clausius 23 June 1868 2843

WÜRZBURG, 23 6/68.

Lieber Tyndall,

Nachdem ich heute Deinen lieben Brief erhalten habe, beeile ich mich noch einmal an Dich zu schreiben, damit mein Brief Dich noch in London trifft, um Dir unsere Wohnung zu bezeichnen. Wir wohnen Hofstrasse No. 80½. Wir würden uns sehr freuen, wenn Du und Hirst uns hier besuchtet, und wir bitten Euch für diesen Fall, von der Eisenbahn gleich in unsere Wohnung zu kommen, und bei uns zu logiren. Wir haben hier eine grössere Wohnung als in Zürich, und können Euch daher sehr gut ein Zimmer zur Disposition stellen, ohne dadurch im Geringsten beschränkt zu sein. Wir haben während der Osterferien dieses Zimmer meinem Bruder und seiner Frau, welche uns besuchten, übergeben, und gleichzeitig ein anderes Zimmer meinem Schwager, welcher ebenfalls zur Taufe gekommen war, eingeräumt, und sind doch nicht genirt gewesen.

In der Hoffnung, Euch bald hier zu sehen, und mit freundlichen Grüssen von meiner Frau und mir

Euer | Clausius.

WURZBURG,[1] 23 6/68.

Dear Tyndall,

After receiving your kind letter[2] today, I am hastening to write to you again so that my letter still reaches you in London, in order to describe our apartment to you. We live at Hofstrasse 80½. We would be very glad if you and Hirst[3] visited us here, and we ask you in this case to come directly from the railway to our apartment, and stay with us. We have a larger apartment here than in Zürich, and thus can very easily put a room at your disposal without being in the least restricted by this. During the Easter holidays, we gave this same room to my brother and his wife,[4] who were visiting us, and at the same time let my brother-in-law,[5] who had likewise come for the christening,[6] have another room, and still were not bothered.

In the hope of seeing you here soon, and with kind regards from my wife[7] and me,

Your | Clausius.

RI MS JT/1/TYP/7/2308
LT Typescript Only

1. *WURZBURG:* Clausius held a professorship at the University of Würzburg.

2. *your kind letter:* letter 2840.
3. *Hirst:* Thomas Hirst.
4. *my brother and his wife:* not identified. Rudolf Clausius was one of eighteen children.
5. *my brother-in-law:* not identified.
6. *christening:* presumably of one of Clausius's children; Alfred was born earlier in 1868 (see letter 2764).
7. *my wife:* Adelheid Clausius.

From Herbert Spencer 25 June 1868 2844

37 Queen's Gardens | Bayswater W.[1] | 25 June 1868

My dear Tyndall

If there is any distribution of slip-copies[2] of the letter from the Pall Mall,[3] pray do not let Mr. Clifford[4] be put to any cost for postage. The distribution of the slips may very well stand over until some abstract of lecture, or other paper, has to be circulated among the members;[5] and then the slips may go in the same envelopes.

ever yours sincerely | Herbert Spencer

RI MS JT/1/S/97
RI MS JT/1/TYP/3/1114

---

1. *37 Queen's Gardens | Bayswater W.:* Spencer's London residence since September 1866.
2. *slip-copies:* individually printed copies of a letter or article for the purpose of distributing to interested parties.
3. *letter from the Pall Mall:* see 'An Explanation,' *PMG,* 24 June 1868, p. 4. William Clifford (see n. 4) wrote to clarify an error he felt he had made when giving a lecture at the RI on 6 March 1868, 'On Some of the Conditions of Mental Development' (*Roy. Inst. Proc.,* 5 (1866–9), pp. 311–28). Clifford wrote that he had not properly acknowledged the originality of the evolutionary theory developed by Spencer in his *Principles of Biology,* 2 vols (London: Williams and Norgate, 1864–7).
4. *Mr. Clifford:* William Kingdon Clifford (1845–79), an English mathematician and philosopher of science. He made contributions to a variety of subjects including ethics, the philosophy of science, and algebra. He died young from tuberculosis (*ODNB*). On the influence of Herbert Spencer on Clifford's work, see J. Petrunic, 'Evolutionary Mathematics: William Kingdon Clifford's Use of Spencerian Evolution', in B. Lightman and M. S. Reidy (eds), *The Age of Scientific Naturalism: Tyndall and His Contemporaries* (London: Pickering & Chatto, 2014), pp. 89–112.
5. *members:* the members of the RI, to whom Clifford had given a lecture on 6 March 1868 (see n. 3).

From Elizabeth Dawson Steuart 25 June 1868 2845

June 25th. 68

My dear John,

I ought to have returned you the half of acc[oun]t., remitted to me, before now, but have been from home and very unwell since my return. With the next half note, I hope to send you Tyndall's receipt, for your generous gift.[1] I am sorry to hear you have given up part of your income,[2] but hope it was not in consequence of health. I do think you should steadily resist all future demands, from your family, for you have already acted with too much liberality to some, unworthy of kindness. What splendid weather for you to run off to your beloved Alps! I wish you could do so.

With kindest regards, ever, | most sincerely yours | E. D. Steuart.

RI MS JT/1/TYP/10/3379
LT Typescript Only

1. *Tyndall's receipt, for your generous gift:* John Tyndall gave William and Fanny Tyndall some money after their cow died (see letters 2828 and 2831).
2. *given up part of your income:* Tyndall had written to Steuart that because of the state of his health he was reducing his workload which would reduce his income (see letter 2828, especially n. 4).

To Jane Barnard 26 June [1868][1] 2846

26th. June

My dear Miss Barnard.

This letter[2] is decisive as to the origin of the machine.[3] Still the memory of my standing at your uncle's[4] side in the model room when I thought he mentioned the name of Perkin[5] is very vivid. Perkin had done much in connexion with high pressure boilers, and I may have confounded his name with that of Armstrong.[6]

I know that the piece referred to you by your brother[7] was missing, and I sought for it with Brown.[8] But Cottrell[9] may know more of the matter; so I have placed your brother's letter[10] in his hands.

Yours very since[re]ly | John Tyndall

RI MS JT/1/T/122
RI MS JT/1/TYP/12/4199

1. *[1868]:* the year is given by a note written on the letter from LT giving this year and by relation to letters 2873, 2879, 2880, and 2881.
2. *This letter:* letter missing.
3. *machine:* as Tyndall noted in letter 2842, a 'little steam engine', but otherwise not identified.
4. *your uncle's:* Michael Faraday.
5. *Perkin:* either Angier March Perkins, or his father Jacob Perkins (see letter 2842, n. 6).
6. *Armstrong:* William Armstrong.
7. *your brother:* either Alfred, John, or Frank Barnard.
8. *Brown:* William Brown.
9. *Cottrell:* John Cottrell.
10. *your brother's letter:* presumably the same letter mentioned in n. 2.

From Heinrich Gustav Magnus 29 June 1868 2847

Berlin 29 Juni 1868

Verehrter Freund!

Gern will ich Ihnen machen lassen was Sie wollen und wenn ich es nicht sollte erhalten können Ihnen meine Apparate senden so weit ich sie noch besitze, denn sie sind theils zerbrochen theils zu andern Zwecken benutzt. Vor Allem müßte ich aber wissen was Sie wünschen, ob den in Pogg. CXII oder den CXVIII beschriebenen. Ich beneide Sie nicht um die Wiederholung dieser Versuche! sie kosten unendliche Zeit und dabei ist es doch sehr schwierig sichere Resultate zu erhalten, man muß die äußersten Vorsichtsmaßregeln ergreifen. Außerdem bin ich ganz davon zurückgekommen einen Raum dadurch auf einer constanten Temp. zu erhalten daß man ihn mit Wasser umgiebt. Luft und schlechte Leiter sind bei weitem vorzuziehn.

Aber warum wollen Sie sich auf diese Versuche einlassen. Wenn Ihnen daran liegt die Sache zur Entscheidung zu bringen, weßhalb wiederholen Sie nicht lieber die in Pogg. CXXX erwähnten Versuche.—Doch ich kann mir denken daß diese Sache besser von mir unerörtert bleibt. Seien Sie überzeugt daß mir nur die Sache am Herzen liegt, das werden Sie auch, wie ich glaube, daraus gesehn haben daß ich die Discussion so objectiv wie möglich geführt habe.—Doch genug davon!

Bis zum Anfang August bleibe ich hier und besorge Ihnen bis dahin was Sie wollen.

Ganz der Ihrige | G Magnus.

Berlin 29 June 1868

Dear friend!

I shall gladly have made for you whatever you want, and should I not be able to obtain it, shall send you my apparatuses as far as I still have them, for they are partly broken and partly being used for other purposes. Above all, though, I would have to know what you wish, whether the one described in Pogg. CXII[1] or the one in CXVIII.[2] I do not envy you for repeating these experiments! They take an endless amount of time and it is very difficult to obtain reliable results as well; one has to take extreme precautions. In addition, I have quite abandoned the idea of maintaining a space at a constant temp[erature]. by surrounding it with water. Air and poor conductors are by far to be preferred.

But why do you want to get involved in these experiments. If it matters to you to settle the issue, why do you not instead repeat the experiments mentioned in Pogg. CXXX.[3]—But I can imagine that this issue remains better undiscussed by me. Rest assured that the issue is just close to my heart. You will have seen that too, I believe, from the fact that I have conducted the discussion as objectively as possible.—But enough of this!

I shall stay here until the beginning of August and shall get you whatever you want until then.

Entirely yours | G Magnus.

RI MS JT/1/M/38

1. *one described in Pogg. CXII:* G. Magnus, 'Ueber die Verbreitung der Wärme in den Gasen', *Poggend. Annal.,* 112 (1861), pp. 497–548. The apparatus is illustrated in figures 1 and 2 on plate 5 at the end of the volume and is described on pp. 504–8, 516–7.
2. *one in CXVIII:* G. Magnus, 'Ueber die Diathermansie trackner und feuchter Luft', *Poggend. Annal.,* 118 (1863), pp. 575–88. The apparatus is illustrated in figures 1 and 2 on plate 8 at the end of the volume and is described on pp. 577–8, 583–4.
3. *experiments that are described in Pogg. CXXX:* G. Magnus, 'Ueber den Einfluss der Vaporhäsion bei Versuchen über Absorption der Wärme', *Poggend. Annal.,* 130 (1867), pp. 207–26.

From François-Napoléon-Marie Moigno 30 June 1868 2848

LES MONDES | Revue Hebdomadaire | DES SCIENCES | ET DE LEURS APPLICATIONS | aux arts et à l'industrie | par | M. l'Abbé MOIGNO | RÉDACTION | 2, RUE D'ERFURTH, 2

Paris, le 30 juin 1868

Mon cher Mr Tyndall

Je suis enchanté de ce qui arrive. Les éditeurs sont des hommes terribles. Surtout quand ils ont la fatale idée de faire des économies de bouts de chandelles.

J'ai lutté tant que j'ai pu, et je ne me suis rendu *[1 word illeg]* qu'on m'eut montré une lettre de vous autorisant la suppression de diverses figures et la substitution des figures de Jamin aux vôtres. Mais je vais insister et vous daignerez vous unir à moi pour qu'on fasse des cartons de manière à rétablir le texte et les dessins originaux. J'aimais beaucoup votre gradation: Sirène de Cagniard de Tour, Sirène Dove, Sirène Helmholtz. Votre sonomètre était aussi très élégant, on pourrait tirer sur le oids, pour montrer que le ton s'élève, parce que la corde était suffisamment longue; on ne le peut pas dans le sonomètre étriqué à trois cordes de Mr Jamin. Si on me fait faire encore des changements je m'y refuserai si on savait ce que je me suis donné de mal pour bien rendre, pour rendre élégamment, votre pensée sans périphrase, on respecterait mieux ma traduction. Quel malheur que la mauvaise foi de Mr Giraud nous ait brouillés.

Je suis désolé de n'avoir pas encore de réponse définitive relativement à ma traduction de votre Faraday inventeur. Je voudrais à tout prix en faire une œuvre populaire à bon marché comme les six leçons sur la chaleur et le froid que vous recevrez au commencement de la semaine prochaine.

Faites donc que Mr Jamin se décide.

Votre humble et dévoué serviteur | l'abbé F. Moigno

Merci de vos compliments, j'ai tout fait pour les mériter.

Le mort de Mr Plucker me désole profondément. Je ne crois pas pouvoir aller cette année en Angleterre.

Il est convenu avec Mr Gauthier-Villars que désormais nous vous enverrions non pas les bonnes feuilles, mais les dernières épreuves, pour que je puisse mettre à profit vos observations. Rien ne remplace l'art du maître. Nous vous prions seulement avec instance de renvoyer les feuilles le plus vite possible à mon adresse

2 rue d'Erfurth | Église St. Germain du pré.

Il est convenu que nous feront les cartons *[1 word illeg]* pour rendre la traduction parfaitement conforme à l'original, et pour qu'il n'y ait s'il est possible aucune faute. Mr Gauthier-Villars regrette sa malheureuse pensée d'économie.

LES MONDES[1] | Revue Hebdomadaire | DES SCIENCES | ET DE LEURS APPLICATIONS | aux arts et a l'industrie | by the | Father MOIGNO | EDITORIAL BOARD | 2 RUE D'ERFURTH, 2

Paris, 30 June 1868

My dear Mr Tyndall,

I am delighted by what is happening. The editors are terrible people, especially when they have the deadly idea of creating a cheeseparing economy.

I struggled as much as I could, and I have not *[1 word illeg]* that I was shown a letter from you authorizing the removal of various figures and the substitution of Jamin's[2] figures for your own. But I am going to insist, and please condescend to join me so that we will do pasteboards so as to restore the original text and drawings.[3] I loved your gradation: Cagniard de Tour's[4] siren, Dove's[5] siren, Helmholtz's[6] siren. Your sound-level meter was also very elegant, we could pull on the weight to show that the tone rises, because the rope was long enough; we cannot do so with Mr Jamin's sound-level meter which is cramped with three ropes. If I am made to make more changes; I will refuse. If they knew what I went through to do it well, to make it elegant, to deliver your thought without circumlocution, they would respect my translation[7] more. How unfortunate it is that Mr Giraud's[8] bad faith made us quarrel.

I am sorry that I do not yet have a definite response in regards to my translation[9] of your Faraday as a Discoverer.[10] I want at all costs to do a popular cheap book such as the six lessons on heat and on cold[11] that you will receive at the beginning of next week.

Therefore, ensure that Mr Jamin decides.

Your humble and devoted servant | Father F. Moigno

Thank you for your compliments; I have done everything to deserve it.

Mr Plucker's[12] death deeply saddened me. I do not think that I can go to England this year.

It was agreed with Mr Gauthier-Villars that from now on we will send you not just good papers, but the latest proofs, so that I can put your observations to good use. Nothing can replace the art of the master. We only ask you to send us back the papers as soon as possible to my address

2 Erfurth Street | Church of Saint-Germain-des-Prés.

It was agreed that we will make the pasteboards *[1 word illeg]* in order to make the translation perfectly true to the original, and so that there is no mistake, if it is possible. Mr Gauthier-Villars regrets his unfortunate thrifty thought.

RI MS JT/1/M/115

---

1. *LES MONDES:* see letter 2706, n. 9.
2. *Jamin's:* Jules Jamin.
3. *original text and drawings: Sound* (1867).
4. *Cagniard de Tour:* Charles Cagniard de la Tour (1777–1859), a French physicist and engineer trained at the École Polytechnique. He studied acoustics and heat, and invented a number of machines included a siren, heat engine, and an air pump (*CDSB*).
5. *Dove's:* Heinrich Wilhelm Dove (1803–79), a Prussian meteorologist and experimental physicist. He was educated at the University of Berlin and became an ordinary professor at Berlin's Friedrich-Wilhelms-Universität in 1845. In his research, he focused on meteorology and was awarded the RS's Copley Medal in 1853 for his contributions to understanding the distribution of heat in the atmosphere (*CDSB*).
6. *Helmholtz's:* Hermann Helmholtz.
7. *my translation:* see letter 2617, n. 2.
8. *Mr Giraud's:* Étienne Giraud.
9. *my translation:* see letter 2751, n. 3.
10. *Faraday as a Discoverer: Faraday as a Discoverer* (1868).
11. *six lessons on heat and on cold:* see letter 2751, n. 2.
12. *Mr Plucker's:* Julius Plücker, who had died on 22 May 1868.

To Thomas Henry Huxley [June 1868][1] 2849

Royal Institution. | Monday night.

My dear Huxley,

I do not understand this matter; and I am the more perplexed inasmuch as a year or two ago he[2] did the same thing to me.[3] Since that time I have made the Institution[4] permanent.

I firmly believe that the man loves us both but whether he is conscious of some great demerit in himself or some surpassing excellence in us, or whether it is that he looks upon himself as a potent grave and reverend Seignor[5] not to be addressed familiarly by young cubs like you and me, I know not—such things have been in the case of weaker men, but I should be sorry to believe that either hypothesis applied to him.

The more I know him the more I like him, the more in fact I become convinced of his true goodness of heart. Perhaps he was in low spirits—he <u>has</u> been low of late, and low spirits ooze out in a comical way sometimes.

That anything is "a-drift" I do not for a moment believe. He and I were never closer friends than we are at present, yet as I have said precisely the same thing occurred in my case. I am sensitive in such matters and at once rendered the relationship rigid which his letter[6] seemed to suggest.

Always yours | John Tyndall

IC HP 8.62
RI MS JT/1/TYP/9/2915
Typed Transcript Only

---

1. *[June 1868]:* a note on the Imperial College typescript states '? June 1868'.
2. *he:* Henry Bence Jones.
3. *year or two ago . . . to me:* see letter 2647 where Thomas Huxley noted: 'I am consoled to find that Bence Jones "Dr's" you again but it is an odd notion. I fear his lease of life is not a long one; and it is a thousand pities, for the man is a good one notwithstanding his way of beginning a letter'. In June 1868, Tyndall resigned from his lecturing position at the Royal School of Mines in order to have more time for original research (see letters 2833, 2834, and 2835), so perhaps Bence Jones had discussed with Tyndall whether or not he was capable of maintaining his lecturing duties with the RI (a decade before, in February 1857, Tyndall had asked Faraday if his annual lecture load could be decreased, which the Managers of the RI agreed to). On Tyndall and the Royal School of Mines, see letter 2601, n. 2.
4. *the Institution:* the RI.
5. *Seignor:* a person high in rank or authority (*OED*).
6. *his letter:* letter missing.

From Henry Bence Jones [June 1868][1] 2850

Dear friend

You will find me for an hour or two at Private box No 15 S[t] James Theatre[2] I will say I expect you at the door. It is very strange to drag you into this matter[3] why does not he[4] write to Clifford[5] himself.

I do not consider the Institution[6] is in any degree bound to know what M[r] Spencer[7] has said. No one knew I suspect

Yours ever | H Bence Jones

RI MS JT/1/J/117
RI MS JT/1/TYP/3/714

---

1. *[June 1868]:* the month and year are given in relation to letter 2844, in which Spencer wrote to Tyndall concerning a letter William Kingdon Clifford had written to the *PMG.*
2. *S[t] James Theatre:* a theatre in London from 1835–1957.

3. *this matter:* see letter 2844, n. 3.
4. *he:* Herbert Spencer.
5. *Clifford:* William Kingdon Clifford (see letter 2844, n. 4).
6. *Institution:* the RI.
7. *Mr Spencer:* Herbert Spencer.

To the Athenaeum Committee[1] 4 July 1868 2851

The Athenaeum | July 4th, 1868

We the undersigned being desirous that there should be no misunderstanding as to the real nature of the dissatisfaction that had been entertained in regard to the original Notice[2] posted up at the door of the Morning room on the 1st. July and most anxious that Mr Longfellow[3] should receive a public testimony of the high respect entertained for his eminent qualities by the members of the Athenaeum Club at large beg to state that in our opinion it would be agreeable to the members of the Club that the Committee should address to Mr Longfellow an invitation to a public dinner—the cost to be defrayed by the Members who give in their names as desirous to attend it.

W. C. Cartwright[4]
Charles Limerick—[5]
C. Wheatstone[6]
J. J. Sylvester[7]
John Bowring[8]
A. Gallenga.[9]
James Heywood.[10]
Charles Wynne Finch—[11]
Walpole[12]
W E H Lecky.[13]
Arthur Russell[14]
Henry Oldham[15]
John Tyndall
Mr Freeland.[16]
William Harness[17]
Robert Browning[18]
G W Hemans[19]
Thomas H. Huxley
W Boxall[20]
G. R. Gleig[21]
Lyttelton[22]
C. G. S. Gleig[23]
John W Prout[24]

Charles Vignoles[25]
John G. Perry[26]
Williams Helps[27]
John Gough Nichols[28]
J. Inglis[29]
Edw. J. Shepherd[30]
J P. Kay Shuttleworth[31]
Edward Thomas[32]
R. D. Thomson[33]
Chas. Landseer[34]
Edward Ryan[35]
C. Ebor[36]
J. S. Stuart Glennie[37]
Fred Leighton[38]
Granville[39]
R C Browne Clayton[40]
E A Darwin[41]
W C Spring Rice[42]
G. Owen Rees[43]
George John Cayley[44]
J. W. Blakesly[45]
F. W. Gibbs[46]
John W Larking[47]
John Edmund Cox[48]
C. J. Bayley[49]
Edw[d] Meryon[50]
Frederick Locker[51]
E C Ravenshaw[52]
T. Erskine May[53]
E B Hamley[54]
53 signatures—

Athenaeum General Committee Minutes, vol. 17 (1865–68)

---

1. *Athenaeum Committee:* while this volume includes two open letters to the General Committee of Athenaeum Club (see letter 2537, n. 8) in which Tyndall was a signatory with other members (this letter and 2909), the meeting minutes for 5 March 1867 state that Tyndall was among the signers of another letter, which is missing. The minutes state: 'Memorial submitted signed by Professor Tyndall and 15 other Members frequenting the Smoking Room "appealing to the humanity of the Committee for a room in which they can enjoy their Tobacco without being compelled to inhale an atmosphere charged with poison & in the highest degree detrimental to health". Memorial read, Resolved, That it be referred to a Sub-Committee to consider whether better accommodation may be afforded

for smoking and to submit any schemes for the object to the General Committee. Further Resolved, That the following be requested to serve on the said Sub-Committee. | Me[m]b[e]rs. D. Burton | J. Fergusson | C. Landseer | W. Pole | E. Thomas. | Three to be a quorum. To fix their own time of meeting. The foregoing Minutes read & confirmed as being correctly entered. | Stanhope'. Stanhope is Philip Henry Stanhope (1805–75), Burton is Decimus Burton (1800–81); for Fergusson, see letter 2582, n. 6; for Landseer, see this letter, n. 34; Pole is possibly William Pole (see letter 2909, n. 7); and for Thomas, see this letter, n. 32. The meeting minutes for 30 April 1867 give an update on this subject: 'Report, with Plan annexed, brought up from the Smoking Room Sub-Committee (appointed March 5th *[1 word illeg]* P. 386 Par. 9) recommending [the] enlargement of the present Room at an outlay not exceeding £400. Report considered, & agreed to'.

2. *dissatisfaction that had been entertained in regard to the original Notice:* the notice presumably announced that the Morning-room of the club would be used for a private dinner with Henry Wadsworth Longfellow on 7 July, who had earlier in 1868 been proposed for an Honorary Membership by Charles Hutton Gregory (1817–98), President of the Institution of Civil Engineers (this was approved by the General Committee and he became an Honorary Member in June). This irked some members, as not all had been invited but were expected to not only give up the use of the room on the day of the planned dinner, but to accept that the club would cover costs for it (see 'A Club Grievance', *Times,* 2 July 1868, p. 5; and 'Club Cliqueism', *London Daily News,* 2 July 1868, p. 3). As this letter shows, many signatories approved of a dinner for Longfellow at the club and suggested that only those who attend would cover the costs.
3. *Mr Longfellow:* Henry Wadsworth Longfellow (1807–82), an American poet, translator, and professor of modern languages. He held professorships at Bowdoin College (1829–35) and Harvard College (1836–54). During his final European tour (1868–9) Longfellow received honourary degrees from Cambridge and Oxford and had a private audience with Queen Victoria (*ANB*).
4. *W. C. Cartwright:* William Cornwallis Cartwright (1825–1915), a British politician and journalist (*ODNB*).
5. *Charles Limerick:* not identified (not listed as a member, according to the Athenaeum Club).
6. *C. Wheatstone:* Charles Wheatstone.
7. *J. J. Sylvester:* James Joseph Sylvester (1814–97), an English mathematician and professor of mathematics at the Royal Military Academy, Woolwich (1855–69). He later held professorships at Johns Hopkins University and Oxford University (*ODNB*).
8. *John Bowring:* John Bowring (1792–1872), an English politician and writer (*ODNB*).
9. *A. Gallenga:* Antonio Carlo Napoleone Gallenga (1810–95), an Italian journalist and writer who had moved to England in 1839 (*ODNB*).
10. *James Heywood:* James Heywood (1810–97), an English politician and philanthropist (*ODNB*).
11. *Charles Wynne Finch:* Charles Griffith Wynne Finch (1815–74), an English politician and Member of Parliament for Carnarvon (1859–65) (*Alum. Oxon.*).
12. *Walpole:* Spencer Horatio Walpole (1806–98), an English politician (*ODNB*).

13. *W E H Lecky:* William Edward Hartpole Lecky (1838–1903), an Irish historian, and author of *History of England in the Eighteenth Century,* 8 vols (1878–90) (*ODNB*).
14. *Arthur Russell:* Arthur John Edward Russell (1825–92), an English politician and Member of Parliament for Tavistock (1857–85) ('Lord Arthur Russell', *Times*, 6 April 1892, p. 9).
15. *Henry Oldham:* Henry Oldham (1815–1902), an English physician who specialized in obstetrics (*ODNB*).
16. *Mr Freeland:* Humphrey William Freeland (1814–92), an English politician and magistrate (E. Walford, *The County Families of the United Kingdom*, 5th edn (London: Robert Hardwicke, 1869), p. 382).
17. *William Harness:* William Harness (1790–1869), an English literary scholar (*ODNB*).
18. *Robert Browning:* Robert Browning (1812–89), an English poet (*ODNB*).
19. *G W Hemans:* George Willoughby Hemans (1814–85), an Irish architect and railway engineer ('George Willoughby Hemans', *Minutes of the Proceedings of the Institution of Civil Engineers*, 85 (1886), pp. 394–9).
20. *W Boxall:* William Boxall.
21. *G. R. Gleig:* George Robert Gleig (1796–1888), a Scottish soldier, author, and priest (*ODNB*).
22. *Lyttelton:* George William Lyttelton (1817–76), an English aristocrat and politician (*ODNB*).
23. *C. E. S. Gleig:* Charles Edward Stuart Gleig (1825–*c.* 1906), a military officer and author.
24. *John W Prout:* John William Prout (*c.* 1817–81), an English lawyer (*Alum. Oxon.*).
25. *Charles Vignoles:* Charles Blacker Vignoles (1793–1875), a civil engineer who surveyed and constructed numerous railway projects in the United Kingdom, Europe, and the Americas (*ODNB*).
26. *John G. Perry:* John George Perry (1802–70), an English surgeon, hospital governor, and Inspector of Prisons of Great Britain for the Southern and Western District (S. McConville, *English Local Prisons, 1860–1900: Next Only to Death* (New York: Routledge, 1995), p. 105).
27. *Williams Helps:* Thomas William Helps (1802–87), an English lawyer (*Athenaeum Rules* (1874), p. 62).
28. *John Gough Nichols:* John Gough Nichols (1806–73), an English publisher and antiquarian (*ODNB*).
29. *J. Inglis:* John Inglis (1810–91), a Scottish politician and judge (*ODNB*).
30. *Edw. J. Shepherd:* Edward J. Shepherd, an English clergyman (*Athenaeum Rules* (1862), p. 80).
31. *J P. Kay Shuttleworth:* James Phillips Kay-Shuttleworth (1804–77), an English physician and civil servant (*ODNB*).
32. *Edward Thomas:* Edward Thomas (1813–86), an English civil servant and historian of India (*ODNB*).
33. *R. D. Thomson:* Robert Dalzell Thomson, an English lawyer (*Athenaeum Rules* (1874), p. 110).

34. *Chas. Landseer:* Charles Landseer (*c.* 1800–79), an English painter (*ODNB*).
35. *Edward Ryan:* Edward Ryan (1793–1875), an English lawyer, judge, and Civil Service Commissioner (*ODNB*).
36. *C. Ebor:* Charles Thomas Longley (1794–1868), Archbishop of Canterbury and York (*ODNB*). The Archbishops of York often signed their name with 'Eboracensis' or 'Ebor', for *Eboracum,* the Roman name for York.
37. *J. S. Stuart Glennie:* John Stuart Stuart-Glennie (see letter 2823, n. 14).
38. *Fred Leighton:* Frederic Leighton (1830–96), 1st Baron Leighton, an English painter and sculptor (*ODNB*).
39. *Granville:* Granville George Leveson-Gower (1815–91), second Earl Granville, an English politician and leader of the Liberal Party in the House of Lords (*ODNB*).
40. *R C Browne Clayton:* Richard Clayton Browne-Clayton (1807–86), a lawyer, judge and magistrate of Irish parentage (E. Walford, *The County Families of the United Kingdom,* 5th edn (London: Robert Hardwicke, 1869), p. 209)
41. *E A Darwin:* Erasmus Alvey Darwin (1804–81), the elder brother of Charles Darwin (*Alum. Cantab.*).
42. *W C Spring Rice:* William Cecil Spring Rice (1823–80), an English lawyer (*Athenaeum Rules* (1874), p. 99).
43. *G. Owen Rees:* George Owen Rees (1813–89), a physician and founding member of the Chemical Society of London (1841) (*ODNB*).
44. *George John Cayley:* George John Cayley (1826–78), an English lawyer, traveller, and writer (*ODNB*).
45. *J. W. Blakesley:* Joseph Williams Blakesley (1808–85), an English clergyman who became Dean of Lincoln in 1872 (*ODNB*).
46. *F. W. Gibbs:* Frederick Waymouth Gibbs (1821–98), an English lawyer and judge (*Alum. Cantab.*).
47. *John W Larking:* John Wingfield Larking (1802–91), a magistrate for Kent and previously British Consul in Egypt (E. Walford, *The County Families of the United Kingdom,* 5th edn (London: Robert Hardwicke, 1869), p. 579).
48. *John Edmund Cox:* John Edmund Cox (1812–90), an English clergyman (G. Storey (ed), *The Letters of Charles Dickens,* 4th edn, 12 vols (Oxford: Clarendon Press, 2002), vol. 12, p. 482).
49. *C. J. Bayley:* Charles John Bayley (*c.* 1817–73), Colonial Secretary of Mauritius (1849) and Governor of the Bahama Islands (1857–64) (*Alum. Cantab.*).
50. *Edwd Meryon:* Edward Meryon (*c.* 1807–80), an English physician (*ODNB*).
51. *Frederick Locker:* Frederick Locker-Lampson (1821–95), an English poet (*ODNB*).
52. *E C Ravenshaw:* Edward C. Ravenshaw (*c.* 1804–77), an English geographer.
53. *T. Erskine May:* Thomas Erskine May (1815–86), Baron Farnborough (1886), clerk of the House of Commons and writer on constitutional theory (*ODNB*).
54. *E B Hamley:* Edward Bruce Hamley (1824–93), an army officer and military historian (*ODNB*).

To John Herschel 11 July 1868 2852

GIESSBACH.[1] | 11th July 1868.

My dear Sir John.

Dr. Bence Jones was good enough to promise me that he would write to you, for I was myself unable to do so. I am still far from well, but not too ill to write a letter.

After you wrote to me I waited to ascertain how my colleagues received my suggestion; for you know I am but one among many. I have long thought Mr. Herschel,[2] to a great extent, a power wasted, and have long wished to see him in a position which would give him a fair opportunity of exercising his powers. I thought the School of Mines an admirable position for him[3] and had the following arrangements in view to promote his success as a teacher in that place. I have had there for many years a trained assistant, who is perfectly acquainted with the entire series of my lecture experiments. During the Autumn, had my suggestion been acted on, it was my intention with the assistance of this man to make Mr. Herschel master of all these experiments, which my peculiar experience at the Royal Institution enabled me to /heap/ up; this I thought would give him a store of matter into which his own intellect would put a soul.

But I soon found that doubts were entertained by some of my colleagues[4] upon the subject; and seeing this I did not think it right to press the matter too keenly. I talked with some of them, and gave my reasons for proposing Mr. Herschel, but they evidently desire a man who has made himself more known. Where they will find such a man I know not.[5]

I think they would have acted wisely had they acted upon my advice. But of course they have an equal right to their own opinions.

I am dear Sir John | most faithfully yours | John Tyndall

Sir John Herschel Bart. | &c. &c. &c.

RS HS 17/402
RI MS JT/1/TYP/2/545

1. *GIESSBACH:* 'Giessbach' is embossed on the stationary; possibly from a hotel situated near Giessbach Falls, near lake Brienz, Switzerland.
2. *Mr. Herschel:* Alexander Herschel.
3. *School of Mines an admirable position for him:* Tyndall advocated for Alexander Herschel to replace him at the Royal School of Mines (see letter 2828, n. 4, and letters 2833, 2834, and 2835). On Tyndall and the Royal School of Mines, see letter 2601, n. 2.
4. *some of my colleagues:* not identified. James Clerk Maxwell wrote to Tyndall recommending Charles Hockin for the position (see letter 2854).
5. *Where they will . . . I know not:* Frederick Guthrie was appointed, see letter 2834, n. 7.

To Thomas Archer Hirst [17 July 1868][1] 2853

Dear Tom

I had a beautiful room[2] and no disturbance, but no sleep. I will walk on before you, leaving it to you to settle as to the Pfarrer.[3] Give the smiling little girl who prepared my room for me something. Jaggi[4] will strap my things together—your flask is there.

J. T.

RI MS JT/2/32d/1833

1. *[17 July 1868]:* the date is given by the inclusion of this letter in Hirst's journal for this date.
2. *beautiful room:* Tyndall and Hirst stayed the night in Gsteig, a village in the Swiss canton of Bern. In his journal entry for 16–17 July 1868, Hirst wrote: 'Left Lenk with our guide and walked over the hills to Gsteig a beautifully situated little village, with a bad inn. By a little maneuvering I got John a bedroom in the Pfarrers house, the best in the village; next morning instead of meeting him at breakfast restored by a night's sleep I found this scrap of paper [this letter]—I followed him to the Diablerets; where we spent half a day lazily' (Brock & MacLeod, *Hirst Journals,* pp. 1832–3).
3. *Pfarrer:* not identified; the parson or minister of Gsteig.
4. *Jaggi:* not identified; possibly a porter that Tyndall and Hirst had hired.

From James Clerk Maxwell 23 July 1868 2854

Glenlair | Dalbeattie[1] | 1868, July. 23rd.

Dear Tyndall

Mr. C. Hockin[2] tells me he is a candidate for your lectureship at the School of Mines.[3] As I have worked with him for five years, I have had opportunities of forming a judgment of his scientific character, in particular, he worked at the determination of the B.A unit of electrical resistance[4] with Jenkin[5] and me and then by himself, and this spring he helped me in finding the ratio of the two electrical units (which according to me gives the velocity of light). You know the kind of difficulties which are always turning up in such investigations, the hunting for sources of discrepancies, the hitches in the working of new apparatus, the mathematical difficulties, and not least the physical difficulty of keeping up the spirit of accuracy to the end of a long and disappointing day's work.

I therefore say no more to you than that I think Hockin's scientific "bottom" and patience such, that united as it is with a power of seeing the essentials of an experiment and securing them first, and with very extensive mathematical knowledge and an unreserved devotion to science, he is safe to do more than any young man I know for the promotion of natural knowledge. Besides this, he has taken care to make himself acquainted with engineering and telegraphy under Jenkin and has studied chemistry with Matthiessen,[6] and in every thing he has done he has reduced his observations and planned his experiments with a mathematical ability and a familiarity with abstruse subjects, which I have seldom seen combined with actual manipulation. If he were placed in any scientific institution like the School of Mines, I am certain that his only object would be to do honour to the institution both by sound teaching, and by experimental researches, on which he seems to have set his heart.

I remain | Yours truly | J. Clerk Maxwell

RI MS JT/1/TYP/9/2922
LT Typescript Only

---

1. *Glenlair | Dalbeattie:* Glenlair House was Maxwell's residence in the town of Dalbeattie, near Corsock in Dumfries and Galloway, Scotland.
2. *C. Hockin:* Charles Hockin (1840–82), an English engineer. In the 1860s he collaborated with Augustus Matthiessen and James Clerk Maxwell in determining the BAAS unit of electrical resistance (*Alum. Cantab.*).
3. *your lectureship at the School of Mines:* Tyndall had resigned his position at the Royal School of Mines (see letter 2828, n. 4, and letters 2833, 2834, and 2835). On Tyndall and the Royal School of Mines, see letter 2601, n. 2.
4. *B.A unit of electrical resistance:* a physical unit of electrical resistance (one ohm) in the form of a platinum-silver coil that was defined, produced, and distributed by the BAAS committee on standards of electrical resistance from 1865 (see G. Gooday, *The Morals of Measurement: Accuracy, Irony, and Trust in Late Victorian Electrical Practice* (Cambridge: Cambridge University Press, 2004), pp. 103–12)
5. *Jenkin:* Fleeming Jenkin (1833–85), a British electrician and engineer. He was the secretary of the BAAS committee on standards of electrical resistance and professor of engineering at University College, London (1866) and the University of Edinburgh (1868) (*ODNB*).
6. *Matthiessen:* Augustus Matthiessen (1831–70), a British physicist and chemist, educated in Germany. During the early 1860s he was engaged in producing standards of resistance for the BAAS (*ODNB*).

From Thomas Archer Hirst 28 July 1868[1] 2855

HOTEL DE LA GRANDE BRETAGNE | VITALO.
BIGANZOLI | MILAN | Tuesday July 28th /69

My dear John

On the point of starting on my return journey to England[2] I feel a desire to send you a line by way of greeting, to express the hope that the weather which has for me been uniformly good has also been favourable for your project of ascending and descending the Matterhorn.[3] By this time I hope that this long merited achievement of yours is safely accomplished and above all that it has been accompanied by a steady improvement in your health. On this side the alps the temperature has been something fearful and many a time I have felt thankful that you were not with me. Thanks to frequent baths and a good hotel I have borne it well enough but I could not have borne it with you as my sleepless suffering companion. Your very absence therefore was a source of satisfaction to me, and after every good night's rest with no clothes at all to cover me I prayed in my own fashion that you were enjoying as good a one at an altitude sufficient to render a feather bed on the top of you tolerable. I did not arrive at Milan until late in the evening of the day after I left you at Chatillon.[4] I received a hearty welcome from Cremona[5] and I have since spent pleasant days with him and Brioschi[6] (who may be described as Cremona's Tyndall) I have often had great pleasure in listening to Cremona's praises of his friend Brioschi, how he has been indebted to him for guidance in the critical moments of his own career, for encouragement in hours of darkness, for severe but friendly admonition when such was needed. I have listened many a time I say to Cremona's account of this and at every moment I have been able to say to myself all this and far more my friend has done for me There is indeed a curious resemblance between Cremona's career and mine and from this fact alone apart from the coincidence of our geometrical studies proceeds, I do believe, a good deal of the thorough sympathy which has existed between us ever since our first acquaintance. When you leaned on my bed a week ago at Chatillon and alluded to the deep interest you had always taken in my welfare and success as a set off against what I had pleased to call your occasional imperiousness when we were discussing your peculiarities I do not think you knew that you had touched on a subject which is to me I may say an ever present one. Hardly a day ever passes at home or abroad that I am not reminded in one way or another of your constancy dear John towards me; such constancy, albeit accompanied with a characteristic *[*aimiability*]* of mood, is so far as my experience reaches unparalled. Certainly the constancy of Brioschi to Cremona does not approach it. It is not for me to compare Cremona's devotion to Brioschi with mine towards you but on this point also I have my own convictions deeply rooted and as Carlyle[7] would say "well covered over".[8] I start home in good health. The intellectual atmosphere here has had a tonic effect on me and I return home with something approaching to a conviction that a better year is in store for both you & me. To you at least will be granted I hope the tranquility of mind necessary to pursue with vigour and effect your own pursuits.

Ever dear John | yours affectionately | "Tom"

Please don't forget to bring home my letters from Zermatt.[9]

You have been elected corresponding member of the Istituto Lombardo.[10] I have seen your books on Heat and Sound[11] several times here. Italy is looking to you for further results and Italy will not be disappointed.

RI MS JT/2/10/479
RI MS JT/2/13c/1382

---

1. *1868:* Hirst mistakenly wrote '69' on this letter; in his journal, he noted having spent time with Cremona from 24–8 July 1868 (see Brock & MacLeod, *Hirst Journals,* p. 1833).
2. *my return journey to England:* Hirst departed Milan on 28 July and arrived in London on 5 August after stopping in Paris and Boulogne (Brock & MacLeod, *Hirst Journals,* 28 July–5 August 1868, pp. 1833–4).
3. *your project of ascending and descending the Matterhorn:* Tyndall made the seventh ascent and first traverse of the Matterhorn, from the Italian side to Switzerland, from 26–8 July 1868 (*Hours of Exercise in the Alps,* pp. 273–96).
4. *day after I left you at Chatillon:* Hirst left Tyndall on 22 July and arrived in Milan on 23 July (Brock & MacLeod, *Hirst Journals,* 22–23 July 1868, p. 1833).
5. *Cremona:* Antonio Luigi Gaudenzio Giuseppe Cremona (1830–1903), an Italian mathematician and professor of geometry first at the University of Bologna and after 1867 at the Istituto Tecnico Superiore of Milan (*CDSB*).
6. *Brioschi:* Francesco Brioschi (1824–97), an Italian mathematician and politician. He was a deputy (1861) and senator (1865) in the Italian parliament and helped to found the Istituto Tecnico Superiore of Milan (1863) where he served as director and professor of mathematics (*CDSB*).
7. *Carlyle:* Thomas Carlyle.
8. *"well covered over":* not identified. This does not appear to be quoted from any of Carlyle's works.
9. *my letters from Zermatt:* possibly referring to letters that were sent to Hirst at Zermatt after he left Tyndall for Italy.
10. *Istituto Lombardo:* Istituto Lombardo Accademia di Scienze, an Italian academy initially founded in Bologna by Napoleon in 1797 and moved to Milan in 1810.

11. *your books on Heat and Sound:* Hirst is referring to translations of Tyndall's *Heat* (1863) with second and third editions in 1865 and 1868, and *Sound* (1867). It appears that there were no Italian-language editions, so this likely refers to French or German editions available in Italy.

To Thomas Archer Hirst [between 28 July and early August 1868][1] 2856

My dear Tom,

I am here at the Riffel[2] where it is snowing fast. I can do nothing out of doors and will therefore try to give my swamp of thoughts an outlet by writing to you. I quitted you very ill at Chatillon;[3] I made with the chanoine[4] up the valley as long as was necessary to read a little book[5] that he had given me, and afterwards I walked. What a splendid appetite he has! He stayed at Breuil[6] for a day or two. I found the guide[7] whom he recommended to me, and had two days exercise with him prior to attacking the Matterhorn.[8] I feared the good weather would pass away through the resolution of the guides to "sanctify" the Sunday. But by putting on a little pressure an early mass (at 2 oC in the morning) was arranged, and we started on Sunday. The weather was much more doubtful than it had been for some days: the air was exceedingly soft and vast masses of fog came swirling up from Italy. The solution and renewal of the fog were rapid, but through all the Matterhorn kept itself tolerably clear. We were off before the dawn at least before day sank firmly into the valley. Ruddy clouds soon announced the rising of the sun, and put me in mind of Christ's expression "it will be foul weather today for the sky is red and lowring".[9] They do not indeed appear to be the harbinger of fine weather. Still after 2 or 3 hours climbing the fog was folded up like a scroll by the south wind and rolled grandly over the Theodule[10] into Switzerland. We came to a coulior[11] which was the track of falling stones; the guides pronounced it dangerous, and prepared for our expeditious passage—while crossing it a few small stones shot down upon us. We gained the rocks at the other side, and while clambering up them, the guide in front suddenly shouted "Cachez vous!"[12] I crouched against the rock, where a huge stone whirred past me struck the mountain side, rebounded with a savage boom and flew down to the glaciers. We moved to an arête[13] to avoid a repetition of this danger. We clambered rocks and ridges which were well known to me. The mountain in my estimation had lost nothing of its grandeur: we halted & lay in the sun for an hour near the tattered residue of one of my old tents. Soon afterwards we came to the precipice down which had hung my cord, now replaced by a stronger one. Up this precipice we all clambered, and were soon afterwards upon a ledge which cuts the mountain in a nearly horizontal direction. This ledge is always loaded with snow, and, looked at from below, suggests the idea of a white necktie: hence it is called the Cravatte.[14] High over this ledge toward the cliffs of gneiss, in some places far overhanging the ledge. Underneath an overhanging portion a hut has been built: it was now nearly filled with snow. This was cleared away; for the cabin was to be our quarters for the night.

Water showered in drops from the overhanging cave above the hut. The guides stuck four axes into the snow: spread a square piece of one of my old tents over the axes, making the cloth sink in the middle: at the middle was a hole. Thus a large surface was spread out to receive the falling drops, and all was directed to the central hole, whence the liquid flowed in a continuous stream. Thus we supplied our bottles, and were enabled to make our coffee and soup without expending our spirits of wine in merely changing the state of aggregation.

I had a blown up India rubber mattress and a blanket: the guides had a large area of sheepskin covered with its wool. When night set in I lay down, so did the guides. Sleep of course is difficult under such circumstances, but I think I should have dozed had it not been for the cold. The mattress was a poor barrier against the frigorific[15] action of the rock underneath. I bore the thing for some hours as I did not wish to disturb the guides; but at length the cold became excruciating and I was obliged to rise. I was stiffened & my feet were very painful. The good fellows wrapped me up for a time in their sheepskin and by degrees restored the proper temperature. Lying as they did side by side they kept each other warm, and my chill was a consequence of my isolation. I should have lain down with them, and perhaps accepted a proferred night cap, had I not seen the neck of one of the porters covered with flea bites. Well day dawned. Provisions were slowly packed, and after the exercise of some patience on my part we made a start. They fed on beefsteak réchauffés,[16] I did not touch food, purposing to reach the top of the Matterhorn in a spirit of prayer and fasting. We got away from the Cravatte and soon reached a pinnacle to which they have given my name, and on which I planted a flagstaff in 1862.[17] From this forward a very formidable piece of work lay before us. A long arête of sharp rocks, against which had been flung by the winds a vast mass of snow which formed over the rock an edge of exceeding steepness. The slope to the left was terribly steep, and the precipices to the right abysmal. Herein consists the unparalled grandeur of the Matterhorn. For savagery in its crags and precipices on the southern side it has no competitor. The guide once turned to me with a smile saying "no room for yielding here Sir". I had gone along the same arête in 1862 but then it was nearly free of snow.

From the arête we passed to the final cap of the mountain. About 700 feet of exceedingly difficult rockwork now lay before us. And it was trebly difficult today through the crust of ice which covered the rocks. Exceeding caution had to be exercised, and where the feet were unsafe the hands and arms were heavily drawn upon. At length we came to the end of another rope stretching down over the iced rocks. The rope had been frayed by the wind and the guides hesitated to trust it. It was in part covered with ice which rendered any firm grasp of it impossible. Notwithstanding the existence of this rope the ascent by it was very difficult. This obstacle once surmounted an easy climb of a few minutes placed us upon the summit of the Matterhorn.

And as we reached it up swirled the fog which had attacked us repeatedly during the day, lowering both the physical and moral tone by its chill and black dimness. But again it cleared away and far below we saw Zermatt & the Vispbach[18] sweeping through the valley. We stood on one end of a wild arete, whence the precipices fell grandly toward Breuil, while a slope of more moderate inclination fell towards Zermatt. We worked along the arête to the other end. At both ends are planted pieces of the old ladder which I had taken with me in 1862. At the distant end of the arête to our great interest & surprise we found foottracks—an ascent had been evidently made from Zermatt the day before. This abolished all wavering as to descending on the Zermatt side, for though none of us knew any thing about this side of the mountain we trusted to the traces to help us. I here took about an ounce of food:—for the first five hours I had eaten nothing, and this ounce of food was all that I consumed in what turned out to be a day of nineteen hours. The descent at first was in no way formidable, but it soon became so. The snow was 'like flour' and rested so loosely upon the steep rocks that the most tender treatment was necessary to keep it attached to the surface of which it rested. At a critical point we lost the traces and betook ourselves to the exceedingly steep arête, still formed of those ice covered rocks. By sheer engineering we got down this. A long rope was doubled at various places: the loop was placed over a projecting hunch of rock, and we let ourselves down as far as the rope reached. Then by a sudden jerk, or a succession of jerks, the rope was detached, to be employed again immediately for the same purpose. But we had to quit the rocks and betake ourselves to the steep mountainside, on which hung with such feeble hold this layer of floury snow. The leading guide behaved with great skill & coolness. No expression of fear or anxiety escaped him though you might <u>see</u> anxiety in his countenance; but his brother was in a state of considerable excitement and at times apparently afraid to put a foot forward. It was on this slope that the poor fellows came to grief in 1865,[19] and no doubt of it the slipping of any one of us would have involved the destruction of all the three. However things came to an end, and by and by, that is to say after nearly three hours of peril, we passed from the snow to practicable rocks. Standing here and permitting the vision to range over the Matterhorn the view was wild and impressive to the last degree. Nothing more difficult can be conceived than the two aspects of the Matterhorn from the top & from the bottom. From below it appears a compact pyramid smooth and defiant of the weather, above it appears torn to pieces by the frosts of ages, while its vast facettes are so foreshortened as to stretch out into the distance like plains. A new danger was here to be avoided. Along the sides of the mountain the discharge of stones was piquant; and at any moment by a detaching a single stone we could let loose a perfect cataract of boulders, which flow with the noise of thunder down the mountain. To avoid this danger we kept on the arête or near it, which you, as a geometer, know is not the line of steepest fall; hence not the line along which the stones will rush. I was deeply impressed by the condition of the mountain, and with the resistless and remorseless character of those small forces which integrating themselves through the ages have thus torn the flanks of the Matterhorn to pieces. Down we went; but at length we went astray, and it was dark before we could free ourselves. Then came the descent to Zermatt. The guides not knowing the way made a terribly long detour: at length we tried to get down to the bottom of the valley of Zmutt,[20] and for nearly two hours were toiling amid trees and brushwood, and along the beds of dried streams till they because too precipitous. For the first time in my life I thought of lying down under a tree and waiting for the day. We jogged however doggedly on, & finally cut the path. We reached Zermatt, & found shelter in Seiler's hotel[21] a little after 1 o'Clock in the morning. Having started at 6 this makes a day of 19 hours.

I wish you to keep this letter for me, as I may like to use it as a substitute for notes. I shall remain at the Riffel probably for 8 or 9 days and write out my address for Norwich.[22] My sleep has not yet come to me; but I am better & more hopeful than I was some time ago.

Yours affectionately | John

I wish you would allow M[rs]. Faraday[23] to read this letter: her address is 12 Barnsbury Villas | Liverpool Road | Islington.

BL Add MS 63092, ff. 45–9
RI MS JT/1/HTYP/546

1. *[between 28 July and early August]:* this letter was written after Tyndall completed his traverse of the Matterhorn on 28 July while at the Riffel hotel in Zermatt. On 3 August, he was still in Zermatt (see letter 2857), and back in London by 15 August (see letter 2859; *Hours of Exercise in the Alps*, pp. 293–4).
2. *Riffel:* the Riffelhaus on the Riffelalp, underneath the Riffelhorn, a mountain in the Pennine Alps, located south of Zermatt in the Swiss canton of Valais. The hotel was conveniently higher up the mountainside towards Gornergrat for early morning ascents of surrounding peaks, and is where Tyndall started for his ascents of Monte Rosa. Opened in 1853, the hotel was managed by a brother of Joseph Seiler of Zermatt who managed many alpine hotels, including the Hôtel du Glacier du Rhône.

3. *I quitted you very ill at Chatillon:* Tyndall left for Breuil on 22 August 1868 while Hirst made his way to Milan. Hirst noted his 'bad feet' on 18 August (Brock & MacLeod, *Hirst Journals,* 18–23 August 1868, p. 1835).
4. *chanoine:* canon (French), here referring to Georges Carrel (see letter 2858, n. 7).
5. *little book:* not identified.
6. *Breuil:* see letter 2622, n. 37.
7. *guide:* Jean Joseph Maquignaz (see letter 2869, n. 3).
8. *attacking the Matterhorn:* see letter 2855, n. 3.
9. *"it will be foul weather today for the sky is red and lowring":* Tyndall here was quoting Jesus (Matthew 16:3)
10. *Theodule:* Col de St. Theodule (see letter 2622, n. 36).
11. *coulior:* a steep, narrow gully on a mountainside (French).
12. *"Cachez vous!":* hide! (French).
13. *arête:* a sharp ascending ridge or 'edge' of a mountain; the local name in French Switzerland, whence it has become a technical term with mountain-climbers (*OED*).
14. *Cravatte:* a neckerchief or broad necktie, usually of linen, cotton, or silk, sometimes fastened with a bow or pinned, and often worn over the top of a high collar (*OED*).
15. *frigorific:* producing cold, freezing; cooling (*OED*).
16. *réchauffés:* a warmed-up dish; a dish made from leftovers (*OED*).
17. *pinnacle to which they have given my name, and on which I planted a flagstaff in 1862:* Pic Tyndall, a lower summit on the western shoulder of the Matterhorn first reached by Tyndall and Johann Joseph Bennen in 1862 (see *Hours of Exercise in the Alps,* pp. 153–65).
18. *Vispbach:* the river that runs through Zermatt to Visp where it meets the Rhône.
19. *poor fellows came to grief in 1865:* on 14 July 1865, four members of the group accompanying Edward Whymper (1840–1911) on their successful first ascent of the Matterhorn fell to their deaths during the descent. They were Francis Douglas (b. 1847), Charles Hudson (b. 1828), Douglas Hadow (b. 1846), and Michel Croz (b. 1830); Whymper and two Zermatt guides, Taugwalder father and son, survived the accident. For Tyndall's account of hearing of the tragedy while he was himself in the Alps, see *Hours of Exercise in the Alps,* pp. 252–4. For Whymper's account, see E. Whymper, 'The Matterhorn Accident', *Times,* 8 August 1865, p. 9; E. Whymper, *Scrambles Amongst the Alps in the Years 1860–69* (London: John Murray, 1871), pp. 395–408; and E. Whymper, *The Ascent of the Matterhorn* (London: John Murray, 1880), pp. 284–98.
20. *valley of Zmutt:* to the west of Zermatt, this valley passed the northern slope of the Matterhorn and ends at the Zmutt Glacier on the Italian border.
21. *Seiler's hotel:* a hotel managed by Alexander Seiler (1819–91), a Swiss hotelier who managed numerous hotels in the Swiss Alps including the Hotel Monte Rosa, the Hotel Riffelalp, and the Hotel Jungfrau (*HLS*).
22. *my address for Norwich:* see letter 2860, n. 4.
23. *M^rs^. Faraday:* Sarah Faraday.

To Elizabeth Dawson Steuart 3 August 1868 2857

Zermatt | Switzerland | 3rd. Aug. 1868.

My dear Mrs Steuart,

Your letter[1] has just reached me. I am quite willing to permit Sarah Tyndall[2] to apply the £50 in the manner you indicate. But marriage at her age[3] is so natural that I doubt whether this application of the money will be a substitute for matrimony.

However if you think this use of the money wise and desirable I am ready to send it to you. I shall leave this place towards the end of the week for England, and shall forward you a cheque for £50 on my return.

Yours ever truly | John Tyndall.

RI MS JT/1/TYP/10/3380
LT Typescript Only

1. *your letter:* letter missing.
2. *Sarah Tyndall:* Sarah Tyndall (not Tyndall's mother; see Biographical Register).
3. *marriage at her age:* Sarah Tyndall would marry David Miller, a farmer, on 2 September 1869 (*Irish Genealogy*).

From François-Napoléon-Marie Moigno 2 August 1868 2858

LES MONDES | Revue Hebdomadaire | DES SCIENCES | ET DE LEURS APPLICATIONS | aux arts et à l'industrie | par | M. l'Abbé MOIGNO | RÉDACTION | 2, RUE D'ERFURTH, 2

Paris, le 3 Aout 1868

Mon cher M^r^. Tyndall

Votre lettre de ce matin me rend bienheureux, car j'étais réellement inquiet. Je ne sais comment le bruit s'était répandu à Paris que vous étiez atteint à Montigny dans le Valais d'une fièvre pernicieuse qui mettait vos jours en danger. En <u>allant</u> aux renseignements, j'ai lu que l'origine de ce bruit néfaste était une lettre écrite de Turyn par M^r^. Hirst votre compagnon de voyage. M^r^. Hirst est venu heureusement à Paris la semaine dernière. J'ai dîné avec lui jeudi chez M^r^. Chasles et il nous a appris que vous étiez près de Zermatt en société avec un vénérable chanoine, dont le nom m'a échappé, mais à qui vous voudrez bien présenter mes respectueux compliments.

Je mets aujourd'hui à la poste les bons à tirer et les bonnes feuilles de la brochure Faraday inventeur. Je la publie d'accord avec M[r]. Gauthier-Villars et M[r]. Jamin. Si ces messieurs ne m'avaient pas autorisé à marcher en avant, nous n'en aurions jamais fini. J'ai fait ce travail avec beaucoup de plaisir, peut être trouverez vous que la traduction est trop littérale, ne vous en plaignez pas. Je suis heureux et fier de l'avoir faite; et je suis sur que cet opuscule dont je fais tous les frais aura un grand succès.

A force d'étudier les leçons sur le son, je suis arrivé à partager votre opinion qu'elles sont vraiment excellentes; de votre côté, croyez bien que ma traduction du son est plus parfaite que celle de la chaleur dont vous avez été si content.

Je puis vous affirmer que j'ai mis à la faire une ténacité extrême, que je n'ai reculé devant aucune difficulté, que je les ai vaincues toutes et que je ne doute en rien de l'honneur que ce travail m'apportera.

Le manuscrit de la chaleur est depuis longtemps entre les mains de M[r]. Gauthier-Villars, mais ces braves éditeurs ne *[1 word illeg]* pas. Et si l'on n'était pas comme je suis armé de courage de pied en cap on perdrait patience.

Il me tarde de savoir que la surprise que je vous cause en vous envoyant tout Faraday, aura été pour vous une surprise très agréable.

Tout à vous d'estime et d'affection vive et sincère

L'Abbé F. Moigno

Avertissez moi, je vous en prie, du jour de votre passage par Paris.

LES MONDES[1] | Revue Hebdomadaire | DES SCIENCES | ET DE LEURS APPLICATIONS | aux arts et à l'industrie | by the | Father MOIGNO | EDITORIAL BOARD | 2 RUE D'ERFURTH, 2

Paris, 3 August 1868

My dear M[r]. Tyndall

Your letter[2] from this morning is making me very happy, because I was very concerned. I don't know how the rumour spread in Paris that, while in Martigny in the Valais,[3] you had come down with a pernicious fever that threatened your life. Looking for some information, I read that this nefarious gossip originated from a letter written from Turyn[4] by M[r]. Hirst,[5] your travelling companion. Fortunately, M[r]. Hirst came to Paris last week. I dined with him at M[r]. Chasles[6] on Thursday, and he informed us that you were near Zermatt in the company of an old canon,[7] whose name has escaped me, but to whom you would be kind enough to present my respectful regards.

I am mailing at the post office today the correct sheets ready for printing for the booklet Faraday inventeur.[8] I am publishing this together with M[r]. Gauthier-Villars and M[r]. Jamin.[9] Had these gentlemen not authorized me to go ahead, we would have never finished it. I undertook this work with a lot of pleasure; you might find the translation is too literal. Do not complain about it. I am pleased and proud to have done it; and I am sure that this opuscule, paid entirely by myself, will meet a great success.

By dint of studying the lessons on Sound,[10] I came to share your opinion that they are really excellent; for your part, rest assured that my translation of Sound[11] is more perfect than that of Heat,[12] with which you were so satisfied.

I can assure you that I put into this an extreme tenacity, that I did not retreat when faced with difficulties, that I conquered all of them, and that I do not doubt this work will do me justice.

The manuscript of La Chaleur[13] has been in the hands of M[r]. Gauthier Villars for a long time, but these brave editors do not *[1 word illeg]*. And those not armed from head to toe with the same courage I have would lose patience.

I long to hear that the surprise I am giving you by sending you the complete Faraday[14] will have been very agreeable.

Yours truly, with my regards and my strong and sincere affection,

Father F. Moigno

Let me know, please, when you are passing by Paris.

RI MS JT/1/M/116

---

1. *LES MONDES:* see letter 2706, n. 9.
2. *Your letter:* letter missing.
3. *rumour spread in Paris that . . . that threatened your life:* Hirst does not mention any particular health problem for Tyndall in his journal, nor does Tyndall in *Hours of Exercise in the Alps.* However, in letter 2853 Tyndall reported to Hirst having insomnia, and in letter 2856 to Hirst he stated that 'I quitted you very ill at Chatillon'.
4. *Turyn:* Turin, Italy.
5. *M[r]. Hirst:* Thomas Hirst.
6. *M[r]. Chasles:* Michel Floréal Chasles (1793–1880), a French mathematician who taught at the École Polytechnique (1841–51) and held the chair of geometry at the Sorbonne from 1846 until his death (*CDSB*).
7. *old canon:* Georges Carrel (1800–70), Canon Carrel, an Italian clergyman, botanist, and mountaineer (*Dizionario Biografico Degli Italiani,* 94 vols (Rome: Istituto della Enciclopedia Italiana, 1960–2019), vol. 20, pp. 728–30).
8. *booklet Faraday inventeur:* see letter 2750, n. 3.
9. *M[r]. Jamin:* Jules Jamin.
10. *lessons on the Sound: Sound* (1867).
11. *my translation of the Sound:* see letter 2617, n. 2.
12. *that of Heat:* the French translation of *Heat* (1863) was published as J. Tyndall, *La Chaleur considérée come un mode de mouvement,* trans. F. Moigno (Paris: Étienne Giraud, 1864).

13. *manuscript of La Chaleur:* Moigno was attempting to publish a second French edition of Tyndall's *Heat,* this time through Gauthier-Villars rather than Étienne Giraud (see letter 2795, n. 7).
14. *complete Faraday:* see letter 2750, n. 3.

To William Thomson [8 August 1868][1] 2859

London, Saturday

My dear Thomson.

I reached London, this morning from the Alps,[2] and the first envelope I broke was that containing the card from Lady Thomson[3] and yourself. Let me say next it would have given me true pleasure to have met you both here, and even the thought that Lady Thomson, has been good enough to call upon me will brighten this day. *[Lovely]* correspondence. Yours ever truly. John Tyndall

MS Kelvin T212, by permission of University of Glasgow Library, Archives & Special Collections

---

1. *[8 August 1868]:* the date is given since Tyndall mentioned in letter 2857 (Monday 3 August) that he would be leaving Zermatt sometime around the end of the week. Given a travel time of a few days from Zermatt to London, the following Saturday (8 August) seems a reasonable date for his return to London.
2. *from the Alps:* Tyndall had left London in early July arriving in Paris on 5 July. He and Hirst then proceeded to the Alps (Brock & MacLeod, *Hirst Journals,* 5 July 1868; *Hours of Exercise in the Alps,* pp. 273–96, p. 1832).
3. *Lady Thomson:* Margaret Thomson (see letter 2786, n. 5).

To Thomas Archer Hirst 17 August 1868 2860

17th. Aug. 1868

My dear Tom.

Your letter[1] from Cromer[2] has come, and has proved very pleasant to me.

Your wishes shall be carried out to the letter: I feel perfectly able to do so. Indeed I look at the whole matter with an infinitely calmer and firmer eye than was mine when we last parted.[3]

Whatever my inner thoughts may be, no outward act of mine shall shew to the world that any change has occurred.

Banish therefore your anxiety and believe me

Yours ever affection[ate]ly | John

It will probably be safer if I run down tomorrow,[4] though I would willingly keep as long as possible away from the hubbub.

You might perhaps send me a telegram tomorrow morning saying whether it would answer if I reached Norwich at 2. for the Committee Meeting[5] on Wednesday

RI MS JT/1/T/667
RI MS JT/1/HTYP/542

---

1. *Your letter:* letter missing.
2. *Cromer:* a coastal town in the English county of Norfolk, near Norwich, location of the thirty-eighth annual meeting of the BAAS from 19–26 August 1868 (*Brit. Assoc. Rep. 1868*). Hirst arrived in Norwich on 13 August, left for Cromer on the 14 August and returned to Norwich on 18 August (Brock & MacLeod, *Hirst Journals,* 13–18 August 1868, pp. 1834–5).
3. *when we last parted:* see letter 2855, n. 4.
4. *run down tomorrow:* from London to Norwich, where Tyndall would give the presidential address to Section A (Mathematical and Physical Section) at the BAAS meeting on 20 August 1868 (*Brit. Assoc. Rep. 1868,* pp. 1–8 of 'Notices and Abstracts of Miscellaneous Communications to the Sections'). Hirst remarked in his journal that Tyndall's address was 'admirable' (Brock & MacLeod, *Hirst Journals,* 20 August 1868, p. 1835). This notable 'Norwich Address' was published as 'Scientific Limit of the Imagination' in *Use and Limit of the Imagination in Science* (London: Longmans, Green, and Co., 1870), pp. 52–65, and as 'Scientific Materialism' in *Fragments of Science,* vol. 2, pp. 75–90.
5. *Committee Meeting:* the BAAS General Committee met on Wednesday, 19 August. Tyndall was a member of the Council and president of Section A (Mathematics and Physics) (see *Brit. Assoc. Rep. 1868,* pp. xxxvi–xxxviii).

From William Armstrong 23 August 1868 2861

Cragside[1] | Morpeth | 23 Aug[us]t 68

My dear Tyndall

In reading the Times this morning my eye fell upon the extract given[2] from your opening address to Section A.[3] I read it twice over & under the feeling of the moment I cannot refrain from writing you a line to say how much I am impressed with the clear vigorous & noble thoughts you have expressed on the subject & how entirely I concur in your views. What a contrast you have presented to the narrow dogmatism both of materialists & spiritualists!

I am sorry I could not arrange to be at Norwich[4] this year. Not knowing the meeting was to be so unusually early I had made engagements which I could not set aside—

Is there any chance of seeing you here this autumn—no one will be more welcome either alone or in company with any of our Scientific friends you might ask to accompany you.

Very truly yours | W. S. Armstrong

RI MS JT/2/10/432

1. *Cragside:* Armstrong's home near Rothbury, and the larger town of Morpeth, Northumberland. It is now a National Trust site.
2. *extract given:* see 'British Association for the Advancement of Science', *Times,* 22 August 1868, p. 4.
3. *your opening address to Section A:* see letter 2860, n. 4.
4. *Norwich:* see letter 2860, n. 2.

From unidentified[1] 23 August 1868 2862

Dolgelley[2] | North Wales | Aug: 23: 1868

My dear Professor,

You will allow me to tell you that I cannot sufficiently express my admiration, and as far as the expression can be used of an impersonal matter, my gratitude for the speech which you have made at Norwich.[3] The report[4] has only this moment reached me. You have uttered what has been my conviction for long years—two sets of phenomena, dependent on each other, it may be, to any extent, but not necessarily identical. Mind and matter remain at bottom what they were: but what a thing to have this said by a man of your vast authority in science, and your brilliancy as a poet and a man of feeling.

I really think this one of the grandest and of the most important utterances made in our day: I feel this most strongly.

You never came to Oxford as you half-promised. Alas,

Very sincerely | /W Price/

RI MS JT/2/10/432
RI MS JT/2/13c/1347

1. *unidentified:* while we first believed this letter to be from the English writer and literary critic Walter Pater (1839–94), Gowan Dawson and Robert M. Seiler have suggested, based on various aspects of the letter, that Pater was not the writer. While the signature might be 'W Price', the writer remains unidentified.

2. *Dolgelley:* now Dolgellau, a town in northwest Wales.
3. *speech which you have made at Norwich:* see letter 2860, n. 4.
4. *report:* a brief abstract of Tyndall's address appeared in the *Times* (see letter 2861, n. 2), and a longer account appeared in the *Standard* ('British Association for the Advancement of Science', *Standard,* 21 August 1868, p. 6).

To [Juliet Pollock][1] 26 August 1868 2863

Norwich[2] 26th. Aug. 1868.

My dear friend.

A friend you are indeed: true, as I always knew you to be, to the very core. I like the blossom for its very freshness but there was no danger, I think. I entirely agree with you as to the advisability of that step.[3] It would be the completion of my life. I have thought much of it, and will think more. But I am in great part deterred by the idea that it is hardly right to offer the tag-end of a life to such a being as you describe.[4] However we will think of it together.

Yours ever & ever | John Tyndall

RI MS JT/1/T/1148
RI MS JT/1/TYP/6/2096

1. *[Juliet Pollock]:* while the identity of the recipient is not known for certain, Juliet Pollock is very likely given the relaxed tone of the letter, discussion of marriage, and that Tyndall often addressed letters to her with 'My dear friend'.
2. *Norwich:* Tyndall was in Norwich for the thirty-eighth annual meeting of the BAAS from 19–26 August 1868 (*Brit. Assoc. Rep. 1868*). See letter 2860, n. 4 and n. 5.
3. *that step:* presumably marriage.
4. *such a being as you describe:* not identified, but presumably someone Juliet suggested for Tyndall to marry.

To John Lubbock 3 September 1868 2864

Royal Institution of Great Britain | 3rd. Sept. 1868.

My dear Lubbock

Mr. Sawyer[1] sent me your photograph with one or two of my own.[2] I do not think I have ever seen anything so good of you. There is sweetness in the face and earnestness in the eye, and there is ease in the position.

I will keep this and like instead of the original. For the want of reverence towards the reverend[3] when you are in my presence has almost banished that background of seriousness which shines forth so beautifully in this picture;

and which I prefer a million times to the brilliancy which dazzles and delights so many of your friends.

But the union of both—this substratum of earnestness and these rippling surface phenomena of brilliancy—is I have little doubt destined to bamboozle the world someday. If you have only strength and patience, and do not turn with too strong a yearning towards quieter joys—I have no doubt whatever as to the final effect of these qualities in the House of Commons.[4]

A gentleman named Sparkes writes to me[5] from Bromley.[6] He appears to be fond of cultivating peaches—and seems a clever observant man—do you know him?

I read myself last night to sleep over the pamphlet of the late Bishop of Norwich.[7] It is a momentous sign of the times. The amount of disintegration in the Church Establishment, and indeed among theological thinkers generally, which it indicates is startling. I should almost like to be as young as Rolfe[8] were it only to see the issue of it all. I dare say even in my brief day great changes may occur. The world seems ready for them.

The mention of Rolfe brings the whole bright troop of children before my mind. What a blessed thing memory is, and how trebly blest am I in the power of "visualizing" which I possess! Every glance & feature of a friend come back to me when I think of that friend; and Amy's[9] beauty & sweetness could hardly be rendered more vivid by her presence to my bodily eye than they are now to the eye of my mind. It is this power of presenting my friends intellectually that renders me so constant to them. I think with all your admirable qualities I best you here. This same power of visualizing would have made me a geometer had I chosen lines & angles instead of molecules as my sphere of intellectual action.

I was much concerned in Norwich to see Lady Lubbock[10] looking weary, and very sorry when she resolved to return home. But the sage people who remained behind assured me that she had acted wisely. I sincerely hope that the issue has proved this to be the case.

Give my kind regards to M^rs^. Birkbeck[11] and to Miss Lubbock.[12] Gould[13] told me a night or two ago that he was quite in love with her.

And now the flask uncorked by your photograph is empty—goodbye.

Yours ever | John <u>Tyndall</u>

BL Add MS 49642, ff. 167–70

---

1. *M^r^. Sawyer:* probably John Robert Mather Sawyer (1828–88), an English photographer who operated as an optician and photographer in Norwich before opening a studio in London in 1871 (J. Hannavy (ed), *Encyclopedia of Nineteenth-Century Photography,* 2 vols (London: Routledge, 2008), vol. 2, p. 1247).
2. *one or two of my own:* the RI holds two photographs of Tyndall taken by Sawyer of Norwich

on 2 September 1868 (RIIC 3618 and RIIC 3620). One of these is the frontispiece for this volume.

3. *the reverend:* not identified.
4. *House of Commons:* see letter 2652, n. 2.
5. *Sparkes writes to me:* George Sparkes (*c.* 1810–78), a British judge who had worked in Madras Civil Service and who had a home in Bromley, Kent ('Deaths', *Standard,* 1 February 1878, p. 1). Letter missing.
6. *Bromley:* a town in Kent, England outside London.
7. *pamphlet of the late Bishop of Norwich:* Samuel Hinds (1793–1872), an Anglican clergyman and bishop of Norwich (1849–57) (*ODNB*; S. Hinds, *Free Discussion of Religious Topics* (London: Longmans, Green, and Co., 1868)).
8. *Rolfe:* Rolfe Arthur Lubbock (1865–1909), the third son of John and Ellen Lubbock ('The Hon. R. A. Lubbock', *Times,* 21 May 1909, p. 11). Tyndall was his godfather.
9. *Amy's:* Amy Lubbock (see letter 2627, n. 26).
10. *Lady Lubbock:* Ellen Lubbock.
11. *M^rs^. Birkbeck:* Mary Birkbeck (née Lubbock, d. 1910), a sister of John Lubbock who married Robert Birkbeck in 1857 (see *The Ascent of John Tyndall,* p. 323).
12. *Miss Lubbock:* probably John Lubbock's sister Harriet who appears to have lived in Canada (H. G. Hutchinson, *Life of Sir John Lubbock,* 2 vols (London: Macmillan and Co., 1914), vol. 1, p. 217, and vol. 2, p. 307). His other sister Diana Hotham Lubbock (1836–1917) had married William Powell Rodney (1829–68) in 1856.
13. *Gould:* possibly John Gould (1804–81), a British ornithologist and publisher (*ODNB*).

From Joseph Dalton Hooker 4 September 1868 2865

Kew[1] | Sept. 4th 1868.

Dear old Tyndall

I have received your vulgar ribaldry,[2] and forthwith took a refresher at the "Guardian",[3] whose praise of the "noble period of Prof. Tyndall's thoughtful address" is contrasted with the unfinished and fragmentary character of the President's address,[4] and I did chuckle to find my beloved friend equally the beloved of the Guardian. I shall send for the Record[5] next and hope to find you there too and still deeper in its affections.

Joking apart, I have read your address, and agree with the Guardian: my wife[6] prepared me for its being very good and I think it is about the completest thing of the kind I ever read, and wondrous subtle. Yours[7] and Berkeley's addresses[8] were much the best of the lot, then came Fergusson's lecture[9] then Huxley's Chalk[10] and then Huxley's paper on races of mankind.[11] These were I think the crowning pieces of the meeting and I do not think any previous meeting could produce 5 such contributions, and if you add Magee's sermon[12] and the music at the Cathedral[13] and the Mayor's déjeuner[14] (which last overcame me altogether) it really makes a splendid total.

We[15] go to Scotland on Monday night, first to Glasgow on business[16] and then to Colvile's.[17]

Willy[18] has had a severe attack of bleeding from the lungs at Oxford during the April week. I brought him home on the Saturday and he has no internal ailment. There were several relapses, he is not to return to school, and I suppose I shall have to send him abroad for the winter. He was going to the International school.[19]

Ever yours aff[ectionate]ly | J. D. Hooker.

RI MS JT/1/TYP/8/2597
LT Typescript Only

---

1. *Kew:* see letter 2554, n. 2.
2. *your vulgar ribaldry:* presumably a letter (missing) from Tyndall to Hooker about Tyndall's presidential address to Section A at the 1868 BAAS annual meeting at Norwich (see letter 2860, n. 4).
3. *"Guardian":* article not identified, although Ruth Barton identifies this as the Anglo-Catholic *Guardian,* a conservative weekly (*X Club*, p. 256).
4. *President's address:* J. D. Hooker, 'Address by the President', *Brit. Assoc. Rep. 1868,* pp. lviii–lxxv. On Hooker's address, see Joseph Hooker to Charles Darwin, [20 August 1868], *Darwin Correspondence,* vol. 16, part 2, pp. 689–90, n. 1–3; this is letter 6326 in the online edition of the Darwin correspondence).
5. *Record:* not identified.
6. *my wife:* Frances Hooker.
7. *Yours:* see n. 2.
8. *Berkeley's addresses:* Miles Joseph Berkeley (1803–89), an Anglican clergyman and botanist (*ODNB*; M. J. Berkeley, 'Address [Biology]', *Brit. Assoc. Rep. 1868,* pp. 83–7).
9. *Fergusson's lecture:* James Fergusson (see letter 2582, n. 6), who delivered an address on 21 August on the 'Archaeology of the Early Buddish Monuments' (*Brit. Assoc. Rep. 1868,* p. lvii).
10. *Huxley's Chalk:* as part of the BAAS annual meeting at Norwich, Thomas Huxley gave a lecture on 'Chalk' to the 'operative classes' on 26 August (*Brit. Assoc. Rep. 1868,* p. lvii).
11. *Huxley's paper on races of mankind:* on 24 August, Thomas Huxley gave a lecture 'On the distribution of the races of mankind' at the International Congress of Prehistoric Archaeology in Norwich (see Thomas Huxley to Charles Darwin, [13–21 September 1868], *Darwin Correspondence,* vol. 16, part 2, pp. 740–1, n. 4; this is letter 6381F in the online edition of the Darwin correspondence).
12. *Magee's sermon:* William Connor Magee (1821–91), an Anglican clergyman who was Bishop of Peterborough (1868–91) and Archbishop of York (1891) (*ODNB*). Magee gave

a sermon on 'The Christian theory of the origin of the Christian life' on the morning of 23 August (W. C. Magee, *The Christian theory of the origin of the Christian life* (London: Simpkin, Marshall, and Co., 1868)).

13. *Cathedral:* the Norwich Cathedral, built in the late eleventh and early twelfth centuries.
14. *Mayor's déjeuner:* Jeremiah James Colman (1830–98), an English mustard manufacturer and politician who was Mayor of Norwich (1867–9). As part of the BAAS annual meeting he held a lunch for 700 people on 26 August (*ODNB*; Joseph Hooker to Charles Darwin, 30 August 1868, *Darwin Correspondence,* vol. 16, part 2, pp. 704–5, n. 4; this is letter 6333 in the online edition of the Darwin correspondence).
15. *We:* with Frances Hooker.
16. *Glasgow on business:* Hooker was travelling to make arrangements for the herbarium of the recently deceased George Walker Arnott (1799–1868) (see Joseph Hooker to Charles Darwin, 30 August 1868, *Darwin Correspondence,* vol. 16, part 2, pp. 702–5; this is letter 6333 in the online edition of the Darwin correspondence).
17. *Colvile's:* probably James William Colvile (1810–80), a lawyer and judge who served in the courts of Bengal in the 1840s and 1850s, and was later appointed to the Judicial Committee of Privy Council (*ODNB*).
18. *Willy:* William Henslow Hooker (1853–1942), the eldest child of Joseph and Frances Hooker. He would be sent to New Zealand (1869–70) for his health.
19. *International school:* the London College of the International Education Society, also known as the Spring Grove School, founded in 1867. Tyndall became a director at this school, and with Thomas Huxley helped plan the science curriculum (see *The Ascent of John Tyndall,* pp. 290–1; and C. Bibby, 'A Victorian Experiment in International Education: The College at Spring Grove', *British Journal of Educational Studies,* 5 (1956), pp. 25–36).

To John Murray 4 September [1868][1] 2866

Glaciers 4th Sept[embe]r

Dear Mr. Murray.

Many thanks to you for the cheque, which is larger than usual.

It is pleasant to find that American publishers possess even the rudiment of a conscience.

And now will you allow me one frank word. Persuaded by my excellent friend Hooker[2] I asked you to publish an edition of my Glaciers of the Alps.[3]

I hate bargaining and therefore left the divisions of profits wholly in your hands. You divided them equally between us and I expressed no discontent.

But I did not imagine your claim to this proportion of the profits extended beyond the edition of that book published by yourself—I did not think that you intended to claim the half profits of the sale in America also.[4]

For reasons which will be immediately evident to you Longman's[5] now publish my books. They allow me two thirds of the profits; they pay me on the day of publication, and they never pretend to the slightest control over my books beyond the editions which are published in this country.

I think it is better to speak out openly in a business matter like the present. To have to do so is, however, far from agreeable to me.

Yours very truly | John Tyndall

John Murray Esq[ui]r[e] | &c &c &c[6]

NLS, John Murray Archives, Ms. 41214

1. *[1868]:* the year is given by a note written on the letter (see n. 6).
2. *Persuaded by my excellent friend Hooker:* see letter 1535, *Tyndall Correspondence,* vol. 6.
3. *Glaciers of the Alps: Glaciers of the Alps.*
4. *the sale in America also:* the American edition was J. Tyndall, *Glaciers of the Alps* (Boston: Ticknor and Fields, 1861).
5. *Longman's:* Longmans, Green, and Co., a London publishing firm with which Tyndall published a number of books.
6. A note on the bottom of the second page, written upside down, reads: '1868 Sepr 4 | Dr Tyndall'.

To Thomas Archer Hirst 5 September 1868 2867

5th. September 1868.

My dear Tom

I have done as you advised[1] and written to Mr. Sawyer[2] for some copies of the photograph[3] which you approved of.

The "sadness" is no doubt there; but time will mellow it. Meantime it is a solace to me that there is one in the world to whom I can speak about it.

I work in the laboratory daily; rather getting my hand in and gathering up my thoughts than anything else. At night I read Lucretius;[4] not Tennyson;[5] but the old Sage himself, a translation of whose work "de rerum natura"[6] fell into my hands a few days ago. To me his poem is intensely interesting—It was he who deduced the world and all that therein is from the "fortuitous concourse of atoms".[7] Of course I am at home among these;—he seems like a great child looked back upon from the vantage of eminence of time.

Wonderfully interesting thoughts flash out from him here and there. He set himself to pull religion to pieces, believing it to be a foul and slavish influence among men. And he attacks the doctrine of the immortality of the soul with a kind of religious fervour. If his efforts were to open the gates of Heaven to men he could not be more ardent in his task. One thought in connexion with this subject struck me much. Referring to the recoil of the soul from extinction he says "Death therefore is nothing and concerns us not a jot ..... in time gone by we felt no distress when the Poeni came together to do battle, and all things shuddered and quaked beneath high heaven";[8] and he contends that we ought to feel just as little distress regarding what is to occur after body and soul have been dissolved.

I saw Sylvester[9] last night; went straight up to him; asked him whether he had seen you at Norwich;[10] his reply was cold & somewhat faltering. I said I would not beat about the bush but tell him frankly & directly that I had noticed his coolness to you & deplored it. That I had spoken with you about it, and that you were as ignorant of its cause as I was. He hesitated & spoke in fragments as a man who touched a delicate subject was likely to do. When I said that you were ignorant of having given him any ground of dissatisfaction he expressed surprise. He said it was a matter which could not be explained; but it was possible that he erred. Or that matters might have been stated to him with an unconsciously erroneous colouring. I then said that were it a matter of his own eyes and ears I should assure him that he had misinterpreted you, but that if any third party had been the cause of the estrangement, I felt doubly certain that error was at the bottom of it all. We talked a good while, and he seemed softened, but still remained dark. He said he would turn the matter over in his mind, and weigh well all that I had said to him. I like the transparent candour of Sylvester: it is a thousand pities that his temper is so sensitive.

Perhaps there are people in the world who say the same of me!!

Yours dear Tom affectionately | John

RI MS JT/1/T/668
RI MS JT/1/HTYP/543

---

1. *as you advised:* if this was in a letter, it is missing.
2. *written to Mr. Sawyer:* see letter 2864, n. 1.
3. *photograph:* see letter 2864, n. 2.
4. *Lucretius:* Titus Lucretius Carus (*c.* 99–*c.* 55 BC), a Roman poet known for his *De Rerum Natura.*
5. *Tennyson:* Alfred Tennyson. Here Tyndall was referring to Tennyson's poem 'Lucretius' published in the May 1868 issue of *Macmillan's Magazine.*
6. *"de rerum natura": On the Nature of Things,* a philosophical poem advancing Epicureanism written by Lucretius in the first century BC.
7. *"fortuitous concourse of atoms":* referring to the view of Lucretius that the world was formed not by divine power but by the 'fortuitous concourse of atoms'. This appears to be a paraphrase of Lucretius's words, and was used in many writings from the seventeenth century

which discussed God, causes, chance, and atheism, especially following Darwin's *On the Origin of Species* (1859). From Titi Lucreti Cari, *De Rerum Natura,* trans. H. A. J. Munro (Cambridge: Deighton Bell and Co., 1864), we find: 'But in what ways yon concourse of matter founded earth and heaven and the deeps of the sea, the courses of sun and moon, I will next in order describe. For verily not by design did the first-beginnings [atoms] of things station themselves each in its right place by keen-sighted intelligence' (pp. 229–30).

8. *"Death therefore is nothing . . . quaked beneath high heaven":* quoted from *On the Nature of Things* by Lucretius. The passage appeared in full on p. 140 of Titi Lucreti Cari, *De Rerum Natura,* trans. H. A. J. Munro (Cambridge: Deighton Bell and Co., 1864): 'Death therefore is nothing and concerns us not a jot, since the nature of the mind is proved to be immortal. And as in time gone by we felt no distress, when the Poeni from all sides came together to do battle, and all things shaken by war's troublous uproar shuddered and quaked beneath high heaven, and mortal men were in doubt which of the two peoples it should be to whose empire must fall by sea and land alike, thus when we shall be no more, when there shall have been a separation of body and soul, out of both of which we are each formed into a single being, to us, you may be sure, who then shall be no more, nothing whatever can happen to excite sensation, not if earth shall be mingled with sea and sea with heaven'. *Poeni* (Latin) refers to the Phoenicians.
9. *Sylvester:* James Sylvester (see letter 2851, n. 7).
10. *Norwich:* see letter 2860, n. 2.

To John Murray 5 September 1868 2868

Cheque returned by JM Sept 7[1] | 5th. Sept[embe]r. 1868

My dear Sir,

I believe you did propose, unprompted by me, to send the early sheets of my book[2] to an American publisher.[3]

You also inform me that you had some trouble in obtaining the money from the said publisher.

If on these two grounds you consider that you had a right to half the profits of the American sale, then I beg to state that as I have left the larger question entirely in your hands I willingly leave this smaller one also to your decision and arrangement.

It would not be at all agreeable to me to accept money regarding my right to which the slightest doubt could be entertained.

Yours very truly | John Tyndall

I return your cheque with thanks. | John Murray Esq[ui]re.

1868. Sept 5 & 8 | John Tyndall[4]

NLS, John Murray Archives, Ms. 41214

1. *Cheque returned by JM Sept 7:* Murray's reply to Tyndall's letter of 4 September (letter 2866) is missing, but must have included a check to Tyndall. As he wrote in this letter, Tyndall returned it, then received it back from Murray, noting on this letter 'Cheque returned by JM Sept 7' (with letter 2872).
2. *my book: Glaciers of the Alps.*
3. *American publisher:* probably Ticknor and Fields of Boston, which published an edition of *Glaciers of the Alps* in 1861.
4. *1868. Sept 5 & 8 | John Tyndall:* written by itself on the reverse side of the letter.

To Mary Egerton 6 September [1868][1] 2869

Royal Institution, | 8th. Sept[embe]r.

My dear Lady Mary,

Of course I should not have mentioned the Matterhorn[2] had I not scrambled over the top.

A very wild top it was—We[3] went up on one side and down on the other; thus completing a pass over the summit. It was hard work, for we went astray on the Zermatt side of the mountain and after a day of 19 hours reached Zermatt half famished between one and two o'clock in the morning.

I thought of going down to you on Sunday week—next Sunday I go to Folkestone[4] to see Bence Jones.[5] He is not so strong as he ought to be. In fact we have been comparing notes and found that neither of us has any "go" in him.

You will find your guest the stupidest of men:—so stupid that I almost hesitate to accept your hospitality.

Yours ever | John Tyndall

RI MS JT/1/TYP/1/388
LT Typescript Only

1. *[1868]:* the year is given based on Tyndall having this year been the first to traverse the Matterhorn.
2. *Matterhorn:* see letter 2855, n. 3.
3. *We:* Tyndall made the traverse in the company of the Italian guides Jean Joseph Maquignaz and Pierre Maquignaz (*Hours of Exercise in the Alps*, p. 288–9; and C. D. Cunningham and W. Abney, *The Pioneers of the Alps*, 2nd edn (London: Low, Marston, Searle and Rivington, 1888), pp. 123–6).
4. *Folkestone:* a port town on the English Channel in Kent, where Bence Jones had a home at 5 Albion Villas since 1851.
5. *Bence Jones:* Henry Bence Jones.

From Mary Egerton 9 September [1868][1] 2870

Sept[embe]r. 9th

Dear Mr. Tyndall,

We shall be delighted to see you on Sunday week, do not give us credit for being quite so selfish as to want you only to be agreeable. The more "stupid" & out of spirits you feel, the more essential it is you should get away, as much as possible from the atmosphere of your work, and if I get you down here, (provided it is not at the expense of something pleasanter,) I shall consider myself to have done a good deed, & "virtue", you know, "is its own reward"!

There is an ominous sentence in your life of Faraday,[2] that makes me fear you made it a sort of point of honour to "work till you drop"—but is a man wise or just who lives on his principal, instead of laying by for his children? & (independent of any other consideration,) is it right to run the risk of depriving your scientific descendants of the fruits of 20 or more years good work, to come by using up your brains now in the middle of life?

Do not think me impertinent, I know I can have no right to advise, but do look at the advice independent of the giver, and ask Dr. Bence Jones[3] if it is not sound.

I hope you can come from Saturday till Sunday; it is a sham rest only to come away on Sunday, when nobody works! Send us word nearer the time.

Y[ou]rs. very truly | M. F. Egerton

19 hours! Do you mean to say you crossed the Matterhorn[4] without spending the night on it? I wonder you are alive to tell the tale!

RI MS JT/1/E/47

---

1. *[1868]:* the year is given by relation to letter 2869, in which Tyndall shared that 'after a day of 19 hours reached Zermatt half famished between one and two o'clock in the morning', and also remarked that if he were to visit Egerton, she would find in her guest 'the stupidest of men'.
2. *life of Faraday: Faraday as a Discoverer.*
3. *Dr. Bence Jones:* Henry Bence Jones.
4. *Matterhorn:* Tyndall shared in letter 2869 about his Matterhorn climb.

From Joseph Dalton Hooker 7 September 1868 2871

Sept. 7th 1868. | Kew.[1]

Dear Tyndall,

Spottiswoode[2] and I intended consulting you about enclosed[3] when at Norwich.[4] If you approve will you get which of the signatures you can and forward it to Spottiswoode, who will send it to the First Comm[issione]r of H. M. Works.[5]

We[6] are off tonight and home again as soon as our visit to Colvile's[7] is over.

Ever yours | J. D. Hooker.

RI MS JT/1/TYP/8/2598
LT Typescript Only

---

1. *Kew:* see letter 2554, n. 2.
2. *Spottiswoode:* William Spottiswoode.
3. *enclosed:* while we do not have the enclosure, it is possibly a memorial in favor of Hooker's 1868 proposal that the Banks herbarium at the British Museum be transferred to Kew (*ODNB*), a suggestion that would not have enthused Richard Owen, superintendent of the natural history department of the British Museum. This was all part of a long-standing argument over which institutions should control national scientific collections, which would erupt in a very public dispute from 1870–2 between Hooker and Acton Smee Ayrton (1816–86), a British Member of Parliament and First Commissioner of Works (1869–73), whom oversaw the Board of Works which funded Kew (*ODNB*). In his Presidential Address at the BAAS annual meeting at Norwich, Hooker discussed the role of museums and the treatment of collections, remarking that the trustees of the museum have neglected the botanical collections (J. D. Hooker, 'Address of Joseph D. Hooker, F.R.S., D.C.L. Oxon.; LL.D. Cantab.; &c. President', *Brit. Assoc. Report 1868,* pp. liii–lxvv). For more on this argument over scientific collections and the Ayrton affair, see *The Ascent of John Tyndall,* pp. 295–9; *X Club*, pp. 155–62, 338; and J. Endersby, *Imperial Nature: Joseph Hooker and the Practices of Victorian Science* (Chicago: University of Chicago Press, 2008), pp. 282–303. For the Ayrton affair as it played out in the journal *Nature,* see M. Baldwin, *Making* Nature*: The History of a Scientific Journal* (Chicago: University of Chicago Press, 2015), pp. 39–41.
4. *Norwich:* see letter 2860, n. 2.
5. *First Comm[issione]r of H. M. Works:* John James Robert Manners (1818–1906), an English politician who served as first commissioner of Her Majesty's Works and Buildings (1858–9, 1866–8). At this time he was Lord John Manners, and would become seventh Duke of Rutland in 1888 (*ODNB*).
6. *We:* with Frances Hooker.
7. *Colvile's:* James Colvile (see letter 2865, n. 17).

From John Murray 7 September 1868 2872

50 Albemarle Street[1] | Sept[embe][r] 7 (1868)

My dear Sir

Thank you for your obliging note[2] of Sat[urda]y but I beg you to keep the cheque which you returned,[3] on my assuring you that it was not sent to you grudgingly or in an unwilling spirit. On the contrary my only regret was that I had laid perhaps too much stress upon the circumstances under which it was obtained & that I had not made it over to the author at once.

I am, Dear sir, yours very faithfully | (signed) John Murray

Prof[esso][r]. Tyndall

NLS, John Murray Archives, Ms. 41214
Transcript Only

1. *50 Albemarle Street:* address of John Murray's home and publishing firm, just down the street from the RI.
2. *your obliging note:* letter 2868.
3. *keep the cheque which you returned:* see letter 2868, n. 1.

From Jane Barnard [8 September 1868][1] 2873

12 Barnsbury Villas[2] | Tuesday

Dear D[r] Tyndall

I am much troubled & do not know what to do, so write to you hoping for your kind assistance & which indeed you have offered. M[r] Vincent[3] left here last night, a copy of the paragraph which appeared in the Mechanics Magazine.[4] When I read it I felt so dismayed & troubled & could not even turn to my Aunt[5] with it, for she is very unwell. I have not yet spoken to her about it for I fear to trouble her in her present state.

I thought I would go to Albemarle St[6] & by to see you but I cannot leave her.

This untrue & unkind report must be contradicted & no one I believe can do it as you will.[7] Still I must tell you honestly what may have given rise to it—; when we were gathering together the apparatus &c and had sorted it there was left some odds & ends which partly you looked over & partly D[r] Frankland[8] did the same for . . . it was pronounced rubbish & I, (but I think I may say we) allowed Copeland[9] to clear it away believing he could make a trifle by it. It consisted of pieces of metal, glass, &c but I do not believe there was any scrap of my dearest Uncle's[10] work among it.

I wish we could see you, if you are not too busy could you call, or any rate send a line to tell me what you think. My Aunt is sleeping under the effects of morphine & I am expecting the doctor immediately, still I trust she will in the course of a few hours be better & able to hear this most vexatious & painful report & see you to comfort us, if you can come. Her indisposition is caused by the heat & over fatigue I believe.

Many little annoyances have met us on our return home & I feel overwhelmed with them having no one to share them with.

Ever yours most sincerely & obliged | Jane Barnard | Dr Tyndall Esq

RI MS JT/1/B/76

---

1. *[8 September 1868]:* the date is given as it appears to precede letter 2874, dated 9 September 1868, a Wednesday. As this letter was written on a Tuesday, it must be 8 September 1868.
2. *12 Barnsbury Villas:* see letter 2695, n. 2.
3. *Mr Vincent:* Benjamin Vincent (1818–99), a British translator and editor. He served as Assistant Secretary and Librarian of the RI for most of the second half of the nineteenth century (*Faraday Correspondence,* vol. 5, p. lvii).
4. *paragraph which appeared in the Mechanics Magazine:* see letter 2874, n. 2.
5. *my Aunt:* Sarah Faraday.
6. *Albemarle St:* the RI.
7. *as you will:* for Tyndall's letter to the editor of *The Mechanics' Magazine,* see letter 2874.
8. *Dr Frankland:* Edward Frankland.
9. *Copeland:* not identified.
10. *my dearest Uncle's:* Michael Faraday.

To the Editor of *Mechanics' Magazine*[1] 9 September 1868 2874

Sir,—In the Mechanics' Magazine for August 28, a short paragraph is published under the heading "Faraday's Royal Institution Apparatus."[2] Will you allow me to assure you, in reference to this paragraph, that Mrs. Faraday's[3] first care was that the apparatus illustrative of the researches of her husband should be worthily distributed. To his most intimate friends portions of it were given—a little to each, as a remembrancer—the most important pieces being bestowed by his own express desire. In the Royal Institution are two cases, one devoted to the relics of Faraday's instruments, and the other to his manuscripts. I do not think that a single piece of apparatus, complete or fragmentary, which could in any way illustrate the genius of Faraday as a constructor, fell into the hands of the porter[4] referred to. What came to his share, through the benevolence of Mrs. Faraday, were the sweepings that remained after all portions that were considered worthy of retention had been selected. I write this note without Mrs. Faraday's knowledge; indeed, I have reason to know that she had not yet seen or heard of the paragraph which causes me to address you; and I would end by expressing an opinion that if you knew Mrs. Faraday, you would not suffer a line to appear in your journal which could give her uneasiness. This subject, I trust, will now be permitted to rest.—I am, Sir, yours, &c.,

JOHN TYNDALL | Royal Institution, September 9.[5]

'Faraday's Apparatus', *The Mechanics' Magazine and Journal of Engineering, Agricultural Machinery, Manufactures and Shipbuilding*, 18 September 1868, p. 234

---

1. *editor of Mechanics' Magazine:* the editors from 1857–72 were R. A. Brooman and E. J. Reed (S. Sheets-Pyenson, 'Popular Science Periodicals in Paris and London: the Emergence of a Low Scientific Culture, 1820–1875', *Annals of Science,* 42 (1985), pp. 549–72, on p. 565).
2. *"Faraday's Royal Institution Apparatus.":* 'Faraday's Royal Institution Apparatus', *The Mechanics' Magazine and Journal of Engineering, Agricultural Machinery, Manufactures and Shipbuilding,* 28 August 1868, p. 174. The text of the paragraph was as follows: 'One of the faculties of this truly great man consisted in his great mechanical ingenuity and constructiveness, as evidence in the apparatus for conducting the original and elaborate experiments by which he arrived at such great results. Their main character was simplicity, which is, indeed, the perfection of ingenuity, and is the distinguishing feature of the work of genius. As has been lately remarked by a good judge, "the practical powers" were never, perhaps, "more strikingly displayed by man than in the various contrivances he adopted whilst conducting his researches—some of them being almost equivalent in ingenuity to the compilation of a steam engine." We regret to have to record the fate of the greater portion of these contrivances. Shortly after Faraday's death they were given by Mrs. Faraday to the porter of the Royal Institution, who, we need not say, could scarcely appreciate them. He accordingly sold them piece-meal, and even parts of the same apparatus, to different buyers, and thus broke up combinations that probably few men besides Faraday himself really understood'. This paragraph also appeared, with some word changes in the final sentences, as 'More Vandalism', *Scientific American,* 30 September 1868, p. 211 (it does not appear that Tyndall replied to it).
3. *Mrs. Faraday's:* Sarah Faraday.
4. *porter:* probably William Brown.
5. The journal included the following after Tyndall's note: 'The paragraph in question did not originate with us, having appeared in other papers besides our own.—Ed. M.M.'.

From Sarah Tyndall[1] 9 September 1868 2875

Bagnalstown[2] | Sep. 9th. 1868.

Professor Tyndall

Sir,

I have the pleasure of informing you that all arrangements has been perfected regarding the farm, all that is to be done now is the payment of the money[3] as the deed is signed.

I have been directed by Mrs Steuart[4] who I regret to say is not very well at present, to apply to you to have the money forthwith directed to her in order to complete the whole affair without further delay, as my father[5] has paid his part, for which I have deposited the receipt in the hands of Mrs Steuart. As soon as the balance is paid I am to get the possession of the house and land, there is to be an auction of the property on the farm on the 23rd Inst.

Waiting your earliest convenience

I am Sir your obedient Servant | Sarah Tyndall.

RI MS JT/1/TYP/10/3381
LT Typescript Only

1. *Sarah Tyndall:* Sarah Tyndall (not Tyndall's mother; see Biographical Register).
2. *Bagnalstown:* see letter 2587, n. 1.
3. *payment of the money:* referring to a dowry from Tyndall (see letters 2634, 2777, 2810, and 2857).
4. *Mrs Steuart:* Elizabeth Dawson Steuart.
5. *my father:* Caleb Tyndall.

From Hector Tyndale 9–10 September 1868 2876

Philadelphia September 9. 1868

My dear John

About a year ago I wrote to you[1] in reply to a good letter[2] I had from you just before. Just after, you sent me a Dundee paper with your address to the Working men of that place.[3] I read it with great pleasure for we Yankees, you know, love Science best when it is brought nearest to the human heart. The address was a fine one and rung out a noble peal. I glory in the Age when great radiates of thought turn again and re-root themselves in the Soul and heart of Man. Not in stretching out in direct lines, but in the spherical growth, which you so well quote from Fichte,[4] in your Norwich address,[5] is the great pleasure or rather is the true enjoyment. *[Realise]* a great eagle-winged Thinker soaring out of skies and finding more beautiful Stars, sailing back, enheartened and lovefull, to bring his mate and all his brood to dwell there! "Not alone but for others". And so I was pleased with what you said when you turned to the Dundee men—"and, being men, mates for all". I regretted, though, that your encouraging words and brave thoughts were diluted by the Nobleman, I believe it was Argyll,[6] who was the Chairman of the meeting—. It is a pity that we of English stock should be so "in the slavery of names".

A month ago, while at the beautiful Mount Desert Island on the Coast of Maine, five or six hundred miles to the N.E. of this, an acquaintance, an excellent man,[7] a Unitarian Clergyman, loaned me your Life of Faraday.[8] I know how tight a hand you bore in your feelings when you wrote that Life. Where you loved the <u>man</u> it was hard to demonstrate his doings merely, on the black board. I could almost do that you were not permitted to do, to cry for you.

Last Spring I laughed over your visit to the "Spirituals"[9]—at your position under the table—"groping after the Truth"! and at your being "the Poet of Science"![10]—I saw the "<u>Seance</u>" in one of our papers. I read your 'Heat'[11] just after writing you last and it was good exercise. In places it seemed like the Q.E.D.[12] of our old Euclid,[13] only it is Euclid brought out of the school room into the air and light, among circumstances and providences. Your experiments, disappointments, balances and counter-balances, the tyrannical conditions, and your neutralizations were all exceedingly interesting apart from the great results. There were one or two verbal expressions I did not recognize as yours, but suppose they were so unimportant as to have escaped your notice—and I mention this only to show you how closely I read '<u>Heat as a Mode of Motion</u>'. '<u>Sound</u>'[14] I have not read except to see that it will be, perhaps, even more interesting than '<u>Heat</u>' to me.

A few days ago came a Norwich paper from you with your Address to Section A of the Science Advancement Asso[c][iatio][n][15] for which I am much obliged to you. It, too, is simple and clear and very interesting. Your showing of the way and the distance that Demonstration may travel is excellent I think, though I do not think that you mark the edges of the chasm (narrowing though it be), always to exist between the Soul and Knowledge, as sharp as we have known them to be in personal conversation. It may be the antennae of the Soul grow so quickly and so rankly out over the Abyss as to hide it, and there growing quicker and stronger from the culture of Science on <u>this</u>, the outer, side, turn back and the Soul is confounded by the vigor of it's own outgrowths—All which having written down, so that my understanding can see it seems to be utter bosh! But that may go too, with the rest—there being, Thank God, more than a plenty of leaves in the forest which <u>seem</u> to be no earthly use!

I have related all this to you to shew you how I have journeyed with you and have been alive with you.

In May last I received a letter from a William Tyndall[16] of Morrisburgh P. O. Co: Dundas Ontario (Canada) who wished to be put into some better way of doing for himself. He is now a School Teacher. He seemed to take pains to tell me of the greatness of the Tyndalls of old, and cited you as an outcropping of the primitive family rock! He scorned to ask a favor, as "any true Tyndale would" but seems to lack the power to force the favors of the World or to gain them. He says he is a Grandson of W[illia]m Tyndall,[17] who was a companion of my Father's[18] in boyhood—he mistook me to be a Son of my Uncle W[illia]m[19]—who left a family here. He had written to my eldest brother Sharon,[20] who was not able to aid him. At the time I happened to be on a visit to Sharon at Springfield Illinois (he is Secretary of State of Illinois)—about a thousand miles West of this—and I wrote to him of my inability and advised that "If you are now gaining means of subsistence in your present place I would respectfully suggest that you retain and grow up rooted and strong in it—giving that place a value by your own worth—for no position in the World, or among men, but can be made of vast importance by the man who, forgetting himself and pushing persistently on for the Right, keeps his life pure, upright and independent. John Tyndall whom I know well and love, is a good illustration of the old truths I have endeavored to speak. You can, perhaps, make him as proud of you as you now are justly proud of bearing his name. Things have changed, are now changing, and names merely, no longer denote qualities, or powers, and Demonstration is the sole means used in judgment to day—I wish you success through the inflexible use of your own powers."—

I had intended to send you his letter and my reply but this will do to inform you of the fact of our having another distant relation, his whereabouts, and the kind of man I take him to be. Can you give me any information about him or tell me of his deserts?[21] I kept a copy of both his letter & my reply, from which I have quoted, to send to you.

Sept 10. Yesterday after writing the above came a pamphlet from you[22]—the Norwich Address which I had read before in the paper you sent me. Many thanks for your kind remembrance.

Some months ago I sent you a paper with the news that I had been nominated by <the> Republican Party of this City it's Candidate for Mayor of the City. A very high honor and one which I sought to avoid by going away from town. The election[23] comes off on the Second Tuesday of October next, the 13th prox.[24] some weeks before the Presidential election.[25] The result is uncertain, tho' it is believed by our Party that we shall succeed. But the Democrats, whom we oppose, talk very assuredly too. For myself personally I would rather not have the office but for the sake of the Party, which holds all of the bright future of this growing Nation within it, and is the Party of true Progress and Conservatism, I hope to succeed. I am not taking any active public steps to be elected and if chosen it shall be without pressure or solicitation on my part. Gen[era]l Grant is our Candidate for President and so far as the relations of this Country extend the World itself is interested in his election as against a repudiating, anarchical, rebellious Democracy. The total vote of Philadelphia is about 110,000 (in a population of about 700,000 to 750,000) and the majority either way will probably be under five thousand (5000). It is this closeness of the political parties at home that made our Rebellion so difficult to conquer—for the Democracy as a Party opposed the War,[26] while the Rebels, more than a third of the population, fought desperately in our front.

In my last year's letter[27] (!) I sent all kinds of remembrances to Tom Hirst and to Mr Debus.[28] How are they getting on? I should like to hear from or of them at least. Please send me their addresses by mail. Mrs Tyndale[29] is well and desires to be remembered to you and to Mr Hirst. Whenever you may feel like it and can spare the time I should like you to write me. Mr Mitchell[30] sends his respects to you.

Please make my best respects and remembrances to your Sister[31] whose kindness to me at Laughlin Bridge[32] I shall never forget. Affectionately your Cousin | Hector Tyndale

RI MS JT/1/T/60
RI MS JT/1/TYP/5/1674–7

---

1. *wrote to you:* letter missing, probably a reply to letter 2627.
2. *good letter:* probably letter 2627.
3. *Dundee paper with your address to the Working men of that place:* see letter 2647, n. 5.
4. *Fichte:* Johann Gottlieb Fichte (1762–1841), a German idealist philosopher (*NDB*).
5. *Norwich address:* see letter 2860, n. 4.
6. *Argyll:* George Campbell, eighth Duke of Argyll. Hector Tyndale appears to have confused Argyll with the fifth Duke of Buccleuch (1806–84) who was President of the BAAS during the annual meeting at Dundee in 1867.
7. *excellent man:* not identified.
8. *Life of Faraday:* probably the American edition, J. Tyndall, *Faraday as a Discoverer* (New York: D. Appleton and Co., 1868).
9. *your visit to the "Spirituals":* as part of Tyndall's controversy with Daniel Home in the *PMG* (see letters 2811, 2812, 2813, 2814, 2815, 2817, 2820, 2822, 2824, and 2825), the paper re-published an account of a séance Tyndall had written in 1864 (J. Tyndall, 'Science and the Spirits', *PMG,* 15 June 1868, p. 3).
10. *"the Poet of Science":* Tyndall reported that the spirits had called him 'the Poet of Science'.
11. *'Heat':* two editions of Tyndall's *Heat* were published by D. Appleton and Co., before 1868, one in 1863 and another in 1867 (J. Tyndall, *Heat Considered as a Mode of Motion* (New York: D. Appleton and Co., 1863)).

12. *Q.E.D.: quod erat demonstrandum* (Latin), meaning 'which was to be demonstrated'; used to convey that a fact or situation demonstrates the truth of a person's theory or claim (*OED*).
13. *Euclid:* Euclid of Alexandria (fl. 295 BC), a Greek mathematician known best for his *Elements* (*CDSB*).
14. *'Sound':* probably the American edition, J. Tyndall, *Sound* (New York: D. Appleton and Co., 1867).
15. *Norwich paper from you with your Address to Section A of the Science Advancement Asso^{c}[iatio]$^{\underline{n}}$:* Norwich paper not identified; for the address, see letter 2860, n. 4.
16. *William Tyndall:* William Tyndall (b. *c.* 1837), an Irish emigrant who settled in Gloucester, Ontario, Canada, and became a school teacher (Census of Canada 1861, 1871).
17. *W[illia]$^{\underline{m}}$ Tyndall:* not identified.
18. *Father's:* Robinson Tyndale (d. 1842), an Irish emigrant who established himself as a china merchant in Philadelphia. He was Hector Tyndale's father and the son of William Tyndall (1742–1846) and Margaret Robinson (1744–1826).
19. *my Uncle W[illia]$^{\underline{m}}$:* William Tyndall (b. 1783), one of Robinson Tyndale's brothers.
20. *my eldest brother Sharon:* Sharon Tyndale (1816–71), a civil engineer and surveyor who was Secretary of State for Illinois (1865–9).
21. *deserts:* an action or quality that deserves its appropriate recompense; that in conduct or character which claims reward or deserves punishment. Usually in plural (*OED*).
22. *a pamphlet from you:* Tyndall's Norwich address (see letter 2860, n. 4) was published separately. In letter 2899 to Henry Acland, Tyndall wrote: 'The printing of 200 additional copies of the Norwich Address enables me to send you half a dozen of them'.
23. *election:* Tyndale lost the mayoral election to Daniel M. Fox (1809–90) by a narrow margin, which was disputed in court (J. R. Young (ed), *Memorial history of the city of Philadelphia, from its first settlement to year 1895,* 2 vols (New York: New-York History Company, 1895), vol. 1, p. 537).
24. *prox.:* next month.
25. *Presidential election:* on 3 November 1868, Ulysses S. Grant was elected the eighteenth President of the United States.
26. *War:* the American Civil War (1861–5).
27. *my last year's letter:* the only other letter from Tyndale in this volume is letter 2620 (dated 19 July 1867), and in it he does indeed write about Hirst, but not Debus.
28. *Mr Debus:* Heinrich Debus.
29. *Mrs Tyndale:* Julia Tyndale.
30. *Mr Mitchell:* Edward Mitchell.
31. *your Sister:* Emma Tyndall.
32. *Laughlin Bridge:* Leighlinbridge, County Carlow, Ireland, where Tyndall was born and his mother had lived.

To Elizabeth Dawson Steuart — 11 September 1868 — 2877

Royal Institution | 11th. Sep. 1868.

My dear Mrs Steuart,

I am very very sorry to hear that you are not well.[1] This year has been a trying one to many and it seems to you and me among the number. I hope sincerely as these intense heats subside that you will recover your strength.

The enclosed note[2] has been sent to me, and I accordingly forward you a cheque for £50.[3] which I intended as a dowry for Sarah Tyndall.[4] But as you think it would be to her benefit to employ it in a farm I have no objection to sending her the sum now instead of waiting for her marriage.

Yours ever faithfully | John Tyndall.

RI MS JT/1/TYP/10/3382
LT Typescript Only

1. *hear that you are not well:* in letter 2875, Sarah Tyndall wrote to Tyndall regarding Steuart, 'I regret to say is not very well at present'.
2. *enclosed note:* letter 2875.
3. *cheque for £50:* see letter 2857.
4. *Sarah Tyndall:* Sarah Tyndall (not Tyndall's mother; see Biographical Register).

From Louis Lortet[1] — 21 September 1868 — 2878

Lyon 21 Sept 1868

Très honoré Monsieur

Combien je vous remercie d'avoir songé à moi au milieu de vos nombreuses occupations et de vous être rappelé de votre disciple du Riffel.

Je ne suis cependant pas seulement votre disciple du Riffel, mais depuis bien des années déjà j'étais en communion avec vous par vos ouvrages et par vos méthodes si claires et si nettes que nous nous efforçons aujourd'hui de transplanter en médicine et en physiologie. Mais j'ai été bien heureux cette année de vous voir par vous-même en chair et en os sur le théâtre de vos travaux. Et aujourd'hui je suis encore plus heureux de posséder cette photographie de vous, Monsieur que j'admirais sans vous connaitre et que maintenant j'ai appris non seulement à admirer mais à estimer.

Veuillez, Monsieur, avec tous mes remerciements bien croire aux témoignages de ma très haute considération.
Doct. L. Lortet

Lyon 21 September 1868

Very honored Sir,

I thank you so much for thinking of me[2] in the midst of your numerous activities and to be reminded of your disciple from Riffel.[3]

However, I am not only your disciple from Riffel, but for many years I was in communion with you through your books and methods that are so clear and neat, and which we strive to transfer today into medicine and physiology. But I was very fortunate to see you this year, appearing before us in the flesh on the scene of your work.[4] And now I am even happier to have this photograph of you, Sir, which I admired without having known you; and now I have learned not only to admire but also to esteem.

Please accept, Sir, with all my thanks, my highest regards.
Doct[or]. L. Lortet

RI MS JT/1/L/35

1. *Louis Lortet:* Louis Charles Émile Lortet (1836–1909), a French physician, botanist, zoologist, and mountaineer (C. Gaillard, 'La vie et les travaux de Louis-Charles Lortet', *Archives du Muséum d'histoire naturelle de Lyon,* 11 (1912), pp. 1–31).
2. *thinking of me:* letter missing; it appears that Tyndall sent Lortet one of the photographs taken in Norwich (see letter 2864, n. 2).
3. *Riffel:* the Riffelhaus. See letter 2856, n. 2.
4. *I was very fortunate to see you this year, appearing before us in the flesh on the scene of your work:* this letter indicates that Tyndall and Lortet saw each other at Riffelalp in the Alps the previous summer.

To Jane Barnard 22 September 1868 2879

22nd. Sept[embe][r]. 1868

My dear Miss Barnard.

I suppose you have seen that my little note[1] was inserted in the last number of the Mechanics Magazine. I hope it met your views.

The Editor of that journal[2] has sent me a note[3] which I now forward to you.

The editor of the Engineer[4] will I feel sure, not hesitate to insert a note from me on the same subject.[5] If you think the writing of the note desirable I will do it at once. I hesitate to do any thing without your permission else I sh[oul][d]. write at once to the Engineer.

I have made some enquiries about the “odds & ends”—If you will allow me to say so I think some of them were unwisely given away. Though nothing I believe that could illustrate your uncle’s power as a constructor was thus bestowed many small scraps with which he experimented were scattered, which would have found their natural place here.

The sword of Davy[6] is a very interesting relic. It would be highly valued by many—more highly I fear than poor William[7] can be expected to value it.

Yours very truly | John Tyndall

RI MS JT/1/T/111
RI MS JT/1/TYP/12/4199

---

1. *my little note:* letter 2874.
2. *Editor of that journal:* see letter 2874, n. 1.
3. *a note:* the enclosure—a letter from the editor of *The Mechanics' Magazine*—is missing.
4. *editor of the Engineer:* while Vaughan Pendred (1836–1912) was from 1865–1905 the editor of *The Engineer,* a London-based weekly technical magazine founded in 1856 by Edward Charles Healey (1822–1906) (‘Mr. Vaughan Pendred’, *Times,* 15 October 1912, p. 9), the letter Tyndall referred to was from one of the magazine’s owners, Charles Chadwyck-Healey (see letter 2891, n. 1).
5. *a note from me on the same subject:* while the paragraph about Faraday’s apparatus in *Mechanics' Magazine* also appeared as ‘Faraday’s Royal Institution Apparatus’, *The Engineer,* 21 August 1868, p. 149, it does not appear that Tyndall’s response was similarly published by the magazine, for its owner wrote asking Tyndall to ‘let the subject rest’ (see letter 2891).
6. *sword of Davy:* a presentation sword belonging to Humphry Davy, who was made baronet in 1818. The sword came into the possession of his younger brother John Davy (1790–1868), an anatomist and physiologist (*ODNB*), who lent it to William Wordsworth when he was installed as Poet Laureate in 1843 (A. Treneer, *The Mercurial Chemist: A Life of Sir Humphry Davy* (London: Methuen & Co., 1963), p. 254). John Davy died on 24 January 1868, so perhaps the sword was offered to the RI soon before or after his death.
7. *William:* William Brown.

From Jane Barnard[1] 22 September 1868 2880

12 Barnsbury Villas[2] | N. | 22 Sept. 1868

Dear Dr Tyndall

Last evening Mr Vincent[3] brought to me the no. of the Mecha[nics’] Mag[azine]. containing your note.[4] It has relieved my mind much, for I need not say how anxious I felt that such a false report[5] should be noticed.

Encouraged by one or two friends I have told my Aunt[6] nothing about it, for I know it would have pained & annoyed her exceedingly; so that I want you to accept my thanks on her behalf, as well as on my own for your friendly assistance in this affair.

Believe me yours | very sincerely obliged | Jane Barnard | John Tyndall Esq | &c &c &c

RI MS JT/1/B/40

1. Barnard enclosed this letter with 2881.
2. *Barnsbury Villas:* see letter 2695, n. 2.
3. *Mr Vincent:* Benjamin Vincent (see letter 2873, n. 3).
4. *your note:* see letter 2874.
5. *false report:* see letter 2874, n. 2.
6. *my Aunt:* Sarah Faraday.

From Jane Barnard 22 September 1868 2881

22nd Sept. 1868. | 12 Barnsbury Villas[1] | Tuesday Evening.

Dear Dr Tyndall

I had written the enclosed note[2] when yours[3] to me came in this afternoon!

I cannot help feeling much concerned about the things that passed into Williams[4] hands; though at the time, both my Aunt[5] and I, assisted by the judgment of those we thought highly of, believed all had been done for the best.

As to the sword,[6] neither she nor I knew that it had belonged to Davy, & how it has been discovered now I cannot tell. I think it very likely did but it had been up in the loft over the kitchen for many many years disregarded & uncared for by my uncle.[7] I could wish we had not done things so hurriedly but now I do not see what further can be done.

I feel much inclined & that it would be a great relief to me to tell all this to my Aunt but I know the paragraph would annoy her, & pain her, exceedingly, so that I must be quiet unless it grows more serious.

Knowing these circumstances as you do I think your note[8] was all we could wish for & I again thank you for doing it. As to your kind offer about the "Engineer"[9] I would ask you to decide what would be best; but for my own part as far as I can judge I should think the matter had better be allowed to drop. As, Dr Bence Jones & others reminded me, anyone who saw or heard of the collection of his apparatus shewn one evening last season at the R.I.[10] would know how false the paragraph was. I could not rest without knowing that my Aunt's name had been cleared, but now you have done so thoroughly, I shall try & be quiet.

I am glad you told me your thoughts about all this,—but can I do anything to repair the error now? I may be at the West end in a day or two & would call if desirable. I feel thankful that we gained nothing by all this.

Believe me yours | very sincerely obliged— | Jane Barnard

John Tyndall Esq | &c &c &c | 22 Sept. 1868

RI MS JT/1/B/41

---

1. *12 Barnsbury Villas:* see letter 2695, n. 2.
2. *enclosed note:* letter 2879.
3. *yours:* letter 2879.
4. *Williams:* William Brown.
5. *Aunt:* Sarah Faraday.
6. *sword:* see letter 2879, n. 6.
7. *my uncle:* Michael Faraday.
8. *your note:* see letter 2874.
9. *"Engineer":* see letter 2879, n. 5.
10. *apparatus shewn one evening last season at the R.I.:* this possibly refers to the evening of Tyndall's first of two Friday Evening Discourses on 'Faraday as a Discoverer' on 17 January 1868. Mention of apparatus on display at the RI can be found in the *Morning Post*'s article about the discourse: 'In the library there were placed a very good portrait of Faraday and a photograph of him taken when delivering one of his Christmas lectures. In addition to many other objects of interest, there was exhibited a magneto-electrical machine, arranged so as to be a test of muscular strength, the power applied to turn the magnets being indicated by the degree of heat imparted to a platina wire lever' ('Royal Institution', *Morning Post,* 18 January 1868, p. 6).

From Charles Darwin 7 October 1868 2882

Down. | Bromley. | Kent S. E. | Oct 7. 1868

Dear Tyndall

Professor G. Hinrichs[1] of Iowa some time ago sent me a letter[2] in which he described how he had worked & sacrificed every thing, almost to the last dollar in getting his chemical & molecular views known. He sent me also a lecture on Religion & Science[3] which seemed to me good, and in some points original.

I answered that I c[oul]d form no judgment on his work, but expressed sincere sympathy for his up hill labour. Now I am punished by receiving copies of all his publications, including the chief one[4] of which only 112 copies were printed, together with some circulars.[5] I write to ask you whether you will accept the whole bundle? and if you will I shall feel truly grateful. W[oul][d] you object to place one of the circulars on the tables of the Royal. Inst[itution]., and Athenaeum;[6] & if any one s[houl][d] call on you who cares for such subjects, give him a copy. I ask all this to ease my own conscience & help a poor devil if that is possible.

Many thanks for your address at Norwich,[7] rec[eive][d] several weeks ago, which I read with the greatest interest, & which I am told excited great admiration at Norwich.

By the way, I must add how much I admired, and how entirely I agreed with you in a paper[8] published a long time ago, I think in the Fortnightly, in which you enlarged on the wonderful power of pondering; I believe you have hit on the whole secret of scientific discovery.

Forgive me for troubling you | & believe me yours very sincerely | Charles Darwin

RI MS JT/1/TYP/9/2805
Darwin Correspondence, vol. 16[9]

---

1. *G. Hinrichs:* Gustavus Hinrichs.
2. *letter:* Gustavus Hinrichs to Charles Darwin, 31 August 1868, *Darwin Correspondence,* vol. 16, part 2, pp. 709–10; this is letter 6337 in the online edition of the Darwin correspondence.
3. *sent me also a lecture on Religion & Science:* a 'Sunday lecture' that Hinrichs delivered to students of Iowa State University in 1867; published in the *Iowa State Press,* 25 December 1867 (Gustavus Hinrichs to Charles Darwin, 13 August 1868, *Darwin Correspondence,* vol. 16, part 2, pp. 671; this is letter 6313 in the online edition of the Darwin correspondence).
4. *chief one:* G. Hinrichs, *Atomechaniks* (Iowa-City, 1867).
5. *some circulars:* see the enclosure with the letter cited in n. 2.
6. *Athenaeum:* see letter 2537, n. 8.
7. *address at Norwich:* see letter 2860, n. 4.
8. *paper:* J. Tyndall, 'Miracles and Special Providences', *Fortnightly Review,* 7 (1867), pp. 645–60. On p. 655, Tyndall wrote: 'There is much in this process of pondering and its results which it is impossible to analyse. It is by a kind of inspiration that we rise from the wise and sedulous contemplation of facts to the principles on which they depend'.
9. *Darwin Correspondence, vol. 16:* our transcription has been checked against that of the Darwin Correspondence Project. See Charles Darwin to John Tyndall, 7 October 1868, *Darwin Correspondence,* vol. 16, part 2, pp. 790–1; this is letter 6413 in the online edition of the Darwin correspondence.

To Charles Darwin 9 October 1868 2883

Royal Institution of Great Britain | 9th. Oct. 1868

My dear Darwin,

Hinrichs[1] is also a correspondent of mine. Had he trusted more to the natural weight of his views if they have any and less to the policy of making a noise about them he would in my opinion have acted more wisely than he has done.

He has published an attack upon Dana[2] which I should not like to circulate as he shows a temper not to be trusted where cool judgment is required. But you can ease your conscience by sending his papers to me,[3] and I will place almost the whole of them on the table of the Royal Institution. You can tell him that you have sent them to me.

I was rejoiced to hear from dear old Hooker[4] on Thursday week that you were able to have a little party of your friends[5] about you, and then again, Bates[6] dashed my happiness by saying that you were not so well as you ought to be. I hardly think that even your naturalists care more about hearing of your improved health than I do.

That charming man Asa Gray,[7] and his still more charming wife[8] are staying with Hooker. I dined there on Wednesday and endeavoured to make it clear to Mrs. Gray that your ill health was a benefit to you inasmuch as it compelled you to ponder a great deal,[9] and this accounted for the extraordinary proportion of thought which your works display.

Goodbye: I hope you will continue to flourish.

Ever Yours | John Tyndall.

RI MS JT/1/TYP/9/2806
Darwin Correspondence, vol. 16[10]

---

1. *Hinrichs:* Gustavus Hinrichs.
2. *attack upon Dana:* Hinrichs accused American geologist James Dwight Dana (1813–95) of plagiarising his *Atomechanik* (see letter 2882, n. 4) in his papers on crystalline form and chemical composition. This was reported in *Quarterly Journal of Science*, 5 (1868): pp. 257–8) (Charles Darwin to Gustavus Hinrichs, 13 August 1868, *Darwin Correspondence*, vol. 16, part 2, pp. 671, this is letter 6313 in the online edition of the Darwin correspondence).
3. *sending his papers to me:* see letter 2882.
4. *old Hooker:* Joseph Hooker.
5. *little party of your friends:* presumably Hooker had told Tyndall of Darwin's naturalist visitors to Down House on 12–13 September 1868 (John Tyndall to Charles Darwin, 9 October 1868, *Darwin Correspondence,* vol. 16, part 2, pp. 792–3, n. 4; this is letter 6414 in

the online edition of the Darwin correspondence). According to Darwin, the visitors were to be Alfred Russel Wallace and his wife Annie Wallace (1845–1916), entomologist and ornithologist John Jenner Weir (1822–94), zoologist Edward Blyth (1810–73), and Henry Walter Bates (see n. 6), although Bates was not able to come (Charles Darwin to Joseph Hooker, [8–10 September 1868], *Darwin Correspondence,* vol. 16, part 2, pp. 731–2; this is letter 6357 in the online edition of the Darwin correspondence).

6. *Bates:* Henry Walter Bates (1825–92), an English naturalist who spent eleven years studying and collecting specimens in Amazonia (1848–59), initially in the company of Alfred Russel Wallace. Bates is known for developing the biological concept of mimicry (*ODNB*).
7. *Asa Gray:* Asa Gray (1810–88), an American botanist and professor of natural history at Harvard University, whose collections formed the Gray Herbarium (*ANB*).
8. *wife:* Jane Lathrop Gray (née Loring, 1821–1909), a daughter of a Boston lawyer, she married Asa Gray in 1848 (L. A. DeCesare, 'Jane Lathrop Loring Gray (1821–1909) and the Archives of the Gray Herbarium', *Harvard Papers in Botany,* 15 (2010), pp. 221–30).
9. *ponder a great deal:* see letter 2882, n. 8.
10. *Darwin Correspondence, vol. 16:* our transcription has been checked against that of the Darwin Correspondence Project. See John Tyndall to Charles Darwin, 9 October 1868, *Darwin Correspondence,* vol. 16, part 2, pp. 792–3; this is letter 6414 in the online edition of the Darwin correspondence.

From Thomas Carlyle 13 October 1868 2884

Chelsea,[1] 13 Oct[obe][r] 1868

Dear Tyndall,

I have not seen Burgess[2] for a good while; cannot discover his address in any of the repositories here, nor quite accurately recollect it;—but think rather confid[en][tly], If you put

"12(?) Park Walk, Chelsea",

it w[oul][d] find him.—By letter perhaps, at the Pall Mall Gazette[3] Office, they are pretty sure to know it with precis[io][n]. Don't neglect any chance for the poor fellow!—

I have been in much dietetic[4] &c botherat[io][n5] of late days (and ending, let us hope), or I sh[oul][d] have been to see you.

Y[ou][rs] ever | T. Carlyle

RI MS JT/1/C/13
CLO, 45 (13 October 1868, TC to John Tyndall)

1. *Chelsea:* Carlyle's London home at 5 Cheyne Row in Chelsea since 1834.
2. *Burgess:* Andrew Hutton Burgess (b. 1835), a Scottish journalist who in 1868 was lodging at 36 Park Walk, Chelsea (*CLO*).

3. *Pall Mall Gazette:* a daily newspaper founded in London in 1865 by George Murray Smith; a number of letters in this volume come from the *PMG.*
4. *diatetic:* of or pertaining to diet (*OED*).
5. *botherat[io]ⁿ:* the state or condition of being bothered; irritation, vexation; inconvenience (*OED*).

To John Stuart Blackie[1] 15 October 1868 2885

Royal Institution of Great Britain | 15th. Oct. 1868.

My dear Prof. Blackie

A sermon[2] which I have received from the Rev[eren]d. J. T. Goodsir[3] reminds me of my duty to you, telling me that I ought long ago to have replied to the long and admirable letter which you sent to me from Oban.[4]

I read that letter with thorough interest. It hit me hard but there was a little padding of affection round the knuckles which so distributed the impact as to prevent it from seriously hurting me.

There is also mauling in this preface of Goodsir's—His warmth carries him a little too far sometimes, but this is easily forgiveable.

One of the objects I had in view in the composition of my Norwich address[5] was to keep open for the faith and hope and pleasure of loving and religious men, a horizon the shutting up of which by science they dreaded.

I really think the effect produced on the great majority of those who heard the address was one of relief. That it should be printed & circulated was proposed by a clergyman of the Church of England, a distinguished Oxford Professor.[6] And on the last day it was referred to by a dissenting minister[7] in terms which seemed to indicate unqualified gratification

I really thought it a very great thing that a man could utter the thought that was in him on such a subject with such a result

It is not for the arrangement of this universe which we should like best that we ought to combat, but for that which the exercise of our best reasons shows us to be the true arrangement. I might fail if I attempted to give you a philosophical reason for it, but my belief is that truth, however uncomfortable in appearance, will in some way or other vindicate itself as man's benefactor in the end.

I should dearly like to visit you in your highland nest.[8] However I might fret in giving structural form to the power which unfolds us all I *[1 word illeg]* your feelings regarding it would not be utterly dark to me Indeed at bottom our difference may be one of "form" rather than of substance.

You refer to Goethe.[9] It was Goethe who said

"Wer darf Ihn nennen | Und wer bekennen | Ich glaub Ihn?"[10]

Goodbye | Yours ever | John Tyndall | Prof. Blackie | &c. &c. &c.

NLS, Blackie Papers, Ms. 2628

---

1. *John Stuart Blackie:* John Stuart Blackie (1809–95), a Scottish classical scholar and Professor of Greek at Edinburgh University (*ODNB*).
2. *sermon:* in September Goodsir published J. T. Goodsir, *The Divine Rule Proceeds by Law; an Old Sermon, Now First Published, with a New Preface, Bearing on the Philosophic and Theological Tendencies of Professor Tyndall's Introductory Address Lately Delivered at Norwich* (London: Williams & Norgate, 1868).
3. *Rev[eren]d. J. T. Goodsir:* Joseph Taylor Goodsir (1815–93), a Scottish theologian and minister at Largo, Fife (C. D. Waterston and A. Macmillan Shearer (eds), *Former Fellows of the Royal Society of Edinburgh 1783–2002,* 2 vols (Edinburgh: Royal Society of Edinburgh, 2006), vol. 1, p. 366).
4. *long and admirable letter which you sent to me from Oban:* letter missing.
5. *my Norwich address:* see letter 2860, n. 4.
6. *clergyman of the Church of England, a distinguished Oxford Professor:* not identified.
7. *dissenting minister:* not identified.
8. *your highland nest:* 24 Hill Street in Edinburgh, Scotland (A. M. Stoddart, *John Stuart Blackie: A Biography* (Edinburgh and London: William Blackwood and Sons, 1896), p. 221).
9. *Goethe:* Johann Wolfgang von Goethe (1749–1832), the outstanding German literary figure of the late eighteenth and early nineteenth centuries. His extensive writings included his theory of colours and a study of the structure and function of plants. During his time in Marburg, Tyndall became familiar with some of Goethe's works (J. Williams, *The Life of Goethe: A Critical Biography* (Oxford: Blackwell, 1998).
10. *"Wer darf Ihn nennen | Und wer bekennen | Ich glaub Ihn?":* from Goethe's *Faust,* which translates to 'Who dares name the nameless? | Or who dares to confess: | 'I believe in him'?' (J. W. von Goethe, *Faust* (München: Friedr. Bruckmann's Verlag, 1847), p. 104).

From Felix Giordano[1] 16 October 1868 2886

CORPO REALE | delle | MINIERE | Florence le 16 octobre 1868

Estimable Professeur Tyndall

Le 4 7bre dernier j'ai eu l'occasion de faire une ascension du Mt Cervin avec beau temps, quoique le pic fut un peu difficile à cause de la neige abondante et nouvelle qui en couvrait la partie supérieure. J'ai répété votre course, c'est-à-dire je suis monté de Breil et descendu à Zermatt. Je portais mon baromètre à mercure avec lequel j'ai pu faire la détermination d'altitude soit du sommet que des principaux épaulements. Je me permets, en signe de respect, de vous envoyer une esquisse de la coupe géologique du pic que j'ai pu relever assez exactement. Les altitudes sont marquées à côté. (a note dernière).

Je vous prie d'agréer mes sentiments d'estime—

Félice Giordano | Inspecteur des Mines

2 Via Venezia | Florence

(a)—L'altitude du sommet 4505 résulte un peu plus forte que celle 4488 donnée par la carte Suisse / Dufour / et qui a été déterminée trigonométriquement. En général les altitudes données par le baromètre des hauts pics se sont trouvées un peu plus fortes que celle déterminées par la trigonométrie. Du reste mon altitude de 4505 résulte soit de la comparaison faite avec la station du Breil, que celle de l'observatoire d'Aoste.

La Baraque de refuge du côté de Breil résulte de 305 *[met.]* plus haut que celle du côté de Suisse.

Quant à la section géologique elle n'a pas beaucoup d'importance si on la considère isolement; mais elle en acquiert beaucoup en la rattachant à celle des montagnes voisines. Cela mériterait un mémoire spécial que je ferai si j'en aurai le temps.

Royal Corps | of | Mining | Florence 16 October 1868

Estimable Professor Tyndall,

On September 4th I had the opportunity to ascend Mt Cervin[2] during good weather, although the peak was a little difficult because of the abundant fresh snow that covered the upper portion. I retraced your path, that is to say, that I climbed from Breil[3] and descended to Zermatt. I was wearing my mercury barometer with which I was able to determine the altitude of the summit as well as of the main escarpments. As a sign of respect, allow me to send you a sketch of the geological section of the peak[4] which I could measure precisely enough. The altitudes are marked on the side. (a final note).

Please accept my regards of esteem—

Félice Giordano | Inspector of Mines

2 Via Venezia | Florence

(a)—The altitude of the summit 4505 results in a slightly higher number than that 4488 which was given by the Switzerland map / Dufour[5] / and which was determined trigonometrically. In general, altitudes of high peaks that are given by barometers, are found to be a little higher than those determined by trigonometry. As for the rest, my altitude of 4505 results from the comparison made with the station of Breil, rather than the observatory of Aoste.[6]

The refuge shack on the Breil side comes to 305 *[meters]* higher than that on the Swiss side.

As for the geological section, it is not very important if we think of it separately; but it gains a lot of significance when we attach it to the section of the neighbouring mountains. That deserves a special report[7] which I will write if I have the time.

RI MS JT/1/G/5

---

1. *Felix Giordano:* Felix Giordano (1825–92), an Italian engineer and mountaineer who worked in the Mining Corps of Italy, becoming Inspector General in 1870 (*Dizionario Biografico Degli Italiani,* 94 vols (Rome: Istituto della Enciclopedia Italiana, 1960–2019), vol. 55, pp. 264–6).

2. *Mt Cervin:* the Matterhorn.
3. *Breil:* see letter 2622, n. 37.
4. *sketch of the geological section of the peak:* missing. The sketch, 'Section Géologique des environs du Mt. Cervin', was included as a foldout at the end of the issue in which Giordano's paper on the geologic structure of the Matterhorn appeared (see n. 7).
5. *Switzerland map / Dufour:* the first topographic map of Switzerland, known as the Dufourkarte, published in 25 leaves (1844–64) under the supervision of Guilliame Henri Dufour (1787–1875), a Swiss army officer and engineer (*HLS*).
6. *the station of Breil, rather than the observatory of Aoste:* meteorological stations situated above the Italian towns of Breuil and Aosta.
7. *special report:* F. Giordano, 'Notice sur la constitution géologique du Mont Cervin', *Archives des sciences physiques et naturelles,* 34 (1869), pp. 255–67.

To George Gabriel Stokes 16 October 1868 2887

16th Oct. 18<u>68</u>

My dear Stokes,

The paper[1] of Mr. Tomlinson[2] is of undoubted value, because it clears away a great deal of the nonsense which has hitherto surrounded the question of supersaturated solutions.

It is however a very difficult paper to report upon. I am strongly inclined to think that if it were again submitted to Mr. Tomlinson with the recommendation to omit no fact or statement of real importance, and still to cut down the paper to at most one half its present volume, and if Mr. Tomlinson could be induced to act upon this recommendation, the Philosophical Transactions, the reader of the paper, and Mr. Tomlinson himself would be gainers.

If I mistake not moreover a good deal of what appears but has already appeared in print in other places.

There is one phrase in the paper of which Mr. Tomlinson appears to be particularly fond, the use of which I think is not unattended by danger. That is the phrase "chemically clean surface" This phrase I believe is meant to sum up Mr. Tomlinson's views as regards the cause of crystallization in supersaturated solutions. It would add I think greatly to the real scientific interest of his paper if he could be induced to omit this phrase totally, making such alterations on the phraseology of the paper as would be rendered necessary by the omission.

"No previous inquirer" says Mr. Tomlinson at page 37 of this manuscript "seems to have had any precise notion as to what was meant by a nucleus". Nor have I after reading Mr. Tomlinson's paper any precise notion of what is meant by a nucleus. The phrase "chemically clean surface" is rather a disguise than an illumination. What is the scientific definition of <u>dirt</u>? It seems exceedingly probable that the organic specks floating in the air exercise a powerful effect in provoking crystallization, but M[r]. Tomlinson seems to think they do this because they are not clean. If so what dirties them? Could we but known this we should be nearer to a precise conception of a nucleus than we now are.

Notwithstanding however the drawback of wordiness, and the vagueness of the writer's statements on the question of dirt, I am of opinion that M[r]. Tomlinson's paper embodies a valuable addition to our knowledge, and it could be readily thrown into a form which would render it well worthy of appearing in the Philosophical Transactions.

Yours very truly | John Tyndall

Prof. Stokes. &c. &c.

1868 | Tomlinson | by Tyndall[3]

RS RR/6/285 Tyndall 16 October 1868

---

1. *paper:* C. Tomlinson, 'On Supersaturated Saline Solutions', *Phil. Trans.* 158 (1868), pp. 659–73.
2. *M*[r]*. Tomlinson:* Charles Tomlinson (1808–97), a British science teacher and prolific author who did research on the surface tension of liquids. With his brother Lewis he ran a day-school in Salisbury, and later became a lecturer on experimental science at King's College School. In 1872 Tomlinson was elected FRS (*ODNB*).
3. *1868 | Tomlinson | by Tyndall:* this is written on the reverse of the letter, probably by Stokes.

From Charles Darwin 20 October 1868 2888

Down. | Bromley. | Kent. S. E. | Oct. 20 1868

My Dear Tyndall

The Asa Grays[1] and Hookers[2] are coming here on Saturday next, and it would give us all very great pleasure if you c[oul][d] spare the time to come to us on that day & stay the Sunday with us.[3]

We dine at 7.30—and your best route will be by S. E. Railway from Charing Cross[4] to Orpington station,[5] which is 4 miles from this house.[6] Very many thanks for your note about Hinrichs[7] whose papers I sent to you.[8]

I am sure you will excuse me in taking the privileges of an invalid & often leaving my guests.

My dear Tyndall | yours very sincerely. | Charles Darwin

P.S. You w[oul][d] find 4.18. or 5.8. from Charing Cross convenient trains & if you will let me know which I will send a fly[9] from here—

If you come to Bromley (6 miles off) flys are plentiful

RI MS JT/1/TYP/9/2807
Darwin Correspondence, vol. 16[10]

---

1. *Asa Grays:* Asa and Jane Lathrop Gray (see letter 2883, n. 7 and n. 8).
2. *Hookers:* Joseph Dalton and Frances Hooker.
3. *stay the Sunday with us:* Tyndall did stay at Down House. According to the Darwin Correspondence Project, 'Emma Darwin reported in her diary (DAR 242) that the Hookers, the Grays and Tyndall, among others, arrived at Down on 24 October 1868' (Charles Darwin to John Tyndall, 20 October 1868, *Darwin Correspondence*, vol. 16, part 2, p. 806, n. 1; this is letter 6424 in the online edition of the Darwin correspondence).
4. *Charing Cross:* a train station in central London.
5. *Orpington station:* a train station in the town of Orpington, southeast of London.
6. *this house:* Down House in the village of Downe, southeast of London, Darwin's home from 1842–82.
7. *Hinrichs:* Gustavus Hinrichs. See letter 2883.
8. *papers I sent to you:* see letter 2882.
9. *fly:* one-horse covered carriage; cab (*OED*).
10. *Darwin Correspondence, vol. 16:* our transcription has been checked against that of the Darwin Correspondence Project. See Charles Darwin to John Tyndall, 20 October 1868, *Darwin Correspondence*, vol. 16, part 2, p. 806 (this is letter 6424 in the online edition of the Darwin correspondence).

To Elizabeth Dawson Steuart 20 October 1868 2889

Royal Institution. | 20th. Oct. 1868.

My dear Mrs Steuart,

Thank you very much for sending me the note[1] to Dr. Trayer[2] (to whom remember me kindly) I have had several letters of the same kind from Gorey.[3] They are what I expected for I know that Emma's[4] usual habits as to food were so calculated to act on the brain through the stomach as to produce the effects observed.

But how these habits were to be overcome I knew not. She would nibble at bread crumbs all day and declare that she could not bear mutton. But my friend Tyndall[5] has succeeded wonderfully in curing her of this. His letters breathe true kindness and it is the greatest possible comfort to me to know that she is in good hands.

I hope Sarah Tyndall[6] will make good use of her farm.[7] I liked what I saw of her very much.

I am working very hard, but happily for some time to come I shall be in a great measure my own master, so that if I feel any symptoms of breaking down I shall be able to run away to the Isle of Wight[8] to restore myself.

My friend Brodie,[9] the present Sir Benjamin is very desirous that I should take a house near him in the country. He lives in an indescribably beautiful place in Surrey.[10] But it is an hour and a quarter from town, and the rent is terribly high.

Goodbye. I am going to begin my experiments for the day

Yours ever | John Tyndall.

RI MS JT/1/TYP/10/3383
LT Typescript Only

---

1. *note:* letter missing.
2. *Dr. Trayer:* James John Trayer.
3. *several letters of the same kind from Gorey:* letters 2734 and 2768.
4. *Emma's:* Emma Tyndall.
5. *my friend Tyndall:* John Tyndall (of Gorey).
6. *Sarah Tyndall:* Sarah Tyndall (not Tyndall's mother; see Biographical Register).
7. *her farm:* Tyndall had been discussing with Steuart arrangements for providing a dowry to Sarah Tyndall (see letters 2634, 2777, 2810, 2857, and 2877).
8. *Isle of Wight:* Tyndall went to the Isle of Wight at the beginning of November. He wrote three letters to Hirst from there between 3–6 November (2895, 2897, and 2898).
9. *Brodie:* Benjamin Collins Brodie (1817–80), second baronet, an English chemist and Aldrichan Chair at Oxford University from 1865–72 (*ODNB*).
10. *indescribably beautiful place in Surrey:* at Broome Park, Brodie's country home in Surrey, a county in southeastern England.

From Carl Schorlemmer[1] 24 October 1868 2890

Owens College | Manchester | Oct[o]b[e]r 24th 1868

Dear Sir

I have sent you to day the following hydrides.

(1) C_5H_{12} Amylhydride boil point. 38–40C prepared from Boghead naphtha. This hydrocarbon is different from the hydride obtained from amylalcohol as its boiling point is 10 higher

(2) C_6H_{14} Hexylhydride boil point 70 from american Rockoil

(3) C_7H_{16} Heptylhydride boil point 99.5–100 from Rockoil

(4) C_8H_{18} Octyhydride boil point 123–125 from Rockoil

The boiling point of this hydride is generally stated to be at 116–119 I found however that on treating this lower boiling fluid with hot concentrated nitric acid to remove all impurities & then subjecting it to continued fractional distillation the boiling point gets raised to 123°–124°

(5) $C_{18}H_{18}$ from Octyl alcohol boil point 124

Appears to be identical with № 4

I shall feel always most happy if I can render you any assistance in your splendid researches[2] and | remain | yours faithfully | C. Schorlemmer

RI MS JT/3/48/379

---

1. *Carl Schorlemmer:* Carl Schorlemmer (1834–92), a German chemist and laboratory assistant to Henry Roscoe at Owens College, Manchester. In 1871 he was elected FRS, and in 1874 became Professor in Organic Chemistry at Owens College (*ODNB*). Along with several chemistry textbooks, in 1879 Schorlemmer published *The Rise and Development of Organic Chemistry* (J. E. Cornish: Manchester).
2. *splendid researches:* scholars interested in the details of Tyndall's experiments on the effect of light on vapours and their connection to the blue colour of the sky, the polarization of light, and the formation of clouds should consult his laboratory notebook (RI MS JT/3/48/99–208). These experiments would result in the following publications: J. Tyndall, 'On a New Series of Chemical Reactions Produced by Light', *Roy. Soc. Proc.,* 17 (1868–9), pp. 92–102 (received 24 October 1868); J. Tyndall, 'On the Blue Colour of the Sky, the Polarization of Skylight, and on the Polarization of Light by Cloudy Matter Generally', *Roy. Soc. Proc.,* 17 (1868–9), pp. 223–33 (received 16 December 1868); and J. Tyndall, 'Note on the Formation and Phenomena of Clouds', *Roy. Soc. Proc.,* 17 (1868–9), pp. 317–19 (received 25 January 1869).

From Charles Chadwyck-Healey[1] 27 October 1868 2891

OFFICE OF "THE ENGINEER"[2] | 163, STRAND, LONDON. W. C. | 27 October 1866

My dear Sir,

You may remember that some conversation passed between us, when I saw you last, respecting the paragraph, which appeared in the Engineer, about Mr Faraday & the apparatus.[3] Since then I have called to task the correspondent from whom we received the announcement, & he still maintains that he was right in the main & further strengthens his position by stating that some relics of Faraday may still be obtained from the Porter[4] of the Institution.[5] That these relics are practically of no importance or are intrinsically worthless I do not for one moment doubt, but it seems to me to be first worth considering whether it would be well to re-call public attention to the matter at all. By this time probably the whole thing is forgotten & although I do not doubt that the paragraph in question was exaggerated & should not have gone in I fancy it would be best to let the subject rest.

What do you think?

I called at Albemarle[6] about ten days ago, to explain matters but was unfortunate in not seeing you.

very truly yours | C E H Chadwyck Healey

D[r] Tyndall FRS | &c &c.

RI MS JT/1/H/73

1. *Charles Chadwyck-Healey:* Charles Edward Heley Chadwyck-Healey (1845–1919), a British lawyer. Before being called to the bar in 1872, he took over ownership of *The Engineer* on the passing of his father Edward Charles Healey (C. Mosley (ed), *Burke's Peerage, Baronetage & Knightage,* 107 edn, 3 vols (Wilmington, Delaware: Genealogical Books, 2003), vol. 1, p. 740).
2. *"The Engineer":* see letter 2879, n. 4.
3. *paragraph, which appeared in the Engineer, about M[r] Faraday & the apparatus:* see letter 2879, n. 5.
4. *Porter:* probably William Brown (see letter 2874, n. 4).
5. *Institution:* the RI.
6. *Albemarle:* the RI was located on Albemarle Street in London.

From Jonathan Frederick Pollock — 27 October 1868 — 2892

Hatton, | Hounslow.[1] | Tuesday | 27th. Oct 1868

My dear Sir

I have just read your "Faraday as a discoverer"[2] and it has given me so much pleasure & delight that I cannot forbear to tell you so—& I hope you will forgive the intrusion—what is sentimental in it has been to me as interesting as what is dry Philosophy—I have been a reader of Encyclopaed[ia][s] & such books since I was 14 or 15 & nothing is more gratifying than to see fact after fact come out in such experiments as Faraday's—the ultimate result is not yet satisfactory—you cannot write such a book about sight as you have written about sound[3]—but why not go as far as you can. I observe (& with pleasure) you mention "the luminiferous" "ether"—which probably exists (as there seems no alternative between the vibration of a fluid & the emanation of particles of light—) & the existence of which seems to me not merely probable but apparently proved by Encke's comet[4] moving apparently in a resisting medium I am told the existence of such a medium is altogether denied by some—it is very singular that not long before Lavoisier[5] & Priestly[6] published their experiments about air & gases[7] Voltaire[8] wrote an article in the French

Encyc<u>lop</u>[aedi][a] entitled l'<u>Air</u>[9] to prove that no such thing as <u>air</u> existed & <u>repudiating the notion</u> that <u>it is the cause of sound</u>!!—as a matter of curiosity (nothing else) it is worth while to look at the article

believe me truly yours | Fred Pollock

RI MS JT/1/P/158

---

1. *Hatton, | Hounslow:* Hatton was Pollock's family estate in Hatton, Hounslow in the county of Middlesex.
2. *"Faraday as a discoverer": Faraday as a Discoverer.*
3. *sound: Sound* (1867).
4. *Encke's comet:* Comet Encke (or 2P/Encke) is a periodic comet that completes an orbit of the Sun once every 3.3 years. For a history of observations of this comet, see Z. Sekanina, 'Encke, the comet', *Journal of the Royal Astronomical Society of Canada,* 85 (1991), pp. 324–376.
5. *Lavoisier:* Antoine Lavoisier (1743–94), a French chemist best known for his identification and naming of oxygen and for his quantitative experiments and reform of chemical nomenclature (*CDSB*).
6. *Priestly:* Joseph Priestley (1733–1804), an English chemist and theologian known as the discoverer of oxygen which he called "dephlogisticated air" (*ODNB*).
7. *published their experiments about <u>air</u> & <u>gases</u>:* for an account of the experiments of Priestley and Lavoisier in the 1770s, see A. Donovan, *Antoine Lavoisier: Science, Administration, and Revolution* (Cambridge: Cambridge University Press, 1996), pp. 138–57.
8. *Voltaire:* François-Marie Arouet (1694–1778), known as Voltaire, a French philosopher and writer who was a central figure in the French Enlightenment and contributed to Denis Diderot's *Encyclopédie* (1751–72) (*ODNB*).
9. *article in the French Encyc<u>lop</u>[aedi][a] entitled l'<u>Air</u>:* while the first volume of the *Encyclopédie* (1751) contains an article on 'Air', this has not been attributed to Voltaire ('Air', *l'Encyclopédie ou Dictionnaire raisonné des sciences, des arts et des métiers,* 17 vols (Paris, 1751–72), vol. 1, pp. 225–36).

From Jonathan Frederick Pollock — 30 October 1868 — 2893

Hatton, | Hounslow.[1] | 30th. Oct[obe][r]. | 1868

My dear Sir

Do not be alarmed—I have not the presumption to propose myself as a scientific correspondent,[2] altho' always a disciple of Science I am aware that I am in my 86th. year & I feel it in this that tho' I fancy I can grasp & deal with a single proposition at least as well as I could 40 years ago—I cannot deal with a group of phenomena & bring many experiments to a single conclusion—but there are some remarks, which I wish to make to you.—

Magnetism seems to be (to borrow an expression from political economy) a self-sustaining power—electricity seeks an equilibrium & disperses itself, but a magnet tho' it distributes magnetism—loses none & at the end of an indefinite period is as magnetic as ever—Is there any difference between the light electricity gives & that given by magnetism—is the latter polarized?—

I remember seeing many years ago a statement that if an indefinite N°. of small particles were moving in all directions at an indefinitely great velocity they w[oul][d] make all atoms of bodies move towards each other as gravity does that is as $\frac{\text{mass}}{\text{dist}^2}$—now electricity is the "Anima Mundi"[3]—it appears in every operation—crystalline chemical or organic whether vegetable or animal—will not the eternal working of Electricity in all directions produce what we call gravity?—Is there not Evidence (my judicial life daily has made me quite aware of the difference between Evidence & proof) enough for a treatise on light as the result of the luminiferous fluid—a stone thrown into a Pond produces an enlarging circular vibration in the water below, it is spherical no doubt—but a Ray of light (as stated in your book[4]) produces everywhere vibrations perpendicular to itself in all directions—I wish Faraday had given us his doubts & suggestions—I fear he was too afraid of being wrong.

Yours truly | Fred Pollock

RI MS JT/1/P/159

1. *Hatton, | Hounslow:* see letter 2892, n. 1.
2. *Do not be alarmed—I have not the presumption to propose myself as a scientific correspondent:* presumably Tyndall had replied to Pollock's letter (2892) for which this letter is Pollock's reply, but Tyndall's letter is missing.
3. *"Anima Mundi":* the soul of the world (Latin); the animating principle, a power or spirit supposed by early philosophers to be present throughout the material universe, organizing and giving form to the whole and to all its parts, and regulating change and movement (*OED*).
4. *your book:* in *Faraday as a Discoverer,* which Pollock acknowledged reading in letter 2892, Tyndall wrote: 'In a beam of ordinary light the particles of the luminiferous ether vibrate in all directions perpendicular to the line of progression; by the act of polarization, performed here by Faraday, all oscillations but those parallel to a certain plane are eliminated' (p. 83).

To George Gabriel Stokes[1] [October 1868][2] 2894

to the beam.

Through practice, I have obtained considerable mastery over the production of delicate blue clouds (if they may be called such) by chemical decomposition. Pure air was led through the liquid iodide of allyl, and allowed to enter the experimental tube till the associated mercury column was depressed two inches. The tube was then filled with air which had been passed through aqueous hydrochloric acid.

In some seconds after the beam was turned on there was nothing seen in the tube. A delicate blue cloud however soon appeared, and immediately vivid selenite colours flashed forth. One inch of Benzol vapour heated in the same manner produced a magnificent effect. In fact I have never seen more brilliant colours even when the best Nicol[3] was used as a polarizer.

When the cloud augments in coarseness the colours become feebler, and in many cases disappear wholly

Liquids are not free from the particles which produce this action; I have examined a considerable number of them and found them all polarizers; but in very different degrees.

Water may be taken as representative of this variability of action. Distilled water when a thin slice of the electric beam is passed through it shows the action feebly, common water vividly, but water in which a little mastic has been precipitated gives a splendid effect. In fact a beam so thin as to produce no sensible action where passed through distilled water, produces brilliant action in the water containing the mastic.

I have used selenites[4] in wedges for these experiments; but now employ a plate there at the centre & gradually thickening outwards. It exhibits a fine system of coloured rings.

The bearing of these experiments upon the polarization of the sky is manifest enough, and it is to enable myself to pursue this question somewhat exhaustively that I write you this somewhat hasty letter.

I am | Yours very truly | John Tyndall

Professor Stokes D.C.L.[5] &c. &c. | Secretary to the Royal Society

RI MS JT/1/T/1552

1. Letter incomplete; fragment only.
2. *[October 1868]:* the month and year are given based on the content of this letter indicating these experiments as those which Tyndall described in a paper he submitted to the RS on 16 December 1868: J. Tyndall, 'On the Blue Colour of the Sky, the Polarization of Skylight, and on the Polarization of Light by Cloudy Matter Generally', *Roy. Soc. Proc.,* 17

(1868–9), pp. 223–33. Also, the same subject is discussed between Tyndall and Stokes in numerous letters they exchanged beginning on 2 December 1868. This letter appears to be an early notice to Stokes on what he was working on.

3. *Nicol:* a Nicol prism, invented in 1828 by the Scottish physicist and geologist William Nicol (1770–1851) (*ODNB*).
4. *selenites:* sulphate of lime (gypsum) in a crystalline or foliated form. Also, a slip or film of this mineral used for the polarization of light (*OED*).
5. *D.C.L.:* Doctor of Civil Law, which Stokes received as an honorary degree from the University of Oxford.

To Thomas Archer Hirst 3 November 1868 2895

3rd. Nov. 1868.

My dear Tom

I have put up at the Royal Hotel Ventnor.[1] I have got a little room which has enabled me already to execute some work.

There is a wild storm in the air. Last night large wisps of cloud were blown in from the southwest, they sped through the lower atmosphere, the soft & snowy clouds of the upper region being motionless. These were beautifully ribbed & interwoven and had a new interest for me in consequence of my recent experiments.[2] The clouds were all wonderfully reddened at sunset. I walked out at night: the sea moaned, and there seemed a responsive moan at the bottom of my life.

But I have been up and acting to day, and already feel much better. I walked over the stony down of St Bonaface:[3] on coming back I looked into the graveyard of Bonchurch[4] and stood over the dust of John Sterling.[5] Then came home, had a cup of tea for luncheon & got some of my South Kensington syllabus[6] off hands.

The X meet to morrow.[7] Had I been in town I sh[oul]d. have joined the herd of candidates & claimed election to a post.[8] The Spectator has made Huxley the Launcelot[9] of scientific men, saying that he unhorses every man & bamboozles every woman that he meets.[10] Now leaving Huxley on this lewd eminence it was my intention to lay claim to the position of Arthur[11]—that is all.

Yours aff[ectionate]ly | John

If you come on Saturday we shall have a breezy walk together. I hope you will.[12]

RI MS JT/1/T/669
RI MS JT/1/HTYP/544

1. *Royal Hotel Ventnor:* a hotel on the Isle of Wight, founded in 1832.
2. *my recent experiments:* see letter 2890, n. 2.
3. *St Bonaface:* St Boniface is a chalk down on the Isle of Wight, near the town of Ventnor.
4. *graveyard of Bonchurch:* at St Boniface Church in Bonchurch, a small village to the east of Ventnor.
5. *dust of John Sterling:* John Sterling (1806–44), a Scottish writer and poet, and friend of Thomas Carlyle (*ODNB*). His headstone reads: 'John Sterling | Died at Ventnor | 18th September, 1844 | aged 38' ('The Grave of John Sterling at Bonchurch', *The Island Quarterly,* 1 (1877), pp. 3–14, on p. 3).
6. *South Kensington syllabus:* since 1859, Tyndall was an Examiner for Science at the Department of Science and Art at the South Kensington Museum.
7. *X meet to morrow:* the X Club (see letter 2774, n. 3).
8. *herd of candidates & claimed election to a post:* possibly a reference to the upcoming elections to the RS Council at the anniversary meeting on 30 November 1868 ('Anniversary Meeting', *Roy. Soc. Proc.,* 17 (1869), pp. 133–54, Council elections listed on p. 151).
9. *Launcelot:* in the Arthurian legend, Lancelot du Lac is one of the Knights of the Round Table, companion to King Arthur.
10. *saying that he unhorses . . . that he meets:* 'Mr. Maurice on the Conscience', *Spectator,* 26 September 1868, p. 19–20. On p. 19, the article reads: 'Professor Huxley is the Sir Lancelot of the English men of science of our generation. As their chosen champion he has broken more lances and unhorsed more doughty enemies than we can at this moment count'.
11. *Arthur:* in legend, King Arthur is the head of the kingdom Camelot and the Knights of the Round Table.
12. *I hope you will:* Hirst was unable to join Tyndall on the Isle of Wight (see letter 2898).

To Mary Lyell[1] [5 November 1868][2] 2896

Ventnor,[3] Thursday

My dear M^rs^. Lyell

I have written to two photographers desiring each of them to send you a dozen Cartes de Visite.[4] I hope they will reach you in time.

I shall be in town on Sunday[5] & should then in case the photographers are lazy be able to dig up something out of my household debris.[6]

Yours very truly | John Tyndall

RI MS JT/1/T/1058
RI MS JT/1/TYP/3/840

1. *Mary Lyell:* see letter 2671, n. 1.
2. *[5 November 1868]:* the date is given by relation to letters 2895, 2897, and 2898 (all to Thomas Hirst and written from the Ventnor). Tyndall arrived on the Isle of Wight on 3

November and hoped to be joined by Hirst to go for walks (2895), but Hirst was unable to go. Tyndall appears to have returned to London by 8 November (2898).

3. *Ventnor:* the Royal Hotel Ventnor on the Isle of Wight.
4. *Cartes de Visite:* visiting card (French); a small photographic portrait mounted on a card, 3½ by 2¼ inches (*OED*).
5. *in town on Sunday:* Tyndall returned to London by Sunday, 8 November (see letter 2898).
6. *my household debris:* Tyndall had moved into Faraday's rooms at the RI in January 1868 (see letter 2707, n. 8 and n. 9), so this may imply that he had not yet organized his new home.

To Thomas Archer Hirst [6 November 1868][1] 2897

I strolled last night to the little graveyard of Bonchurch[2] where John Sterling[3] sleeps. It was very solemn. The wind had subsided in the lowlands though it was still gusty on the heights. The moon threw herself backward in the sky, while at the sea horizon bars of lightning shot upward like the explosions of a volcano. And up to the dead came the low musical murmur of the sea. You remember writing to me about John Sterling[4]—indeed it was your letter that caused me to buy his Life[5] when I could not well afford the outlay. I lay hold of you my boy by many memories. As we grow older such memories increase in value, and make even more intense but later friendship feeble through want of depth of recollection. It gives me pleasure to think that I have a strong hold of Huxley in the same way. I was the only friend behind him as he walked to the grave of his firstborn.[6] I still remember one shining gray hair on his uncovered head on that occasion. If he should ever prove bumptious I can bring him down.

RI MS JT/1/T/670
RI MS JT/1/HTYP/544

1. *[6 November 1868]:* 'Nov. 6/68' is written on the handwritten letter but is perhaps not in Tyndall's hand. Given that Tyndall wrote to Hirst from the Isle of Wight on 3 November and mentioned visiting the graveyard of Bonchurch (see letter 2895), this date seems reliable. Tyndall wrote another letter to Hirst on 6 November 1868 (see letter 2898).
2. *graveyard of Bonchurch:* see letter 2895, n. 4.
3. *John Sterling:* see letter 2895, n. 5.
4. *writing to me about John Sterling:* Tyndall here refers to Hirst's letter of 9[–10] November 1851 (letter 0565, *Tyndall Correspondence,* vol. 3), but Hirst also wrote about Sterling in a letter of 31 August 1850 (letter 0341, *Tyndall Correspondence,* vol. 3).
5. *Life:* T. Carlyle, *The Life of John Sterling* (London: Chapman and Hall, 1851).
6. *walked to the grave of his firstborn:* see letter 1729, n. 1, *Tyndall Correspondence,* vol. 7.

To Thomas Archer Hirst [6 November 1868][1] 2898

Friday

My dear Tom,

The whole world is proving false to me, and you go along with the world. I have been planning the excursions that we were to have together, but now all my plans are wrecked.[2] Never mind; there will be an end to it all some day.

The weather is splendid here, bright skies but intensely cold air. I give myself about four hours exercise daily. I find I am still good for 20 miles in four hours. I can walk & trot 5 miles an hour without much labour to myself. My legs are good: but it is my head that wearies me.

I will return on Sunday. By the way you will be dining with the Busk's[3] today: had I thought of it I sh[oul]d. have asked you to try & drive her nonsensical notions regarding me[4] out of Mrs. Busk. I know my own defects a thousand times better than she does, but her method of curing them would indefinitely exaggerate them.

Yours affectionately | John

RI MS JT/1/T/671
RI MS JT/1/HTYP/545

1. *[6 November 1868]:* 'Nov. 6/68' is written on the handwritten letter but is perhaps not in Tyndall's hand. Given that Tyndall wrote to Hirst from the Isle of Wight on 3 November and mentioned visiting the graveyard of Bonchurch (see letter 2895), this date seems reliable. Tyndall wrote another letter to Hirst on 6 November 1868 (see letter 2897).
2. *all my plans are wrecked:* Tyndall had hoped that Hirst would join him on the Isle of Wight so they could go on walking excursions but Hirst was unable to go (see letter 2895).
3. *Busk's:* George and Ellen Busk.
4. *her nonsensical notions regarding me:* not identified.

To Henry Acland 9 November 1868 2899

9th Nov. 1868.

My dear Dr. Acland.

Very many thanks for your two Addresses[1]—as usual high in tone & aspiration.

The printing of 200 additional copies of the Norwich Address[2] enables me to send you half a dozen of them.

Your allusion[3] is very very nice, and you are very kind to put it thus nicely.
Yours ever | John Tyndall

Oxford, MSS.Acland d.81 f.116

1. *your two Addresses:* possibly the following two addresses which were delivered and published in 1868: H. W. Acland, *Address to the British Medical Association, delivered in the Hall of Christ Church, August 4, 1868* (Oxford, 1868); and H. W. Acland, *Address to the Students of St. George's Hospital on the opening of the new school, October 1, 1868* (London: Macmillan and Co., 1868).
2. *Norwich Address:* see letter 2860, n. 4.
3. *Your allusion:* not identified; Acland's letter to Tyndall is missing.

To James Croll 9 [November 1868][1] 2900

9th.

My dear Sir,

If had a separate copy of that article[2] I would gladly send it to you. I will write a note to the sub editor[3] & ask him to send you the date.

It gave me pleasure to hear Mr. Darwin express the delight he experienced in reading one of your recent papers.[4] I had been staying with him for a day or two.[5] He is for him, exceedingly well

Yours very truly | John Tyndall

I am glad the notice in the Proceedings[6] pleased you—are you a Fellow of the Royal Society?[7]

BL Add MS 41077, f. 39
J. C. Irons (ed), *Autobiographical Sketch of James Croll with Memoir of his Life and Work* (London: Edward Stanford, 1896), p. 203

1. *[November 1868]:* the month and year are given by relation to letter 2888 (see n. 3).
2. *article:* not identified.
3. *sub editor:* not identified.
4. *one of your recent papers:* according to a letter from Charles Darwin to Croll, Croll had sent the first two parts (of three) of: J. Croll, 'On geological time, and the probable date of the Glacial and the Upper Miocene Period', *Phil. Mag.*, 35 (1868), pp. 363–84; 36 (1868), pp. 141–54, 362–86. See Charles Darwin to James Croll, 19 September 1868, *Darwin Correspondence*, vol. 16, part 2, p. 775–6, n. 1; this is letter 6380 in the online edition of the Darwin correspondence.

5. *staying with him for a day or two:* see letter 2888, n. 3.
6. *notice in the Proceedings:* possibly J. Tyndall, 'On a New Series of Chemical Reactions produced by Light', *Roy. Soc. Proc.,* 17 (1868–69), pp. 92–102, which was received on 24 October 1868.
7. *are you a Fellow of the Royal Society?:* James Croll was not elected FRS until 1 June 1876. Tyndall signed his certificate 'from general knowledge', i.e. not 'from personal knowledge' (RS, Certificates of Election and Candidature, EC/1878/08).

To Edward Frankland 14 November 1868 2901

14th. Nov. 1868.

My dear Frankland.

I am perfectly ashamed of the trouble I have given you. But you must ascribe some of it to my ignorance. I shall need your help for some time to come, but I will do my best to bother you as little as possible.

These light waves, as I thought, must be a powerful means of shaking molecules to pieces.[1] I am sure they will be a powerful agent in the hands of the chemist, and you must for some time to come help me to make it so.

I have been trying a few mixtures, and they reveal actions which I think are entirely unknown to Chemists. You must let me come and talk to you someday about the matter.

Yours ever | John Tyndall

Hirst told me last night that he fears Mrs. Frankland[2] is angry with both of us. I fear all my friends are angry with me. I have not seen Mrs. Huxley[3] for 8 months, and others of my friends have the same cause of complaint against me.

Professor Frankland | 42 Park Road | Haverstock Hill.[4]

John Rylands Library, University of Manchester, Frankland Letters, 2183

1. *These light waves . . . molecules to pieces:* see letter 2890, n. 2.
2. *Mrs. Frankland:* Sophie Frankland (née Fick, 1821–74), of Cassel, near Marburg in the state of Hesse-Cassel. Edward Frankland met her in 1847, on his first visit to Marburg. They became engaged in 1849, and married in 1851. Two of her brothers were professors in the University of Marburg, and her father, Friedrich W. Fick (1783–1861), was chief engineer of the state (C. A. Russell, *Edward Frankland: Chemistry, Controversy and Conspiracy in Victorian England* (Cambridge: Cambridge University Press, 1996), pp. 83–91).
3. *Mrs. Huxley:* Henrietta Huxley.
4. *Professor Frankland | 42 Park Road | Haverstock Hill.:* Frankland's residence in London; written as the address on the envelope, with a postmark given as 'London W | Z | NO[vember] 14 [18]68'.

From Edward Frankland 16 November 1868 2902

R. C. C.[1] Nov. 16/68.

My dear Tyndall

I shall be glad to have a talk with you at any time when you can call.[2] You can scarcely find me out between 11 & 5.

You will not make much impression upon chemists unless you can get quantitative results. Herewith I send you a couple of papers,[3] written soon after my return from Germany,[4] in which you will see some facts & prophecies in connection with the subject. It has only been touched since by Bunsen[5] & Roscoe[6] so far as I know.

Yours sincerely | E. Frankland.

RI MS JT/1/F/49

---

1. *R. C. C.:* the Royal College of Chemistry within the Royal School of Mines, where Frankland taught from 1865–85.
2. *have a talk . . . when you can call:* this letter is a reply to 2901.
3. *couple of papers:* possibly E. Frankland, 'Researches on the organic radicals: Part 3. On the action of solar light upon Iodide of Ethyl', *Quarterly Journal of the Chemical Society*, 3 (1851), pp. 322–47; and E. Frankland, 'Observations, economical and sanatory, on the employment of Chemical Light for artificial Illumination', *Roy. Inst. Proc.*, 1 (1851–4), pp. 319–25.
4. *my return from Germany:* in 1850 Frankland left the University of Marburg without completing his studies, returning to England to take on the Professorship of Chemistry at the Putney College of Civil Engineers in London left vacant by Lyon Playfair (1818–98), but soon after in 1851 he became the Professor of Chemistry at Owens College, University of Manchester (1851–7).
5. *Bunsen:* Robert Bunsen.
6. *Roscoe:* Henry Roscoe.

From Anna Helmholtz 17 November 1868 2903

Heidelberg[1] 17. Nov. 68.

Dear Sir,

If I had been well enough, I should have written a month ago to thank you with all my heart for the friendly welcome you sent our little boy[2] on his arrival in the world. He is so strong and big that I have no doubt he will be able very soon to drink, and of his beautiful mug which is everybodys admiration. I really don't know how to tell you the pleasure it gave me—and how proud I shall be to explain one day to baby of the kind and great friend, who sent this encouragement to his natural propensities towards drinking. He really seems to have some inclinations that way—as every kind of liquid is very welcome to him.

The little man must be my excuse for not having been quicker in my translation[3] of Faraday's life.[4] The translation was all done and ready so far as my part went, at the end of Mai[5]—but I was sent to the sea side with our children who needed it very much and when I came back in August my husband[6] went away—and we had but half finished the correcting part when the baby made his appearance and I have been ill ever since.

But now I hope to get on very quickly both with my health and with my work—so as to be able to send the manuscript to Vieweg[7] before Christmas. Pray do not think me lazy or ungrateful for the great privilege you conferred on me. I really could not help it. My aunt, Madame Mohl[8] from Paris, who had the pleasure of meeting you at the duc d'Aumale's,[9] if I remember rightly, gave us some hope of a visit from you during the Summer—but we never had this great pleasure. I hope you will "dédommager"[10] us next year—I am very glad, you did not come last summer, because I should have missed you.

My husband,[11] who is not at home to day, told me to send you his most friendly regards. He has been quite as much touched by your kindness as myself. Pray dear Sir, let me thank you once more and believe me very sincerely yours | Anna Helmholtz.

RI MS JT/H/32

---

1. *Heidelberg:* see letter 2677, n. 1.
2. *our little boy:* Hermann and Anna Helmholtz's son Friedrich Julius was born on 15 October 1868 (*Helmholtz,* p. 312).
3. *my translation:* see letter 2686, n. 4.
4. *Faraday's life: Faraday as a Discoverer.*
5. *Mai:* May (German).
6. *my husband:* Hermann Helmholtz.
7. *Vieweg:* Heinrich Rudolf Vieweg.
8. *Madame Mohl:* Mary Elizabeth Mohl (née Clarke, 1793–1883), an English writer, feminist, and salon hostess in Paris. In 1847, she married the orientalist Julius Mohl (*ODNB*).
9. *duc d'Aumale's:* the home of Henri Eugène Philippe Louis d'Orléans, Duke of Aumale (1822–97), a leader of the Orleanists, a political faction in France which supported a constitutional monarchy (M. Prevost and R. D'Amat (eds), *Dictionnaire de biographie française,* 21 vols (Paris: Letouzey et Ané, 1932–2016), vol. 4, pp. 611–24). Tyndall's journals and correspondence do not mention such a visit in Paris, but as Tyndall travelled to or through Paris many times during his life, it is quite possible he did.
10. *"dédommager":* compensate (French).
11. *My husband:* Hermann Helmholtz.

From Joseph Plateau[1] 17 November 1868 2904

Gand, 17 Novembre 1868.

Mon cher Monsieur Tyndall

J'ai l'honneur de vous adresser trois exemplaires du mémoire que je viens de publier; l'un d'eux vous est naturellement destiné, et j'espère que vous voudrez bien offrir, de ma part, le second à la Société Royale et le troisième à l'Institution Royale. Je viens de présenter à notre Académie trois nouvelles Séries, ce qui complète mon ouvrage; celui-ci, comme je vous l'ai déjà fait remarquer, constitue une Statique expérimentale et théorique des liquides soumis aux seules forces moléculaires.

A propos de la Société Royale, je me décide, après avoir longtemps hésité, à vous exprimer un sentiment douloureux: j'ai consacré ma vie à l'étude et aux progrès de la science, et, depuis 1842, malgré une cécité complète, je n'en continue pas moins mes travaux, luttant sans cesse contre mon infériorité; j'ai la conscience de l'importance de ces travaux, et il me semble qu'on doit me tenir compte de mes efforts et de mes résultats. Deux grandes sociétés savantes, l'Institution royale de Londres et l'Institut de France l'ont compris et m'ont placé au nombre de leurs correspondants; pourquoi donc la Société royale paraît-elle me dédaigner? Pourquoi ne m'a-t-elle pas encore nominé correspondant ou ne m'a-t-elle pas décerné l'une de ses médailles? Peut-être fallait-il, comme pour l'Institut de France, se mettre sur les rangs; si cela est, je vous prie de vouloir bien me considérer comme candidat, et transmettre ma demande à la Société royale. Dans ce cas aussi, voici l'énumération de mes principales publications:

1°. Dissertation sur quelques propriétés des impressions produites par la lumière sur l'organe de la vue, Liège, 1829.

Ce mémoire, dans lequel j'ai étudié spécialement la persistance des impressions, n'a été imprimé qu'à un petit nombre d'exemplaires; je compte reproduire, dans une publication ultérieure, les principaux résultats qu'il renferme.

2°. Essai d'une théorie générale comprenant l'ensemble des apparences visuelles qui succèdent à la contemplation des objets colorés et de celles qui accompagnent cette contemplation.

Cette théorie n'a pas été acceptée en Angleterre, à cause d'une théorie différente avancée par Sir D Brewster, mais elle est généralement adoptée en France.

3°. Mémoire sur l'Irradiation. C'est une étude détaillée de toutes les particularités et de toutes les lois du phénomène.

4°. Sur un problème curieux de magnétisme. Dans ce mémoire, je démontre par l'analyse l'impossibilité absolue de soutenir en l'air une aiguille aimantée dans un état d'équilibre stable, au moyen d'un système quelconque de barreaux aimantés.

5°. Mes onze séries intitulées: Recherches expérimentales et théoriques sur les figures d'équilibre d'une masse liquide sans pesanteur. Dans ce long travail, dont je vous ai rappelé plus haut la partie générale, j'ai fait connaître deux procédés essentiellement différents au moyen desquels on place un liquide présentant une surface considérable dans la même condition que s'il n'avait pas de pesanteur; j'ai étudié les phénomènes résultant d'un mouvement de rotation *[1 word illeg]* à la masse ainsi non pesante, et les figures d'équilibre qu'elle peut prendre à l'état de repos. J'ai fait également une étude détaillée des lames liquides, de leur génération, de leurs assemblages, et de toutes leurs propriétés; enfin j'ai cherché les conditions de stabilité d'un certain nombre des figures d'équilibre dont il s'agit. Ces recherches m'ont conduit à l'explication de certains faits que l'on envisageait d'une manière erronée: elles m'ont conduit, par exemple, à la vraie théorie de la constitution des masses liquides.

Je parle ici de mes onze séries, bien que les trois dernières ne soient pas encore imprimées; mais, ainsi que je vous l'ai dit plus haut, elles sont présentées à notre Académie et l'impression en a été décidée.

J'omets ici un grand nombre de notes sur différents sujets de physique; mais j'ajoute à l'énumération ci-dessus l'invention de l'Anorthoscope, et celle du Phénakistiscope; je partage, du reste, cette dernière avec M. Stampfer de Vienne.

Et maintenant, si vous jugez mes titres insuffisants, considérez comme non avenue toute la partie de ma lettre qui les concerne; dans le cas contraire, j'espère que vous voudrez bien les faire valoir.

Agréez, mon cher Monsieur Tyndall, l'assurance de tous mes sentiments de haute considération.

J. Plateau.

Ghent,[2] 17 November 1868

My dear Mr Tyndall

I have the honour of sending you three copies of the memoir that I have just published.[3] One of them is naturally for you, and I hope that you would kindly offer on my behalf the second one to the Royal Society, and the third one to the Royal Institution. I have just presented three new series in our Academy,[4] which completes my book.[5] This one, as I have already mentioned, is the Statique expérimentale et théorique des liquides soumis aux seules forces moléculaires.[6]

About the Royal Society, I decided after lengthy hesitation, to express to you a painful emotion: I dedicated my life to studies and to the progress of science, and since 1842, despite complete blindness I nevertheless continued my work, fighting ceaselessly against my disadvantage. I am aware of the importance of these works, and I think that my efforts and my results should be taken into account. Two major scientific societies, the Royal Institution of London and the Institute of France,[7] have understood that and placed me among their correspondents. So why does the Royal Society seem to despise me? Why has it not yet nominated me as a correspondent, or has not awarded me with one of its medals? Perhaps it was necessary, like in the Institute of France, to get into the ranks. If it is so, I ask you to kindly consider me as a candidate, and to forward my request to the Royal Society.[8] In that case also, here is the list of my main publications:

1°. Dissertation sur quelques propriétés des impressions produites par la lumière sur l'organe de la vue, Liege, 1829.[9]

This memoir, in which I specifically studied the persistence of impressions, had only a small amount of copies printed. I intend to reproduce, in the subsequent publication, the main results that it contains.

2°. Essai d'une théorie générale comprenant l'ensemble des apparences visuelles qui succèdent à la contemplation des objets colorés et de celles qui accompagnent cette contemplation.[10]

This theory had not been accepted in England, due to a different theory that was advanced by Sir D. Brewster,[11] but it is generally adopted in France.

3°. Mémoire sur l'Irradiation.[12] It is a detailed study about all the particularities of all the laws of this phenomenon.

4°. Sur un problème curieux de magnétisme.[13] In this memoir, I demonstrate through analysis the absolute impossibility of keeping a magnetic needle in the air in a state of stable equilibrium, using any system of magnetized bars.

5°. My eleven series, entitled: Recherches expérimentales et théoriques sur les figures d'équilibre d'une masse liquide sans pesanteur.[14] In this long work, which I reminded you above in general, I got to know two essentially different procedures through which we place a liquid and present a considerable surface in the same condition as if there was no gravity. I studied the phenomena resulting from a rotational movement imparted to the mass thus weightless, and the figures of equilibrium that it could take up in a resting state. I also did a detailed study of liquid layers, of their generation, their assemblies, and all their properties. Finally, I researched the conditions of stability of a certain number of figures of equilibrium in question. This research led me to the explanation of certain facts that were contemplated in an erroneous way: they led me, for example, to the true theory of the constitution of liquid masses.

I am speaking here about my eleven series, even though the last three have not yet been printed. But, as I told you above, they were presented at our Academy and their printing was decided.

I omit here a lot of notes on various topics of physics, but I add to the aforementioned list the invention of the Anorthoscope[15] and of the Phenakistiscope;[16] I am sharing the latter one with Mr Stampfer[17] from Vienna.

And now, if you think my titles are insufficient, consider void the entire part of my letter that concerns them. Otherwise, I hope that you would kindly present them.

Please accept, my dear Mr Tyndall, the assurance of my best regards.
J. Plateau.

RI MS JT/1/P/102

---

1. *Joseph Plateau:* Joseph Antoine Ferdinand Plateau (1801–83), a Belgian physicist and professor at the University of Ghent who studied optics. In 1832 Plateau invented a device to illustrate moving images called the 'phenakistoscope' (*CDSB*).
2. *Ghent:* a city in Belgium.
3. *the memoir that I have just published:* J. Plateau, 'Recherches expérimentales et théoriques sur les figures d'équilibre d'une masse liquide sans pesanteur [series eight]', *Mémoires de l'Académie royale des sciences, des lettres et des beaux-arts de Belgique,* 37 (1869), pp. 1–102 (presented on 4 July 1868).
4. *three new series in our Academy:* Plateau presented series nine, ten, and eleven on 11 October 1868 at the Royal Academy of Science, Letters and Fine Arts of Belgium.
5. *my book:* see n. 6.
6. *Statique expérimentale et théorique des liquides soumis aux seules forces moléculaire:* J. Plateau, *Statique expérimentale et théorique des liquides soumis aux seules forces moléculaires,* 2 vols (Gand: F. Clemm, 1873). The book was a revision of Plateau's series of eleven papers cited in n. 13.
7. *Institute of France:* the Institute of France (Institut de France), a French learned society founded in 1795 and made up of five member societies including the Académie des Sciences.
8. *Royal Society:* Plateau would be elected a foreign member of the RS on 24 November 1870, with Tyndall among those who signed the certificate (RS, Certificates of Election and Candidature, EC/1870/18).
9. *Dissertation sur quelques propriétés des impressions produites par la lumière sur l'organe de la vue,* Liege, 1829: J. Plateau, *Dissertation sur quelques propriétés des impressions produites par la lumière sur l'organe de la vue* (Liège: H. Dessain, 1829).
10. *Essai d'une théorie générale comprenant l'ensemble des apparences visuelles qui succèdent à la contemplation des objets colorés et de celles qui accompagnent cette contemplation:* J. Plateau, 'Essai d'une théorie générale comprenant l'ensemble des apparences visuelles qui succèdent à la contemplation des objets colorés, et de celles qui accompagnent cette contemplation', *Nouveaux mémoires de l'Académie royale des sciences et belles-lettres de Bruxelles,* 8 (1834), pp. 1–68.
11. *Sir D. Brewster:* David Brewster.
12. *Mémoire sur l'Irradiation:* J. Plateau, 'Mémoire sur l'Irradiation', *Nouveaux mémoires de l'Académie royale des sciences et belles-lettres de Bruxelles,* 11 (1838), pp. 1–112.
13. *Sur un problème curieux de magnétisme:* J. Plateau, 'Sur un problème curieux de magnétisme', *Mémoires de l'Académie royale des sciences, des lettres et des beaux-arts de Belgique,* 34 (1864), pp. 1–37.

14. *Recherches expérimentales et théoriques sur les figures d'équilibre d'une masse liquide sans pesanteur:* J. Plateau, 'Recherches expérimentales et théoriques sur les figures d'équilibre d'une masse liquide sans pesanteur', *Mémoires de l'Académie royale des sciences, des lettres et des beaux-arts de Belgique,* 16 (1843); 23 (1849); 31 (1859); 33 (1861); 36 (1867); and 37 (1869). The eleventh and final paper was read to the Academy on 11 October 1868.
15. *Anorthoscope:* an optical toy for viewing distorted figures drawn on a rotating disk (*OED*).
16. *Phenakistiscope:* a toy consisting of a disc or drum with figures representing a moving object in successive positions arranged radially on it, to be viewed in such a way (e.g., through a fixed slit) that the persistence of the successive visual images produces the impression of actual motion when the disc or drum is rapidly rotated (*OED*).
17. *Mr Stampfer:* Simon Stampfer (1792–1864), an Austrian mathematician and surveyor who in 1832 simultaneously invented a device identical to Plateau's Phenakistiscope which Stampfer named the Stroboscope (*ADB*).

To Rudolf Clausius 19 November 1868 2905

19th Nov. 1868

My dear Clausius

Some time ago you wrote to me[1] making enquiries about a work of Tait[2] on the Dynamical Theory of Heat.[3]

I have this day received the work from Tait himself. He will probably send you a copy, otherwise I should purchase the book and send it to you at once.

I have been long thinking of writing to you. Your kind letter[4] asking Hirst[5] & myself to visit you came to me before I went to Switzerland. It would have given both of us real pleasure to have gone to you; but I was terribly ill:—so ill that I was obliged to break the journey to Switzerland several times. After some weeks among the mountains I recovered my strength; but then I was compelled to return in time for the Norwich meeting of the British Association; for I had accepted the presidency of the physical section.[6]

Since my return I have been hard at work. I suppose the Royal Society sends its Proceedings to its Foreign members; the last number gives an account of what I have been doing.[7] I dare say you remember the conversation[8] to which I there allude. It was in the Hotel Bauer au Lac.[9]

In Switzerland I did but little climbing this year. I however completed a piece of work which has been on my mind for some years. I climbed the Matterhorn from the Italian side:[10] reached the western end of the ridge which forms the summit; went along the summit to its eastern end. And then though my friends did not know anything about the Northern slopes of the mountain, we descended together to Zermatt. I found the slope on which those poor fellows slipped in 1865,[11] exceedingly dangerous.

I am now hard at work experimenting upon the chemical actions of light.[12] I had no idea of the power of light in shaking molecules asunder. The experiments reveal energetic action, where chemists have hitherto supposed there was no action. I have added a great number of substances to the list already published, and I should not wonder if the cases in which decomposition is not produced by light should prove the exception instead of the rule.

But the cloud forms which accompany the decomposition are the most wonderful looking things imaginable. I think they will also prove to be characteristic of the substances from which they are derived.

<u>You</u> will see that I am working on towards a solution of the differences between Magnus and myself.[13] His paper[14] was to me in no way convincing, still it is a paper calculated to make an impression on the outside world. I think he is wrong, but whether he is so or not, I hope to set the question for ever at rest during my present investigation.

Give my kindest regards to M^rs^. Clausius;[15] give Johnny[16] my love & believe me dear Clausius | always yours | <u>John Tyndall</u>

RI MS JT/1/T/201

---

1. *you wrote to me:* letter 2839.
2. *Tait:* Peter Guthrie Tait.
3. *Dynamical Theory of Heat:* see letter 2742, n. 4.
4. *your kind letter:* letter 2843.
5. *Hirst:* Thomas Hirst.
6. *Norwich meeting . . . the physical section:* see letter 2860, n. 4.
7. *an account of what I have been doing:* see J. Tyndall, 'On a New Series of Chemical Reactions Produced by Light', *Roy. Soc. Proc.,* 17 (1868–9), pp. 92–102 (received 24 October 1868).
8. *conversation:* in his paper, Tyndall provided the following footnote for the passage 'A question of extreme importance in molecular physics here arises:—What is the real mechanism of this absorption, and where is its seat?': 'My attention was very forcibly directed to this subject some years ago by a conversation with my excellent friend Professor Clausius' (p. 95). It appears from previous Tyndall correspondence between Tyndall and Clausius that they saw each other in Zürich in the summer of 1858 and in the fall of 1864.
9. *Hotel Bauer au Lac:* a hotel in Zürich, Switzerland, founded in 1844.
10. *climbed the Matterhorn from the Italian side:* see letter 2855, n. 3.
11. *those poor fellows slipped in 1865:* see letter 2856, n. 19.
12. *experimenting upon the chemical actions of light:* see n. 7 of this letter, and letter 2890, n. 2.
13. *differences between Magnus and myself:* see letters 2578, 2580, and 2598.
14. *His paper:* see letter 2578, n. 1.
15. *M^rs^. Clausius:* Adelheid Clausius.
16. *Johnny:* Rudolf John Clausius.

To Thomas Henry Huxley 19 November [1868][1] 2906

Royal Institution. | 19th Nov.

My dear Huxley,

Mill deserves what he got[2]—but if I had a pocket borough[3] at my disposal he is the first man that I should choose to represent it.

The parcels delivery company has I fear made a mess of the eggs. Only think of that little damsel hazarding the hypothesis that the eggs were hatched and flew away! Her mother has much to answer for.

Thanks as regards Barrett[4] I always thought him an exceedingly clever fellow—He is sure to be upright—But why he goes in for this thing I do not understand[5]

Yours ever | John Tyndall.

IC HP 8.67
RI MS JT/1/TYP/9/2924
Typed Transcript Only

---

1. *[1868]:* the year is given by the reference to John Stuart Mill (see n. 2).
2. *Mill deserves what he got:* John Stuart Mill lost his bid for reelection as MP for Westminster in 1868, in part due to his role in the Jamaica Committee (see letter 2571, n. 6 and n. 8).
3. *pocket borough:* a borough in which the election of political representatives was controlled by one person or family (*OED*). These were abolished by the Reform Acts of 1832 and 1867.
4. *Barrett:* William Fletcher Barrett (1844–1925), a physicist and psychical researcher who was an assistant to Tyndall at the RI from 1863–6. Following his time at the RI, he held various teaching positions at the International College, London, the Royal School of Naval Architecture, and the Royal College of Science for Ireland, and was elected FRS in 1899 (*ODNB*).
5. *But why he goes in for this thing I do not understand:* Huxley possibly sent Tyndall notice of Barrett's 1868 lecture at the Royal Dublin Society on sensitive flames and spiritualism ('Royal Dublin Society', *Irish Times,* 4 January 1868, p. 3; 'On Musical and Sensitive Flames. Abstract of a Lecture delivered, at the invitation of the Council, before the Royal Dublin Society, on January 3, 1868, by W. F. Barrett', *Chemical News,* 17 (1868), pp. 220–2). Barrett had published a paper on sensitive flames (W. F. Barrett, 'Note on "sensitive flames"', *Phil. Mag.,* 33 (1867), pp. 216–22), and Tyndall became interested in the phenomena as well, incorporating sensitive flames in his lectures on sound and heat, and giving on 18 January 1867 a Friday Evening Discourse at the RI on sensitive flames (see letter 2535, n. 3). Barrett presumably felt that Tyndall was co-opting the phenomena from him, and Barrett resigned as Tyndall's assistant at the RI in 1866 because of growing tensions between

the two. However, Tyndall was also unenthusiastic about Barrett's increasingly spiritualist leanings, which began in the 1860s when Barrett had an experience with mesmerism, and he included psychic effects in explaining sensitive flames. Following controversy over publishing spiritualist investigations in the journal *Nature* and presenting a paper on spiritualism and mesmerism at the 1876 annual meeting of the BAAS in Glasgow, Barrett founded both the Society for Psychical Research and the American Society for Psychical Research in the early 1880s. On Barrett, Tyndall, and spiritualism, see R. Noakes, 'The "Bridge Which Is Between Physical and Psychical Research": William Fletcher Barrett, Sensitive Flames, and Spiritualism', *History of Science,* 42 (2004), pp. 419–64.

To Thomas Archer Hirst 20 November 1868 2907

Nov. 20th /68

My dear Tom,

I must ask you not to press me regarding the lecture at University College.[1] I cannot, without damage to my work,[2] get up such a lecture as I should deem worth giving. Wh[en]: I have anything to say I will let you know: but at present if I attempted what you require of me, it would necessitate a wilfull deflection of my thoughts from the subjects which ought to occupy them. The very thought of my being pledged to such a thing would have a prejudicial influence upon me.

Yours ever | John Tyndall

RI MS JT/1/T/672
RI MS JT/1/HTYP/545

1. *lecture at University College:* possibly referring to a lecture Tyndall would give on 25 June 1869 at University College London, where Hirst was Professor of Pure Mathematics (1867–70; he was at the college in total from 1865–73). See J. Tyndall, 'Address to the Students, University College London on the Distribution of Prizes in the Faculty of Arts Session 1868–69', in *Fragments of Science for Unscientific People: A Series of Detached Essays, Lectures and Reviews* (London: Longmans, Green, and Co., 1871), pp. 95–106. On the day of the lecture, Hirst wrote that 'Tyndall's address was greatly admired I thought it too discursive' (Brock & MacLeod, *Hirst Journals,* 25 June 1869, p. 1848).
2. *my work:* see letter 2890, n. 2.

From Jonathan Frederick Pollock 21 November 1868 2908

Hatton, | Hounslow.[1] | 21st. Nov[embe][r]. 1868

My dear Sir!

I am much obliged and most thankful for your letter[2] and the Paper[3] which I found very interesting—I should like to be present if any similar experiments take place in the course of the spring lectures[4]—

I owe my handwriting to my Cambridge-Exam[inatio][n]. in 1806[5]—most of the examination—indeed all the answers are in writing—& I took great pains with mine, for that reason—in the present day it is not easy to read a letter & frequently the name of the writer (if a Stranger) is the most difficult part of it to read—what right has any man to give to others so much trouble by taking none himself

Yours very sincerely & gratefully | Fred Pollock

RI MS JT/1/P/160
RI MS JT/1/TYP/6/2101

1. *Hatton, | Hounslow:* see letter 2892, n. 1.
2. *your letter:* letter missing.
3. *Paper:* see letter 2905, n. 7.
4. *spring lectures:* Tyndall gave nine lectures 'On Light' at the RI on Thursdays from 8 April to 3 June 1869 (*Roy. Inst. Proc.,* 5 (1866–9), pp. 428).
5. *my Cambridge-Exam[inatio][n]. in 1806:* Pollock graduated BA in 1806 from Trinity College, Cambridge (and MA in 1809).

To the Athenaeum Committee[1] 24 November 1868 2909

106 Members | Urging Appointment of Mr John Bennett now the Butler to the Vacant Stewardship.

The Athenaeum

We, the undersigned, hearing that a vacancy has occurred in the office of the Steward, beg to ask the attention of the Committee of the Athenaeum Club, to the long continued & faithful services of the Butler, Mr Bennett,[2] who has been associated with this Club from its Establishment in 1824.

The undersigned believe that no selection would be more acceptable to the majority of those frequenting the Club than the promotion to the office of Steward of this courteous and experienced servant; so well acquainted with the requirements of its Members.

The undersigned feel that such a recognition of Mr Bennett's services should not rest on the mere length of those services, but they may reasonably assume that the manner in which he has assisted to fulfil the duties of Steward during the absence of Mr Peters may be accepted as a strong proof of his fitness to be his successor.

Tho. Graham[3]
John Tyndall
John Hilton[4]
T Fred^k^ Elliot[5]
H A Ouvry[6]
H. K. Storks.[7]
Fred. P. Maude[8]
G W Hemans[9]
Charles Hawkins[10]
W. Bovill[11]
Wilkie Collins[12]
Warington W. Smyth.[13]
J. W. Blakesley.[14]
G O Rees—[15]
Thos H Wyatt[16]
Alexander Brodie[17]
Charles Vignoles[18]
S. A. Hart[19]
Robert Browning[20]
Sydney Smirke[21]
Montague Smith[22]
Wm Smith[23]
A. Taylor.[24]
F. D. Massy Dawson[25]
A. S. Waugh[26]
Edwd Meryon[27]
Matthew Arnold.—[28]
T H Huxley.
Christopher Erle[29]
[1 signature illeg]
(over)
John Gould[30]

T C Paris[31]
George Long[32]
Alex[r] W. Williamson[33]
H. Sandwith[34]
Duncan MacGregor[35]
Tho[s]. Sopwith[36]
T. Tatum[37]
[1 signature illeg]
Geo. Richmond.[38]
John Edmund Cox.[39]
W R Basham[40]
R Malcolm Kerr[41]
E M Ward[42]
D Salomons[43]
G Lathom Browne[44]
W. Ebor[45]
Richard Quain[46]
Francis Sibson[47]
P. F. Campbell Johnston[48]
Mark Pattison[49]
Charles Manby[50]
A. R. Johnston[51]
Geo Capel[52]
Holdsworth Hunt[53]
John W Prout[54]
W. J. Radford.[55]
Alex[r]. Dyce[56]
Robert Whiston[57]
J. Palgrave Simpson[58]
<u>Charles Dickens</u>[59]
E A Darwin[60]
C. Wheatstone
Dan[l]. Maclise[61]
C. J. Herries[62]
Nelson Matcham[63]
H. Hunt[64]
G. H. A. Forbes[65]
Sp Perceval[66]
John Hawkshaw[67]
Revd Tulloch[68]
Henry Wyndham Phillips[69]

L L Dillwyn[70]
Thomas Debary[71]
W. Frith[72]
W. Boxall[73]
E. Chisholm Batten[74]
John Everett Millais[75]
Frederic Quvry[76]
William Pole[77]
A H Layard[78]
Henry Vaughan[79]
Fred^k^. Marrable[80]
Geo T Clark[81]
C. Austen Leigh[82]
C Wigram[83]
Henry Oldham[84]
Chas. Landseer[85]
T B Curling[86]
John E. Holdsworth[87]
A. W. Fonblanque[88]
Joseph Whitworth[89]
de Strzelecki[90]
William J. Thoms.[91]
Geo. Busk
George Denman[92]
Trenham Reeks[93]
E. Landseer[94]
C. M. Roupell[95]
R W Mylne[96]
H Weekes[97]
J. J. Sylvester[98]
John Percy[99]
J. F. Lewis[100] over
J. A. Jeremie[101]
106
20 or more of the above have served on the Committee

Athenaeum General Committee Minutes, vol. 17 (1865–68)

---

1. *Athenaeum Committee:* see letter 2537, n. 8.
2. *M^r^. Bennett:* the petition was successful; John Bennett became the Steward of the Athenaeum Club.
3. *Tho. Graham:* Thomas Graham (1805–69), a Scottish chemist who from 1851 was an

assayer of the Royal Mint. He succeeded John Herschel as Master of the Mint in 1855 holding the post until his death (*ODNB*).

4. *John Hilton:* John Hilton (1805–78), an English anatomist and surgeon at Guy's Hospital (1849–70) (*ODNB*).
5. *T Fred[k] Elliot:* Thomas Frederick Elliot (see letter 2721, n. 1).
6. *H A Ouvry:* Henry Aimé Ouvry (1813–99), a British army officer who served in India (*Athenaeum Rules* (1874), p. 65).
7. *H. K. Storks:* Henry Knight Storks (1811–74), a British army officer and colonial official (*ODNB*).
8. *Fred. P. Maude:* Frederick Philip Maude (dates unknown), an English barrister (*Athenaeum Rules* (1862), p. 66).
9. *G W Hemans:* George Hemans (see letter 2851, n. 19).
10. *Charles Hawkins:* Charles Hawkins (1812–92), a surgeon who served as Inspector of Anatomical Schools for the Metropolis ('Obituary', *Lancet,* 30 April 1892, pp. 1004–5).
11. *W. Bovill:* William Bovill (1814–73), an English lawyer and chief justice (*ODNB*).
12. *Wilkie Collins:* William 'Wilkie' Collins (1824–89), an English writer (*ODNB*).
13. *Warington W. Smyth:* Warington Wilkinson Smyth (1817–90), a British mining engineer and geologist (*ODNB*).
14. *J. W. Blakesley:* Joseph Blakesley (see letter 2851, n. 45).
15. *G O Rees:* George Rees (see letter 2851, n. 43).
16. *Tho[s] H Wyatt:* Thomas Henry Wyatt (1807–80), an English architect (*ODNB*).
17. *Alexander Brodie:* Alexander Brodie (b. *c.* 1826) (*Alum. Oxon.*).
18. *Charles Vignoles:* Charles Vignoles (see letter 2851, n. 25).
19. *S. A. Hart:* Solomon Alexander Hart (1806–81), an English painter (*ODNB*).
20. *Robert Browning:* Robert Browning (see letter 2851, n. 18).
21. *Sydney Smirke:* Sydney Smirke (1798–1877), a British English architect (*ODNB*).
22. *Montague Smith:* Montague Smith (1806–91), an English judge (*ODNB*).
23. *W[m] Smith:* William Smith (1813–93), a British lexicographer and from 1853–69 he was classical examiner to the University of London (*ODNB*).
24. *A. Taylor:* probably Alexander Taylor (*c.* 1824–84), Michel Fellow at Queen's College, Oxford, and Reader and Afternoon Preacher to the Honourable Society of Gray's Inn (*Alum. Oxon.*).
25. *F. D. Massy Dawson:* Francis Dennis Massy-Dawson (1803–70), an English barrister (*Alum. Cantab.*).
26. *A. S. Waugh:* Andrew Scott Waugh (1810–78), a British army officer who served as Surveyor General of India (*ODNB*).
27. *Edw[d] Meryon:* Edward Meryon (see letter 2851, n. 50).
28. *<u>Matthew Arnold</u>:* Matthew Arnold (see letter 2588, n. 1).
29. *Christopher Erle:* Christopher Erle (*c.* 1790–1870) (*Alum. Oxon.*).
30. *John <u>Gould</u>:* John Gould (see letter 2864, n. 13).
31. *T C Paris:* Thomas Clifton Paris (1818–1909), an English lawyer (*Alum. Cantab.*).
32. *George Long:* George Long (1800–79), an English lawyer and classicist (*ODNB*).

33. *Alex^r W. Williamson:* Alexander William Williamson (1824–1904), an English chemist (*ODNB*).
34. *H. Sandwith:* Humphry Sandwith (see letter 2723, n. 3).
35. *Duncan MacGregor:* Duncan MacGregor (*Athenaeum Rules* (1862), p. 64)
36. *Thos. Sopwith:* Thomas Sopwith (1803–79), a surveyor and engineer (*ODNB*).
37. *T. Tatum:* Thomas Tatum (*Athenaeum Rules* (1862), p. 84)
38. *Geo. Richmond:* George Richmond (1809–96), a British portrait painter (*ONDB*).
39. *John Edmund Cox:* John Edmund Cox (see letter 2851, n. 48).
40. *W R Basham:* William Richard Basham (1807–77), an English physician (*ODNB*).
41. *R Malcolm Kerr:* Robert Malcolm Kerr (1821–1902), a British judge and barrister (G. Pitt Lewis, *Commissioner Kerr—An Individuality* (London: T. Fisher Unwin, 1903)).
42. *E M Ward:* Edward Matthew Ward (1816–79), an English painter (*ODNB*).
43. *D Salomons:* David Salomons (1797–1873), an English banker and politician who served as the first Jewish mayor of London (1855), and Member of Parliament for Greenwich (1859–73) (*ODNB*).
44. *G Latham Browne:* George Lathom Browne (1815–92), an English lawyer (*Alum. Oxon.*).
45. *W. Ebor:* William Thomson (1819–90), Archbishop of York (*ODNB*). The Archbishops of York often signed their name with 'Eboracensis' or 'Ebor', for *Eboracum,* the Roman name for York.
46. *Richard Quain:* Richard Quain (1816–98), an Irish physician who in 1863 was appointed as a Member of the General Medical Council (*ODNB*).
47. *Francis Sibson:* Francis Sibson (1814–76), a physician and mountaineer who had a practice in London (*ODNB*).
48. *P. F. Campbell Johnston:* Patrick Francis Campbell-Johnston (1802–92), a British diplomat (*Athenaeum Rules* (1862), p. 58; B. Burke, *A Genealogical and Heraldic Dictionary of the Landed Gentry of Great Britain and Ireland,* 2 vols, 4th edn (London: Harrison, Pall Mall, 1862), vol. 1, p. 788).
49. *Mark Pattison:* Mark Pattison (1813–84), an English scholar and clergyman (*ODNB*).
50. *Charles Manby:* Charles Manby (1804–84), a civil engineer (*ODNB*).
51. *A. R. Johnston:* Alexander Robert Campbell Johnston (1812–88), a British colonial official (*ODNB*).
52. *Geo Capel:* George Capel (dates unknown), an Anglican clergyman.
53. *Holdsworth Hunt:* Holdsworth Hunt (1806–83), an English barrister (F. Boase, *Modern English Biography,* 3 vols (Truro: Netherton and Worth, 1892–1901), vol. 1, p. 1590).
54. *John W Prout:* John Prout (see letter 2851, n. 24).
55. *W. J. Radford.:* William Radford, MD (*Athenaeum Rules* (1862), p. 76).
56. *Alex^r. Dyce:* Alexander Dyce (*c.* 1799–1869), an Anglican clergyman (*Alum. Oxon.*).
57. *Robert Whiston:* Robert Whiston (1808–95), an Anglican clergyman and headmaster (*ODNB*).
58. *J. Palgrave Simpson:* John Palgrave Simpson (1807–87), a novelist and playwright (*ODNB*).

59. *Charles Dickens:* Charles Dickens (1812–70), the famed English writer and critic (*ODNB*).
60. *E A Darwin:* Erasmus Darwin (see letter 2851, n. 41).
61. *Dan[l]. Maclise:* Daniel Maclise (*c.* 1806–70), an Irish painter (*ODNB*).
62. *C. J. Herries:* Charles John Herries (1815–83), an English financier, chair of the Board of Inland Revenue, and cousin of Juliet Pollock (*ODNB*).
63. *Nelson Matcham:* Nelson Matcham (1811–86), an English lawyer (*Alum. Cantab.*).
64. *H. Hunt:* Henry Hunt (*c.* 1802–1877), a London physician and author of medical books (*Athenaeum Rules* (1874), p. 65; F. Boase, *Modern English Biography,* 3 vols (Truro: Netherton and Worth, 1892–1901), vol. 1, p. 1590).
65. *G. H. A. Forbes:* G. H. A. Forbes (dates unknown), a British military officer (*Athenaeum Rules* (1862), p. 55).
66. *Sp Perceval:* Spencer Perceval (1828–90), an English lawyer (*Alum. Cantab.*).
67. *John Hawkshaw:* John Hawkshaw (1811–91), a British civil engineer (*ODNB*).
68. *Revd Tulloch:* John Tulloch (1823–86), a Scottish clergyman (*ODNB*).
69. *Henry Wyndham Phillips:* Henry Wyndham Phillips (1820–68), an English painter and son of the painter Thomas Phillips (1770–1845) (*ONDB*).
70. *L L Dillwyn:* Lewis Llewelyn Dillwyn (1814–92), a British industrialist and member of parliament (*ODNB*).
71. *Thomas Debary:* Thomas Debary (b. *c.* 1817), an Anglican clergyman who was vicar of Hatfield Peverell Essex (*Alum. Oxon.*).
72. *W. Frith:* William Powell Frith (1819–1909), an English painter (*ODNB*).
73. *W. Boxall:* William Boxall.
74. *E. Chisholm Batten:* Edmund Chisholm Batten (1817–97), an English barrister and antiquarian (C. D. Waterston and A. Macmillan Shearer (eds), *Former Fellows of the Royal Society of Edinburgh 1783–2002,* 2 vols (Edinburgh: Royal Society of Edinburgh, 2006), vol. 1, p. 180).
75. *John Everett Millais:* John Everett Millais (1829–96), an English painter (*ODNB*).
76. *Frederic Quvry:* Frederic Ouvry (1814–81), an English lawyer and antiquarian (*ODNB*).
77. *William Pole:* William Pole (1814–1900), an engineer and musician (*ODNB*).
78. *A H Layard:* Austen Henry Layard (1817–94), an English archeologist and politician (*ODNB*).
79. *Henry Vaughan:* Henry Vaughan (dates unknown), an English lawyer (*Athenaeum Rules* (1862), p. 88).
80. *Fred[k]. Marrable:* Frederick Marrable (*c.* 1819–72), an English architect for the Metropolitan Board of Works (*ODNB*).
81. *Geo T Clark:* George Thomas Clark (1809–98), an English engineer and antiquarian (*ODNB*).
82. *C. Austen Leigh:* Cholmeley Austen Leigh (b. *c.* 1830), an English lawyer (*Alum. Oxon.*).
83. *C Wigram:* Clifford Wigram (dates unknown), an English banker (*Athenaeum Rules* (1862), p. 91).
84. *Henry Oldham:* Henry Oldham (see letter 2851, n. 15).

85. *Chas. Landseer:* Charles Landseer (see letter 2851, n. 34).
86. *T B Curling:* Thomas Blizard Curling (1811–88), an English surgeon (J. F., 'Obituary Notices of Fellows Deceased', *Roy. Soc. Proc.,* 44 (1888), pp. xxv–xxvi).
87. *John E. Holdsworth:* John Eastabrooke Holdsworth (1811–88), an English lawyer (*Athenaeum Rules* (1874), p. 64; J. Foster, *Men-at-the-Bar: A Biographical Hand-list of the Members of the Various Inns of Court, including Her Majesty's Judges, etc.,* 2nd edn (London: Hazell, Watson, and Viney, 1885), p. 222).
88. *A. W. Fonblanque:* Albany William Fonblanque (1793–1872), a political writer and editor (*ODNB*).
89. *Joseph Whitworth:* Joseph Whitworth (1803–87), an English engineer (*ONDB*).
90. *de Strzelecki:* Paul Edmund de Strzelecki (1797–1873), a Polish nobleman and explorer (*ODNB*).
91. *William J. Thoms.:* William John Thoms (1803–85), an antiquarian and librarian for the House of the Lord (*ODNB*).
92. *George Denman:* George Denman (1819–96), a judge and Member of Parliament for Tiverton (*ODNB*).
93. *Trenham Reeks:* Trenham Reeks (1823–79), a mineralogist and registrar of the Royal School of Mines ('Obituary', *Geological Magazine,* 6 (1879), p. 288).
94. *E. Landseer:* Edwin Henry Landseer (1802–73), an English painter (*ODNB*).
95. *C. M. Roupell:* Charles Morris Roupell (1813–95), an English Lawyer (*Alum. Cantab.*).
96. *R W Mylne:* Robert William Mylne (1817–90), an architect, civil engineer, and geologist (*ODNB*).
97. *H Weekes:* Henry Weekes (1807–77), a sculptor and professor at the Royal Academy (*ODNB*).
98. *J. J. Sylvester:* James Joseph Sylvester.
99. *John Percy:* John Percy (1817–89), a metallurgist and professor at the Royal School of Mines (*ONDB*).
100. *J. F. Lewis:* John Frederick Lewis (1804–76), an English Orientalist painter (*ODNB*).
101. *J. A. Jeremie:* James Amiraux Jeremie (1802–72), a professor of Divinity at University of Cambridge (1850–70) and Dean of Lincoln (1864–72) (*ODNB*).

From John Herschel 26 November 1868 2910

Collingwood.[1] Nov 26/68

Dear Sir

Accept my thanks for your most concise and interesting paper on the action of Solar and Electric light on Vapours.[2] It is a class of reactions eminently calculated to set one thinking, and it seems to have had that effect on you to excellent purpose. I am glad to see that it has brought you into contact with the blue colour of the sky.—Still more if it should lead you to any satisfactory explanation of the polarization of sky-light. We are told that it is due to the reflexion of the ⊙'s[3] light on the molecules of air as the plane of pol[arizatio]ⁿ. passes through the Sun, I suppose that reflexion is somehow or other the cause of it and as the max[imu]ᵐ. of polarization is at 90° from the Sun—Brewsters law[4] would lead to the conclusion of the relative refractive index of the reflecting matter being 1.—That is a reflexion on air in air!! a conclusion which has always to me been a stumbling block.

The polarization of sky-light continues long after sunset—indeed so long as there is light enough in the sky to admit of the fact of its partial polarization being brought to the test of experiment—& the plane (in absence of the Moon) of pol[arizatio]ⁿ still continues to pass through the Sun as in the day time—how this clearly shows that the pol[arizatio]ⁿ is not effected by a single reflexion at the polarizing angle (90°—corresponds to μ=1) but has been induced by several successive reflexions—and this it is which leads me to conclude that the light of the corona in a total eclipse emanates (ultimately) in at least a very large proportion from the hinder part of the Sun's disc[5]—the fact of its polarization (assuming that fact established) proving the corona to be a true Solar atmosphere and the appearance not an optical effect originating in our atmosphere in which case (being clear to the ⊙) the polarization would be nil.

Is it conceivable that in passing out of a molecule of oxygen into one of nitrogen any percentage of light could be reflected? and in that case what is to be understood by the index of refraction as from one molecule into another?

In your luminous conversion of gas into visible vapour have you had any reason to suspect the formation of Vesicles?[6]—and—apropos of Vesicles did you ever see Saussure's[7] phenomenon of visible floating Vesicles in the Alpine clouds[8] when surrounding you in any of your excursions?

Excuse this rambling letter and believe me My dear Sir

Yours very tr[ul]y | J F W Herschel

P.S. Alex[ande]r[9] is working excessively hard at Glasgow and his classes of all kinds are increasing on him. I hope his health will stand it.

RI MS JT/1/H/110
RI MS JT/1/TYP/2/547–8
RS HS 24/241

---

1. *Collingwood:* see letter 2669, n. 1.
2. *paper on the action of Solar and Electric light on Vapours:* see letter 2905, n. 7.
3. ⊙*:* Herschel used this symbol for the sun twice in this letter.
4. *Brewsters law:* the law that the tangent of the Brewster angle is equal to the refractive index of the dielectric (in a vacuum), or to the ratio of the refractive indices of the media either

side of the interface (*OED*). See D. Brewster, 'On the laws which regulate the polarisation of light by reflexion from transparent bodies', *Phil. Trans.*, 105 (1815), pp. 125–159.

5. *light of the corona in a total eclipse . . . part of the Sun's disc:* possibly a stimulus to Tyndall's visit to Algeria in 1870 to observe the solar eclipse.
6. *suspect the formation of Vesicles?:* possibly a reference to Newton's idea, given in *Opticks* (1704), that the blue colour of the sky was due to the effects of minute vesicles of water with walls of different thickness. In his paper (see n. 2), Tyndall wrote: 'Newton conceived the colour to be due to exceedingly small water particles acting as thin plates' (p. 231).
7. *Saussure's:* Horace-Bénédict de Saussure (1740–99), a Swiss geologist, botanist, and mountaineer, educated at the University of Geneva. He published his research in his four-volume *Voyages dans les Alpes* (Neuchâtel: Samuel Fauche, 1779–96), and made the third ascent of Mont Blanc in 1787, conducting a number of scientific experiments upon the summit (*CDSB*).
8. *phenomenon of visible floating Vesicles in the Alpine clouds:* Saussure wrote in his *Voyages dans les Alpes: précédés d'un essai sur l'histoire naturelle des environs de Genève* (Genève: Barde, Manget, & Comp., 1886): 'Nous avons même une variété de Petrosilex, qui s'est complettement fondue en un verre brun, demi-transparent, compacte dans le fond du creuset; mais cellulaire à la surface. Cette variété est remarquable par des especes de tubercules arrondis, un peu plus petits que des pois, dont sa surface est couverte en quelques endroits. Ces tubercules sont gris, comme le reste de la pierre; quelques-uns d'entr'eux blanchissent vers le centre. Je pris d'abord ce caillou pour une Variolite; mais il a la cassure, la dureté, le degré de densité, & tous les autres caracteres du Petrosilex' (pp. 66–7). This translates to: 'We even have a variety of Petrosilex, which has completely melted into a brown, semi-transparent, compact glass in the bottom of the crucible; but cellular on the surface. This variety is remarkable for the species of rounded tubercles, a little smaller than peas, whose surface is covered in some places. These tubercles are gray, like the rest of the stone; some of them are bleaching towards the center. I took this pebble first for a Variolite; but it has the fracture, the hardness, the degree of density, and all the other characters of Petrosilex'. Petrosilex is 'any of various hard, compact, igneous or metamorphic rocks rich in silica', such as chert or felsite (*OED*). Saussure's observation is recounted in other works, some seemingly taking Saussure's claim with some skepticism. An 'elder Saussure saw in the high Alps, these vesicles floating around him, and although many of them were very small, some were as large as a pea, and of course large enough to bear ocular examination' ('The Wisdom of God as displayed in the formation of water', *Greenbank's Periodical Library*, 3 (1833), p. 390–9, on p. 395). 'Mists and clouds seem to consist of a multitude of hollow vesicles, or bladders, the coatings of which are inconceivably thin, and similar in structure to those usually blown from soap-suds. These vesicles vary in size, according to the measurement of de Saussure, from 1/4222 to 1/2620 of an English inch in diameter. M. de Saussure, senior, while travelling in the Alps, happened to be enveloped in mist, the vesicles of which he described being as large as peas. The remarkable magnitude of these vesicles compared with those seen by other observers, throws doubt upon the truth of the statement' (G. Hutchinson, *A Treatise on the Causes and Principles of Meteorological*

Phenomena (Glasgow: Archibald Fullarton and Co., 1835), p. 152). 'This last theory was supported by M. de Saussure, who affirms that in the dense fogs which prevail among the Alps, he saw vesicles as large as pease float past him, and covered with a coating as thin as it is possible to conceive' ('Visible Vapour', *Chambers's Journal of Popular Literature, Science, and Art,* 8 July 1865, pp. 422–5, on p. 424). 'Meteorologists have generally assumed, as an unsupported dogma, that cloud particles are small vesicles or watery bubbles. Saussure, indeed, is said to have examined these vesicles in the mists of high mountains, and to have found them vary in size from 1/4500 inch to the 1/2780 inch, occasionally however attaining the size of a pea. Now although no observer is in general more deservedly trusted than Saussure, his unsupported announcements must not always be adopted as conclusive, and we are quote unaware of any one else having witness vesicles of appreciable size floating in the air and forming clouds or mists' (H. Watts, *A Dictionary of Chemistry and the Allied Branches of Other Sciences,* 5 vols (London: Longmans, Green, and Co., 1866), vol. 1, p. 1028).

9. *Alex[ande]r:* Alexander Herschel.

To John Herschel 27 November 1868 2911

Royal Institution of Great Britain | 27th. Nov. 1868.

My dear Sir John

Very many thanks for your most suggestive letter[1] to which I shall pay due attention.

I have <u>never</u> seen those vesicles of Saussure[2] though I have anxiously sought them. Faraday once told me that he had seen them; but I have never been so fortunate.

I <u>am</u> devoting attention to the question of vesicles. The proofs hitherto adduced of their being vesicles are I think valueless. They float it is true in my tube for 24 hours without subsidence — but so do <u>solid</u> particles of asbestos.

I trust Mr. Herschel's[3] present work will give us a "purchase" to aid in getting him to London.[4] Nothing that I know has as yet been decided at the School of Mines.

Yours most truly | <u>John Tyndall</u>

RS HS 17/403
RI MS JT/1/TYP/2/549

---

1. *your most suggestive letter:* letter 2910.
2. *vesicles of Saussure:* see letter 2910, n. 7 and n. 8.
3. *Mr. Herschel's:* Alexander Herschel.
4. *getting him to London:* see letter 2833, 2834, and 2836.

To John Herschel 30 November 1868 2912

Royal Institution of Great Britain | 30th Nov. 1868

My dear Sir John.

I am this moment subjecting carbonic acid to the concentrated beam of the electric lamp. The action has been going on for some hours.[1]

A cloud, poor in comparison with many that I have obtained, though blue, or bluish, is formed in the tube. I looked at it this day through a Nicol's prism[2] and through tourmaline and observed the following:—

1°. Different quantities of light pass through the prism in different positions.

2° The tourmaline (cut parallel to the axis) shows the same difference.

3°. The difference is greatest when the line of vision is at right angles to the beam which illuminates the blue cloud.

4. A plate of selenite placed between the Nicol and the cloud shows vivid colours. A wedge of the substance shows in the proper position bright bars of colour. Unannealed glass and rock crystal also show the colours of polarized light.

5°. The colours are brightest when the line of vision is at right angles to the beam.

6°. A plate of tourmaline and one of selenite also show colours.

7°. The plane of vibration of the polarized light when the polarization is a maximum, is at right angles to the beam. In other words the plane of polarization passes through the source of light. When the axis of the tourmaline is perpendicular to the beam more light passes than when both are parallel.

Excuse these hasty lines—I thought the observations might interest you. I shall pursue the subject with finer clouds.

Yours very faithfully | John Tyndall

The dust particles of our laboratory, strongly illuminated by the electric light show the same effects with great splendour. I am not yet sure whether they are due to the visible spores & germs floating in the air, or to a residue of vapour fumes. The track of the beam is distinctly blue.

Has ordinary London dust with the sun shining upon it been examined?

As the cloud in the carbonic acid tube became coarser and whiter the polarization diminished materially.[3,4]

RS HS 17/404
RI MS JT/1/TYP/2/550–50a

1. *I am this moment . . . for some hours:* this experiment was for the research published as J. Tyndall, 'On the Blue Colour of the Sky, the Polarization of Skylight, and on the

Polarization of Light by Cloudy Matter Generally', *Roy. Soc. Proc.,* 17 (1868–9), pp. 223–33. Also see letter 2890, n. 2.

2. *Nicol's prism:* see letter 2894, n. 3.
3. *dust with the sun shining . . . polarization diminished materially:* this text is missing from the handwritten letter and is taken from the typescript.
4. *As the cloud in the carbonic acid tube became coarser and whiter the polarization diminished materially:* Tyndall would explain this in terms of the white colour developing (from blue) and polarization reducing because of the increasing size of the cloud particles (see n. 1).

From Alexander Buchan[1,2] [November 1868][3] 2913

Scottish Meteorological Society[4] | General Post Office Buildings | Edinburgh ____________18__

your experiments. I very sincerely hope your experiments[5] will enable you to expound the diathermic differences of the different states of vapour in all its transitions from pure to visible vapour.

I send you a copy of last № of our Journal containing an Article on certain American and European Storms,[6] with three synoptic charts representing a very large portion of the earth's atmosphere in three mornings of March 1859. This discussion has led me to consider more closely a question which I have ventured to dispute at page 347 et seq.[7] of my Book[8], viz that storms are not brought about by the mutual action of polar and equatorial currents. An equatorial current, moist and warm, always precedes and occupies the front part of storms; but a true polar current, rarely and only by accident, follows. There always follows storms an irregularly shaped patch marked off by a high Barometer, low (as compared with what precedes) temperature and dry air. This patch is seldom connected with the polar, or northern regions, but is with regard to European storms, about always to the SW or WSW, of the atmospheric disturbance; and out of it the wind blows in all directions as is determined by differences of pressure; that is gravitation acting on the inert matter of the atmosphere. Though the wind blows everywhere out of this patch, the pressure is not lessened, but on the contrary it increases provided that the storm in its vicinity increase in intensity and so diminish the pressure about their centres. The conclusion seems to me inevitable that the high pressures which succeed storms are produced by flow of air upon them from the storms in their vicinity, by upper currents.

Now if the lower temperature in the lower beds of the atmosphere where the high pressure prevails could be accounted for, the theory of storms is solved; at least the materials of solution are obtained. I present this problem of the low temperature to you,—to be solved, unless I greatly mistake, by a discussion of the changes which take place in the temperature and amount of vapour of a mass of air ascending, or being buoyed up by a continuous inflow of warm moist currents from below into the higher regions of the atmosphere, and then descending to the surface by being pressed down by fresh accessions constantly poured upon it from above. The facts are, that air is poured in, vorticosely, upon the space of low pressure, necessitating an ascending current there; and that air flows out of the space of high barometer, necessitating a descending current there. The question cannot be decided by observations, because the clouded state of the atmosphere does not admit of observations being made of the upper currents. It must therefore be decided by inference; and by an exposition of the physics of the questions which in my opinion is now the desideration.

I am working hard at my Isobaric Lines and Prevailing Winds over the Globe;[9] and have now collected or calculated good averages of Winds for upwards of 180 Places in all parts of the Globe. I hope to read the paper before our Royal Society early in March and am in hopes it will be published in the Transactions, Charts and Tables of Mean Monthly Pressure and Wind in extenso. The results are very interesting, and the Isobaric curves and arrows of mean direction of Wind for the months present the surface movements of the atmosphere in a simple and striking light.

The result is such as should be anticipated by all that regard the winds as masses of inert matter, and not as living agencies,—'chartered libertines',[10] in fact, as they are often represented.

Excuse the length of this note and believe me
very truly yours | Alexander Buchan
John Tyndall, Esq[ui]r[e]., L.L.D. &c. &c.

RI MS JT/1/B/139
RI MS JT/1/TYP/1/166–7

1. *Alexander Buchan:* Alexander Buchan (1829–1907), a Scottish meteorologist and oceanographer known for established the use of maps of weather patterns for weather forecasting. He was elected a Fellow of the Royal Society of Edinburgh in 1868, and FRS of London in 1898 (*ODNB*).
2. Letter incomplete; fragment only.
3. *[November 1868]:* the month and year are given based on the mention of a recently published article which appeared in October 1868 (see n. 6).
4. *Scottish Meteorological Society:* Buchan served as Secretary of the Scottish Meteorological Society from 1860–1907.
5. *your experiments:* see letter 2890, n. 2.
6. *Article on certain American and European Storms:* A. Buchan, 'On Two Storms which passed over the United States between the 13th and 22d March 1859', *Journal of the*

Scottish Meteorological Society, 2 (1868), pp. 198–213. Buchan was editor of this journal from 1860 until his death.

7. *et seq.:* and in what follows (Latin).
8. *my Book:* A. Buchan, *Handy Book of Meteorology,* 2nd edn (Edinburgh: William Blackwood and Sons, 1868).
9. *Isobaric Lines and Prevailing Winds over the Globe:* A. Buchan, 'The Mean Pressure of the Atmosphere and the Prevailing Winds over the Globe, for the Months and for the Year. Part II', *Transactions of the Royal Society of Edinburgh,* 25 (1869), pp. 575–637 (3 plates of charts precede p. 575). Read at the Royal Society of Edinburgh on 19 April 1869. The first part of this article, read 16 March 1868, was published as: A. Buchan, 'The Mean Pressure of the Atmosphere over the Globe for the Months and for the Year. Part I.—January, July, and the Year', *Proceedings of the Royal Society of Edinburgh,* 6 (1868), pp. 303–7.
10. *'chartered libertines':* freethinkers or nonconformists.

To John Herschel 1 December 1868 2914

Royal Institution of Great Britain | 1st. December. 1868

My dear Sir John

I thought the polarization effects that I mentioned to you yesterday[1] might be produced by the residue of vaporous fumes floating in the laboratory.[2] I therefore moved my lamp to the top of the house this morning far away from the laboratory, and found the air there well charged with floating dust which was illuminated by the cylindrical beam of the electric light.

I had a length of 12 or 13 feet of room thus illuminated.

Under these conditions I obtained all the effects which I described to you yesterday. A plate of tourmaline cut parallel to the axis allowed more light to pass when it was perpendicular to the beam than when it was parallel to it.

I have a plate of gypsum hollowed out great thinness in the centre, and gradually increasing in thickness all round. In the proper positions between the polarizer & analyzer of the polariscope, this plate gives a splendid system of rings.

Using the beam as polarizer & a Nicol[3] as analyzer the rings were also very vivid.

Doubtless this cloud of dust was mainly composed of germs and spores. I had great trouble with it sometime ago, and then proved it to be almost wholly organic.

But blue wood smoke, the smoke of a cigar, or even the fumes of chloride of ammonium bring out the effect splendidly.

But now comes a point that will interest you. As I have said the polarization is a maximum and the chromatic effects are most brilliant when you look straight across the beam. Supposing a very brilliant system of rings to be thus obtained and suppose the line of vision to pass gradually from the perpendicular, still keeping the Nicol steady before the eye, and the selenite steady between the Nicol and the beam, the colours gradually fade & at length entirely vanish. Continue to augment the obliquity, and they start into existence again, But now they are complementary to what they were before.

The line from the eye to the "neutral point" encloses roughly estimated an angle of 60° with the perpendicular to the beam.

This as I have said is produced by diverse objects, if they are only small enough to produce a blue tinge. Nor is it necessary to converge the beam to produce the effect; the uncondensed rays of the electric light acting on the floating atmospheric matter produce the same effect.

At all points of a beam from 12 to 15 feet long polarized light was copiously emitted.

What is true of particles floating in the air is also true of particles floating in water.[4] I placed today a beaker of water rendered bluish by precipitated mastic on the path of my beam. Looked at in a direction perpendicular to the beam, with the Nicol and the selenite in their proper positions I have never seen any thing more brilliant than the system of rings exhibited. You could produce nothing better with the best polariscope.

I hope I do not weary you with these details. I think we shall explode utterly the idea that the polarization of the atmosphere is due to reflection by the particles of air.[5]

Yours very faithfully | John Tyndall

I may add that when the neutral point was passed the horizontal tourmaline allowed most light to pass.

RS HS 17/405a
RI MS JT/1/TYP/2/551–2

1. *mentioned to you yesterday:* letter 2912.
2. *in the laboratory:* see letter 2890, n. 2.
3. *Nicol:* see letter 2894, n. 3.
4. *What is true of particles floating in the air is also true of particles floating in water:* this effect seen in colloids is now known as the Tyndall Effect.
5. *I think we shall explode . . . particles of air:* interestingly, in his paper on the colour of the sky (see letter 2894, n. 2), Tyndall never stated that it was molecules ('particles of the air') that were responsible, but merely particles with diameters 'but a very small fraction of the length of a wave of violet light' (p. 223). See also *The Ascent of John Tyndall,* pp. 245–8.

From William Fairbairn[1] 2 December 1868 2915

Manchester | Dec. 2nd 1868.

My dear Tyndall,

Many thanks for your interesting pamphlet on your new series of Chemical Reactions produced by light.[2] The results are very curious and full of instruction. I am not sufficiently informed on chemistry to follow you through all your experiments, but those with nitrite vapour and the light of the electric lamp and the light of the sun appear to produce the same results in the decomposition of the nitrite vapour and with the same rapidity and intensity

I trust your intentions of setting others to work will be fulfilled and with many thanks for thinking of me in these researches

I am always yours faithfully | W. Fairbairn

RI MS JT/1/TYP/12/3986
LT Typescript Only

1. *William Fairbairn:* William Fairbairn (1789–1874), a Scottish civil engineer who worked on bridges, shipbuilding, and railway locomotives. He was elected FRS in 1850, was third president of the Institute for Mechanical Engineers (1854–5), and President of the BAAS in 1861 (*ODNB*). Fairbairn published widely on construction and manufacturing, including *Useful Information for Engineers* (London: Longmans, 1856).
2. *pamphlet on your new series of Chemical Reactions produced by light:* see letter 2905, n. 7.

From John Herschel 2 December 1868 2916

Collingwood[1] | Dec 2/68

My dear Sir

Go on & prosper! It seems to me very clear that you are on the high road to something very remarkable.

I have often noticed the polarization produced by very moderate thicknesses of the ordinary atmosphere between the eye and surrounding objects (not looking up into the sky.)

A <u>spore</u> or a particle of <u>dust</u> can of course reflect light <u>in</u> air since it consists of material of greater refractive index than air. Still its index can hardly be inferior to that of water—so that a <u>single reflexion</u> at 45° incidence would not operate more than a very partial polarization & could give rise to no such neutral point & conversion to the opposite state.

When considering the nonproduction of a rainbow or cloud—Kamtz's[2] explanation has always appeared to me unsatisfactory as it assumes the vesicular state of the cloudy particles—and I have been driven to the conclusion that the absence of that internal reflexion on which the rainbow depends* may depend on the fact that the diameter of the drop may be smaller than a fraction of the undulation-length of the more refrangible rays, and that a refined analysis applied to the undulatory theory of reflexion taking into consideration this element, might possibly lead to conclusions as to reflexion from very thin plates parallel (of course not analogous) to Newton's[3] explanation of the black spot in a soap bubble. The excessive sharpness of termination of this black spot, or irregular black area, and its immediate transition to brilliant white without any gradation is a thing for which I see no explanation in the ordinary way of applying the undulatory theory to reflexion & points I think to something in that theory requiring to be approfondi[4]—Just as the separation into colours by the prism which at one time seemed to be an insuperable difficulty yielded to the conception of the sphere of action of the etherial particles not being infinitely smaller than the length of undulation. If a reflecting particle were less in diameter than 1/500000th of an inch (1 undul[ation] of a red ray) might not its reflexion of light be modified by this very same cause?

One thing puzzles me in your experiments. What are the nebulous particles which light produces in a gas or vapour? Are they the gas or vapour converted simply from the vaporous state to the liquid—If so they must be denser—& if so their production must produce a partial vacuum. Is this the case? as tested by a mercurial gauge.—You speak of such nebula being produced in Carbonic Acid! Can it be that the acid is really liquified or solidified—or is it decomposed and is the cloud infinitesimally divided Carbon?

Or—have we here that intermediate state between vapor & cloud that Mr Baxendell[5] suggests to account for the increase of the quantity of rain in the lower strata of the air—and for the enormous size of the drops on reaching the bottom of a deep shaft, which quitted their hold on a dropping beam or cross tie no bigger than ordinary drops?

Such rapprochemens[6] I think are not to be neglected—they are much easier suggested than verified or disproved which is my only excuse for writing you this kind of rambling talk.

One such may be worth mentioning—Query? is not the tumefaction[7] of Mercury when electrified in a sol[utio]^{n}. of Chloride of Ammon[iu]^{m} due to a metallic absorption of hydrogen analogous to Graham's strange facts of its absorption by Palladium &c? and not to the formation of a very light amalgam of ammonium? (though such an amalgam may still be present—whether light or heavy)

Mr. Huggins[8] speaks of the spectrum of Carbon as exhibited in the tail of a Comet.[9]—Can this be a case analogous to your phenomena produced by light in Carbonic Acid?

Believe me | My dear Sir | Yours very faithfully | J F W Herschel

*or two for the secondary Bow, 3 for a tertiary, &c. | Mr. Billet[10] has succeeded in obtaining ocular evidence of the existence of Bows as far as the 26th order!![11]

RI MS JT/1/H/111
RI MS JT/1/TYP/2/553–4
RS HS 17/405b

---

1. *Collingwood:* see letter 2669, n. 1.
2. *Kamtz's:* Ludwig Friedrich Kämtz (1801–67), a German physicist, glaciologist, and meteorologist who was appointed Director of the Central Physical Observatory of St. Petersburg in 1865 (*NDB*).
3. *Newton's:* see letter 2781, n. 6.
4. *approfondi:* studied in depth (French).
5. *Mr Baxendell:* Joseph Baxendell.
6. *rapprochemens:* rapprochement (French), a coming or bringing together; an establishment or resumption of harmonious relations (*OED*).
7. *tumefaction:* the action or process of tumefying, or state of being tumefied; swelling; swollen condition (*OED*).
8. *Mr. Huggins:* William Huggins.
9. *spectrum of Carbon as exhibited in the tail of a Comet:* this may have been a stimulus for Tyndall to think about comets. See J. Tyndall, 'On Cometary Theory', *Phil. Mag.,* 37 (1869), pp. 241–5.
10. *Mr. Billet:* Félix Billet (1808–82), a French physicist who studied optics, including the phenomena of rainbows. He was a professor at the Faculté des sciences, Dijon from 1843 until his death (*DBF*).
11. *26th order:* in the handwritten copy of this letter from the RS, '26th' is crossed out and '19th' added. In a later letter (2929), Herschel wrote, 'I think I mentioned in one of my letters that M. Billet had succeeded in rendering visible rainbows as far as the 26th order. It was a slip of memory. I should have said 19th'. Billet had recently published a paper on the 19th order of rainbows. See F. Billet, 'Mémoire sur les dix-neuf premiers arcs-en-ciel de l'eau', *Annales scientifiques de l'École Normale Supérieure,* 5 (1868), pp. 67–109.

To George Gabriel Stokes 2 December 1868 2917

Royal Institution 2nd December 1868

My dear Stokes

I trust you will consider the following facts of sufficient interest to be brought before the Royal Society. I may say that I have obtained them sooner than I otherwise should have done because of the great interest which Sir John Herschel manifested in a letter[1] which he had wrote to me a few days ago, in the question of atmospheric polarization.

I had been endeavouring to decompose carbonic acid by light; a bluish cloud was in my experimental tube; I looked at this cloud through a Nicol's prism.[2] Different qualities of light passed through the prism during a complete rotation around its axis.

A plate of tourmaline cut parallel to the axis of the crystal, and held between the eye and the experimental tube allowed neon light to pass when the axis of the crystal was perpendicular to the axis of the illuminating beam than when both of them were parallel.

The quantity of polarized light emitted by the beam in a direction at right angles to itself is a maximum.

Plates of selenite of suitable thickness placed between the Nicol and the illuminated cloud showed the colours of polarized light.

The colours were most vivid when the direction of vision was a right angles to the beam. When a plate of tourmaline was employed instead of the Nicol the effect was the same.

The beam here employed also illuminated the floating dust in the air of the laboratory. The track of the beam through this dust was distinctly blue. The dust must in part have been very attenuated.

All the effects obtained with the cloud in <*1 or more pages missing*>

It is the matter thus diffused which produce the effect; for when pure clear air is admitted into the experimental tube it shows no signs of polarization: the common air admitted into the tube shows vividly the colours of polarized light.

RI MS JT/1/S/244

1. *letter:* letter 2910. Also see letter 2890, n. 2.
2. *Nicol's prism:* see letter 2894, n. 3.

To John Herschel 3 December [1868][1] 2918

Royal Institution of Great Britain | 3rd Dec.

My dear Sir John,

You have set me to work upon this subject, and your notes[2] instead of "rambling" are of the most essential service to me. Carbonic acid as yet has yielded no definite results as regards its decomposition, though I have subjected it for about 15 hours to the most intense light. It however has nothing essential to do with our present subject.

I would give a great deal to show you the splendid effects which I obtained yesterday. I employed in the experiments iodine of Allyl, Benzol and bisulphide of Carbon; allowing a small quantity of the vapour of each to enter my tube, and then filling the tube with air which had been passed through aqueous hydrochloric acid.

On first turning on the light the tube seems absolutely empty. Looked at through the Nicol's prism[3] and the plate of selenite nothing is to be seen. In a few seconds however a cloud as purely blue as the firmament itself appears, and instantly the colours flash forth. They increase for a time in intensity and then begin to grow feebler; after a time the blue cloud(?) has changed to a white one and all traces of polarization disappear.

This blue cloud is undoubtedly the first step towards the formation of the real cloud. It is a case of true condensation, and the blue colour is due to the smallness of the particles. These particles are undoubtedly at first far less in diameter than length of a wave of red light.[4] Huxley tells me he has measured the spores floating in the air when they were less than 1/30000th of an inch in diameter.

May not interference have something to do with the matter? and may not the continuous accession of new small particles, as the old ones grow bigger be the cause of the whitening of the light?[5] A mixture of particles of different sizes would produce this effect—would it not?

Yours most truly | John Tyndall

RS HS 17/412
RI MS JT/1/TYP/2/555

---

1. *[1868]:* the year is given by reference to the experiments Tyndall was conducting toward the end of 1868 (see letter 2890, n. 2).
2. *your notes:* letter 2916.
3. *Nicol's prism:* see letter 2894, n. 3.

4. *the blue colour is due . . . wave of red light:* in his paper resulting from these experiments (see letter 2894, n. 2), Tyndall went further and stated that it was less in diameter than the length of a wave of violet light (p. 223).
5. *may not the continuous . . . whitening of the light?:* this was a conclusion Tyndall made in the paper resulting from these experiments (see letter 2894, n. 2)

From William Odling[1] 3 December 1868 2919

December 3rd 1868

My dear Tyndall

The chemical laboratory here[2] is sadly deficient in apparatus for lecture illustration I am without gas-holders, gas-bags for oxyhydrogen blowpipe,[3] gas-bottles for deflagrations,[4] suitable test glasses, tall gas-cylinders, glass jars of different kinds and other miscellaneous apparatus that would be in constant requisition for lecture purposes

As the cost of this apparatus would not probably fall within the regular working expenses of the laboratory I have to ask you kindly to satisfy yourself of the want of these things and then to get me a special permission from the Managers[5] to procure them. I will only add with regard to most of them that time presses as my course[6] is to begin in another three weeks

There is one other point on which I have to ask your assistance I want some place (independent of the room used by you) in which to put aside the apparatus got ready for the particular experiments of my course You will I am sure appreciate the necessity of having every thing properly put together tested as to functions &c. and ready to hand before the actual period of lecturing.

I am | My dear Tyndall | Yours very truly | W Odling

RI MS AD/03/D/04

---

1. *William Odling:* William Odling (1829–1921), an English chemist who worked on classifying the elements and refining Dmitri Mendeleev's periodic table of elements. Before becoming Fullerian Professor of Chemistry at the RI in 1868, he was a lecturer in chemistry at St Bartholomew's Hospital Medical School, and from 1872–1912 he was a professor of chemistry at Worcester College, Oxford. Odling was elected FRS in 1859 (*ODNB*).
2. *chemical laboratory here:* at the RI.
3. *oxyhydrogen blowpipe:* a type of blowpipe in which streams of oxygen and hydrogen meet as they emerge, producing an extremely hot flame by the burning of the hydrogen in the oxygen (*OED*), developed by the French chemist Jean Baptiste Gaspard Bochard de Saron (1730–94). See E. L. Gonzalez, 'Bochard de Saron and the Oxyhydrogen blowpipe', *Bulletin for the History of Chemistry,* 4 (1989), pp. 11–15.

4. *deflagrations:* rapid, sharp combustion with sudden evolution of flame; esp. the sudden combustion of a substance for the purpose of producing some change in its composition by the joint action of heat and oxygen (*OED*).
5. *Managers:* see letter 2833, n. 6.
6. *my course:* Odling delivered the RI's annual Christmas Lectures for children in 1868, the subject being 'On the Chemical Changes of Carbon'. The series of lectures were given on 29 and 31 December 1868, and 2, 5, 7, and 9 January 1869 (see *Roy. Inst. Proc.,* 5 (1866–69), pp. 427).

From George Gabriel Stokes 3–4 December 1868 2920

Dec 3/68

My dear Tyndall,

I received your letter[1] this morning.

The existence and the polarization of beams such as you describe is a fact with which I have long been familiar. They forced themselves upon my attention at every turn in experiments on fluorescence,[2] though I know I regarded them as enemies rather than friends, as things tending to mask what I was interested about than as things claiming attention for their own sake. You will find several references to them under the head "dispersion, false," at Phil. Trans. for 1852 Part 2. p. 562 of the vol.[3]

That the light of the clear sky was in part at least produced in this way is an idea with which I have so long been familiar that I cannot say when I first thought about it. I say in part at least for it has long been a problem in my mind, If we could absolutely get rid of all motes (including haze &c.) would we or would we not get rid of all the reflected light which we observe in the sky? Would the ultimate particles of matter—the diffused chemical molecules—be absolutely non-reflective? Again seeing that the density changes, would the fact that the change is continuous—that it takes place by an infinite number of infinitely small steps—prevent all reflexion?

I was surprised by the way in which Sir John Herschel in a letter I got from him lately attempted to explain the polarization of the corona in the late eclipse—by a series of reflections. I wrote back to him saying that I did not see the necessity for more than one reflexion at an incidence (mainly) of 45°, and adduced the case of the blue sky 90° from the Sun. I have not heard from him since. It was only on Monday I wrote.

I told you I regarded these beams as enemies rather than friends, and I paid no great attention to them.

Dec. 4. I recollect about the time of the appearance of Faraday's paper on finely divided gold[4] preparing his ruby fluid[5] and noticing that the scattered beam (viewed perpendicularly as I was in the habit of viewing it) showed less polarization than usual, which I connected in my mind with the metallic or quasi-metallic nature of the precipitate.

The polarization was evident, only I thought less than usual. I said quasi-metallic, because after all that Faraday has done I *[strike]* at the colour, and can't help thinking that the suspended body is some yet unknown compound of gold and not gold itself.

To return to the corona in a total eclipse, Prazmowski[6] who observed the strong polarization of that of the eclipse of 1860 also rightly interpreted it to indicate a reflexion at 45°; but how he conceived this reflexion produced is not very clear from the description.

The common phenomenon of sun beams recalls so completely the scattering of a sunbeam admitted into a darkened room by the dust in air that I can hardly imagine anyone not connecting the two. The polarization of the light of such beams I should imagine to be a familiar phenomenon—certainly my work at fluorescence rendered it so to me and Sir D. Brewster speaks of it in his paper on internal dispersion written at a time when the nature of the phenomenon shown by quinine solution &c. was still unknown. I recollect many years ago on the Faulhorn[7] endeavouring to make out by a differential observation whether the light reflected from a cloud was polarized. I turned a Nicol's prism[8] capped with calcspar alternately on to a cloud rather below my level, and to a point just behind it where I had for background a dark pine wood. The light scattered by the intervening air (air that is taken as a whole) was strongly polarized, and the pol[arizatio]n observed on the cloud seemed about as much as would be produced by this light superposed on white unpolarized light. I merely mention this to show that the polarization of a sunbeam was an idea with which I was familiar though I did not regard it as any novelty. In looking up, say from an ordinary low level, on the clear sky we have the light thus scattered, from the lower regions especially, superposed on the light, if any, which would come from pure air. Whether any light would under these circumstances come from pure air, whether independently of or in consequence of variations of density, is what I have long regarded as an open question.

In what you wrote to me on Dec 2 there is one thing quite new to me namely the existence of a neutral point in such a beam and the reversal of the polarization beyond it. This well deserves following out and applying to the neutral point or points in the sky.

I heard from Herschel[9] this morning. It seems the point he struck at was precisely how there could be reflexion from air in air. If the sun's atmosphere be constituted like our own there must be reflexion as a fact account for it as we will

Yours very truly | G. G. Stokes

RI MS JT/1/S/245
RI MS JT/1/TYP/4/1405–6

1. *your letter:* letter 2917.
2. *my experiments on fluorescence:* see G. G. Stokes, 'On the Change of Refrangibility of Light', *Phil. Trans.*, 142 (1852), pp. 463–562; on p. 479, in a note, Stokes coined the term *fluorescence.*
3. *"dispersion, false," at Phil. Trans. for 1852 Part 2. p. 562 of the vol.:* in the paper cited in n. 2, p. 562 is an index.
4. *Faraday's paper on finely divided gold:* M. Faraday, 'Experimental Relations of Gold (and other Metals) to Light', *Phil. Trans.*, 147 (1857), pp. 145–81.
5. *ruby fluid:* the ruby colour of colloidal gold was the main topic of Faraday's paper. On this, see D. Thompson, 'Michael Faraday's Recognition of Ruby Gold: the Birth of Modern Nanotechnology', *Gold Bulletin,* 4 (2007), pp. 267–9.
6. *Prazmowski:* Adam Józef Ignacy Prażmowski (1821–85), a Polish astronomer known for discovering the polarized emissions from the sun's corona during the total solar eclipse of 18 July 1860. See A. Prazmowski, 'Causes des rayons courtes de la couronne des éclipses solaires', *Cosmos,* 17 (1860), pp. 748–9; and A. Prazmowski, 'Observation de l'éclipse totale de Soleil du 18 Juillet 1860', *Comptes Rendus,* 51 (1860), pp. 195–7.
7. *Faulhorn:* a mountain in the Bernese Alps.
8. *Nicol's prism:* see letter 2894, n. 3.
9. *Herschel:* John Herschel.

To George Gabriel Stokes [7 December 1868][1] 2921

Monday

My dear Mr Stokes.

Would you kindly give me your opinion on a little optical point. I have written a short addendum to my paper on aqueous vapour having reference to the colour of the sky.[2] The vapour *[attracts]* the same class of rays as the liquid from which it is derived:—in other words it shares the colour of water. We are therefore surrounded by a blue medium. May not the colour of the sky be due to light transmitted through it? Were the vapour of the air condensed to a liquid shell it would not be competent to colour sensibly the light passing through it; but you perhaps remember those observations of mine on the colour of newly fallen snow; a little crack or hole in such snow is filled with pure blue light. This is no doubt caused by the innumerable reflections that occur on the limiting planes of the granules, by which what I may call the virtual depth of the snow traversed by the light is enormously increased. Thus while the light of the snow is transmitted light, reflection is the agency which renders the absoption[3] effective.

Similar reflections must occur in the atmosphere; you cannot whip air & aqueous vapour to a homogeneous whole; the atmosphere is an aggregate of air parcels in different states of saturation. Reflection must take place at their limiting surfaces, and in this way the vertical depth of vapour is augmented & the light finally reaches us from the atmosphere as blue.

The polarization of the sky cannot be accounted for by reflection or water particles or vesicles. The angle of maximum polarization does not admit of this assumption. Sir J. Herschel found that angle to be about 90°—the polarizing angle therefore 45°, & hence the index of refraction close to unity—Would not reflection at the limiting surfaces of differently saturated air parcels meet this case? I should be much obliged to you if you would give me your opinions on these points before I send in the addition to my paper. I do not forget the light of sunset. It is a difficulty, but I think one to be got over.

Yours very since[rel]y | John Tyndall

RI MS JT/1/T/1429
RI MS JT/1/TYP/4/1482

---

1. *[7 December 1868]:* based on content, this letter (written on a Monday) follows letters 2917 (2 December, a Wednesday) and 2920 (December 3, a Thursday) and precedes but is written on the same day as 2922 (7 December).
2. *my paper on aqueous vapour having reference to the colour of the sky:* see letter 2894, n. 2.
3. *absoption:* 'absorption' misspelled by Tyndall.

From George Gabriel Stokes 7 December 1868 2922

Dec 7/68

My dear Tyndall,

I think your paper[1] certainly worthy of being communicated to the Royal Society. If the idea is new to so distinguished a physicist as Sir John Herschel,[2] and has not been put in print by any one there is reason enough for producing it. The idea of accounting in this way for the blue of the sky was one which could hardly help forcing itself upon me, familiar as I was from my experiments on fluorescence[3] with the beams produced by finely divided colourless precipitates, with the bluish appearance of such a mixture by reflected light, and with the polarization of the scattered beam. But I never followed out the subject either theoretically (beyond giving the obvious explanation of the bluishness) nor experimentally, and I have long regarded it as an undecided question whether this was the true account of the blue colour or whether there might not possibly be reflexion even within homogeneous air (i.e. air free from motes or haze) whether by reason of a minute reflexion even from the ultimate molecules, or from variations of density giving surfaces on opposite sides of which the medium had slightly different values of the refractive index. To disprove the existence of such reflexion in homogeneous (or mote[4]-less) air is a thing hardly within the reach of experiment considering how small are the thicknesses of air looked through in our experiments compared with the miles through which we look in regarding the sky. Still if you can show by experiment that you can by a sufficiently fine precipitate get a blue as deep as that of an Italian sky, and that air can be so purged as to show no sensible beam that goes far to render improbable reflexion otherwise than by motes.

I have marked your letters[5] with the dates of reception not knowing whether you mean the first of them[6] for a communication to the Society or merely a sketch of a somewhat longer communication you meant to write on hearing from me.

Yours very truly | G. G. Stokes

RI MS JT/1/S/246
RI MS JT/1/TYP/4/1407

1. *your paper:* see letter 2894, n. 2. As the final paper was received by the RS on 16 December, it appears that Tyndall revised the paper he initially sent to Stokes prior to 7 December.
2. *if the idea is new to so distinguished a physicist as Sir John Herschel:* see previous letters from Herschel on this subject, 2910, 2911, 2912, 2914, 2916, and 2918. Letters to Herschel following this are 2927 and 2929.
3. *my experiments on fluorescence:* see letter 2920, n. 2.
4. *mote:* a particle of dust, esp. one of the innumerable minute specks seen floating in a beam of light (*OED*). While Tyndall referred to 'motes' in his paper (see letter 2894, n. 2), Stokes here referred to 'molecules' as he does in his paper that Tyndall cited (see letter 2920, n. 2).
5. *your letters:* letters 2917 and 2921.
6. *first of them:* letter 2917.

To Thomas Archer Hirst 8 December 1868 2923

8th Dec. 1868

Dear Tom

I have got a pheasant & a hare: If you dine at the Athenaeum[1] today send up for them & give orders about the cooking of the pheasant.

Yours affect. | John

RI MS JT/1/T/673
RI MS JT/1/HTYP/545

1. *Athenaeum:* see letter 2537, n. 8.

From George Gabriel Stokes 9 December 1868 2924

Dec 9/68

My dear Tyndall,

I return you your letters[1] as you wish. I would make these suggestions

(1) The air of any dwelling house, and indeed any air about noisy bustling London is hardly a fair specimen of ordinary pure air. You remember the dust-stratum a few thousand feet thick that Piazzi Smyth[2] speaks of (Phil. Trans. Teneriffe experiment[3]) I think it would be worth while in one of your mountain tours to rig up a dark box in such a manner that the pure mountain air should blow freely through it, at least about the focus of the condensing lens turned to the sun, while the observation was being made.

(2) Please to remember that with reference to the point I have mentioned in Art 183 of my paper in the Phil. Trans. for 1852 p. 530[4] it would be very interesting to me if you would kindly notice whether it appeared to be a general rule when the suspended particles were the very finest that the polarization at 90° was sensibly perfect. Also whether it makes a difference in this respect whether the suspended particles are transparent or sensibly opaque or rather endowed with an opacity comparable with that of metals.

(3) It would be worth your while to observe with light (the incident light I mean) polarized in a plane ⊥ to that passing through the eye and the axis of the beam. You would probably thus get at a glance phenomena similar to those otherwise obtained piecemeal.

I can't find the reference[5] to Arago[6]

Yours very truly | G. G. Stokes

RI MS JT/1/S/247
RI MS JT/1/TYP/4/1408

1. *your letters:* letters 2917 and 2921. Also see letter 2922, n. 1.
2. *Piazzi Smyth:* Charles Piazzi Smyth (1819–1900), an English astronomer who was Astronomer Royal for Scotland and Professor of Practical Astronomy at the University of Edinburgh, both from 1846–88. Beyond his many astronomical contributions, Piazzi Smyth also studied the Great Pyramid of Giza (*ODNB*).
3. *Phil. Trans. Teneriffe experiment:* C. P. Smyth, 'Astronomical Experiment on the Peak of

Teneriffe, carried out under the sanction of the Lords Commissioners of the Admiralty', *Phil. Trans.,* 148 (1858), pp. 465–533.

4. *my paper in the Phil. Trans. for 1852 p. 530:* G. G. Stokes, 'On the Change of Refrangibility of Light', *Phil. Trans.,* 142 (1852), pp. 463–562, article 183 on pp. 530–2.
5. *reference:* F. Arago, 'Sur la polarisation de la lumière atmosphérique et de la lumière de la lune', in J.-A. Barral (ed), *Oeuvres complètes de François Arago,* 7 vols (Paris: Gide et J. Baudry, 1854–8), vol. 7, pp. 435–6.
6. *Arago:* Dominique François Jean Arago (1786–1853), a French mathematician, physicist, and astronomer. Arago made significant contributions to the study of optics and was awarded the RS's Copley Medal (1825) for his work on magnetism. He was appointed astronomer at the Paris Observatory in 1809 later serving as Director (1843–53) (*CDSB*).

From Rudolf Clausius 11 December 1868 2925

WURZBURG, 11 12/68.

Lieber Tyndall,

Es hat mir grosse Freude gemacht, durch Deinen lieben Brief einmal wieder etwas von Dir und Hirst zu hören, zugleich aber hat es mir leid gethan zu erfahren, dass Du so unwohl gewesen bist. Ich glaube Du arbeitest zu viel, und zu angestrengt. Dass Du aber, wenn Du unter solchen Umständen zur Erholung nach der Schweiz reisest, auch noch so anstrengende und gefährliche Touren machst, wie die Besteigung des Matterhorns, hat mich in Erstaunen gesetzt. Ich will hoffen, dass der Aufenthalt in den Bergen Deine Gesundheit nicht nur vorübergehend, sondern dauernd gekräftigt hat.

Für die Mittheilung über das Buch von Tait danke ich Dir bestens. Ich habe es noch nicht erhalten, aber ich denke doch dass er es mir schicken wird, da er mir sonst alle seine Bücher geschickt hat. Vielleicht hat er es nach Zürich geschickt, von wo es dann etwas verspätet hierher kommt. Sollte es nicht kommen, so kann ich es mir leicht durch einen hiesigen Buchhändler besorgen lassen, ohne dass Du Dich deshalb zu belästigen brauchst.

Die Proceedings der Royal Society erhalte ich zwar, aber nicht die einzelnen Hefte, wie sie erscheinen, sondern immer eine grössere Anzahl von Heften zusammen. Ich habe daher das Heft in welchem sich Deine Arbeit befindet, noch nicht erhalten, was mir sehr leid thut. Sollte es sich ohne Unbequemlichkeit machen lassen, dass mir die einzelnen Hefte gleich nach ihrem Erscheinen auf meine Kosten zugeschickt würden, vielleicht durch die Buchhandlung welche sie herausgiebt, so würde es mir lieb sein. Ich wünsche aber durchaus nicht, dass der Secretär der R.S. durch mich irgend wie belästigt wird.

Was Du mir über Deine Resultate in Bezug auf die chemische Wirkung des Lichts schreibst, interessirt mich sehr. Es wird dadurch ein ebenso wichtiger als dunkler Punct der Physik und Chemie aufgeklärt und vielleicht ein neues Gebiet der Forschung eröffnet werden. Zugleich spielen die chemischen Wirkungen des Lichtes in der Natur eine so hervorragende Rolle, dass kaum irgend ein Gegenstand mehr geeignet ist, allgemeines Interesse zu erregen.

Ich weiss gar nicht ob Du noch mit Hirst zusammen an Waverly Place wohnst, oder ob Du jetzt in Faraday's Räumen im Gebäude der Royal Institution wohnst, und Dich von Hirst getrennt hast. Schreibe mir doch gelegentlich auch darüber etwas.

Uns geht es hier in Würzburg ganz gut. Ich hoffe jetzt da das erste Jahr vorüber ist, und ich im physikalischen Cabinet nicht mehr so viel zu ordnen habe, auch wieder mehr zu eigenen Arbeiten zu kommen. Die Kinder sind wohl und wachsen frisch heran. Die beiden ältesten Mädchen fangen schon an zu arbeiten, während die Kleinen ihre Zeit mit Spielen verbringen, wobei sie in Lärmen und Toben das Nöthige zu leisten verstehen.

Nun, lieber Tyndall, lebe wohl, und nimm für dich und Hirst freundliche Grüsse von meiner Frau und mir,

Dein | Clausius.

WÜRZBURG,[1] 11 12/68.

Dear Tyndall,

I was very pleased to hear through your kind letter[2] something about you and Hirst[3] again, but at the same time I was sorry to learn that you have been so unwell. I believe you are working too much and too hard. The fact, though, that when you travel under such circumstances to Switzerland to relax, you also go on such strenuous and dangerous excursions as climbing the Matterhorn,[4] astonished me. I want to hope that your stay in the mountains has strengthened your health not just temporarily, but permanently.

I thank you very much for the information about the book by Tait.[5] I have not received it yet, but I do think that he will send it to me, as he has sent me all his books before. Perhaps he has sent it to Zürich, from where it will then come here with somewhat delayed. Should it not arrive, then I can easily have it provided to me through a local bookseller without you needing to bother yourself about it.

I do receive the Proceedings of the Royal Society, though not the individual issues as they appear, but always a larger number of issues together. I have therefore not yet received the issue in which your article[6] is found, which I very much regret. If it can be done without inconvenience that the individual issues were sent to me <u>at my expense</u> immediately after they appear, perhaps by the bookshop which publishes them,[7] then I would be glad. I definitely do not wish, though, that the secretary of the R.S.[8] is in any way bothered because of me.

What you write to me about your results regarding the chemical effect of light interests me very much. Through it, a point of physics and chemistry that is as important as it is obscure is cleared up and perhaps a new field of research will be opened. At the same time, the chemical effects of light in nature play such an outstanding part that hardly any topic is more suitable to arouse general interest.

I do not know at all whether you are still living at Waverly Place[9] together with Hirst, or whether you are now living in Faraday's rooms in the building of the Royal Institution,[10] and have parted company from Hirst. Do write something about this to me sometime or other too.

We are quite well here in Würzburg. I am hoping now, as the first year is over and I no longer have so much to put in order in the physics laboratory, that I shall return more to my own work also. The children are well and growing healthily. The two oldest girls[11] are already beginning to do some work, while the little ones[12] spend their time at play, in which they know how to produce the requisite noise and do the requisite stomping around.

Now, dear Tyndall, fare well and accept kind regards from my wife[13] and me to you and Hirst.

Your | Clausius.

RI MS JT/1/TYP/7/2309–10
LT Typescript Only

---

1. *WÜRZBURG:* Clausius held a professorship at the University of Würzburg.
2. *your kind letter:* letter 2905.
3. *Hirst:* Thomas Hirst.
4. *climbing the Matterhorn:* see letter 2855, n. 3.
5. *book by Tait:* see letter 2742, n. 4.
6. *your article:* J. Tyndall, 'On a New Series of Chemical Reactions Produced by Light', *Roy. Soc. Proc.* 17 (1868–9), pp. 92–102.
7. *bookshop which publishes them:* the *Roy. Soc. Proc.* and the *Phil. Trans.* were both printed by Taylor and Francis.
8. *secretary of the R.S.:* in 1868 the secretaries of the Royal Society were George Gabriel Stokes who held the position from 1854–85 and the physiologist William Sharpey (1802–80) who held the position from 1854–72.
9. *Waverly Place:* see letter 2707, n. 8.
10. *Faraday's rooms in the building of the Royal Institution:* see letter 2707, n. 8.
11. *two oldest girls:* Mathilde (1861–1907) and Helene (1863–1919).
12. *little ones:* Rudolf John (1864–1915?), Hedwig (1866–1946), and Alfred (1868–1950).
13. *my wife:* Adelheid Clausius.

From George Gabriel Stokes 11 December 1868 2926

Dec 11/68

My dear Tyndall,

I thought I recollected talking to Roscoe[1] about the cause of the blue of the sky, and stating my belief that it arose, in part at least, from suspended particles—whether over and above that there is such a thing as molecular reflexion, to which some portion of the deeper blue may be due, I regard as a point which it is hardly in the power of experiment to settle. Yesterday I casually came across Roscoe's lecture at the Royal Institution (Proc. Vol. IV) on the opalescence of the atmosphere[2] in which precisely that explanation is given. If I am right in supposing that you regarded this as the capital point of your communication, the discovery that it has already been given in print may modify the form in which you may think it right to bring forward your communication.[3] As to the theory of the production of blue—why it is that finely suspended particles give a bluish reflexion—I think Brücke[4] altogether wrong.

I enclose a glass[5] which it may interest you to examine. Please to let me have it again when you have done with it, but there is no hurry.

Yours very truly | G. G. Stokes

RI MS JT/1/S/248
RI MS JT/1/TYP/4/1409

1. *Roscoe:* Henry Roscoe.
2. *Roscoe's lecture at the Royal Institution (Proc. Vol. IV) on the opalescence of the atmosphere:* H. Roscoe, 'On the Opalescence of the Atmosphere', *Roy. Inst. Proc.,* 4 (1862–66), pp. 651–9. Delivered as a Friday Evening Discourse at the RI on 1 June 1866.
3. *may modify the form . . . forward your communication:* indeed Tyndall did mention Roscoe in his paper (see letter 2894, n. 2), on p. 232: 'Roscoe, in connexion with his truly beautiful experiments on the photographic power of sky-light, has also given various instances of the production of colour by suspended particles. In the foregoing experiments the azure was produced in *air,* and exhibited a depth and purity far surpassing anything that I have ever seen in mote-filled liquids. Its polarization, moreover, was *perfect*'.
4. *Brücke:* Ernst Wilhelm Ritter Brücke (1819–92), a German physiologist and professor at the University of Vienna from 1849. Educated in Berlin he was one of the founders of the Physikaleische Gesellschaft (1845) along with Emil du Bois-Reymond and Hermann Helmholtz (*CDSB*; E. Brücke, 'Ueber die Farben, welche trübe Medien im auffallendedn un durchfallenden Lichte zeigen,' *Poggend. Annal.,* 88 (1853), pp. 363–85).
5. *glass:* not identified.

To John Herschel 12 December [1868][1] 2927

Royal Institution of Great Britain | 12th Dec.

My dear Sir John.

If you would kindly glance at this paper[2] & return it to me as soon as you have done so with any remarks that strike you as bearing on it I would thank you much.

Yours very truly | John Tyndall

RS HS 17/413
RI MS JT/1/TYP/2/556

1. *[1868]:* the year is given based on mention of Tyndall's paper on the blue colour of the sky (see n. 2), which was received by the RS on 16 December and read at the RS on 14 January 1869.
2. *this paper:* J. Tyndall, 'On the Blue Colour of the Sky, the Polarization of Skylight, and on the Polarization of Light by Cloudy Matter Generally', *Roy. Soc. Proc.,* 17 (1868–9), pp. 223–33.

From George Gabriel Stokes 12 December 1868 2928

Dec 12/68

My dear Tyndall,

I forgot when I was with you to give you a reference to Arago's[1] work on Atmospheric polarization.[2] It is to be found in Vol VII of his works p. 435 and thereabouts. He appears to have written short notices which were read to the Academy[3] and preserved in the Archives, which accounts for their not appearing in our catalogue.[4]

I doubt if I ever saw in any country so pure and deep a blue in the sky as yesterday about 3h30m P.M.[5] I can tell you the circumstances when we meet.

Your observation that the polarizing angle becomes 45° when the suspended particles are of the most extreme fineness is a capital point for me. It agrees perfectly with the theoretical statement which I have made in Phil. Trans. 1852 p. 531, l. 4–6.[6] I did not examine my beams with a view to determine the changes dependent upon a varying inclination, but received them ⊥ly or nearly so, and when the suspended particles were very fine I found the pol[arizatio]n perfect, on all but (art. 182[7]) with suspended gold (?) ruby gold it was somewhat different.

Yours very truly | G. G. Stokes

RI MS JT/1/S/249
RI MS JT/1/TYP/4/1410

---

1. *Arago's:* see letter 2924, n. 6.
2. *work on Atmospheric polarization:* see letter 2924, n. 5.
3. *Academy:* l'Académie des sciences, or French Academy of Sciences, founded in 1666. Arago was elected a member of the academy in 1809.
4. *our catalogue:* Stokes was referring to the *Catalogue of Scientific Papers (1800–1863), Compiled and Published by the Royal Society of London,* 6 vols (London: Eyre and Spottiswoode, 1867–72). Arago's papers are listed in vol. 1 (1867), pp. 80–4.
5. *I doubt if I ever saw . . . about 3^h30^m P.M.:* Stokes was in Cambridge.
6. *Phil. Trans. 1852 p. 531, l. 4–6:* see letter 2920, n. 2. Lines 4–6 on p. 531 read, 'But if the diameters of the particles be small compared with the length of a wave of light, it seems plain that the vibrations in a reflected ray cannot be perpendicular to the vibrations in the incident ray'.
7. *art. 182:* article 182 in Stokes's paper is on p. 530.

From John Herschel 14 December 1868 2929

Collingwood[1] | Dec 14/68

My dear Sir,

I return your paper[2] (per Book Post registered) with many thanks for permitting me to read it.[3] I am exceedingly delighted with the way in which you have followed out this most curious and interesting problem and verified so completely my prediction that the true explanation of the blue colour of the sky would be found to carry with it that of the polarization of skylight. You have made it clear I think that they are part & parcel of one and the same ultimate fact—an ultimate fact which in the action of light itself on vapours, is allied to chemistry and in the relation of the material particles of all but infinitesimal magnitude to the luminiferous Ether belongs to the department of analytical dynamics—an ultimate fact for which your experiments place us on the high road but which it will require a good deal more of experiment in such hands as your own to present in a perfectly definite & intelligible form

Among the remarks which have occurred to me in reading the paper there is only one which seems to me worth mentioning, it is in reference to the gradual change of the polarizing angle, and the appearance of a neutral point and thence going over to the opposite (a negative) polarization. This change it appears accompanies & is commensurate with the gradual increase of the density of the cloud. The neutral points in sky-polarization have never so far as I know had any attempt made to explain them. But I think your exp[erimen]ts give a clue to one. Why is it that a white cloud viewed at the polarizing of water (of which it no doubt consists) shows no sign of polarization? Obviously because the rays it sends to the eye consist of a mixture of rays reflected, rereflected, refracted, rerefracted &c &c in all sorts of planes from all sorts of depths within the clouds.—Again why does your artificial blue sky polarize "positively" the light it sends to the eye?—Because only the light immediately sent off by a first reflexion is strong enough to affect the eye. But as the density of the cloud increases, light which has undergone refraction (and of which therefore a percentage equal to the reflected light has become negatively polarized) in greater and greater abundance becomes deflected by successive reflexions not at the polarizing angle & begins to mix with the first reflected rays in sufficient abundance 1st to neutralize the character of the total beam and then to predominate somewhat, till at length the increasing quantity of light that has undergone many cross reflexions & refractions prevails over all and confounds all sign of pol[arizatio]n.

I am aware that this explanation is somewhat confused and that there will arise against it objections to which I do not yet see a very clear answer, but to which I am inclined to think more consideration will furnish a reply. Any how one thing seems clear. As positive Polariz[atio]n arises from reflexion so negative looks, naturally, for its origin to that portion of each ray which (escaping reflexion) acquires ipso facto what is called a partial negative pol[arizatio]n. i.e. a true neg[ative]. pol[arizatio]n. of a percentage of its total amount equal to the percentage reflected. All I want to postulate is that somehow or other this portion, preserving its negative character shall somehow or other get directed towards the observer—and in sufficient abundance to exceed the other—without prejudice to its being reinforced by refracted light from the general mass.

As regards the gradual change of the angle of polarization—that seems to have relation to the gradual increase of size of the particles and to be connected with the dynamics of the subject.[4] It seems to me almost self evident that if a material reflecting or refracting particle be so small as that its dimension shall stand in a commensurable ratio to the extent of the sphere of action of the repulsive force of the ethereal particles constituting the luminous Ether, its total action will not be identical with that of a mass of such dimensions as shall infinitely exceed that sphere of action.

Of course all this is very crude.

I think I mentioned in one of my letters[5] that M. Billet[6] had succeeded in rendering visible rainbows as far as the 26th order. It was a slip of memory. I should have said 19th.—

I remain | my dear Sir | Yours very truly | J F W Herschel

PS. How about the blue colour of water? Does that too arise from disseminated particles? or is the blue colour absorption?[7]

RI MS JT/1/H/112[8]
RI MS JT/1/TYP/2/557–8
RS HS 24/243[9]

---

1. *Collingwood:* see letter 2669, n. 1.
2. *your paper:* see letter 2927, n. 2.
3. *permitting me to read it:* in letter 2927, Tyndall had asked Herschel to read the paper and comment on it.
4. *As regards the gradual change . . . dynamics of the subject:* Tyndall picked up on this in his paper (see letter 2927, n. 2), writing: 'But as the sky-blue is gradually rendered impure by the introduction of particles of too large a size, in other words, as real clouds begin to be formed, the polarization begins to deteriorate, a portion of the light passing through the prism in all its positions' (p. 226).
5. *one of my letters:* letter 2916.
6. *M. Billet:* see letter 2916, n. 10.
7. *How about the . . . blue colour absorption?:* Tyndall would later explain this as due to the absorption of red and reflection back of blue, the same reasons for the blue colour he observed inside a hole in the snow made by an ice axe shaft while climbing in the Alps.
8. A draft of this letter, dated 13 December 1868, is held by the RS (RS HS 19/160). There are only minor variations in text.
9. This is a handwritten transcript.

To Harriet Grote[1] 15 December 1868 2930

15th. December <u>1868</u>

My dear Mrs. Grote,

I fear very much that the perfume of Arabia will not breathe from this note, for it is written in the laboratory in the midst of doubtful odours. At all events it is not addressed to you from a Castle of Indolence.[2]

It would give me exceeding pleasure to join you on Saturday, but I am pledged to dine in this Babylon[3] on that day, and am also pledged to quit it with a friend on Sunday morning.

I am right glad to hear of your welfare; and if the inexorable Gods would but open their ears to my wishes your welfare would be always secured. But the Gods are stupid and cannot hear.

Last Sunday evening I was talking about you with a clear headed quick witted, and altogether decided friend of yours—Lady Seabright.[4] I went to chapel, heard a hymn, and fancied myself like Faust[5] when the easter chaunt[6] charmed him. The law of association gives these old hymn tunes a fascination for even an unbeliever like me. I had also a very quiet, but very earnest discussion with Sir Roundel Palmer.[7] As long as they are not dogmatic, and do not talk about Hellfire, I am very tender to the holders of the old faith. It warms them inwardly and does me no harm. There are things in optics called "imaginary foci", from which light & heat <u>seem</u> to come, though the <u>real</u> focus is a very different thing. Now I take it Sir Roundel Palmer & such people get their light & heat from imaginary foci, thinking them <u>real</u>. They simply make a mistake as to its origin.

Health & loyal service to the Historian.[8]

Most truly yours | <u>John Tyndall</u>

MS ADD 266, UCL Library Services, Special Collections

---

1. *Harriet Grote:* Harriet Grote (née Lewin, 1792–1878), an English biographer who also supported her husband George Grote's scholarly work (*ODNB*).
2. *Castle of Indolence:* a reference to 'The Castle of Indolence', a 1748 poem by Scottish poet James Thomson (1700–48).
3. *this Babylon:* London.
4. *Lady Seabright:* possibly Marie, Baroness d'Everton, the first wife of Charles Sebright (1807–84), Consul-General of the Ionian Islands ('Obituary for 1884. October. Sir Charles Sebright', *Times Register of Events in 1884* (London: Times, 1884), p. lxvi). Following Marie's death, he married his second wife Georgina Mary Muir Mackenzie (1833–74) in 1871 (*ODNB*).
5. *Faust:* a reference to the main character in Goethe's novel *Faust* (1847). For Goethe, see letter 2885, n. 9.
6. *Chaunt:* alternative spelling of 'chant'.
7. *Sir Roundel Palmer:* Roundell Palmer (1812–95), 1st Earl of Selborne, an English lawyer and politician (*ODNB*).
8. *the Historian:* Harriet Grote's husband George Grote (1794–1871), an English classical historian and politician, known for his 12-volume *A History of Greece* (1846–56) (*ODNB*). He married Harriet Grote on 6 April 1820 ('Marriages', *London Magazine,* 1 (1820), p. 601).

To Rudolf Clausius 17 December 1868 2931

17th Dec. 1868

My dear Clausius

I have this day given directions to have a copy of Tait's[1] book[2] sent to you without delay. I have not yet read it myself but I have glanced at the preface & at the letter of Helmholtz[3] to Tait. Helmholtz does not agree with Tait, he rather expresses dissent from his views regarding Mayer,[4] but he uses comparisons and illustrations which I hardly think posterity will ratify.

When I have a little more time I shall translate the collected works of Mayer,[5] so that the world may know what the man has done.

Tait's confession that he acts as an advocate is very amazing. It is to me a strange way of dealing with a scientific question.

I have been working very hard of late at a subject which you touched upon some years ago, namely, the colour of the sky & the polarization of skylight; also the polarization of light by cloudy matter generally.[6] I have established the following points:—

1. A blue as deep as that of the purest Italian sky can be produced by the fine matter of an incipient cloud.

2. This matter, as long as it continues fine enough to give the blue, polarizes light perfectly. The direction of perfect polarization is at right angles to the track of the beam which illuminates the incipient cloud.

For example, supposing the cloud to be in a wire glass tube through which a beam of light is passing; when you look at the illuminated cloud horizontally through a nicol[7] with its long diagonal vertical, the light emitted by the cloud is absolutely quenched; when the short diagonal is vertical the light from the cloud is transmitted.

3. As the particles of the cloud grow larger the polarization becomes less perfect; and when at length a texture is obtained approaching in coarseness to that of an ordinary cloud the polarization along the normal disappears.

4. But such a course cloud polarizes light in oblique directions. For example in many cases, where the cloud had ceased to discharge polarised light normally I found the light emitted at an angle of 12° or 14° with the beam, strongly polarized.

5. If the fine blue cloud, be illuminated by a beam of polarized light instead of by common light, the cloud prove itself capable of reflecting light in two directions only Speaking more strictly: suppose the plane of vibration of the polarized beam to be vertical the cloud emits light right & left horizontally; but is absolutely incompetent to emit light vertically upwards or downwards.

This continues only so long as the cloud remains in the fine condition which yields a blue colour.

6. The polarization of light by matter in this condition is absolutely independent of the polarizing angle. No matter what the refractive index of the substance may be when the matter is fine enough the colour by reflected light is blue, and the polarization along the normal perfect.

7. When a strong beam of light is sent through ordinary London air it is polarized, though not so vividly and perfectly as by my incipient cloud. The direction of most perfect polarization is also normal to the beam.

8. But when the fumes of resin, gunpowder, incense, chloride of ammonium &c fill the laboratory the maximum polarization is no longer normal, but oblique. The beam regarded obliquely also divides itself into two parts in one of which the light is polarized in a plane at right angles to the plane of polarization of the other portion. A neutral point exists between both parts.

Wheatstone[8] has informed me that Govi of Turin[9] noticed this some years ago, but he did not pursue the subject further.

9. The position of the neutral point I find to be a function of the density of the fumes. For example, if after the laboratory has been well filled with fumes, and the position of the neutral point determined, the ventilators be spaced so as to gradually dissipate the fumes; the angle enclosed between the line of vision to the neutral point, and the axis of the beam becomes less & less. In many cases it has gradually diminished from 66° to 33°.

10. Supposing you look at the beam passing through the ordinary air of the laboratory through a nicol & a selenite plate thinnest at the centre, you obtain the usual ring system when the nicol is in its proper position. If the fumes of tobacco, phosphorus, sulphur or chloride of ammonium be now thrown into the beam at the proper degree of density, the ring system becomes much more brilliant, but the polarization remains unchanged. Sulphur shows this with particular beauty & intensity.

11. But if an ordinary aqueous cloud, the fumes of hydrochloric, hydrobromic, or nitric acid, be thrown into the beam, the plane of polarization turns suddenly 90° and the selenite tints are all rendered complementary.

These are a few of the principal points which I have established. I think they will give new work to the undulatory theory.

A happy new year to you all.

Yours ever | John Tyndall

RI MS JT/1/T/202

---

1. *Tait's:* Peter Guthrie Tait.
2. *book:* see letter 2742, n. 4.
3. *Helmholtz:* Hermann Helmholtz.
4. *Mayer:* Julius Mayer (see letter 2840, n. 12).
5. *collected works of Mayer:* Tyndall had previously translated several of Mayer's papers and had them published in the *Phil. Mag.* in 1872 and 1873, but he never published a collected works.
6. *I have been working . . . by cloudy matter generally:* see letter 2890, n. 2.
7. *nicol:* see letter 2894, n. 3.
8. *Wheatstone:* Charles Wheatstone.

9. *Govi of Turin:* Gilbert Govi (1826–89), an Italian physicist who studied optics and electricity as well as the works of Leonardo da Vinci. From 1861 he held the professorship of experimental physics at the University of Turin (*Dizionario Biografico Degli Italiani,* 94 vols (Rome: Istituto della Enciclopedia Italiana, 1960–2019), vol. 58, pp. 174–7).

To Elizabeth Dawson Steuart 19 December 1868 2932

Royal Institution. | 19th. Dec. 1868.

My dear Mrs Steuart,

Much against my will I trouble you with a letter written to me by a very troublesome person of whom I have no knowledge.[1] I also enclose my reply to this man.[2]

I think it would be best to pay him his paltry account unless it is evidently a fraud.

I am going out of town to-morrow for a few days[3] to rest my brain and cool my blood. I have been working exceedingly hard of late, rising very early, and experimenting all day long. I am happy to say that it is not without result. I have been able by my work to add materially to our knowledge.

I have had some exceedingly interesting correspondence with Sir John Herschel,[4] who though I believe over 80, has a wonderfully fruitful and elastic mind.

Speaking of ages I have had letters from Mrs Somerville[5] who is now living at Naples; and who is in her 88th year. She writes without the slightest tremor.

I have also had two letters from the late Chief Baron Pollock[6] which might be mistaken for the letters of a boy; they are so plainly and firmly and roundly written. He is in his 86th year.

The President of this Institution[7] is over 80, but he runs up three flights of stairs to my rooms like a youth.

The President of the Royal Society[8] is over 80, but he is able to attend its evening meetings, its dinners, and to deliver its addresses.

A member[9] of the Royal Society Club,[10] where I sometimes dine and of which I am a member, is 95. He is quite brisk, only a little deaf.

But I do not know what has caused me to write about these things. They may amuse you.

Last Sunday I met a remarkable group of men under remarkable circumstances. A number of liberal Statesmen surrounding a Tory nobleman. The Marquis of Salisbury[11] was our host, and among the guests were Gladstone,[12] Cardwell,[13] Lowe[14] and Roundell Palmer.[15]

Yours ever | John Tyndall

A happy New Year.

RI MS JT/1/TYP/10/3384
LT Typescript Only

---

1. *letter written to me . . . I have no knowledge:* letter missing; man not identified.
2. *my reply to this man:* letter missing; man not identified.
3. *out of town to-morrow for a few days:* on 20 December, Tyndall and Thomas Hirst left London for a trip to Malvern and surrounding areas, going for walks and visiting sites. They separated on 23 December, Hirst heading to Bristol while Tyndall remained in Malvern (Brock & MacLeod, *Hirst Journals,* 20–23 December 1868, p. 1838). From letter 2933 to Herbert Spencer we know Tyndall was still in Malvern on 24 December, but when he returned to London is unknown.
4. *correspondence with Sir John Herschel:* see letters 2910, 2911, 2912, 2914, 2916, 2918, 2927, and 2929.
5. *letters from Mrs Somerville:* Mary Somerville (see letter 2803, n. 1); letters missing.
6. *had two letters from the late Chief Baron Pollock:* letters 2892 and 2893. Jonathan Frederick Pollock had turned 85 on 23 September 1868. By 'late', Tyndall was referring to the fact that Pollock was no longer Lord Chief Baron of the Exchequer, an appointment he served from 1844–68.
7. *President of this Institution:* Henry Holland, who had turned 80 on 27 October 1868.
8. *President of the Royal Society:* Edward Sabine, who had turned 80 on 14 October 1868.
9. *member:* not identified.
10. *Royal Society Club:* the Philosophical Club of the RS, founded in 1847 by William Robert Grove in the aftermath of the reform of the RS's governance. Tyndall was elected a member on 10 May 1855 (Journal, JT/2/13c/745). See also T. G. Bonney, *Annals of the Philosophical Club of the Royal Society, Written from Its Minute Books* (London: Macmillan and Co., 1919).
11. *Marquis of Salisbury:* Robert Gascoyne-Cecil (1830–1903), 3rd Marquess of Salisbury, an English conservative politician who would serve as Prime Minster three times (1885–6, 1886–92, and 1895–1902) (*ODNB*).
12. *Gladstone:* William Gladstone (1809–98), a British liberal politician who would serve as Prime Minister four times (1868–74, 1880–5, 1886, and 1892–4) and as Chancellor of the Exchequer four times (1852–5, 1859–66, 1873–4, and 1880–2) (*ODNB*).
13. *Cardwell:* Edward Cardwell (1813–86), the Peelite politician who had served under Lord Aberdeen as President of the Board of Trade from 1852–5 (*ODNB*).
14. *Lowe:* Robert Lowe (1811–92), 1st Viscount Sherbrooke, a British liberal politician who served as Chancellor of the Exchequer (1868–73) and later Home Secretary (1873–4) (*ODNB*).
15. *Roundell Palmer:* see letter 2930, n. 7.

To Herbert Spencer 24 December [1868][1] 2933

Great Malvern[2] | Xmas-eve

My dear Spencer.

I have read your proof[3] from beginning to end. It is excellent on the whole and as clear for the most part as mud itself.

The unmarked parts pleased me exceedingly; you have cleared away what I always thought to be a blemish both in Darwin's and Huxley's view of Evolution; and have stated your case with exceeding clearness.

I have made one or two trifling remarks on the margin of the marked portion. I would break up the latter part of the article into smaller paragraphs.

It is perfectly manifest that your reviewer[4] is not in the least fit to execute the task that he has undertaken. His limitation of the "mechanical" caused me to stare.[5]

Yours ever | John Tyndall

University of London—GB 96 MS 791/77
RI MS JT/1/TYP/3/1030

---

1. *[1868]:* the year is given by reference to Spencer's reply to a review of a book of his (see n. 3).
2. *Great Malvern:* part of the spa town of Malvern, Worcestershire, England, to the north-west of London. See letter 2932, n. 3.
3. *your proof:* Spencer's reply to a review of an American edition of his *The Principles of Biology,* 2 vols (New York: D. Appleton and Co., 1866–7) by Francis Ellingwood Abbot. The review had been published in October in *The North American Review,* 107 (1868), pp. 377–422. Spencer sent his reply to the same journal, but as he later wrote it was 'declined by the Editor in pursuance of a general rule' (R. G. Perrin, *Herbert Spencer: A Primary and Secondary Bibliography* (New York: Routledge, 2014), pp. 162–3). Spencer then sent it to *Appleton's Journal of Literature, Science, and Art,* which published it in two parts the following summer: H. Spencer, 'Spontaneous Generation', *Appleton's Journal,* 31 July 1869, pp. 562–3; and 7 August 1869, pp. 597–8. Appleton later published it separately as: H. Spencer, *Spontaneous Generation, and the Hypothesis of Physiological Units: A Reply to the North American Review* (New York: D. Appleton and Co., 1870). Spencer included his reply in later editions of *The Principles of Biology.*
4. *your reviewer:* Francis Ellingwood Abbot (1836–1903), a controversial American philosopher and theologian who in a 1868 court trial was deemed not fit to serve his congregation because of his support for 'free religion' in which individuals were encouraged to find their own truth and not rely on dogma or scripture (*ANB*; H. W. Bowden, *Dictionary of American Religious Biography,* 2nd edn (Westport, CT: Greenwood Press, 1993), pp. 2–3).

5. *His limitation of the "mechanical" caused me to stare:* on p. 381 the reviewer wrote: 'But in the adoption of a false method, namely, the interpretation of evolution as a purely mechanical process, we find the cardinal demerits of his attempt'.

From Alexander Herschel 28 December 1868 2934

57 N. Hanover Street | Glasgow NB.[1] |
28th Dec[embe]r 1868.

Dear Professor Tyndall

I should have thanked you long ago for the Paper[2] which you were so good as to send me, on a new Action of Light on Certain Vapours,—Nitrite and Iodide of Amyl and Isopropyl, and the three Hydracids,—in the "Experimental Tube".—But my attention was taken up with my classes too constantly to allow me to acknowledge it until the Xmas holidays, which give me a little time and leisure for following your very remarkable and interesting experiments more closely.

I did not perceive their importance in the first place, in relation to what Professor Roscoe[3] calls the "opalescence" of the sky[4], and notice the production of sky blue in certain cases; but was more struck with the chemical changes produced by the action, to which you particularly draw attention in order to determine the nature of the action, whether the absorption of energy takes place internally within a molecule or without it; and by the extraordinary symmetry of the figures or "Nebulae" produced by the beam! Add to which the iridescence and the suite of colors succeeding each other in them.—I have more lately been reflecting on the examples of "opalescence" or production of sky blue which is one of the chief features of the precipitations; and have tried Brücke's[5] experiments with solution of gum-mastic in Alcohol,[6] with the help of the Magnesium light. This light shows the bluish tint of the diffuse light of the solutions very distinctly, and beautifully, but probably not more so than the electric light would do.—

Some time ago, Professor Roscoe was so good as to send me a Paper, which he read to the Manchester Philosophical Society,[7] describing the result of my Brother's[8] observations on the Photographic intensity of the Sun's light during the eclipse;[9] in which he shows that the photographic intensity of the diffuse day-light fell more rapidly at the beginning of the eclipse, than corresponded with the diminution of the visible area of the Sun's disc, and rose again after the eclipse in the same manner. He attributes the circumstance to the moon concealing, first, the Corona and the red-flames, which have a certain photographic intensity, before beginning to encroach on the visible disc, and continuing to do so during the beginning, and <u>un</u>covering them in the same way at the end of the eclipse, more rapidly and completely than in the middle, of it, about the total phaze; (as the figures show more clearly, in this sketch).

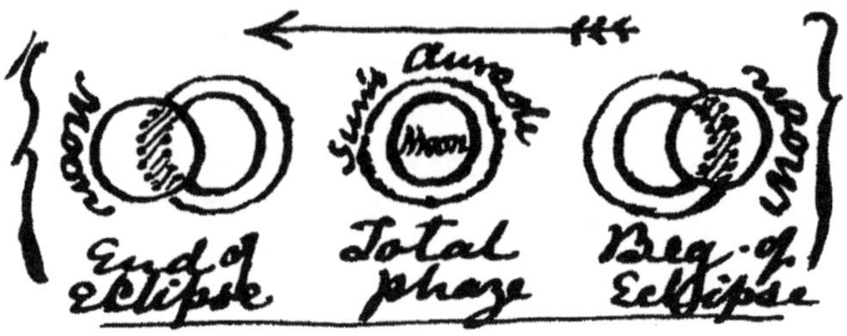

The discovery of a means of measuring the photographic intensity of the Sun's light, either direct, or as diffuse daylight is due to the Professors Roscoe and Bunsen,[10] and leads to the curious result, that the chemical power of the direct rays of the Sun's light is sometimes nil in comparison with that of diffuse daylight, when the sun is even 120 above the horizon; owing to the "opalescence" of the atmosphere absorbing the chemical and transmitting only the rays of low refrangibility.—The phenomenon of diffused skylight close to the Sun which prevents us from seeing the red prominences without a total eclipse, may, I suppose be classed under the head of "opalescence" and as its amount is sometimes so extraordinary as the above experiments describe I would like to allude to Professor Roscoe's researches[11] in speaking of the great discovery of Mr. Lockyer,[12] that the red-flames may be seen with the spectroscope,[13] even without an eclipse of the Sun; and of Mr. Janssen's[14] observations[15] of them for 17 successive days after the total eclipse in August last![16]

The turbid liquid which I will use will be Brücke's solution of gum-mastic dropped into water; and the experiment to illustrate the energetic absorption of the chemical rays, by this opalescent liquid, will be one which Prof. Roscoe showed at the Royal Institution on the 1st of June 1866;[17] but in which he says that he only partly succeeded, because the Sulphur-tinged liquid which he use coagulated, and the bulbs exploded more freely than he expected.—

He has sent me the troughs and bulbs which he used in that experiment; and with a little care, and the use of Brücke's liquid instead of the mixture of Sulphur and water, I hope to make it perfectly succeed.—In his letter (which I enclose herewith) he speaks of the gold solutions prepared and used by Dr Faraday in his experiments on the same remarkable peculiarities of light;[18] which are still at the Royal Institution; and I hope that you will allow them to be placed on the table, where they will occupy such a very conspicuous and important position in my lecture, on Friday evening, the 22nd of next month?[19]—

I am writing to Mr Ladd[20] to make ready a few demonstrations of the complementary colours in quartz, and other properties of polarized light illustrating the circumstance of its "having sides", by which the planes of reflection of light from the sun, reflected to us from the corona, and from all parts of the clear sky on a cloudless day, are shown to pass through the Sun's centre, and accordingly to derive their light from his source.—

I have, here, a very fine Dubosq's Polariscope[21] with large Nicoll's prism[22] and two large doubly refracting prisms of Iceland spar, which I have used for this purpose very successfully with the electric lamp, and there is besides a Nüremburg Apparatus,[23] and a great collection of Selenite Slides which will not require any additional apparatus to be prepared to illustrate the subject of polarization. Mr Ladd will however provide me a thin plate of Iceland spar, showing the black cross; and perhaps one, also, of Nitre for the same purpose:—He will I hope be able to show some of Mr Huggin's[24] solitary line spectra of the Nebulae,[25] and of the new star in corona, to which he will add one of the red-flames as observed during the eclipse, and, as more recently measured, again, by Mr Lockyer ~—Mr. Warren DelaRue,[26] to whom I wrote, a few days since, asking him for the loan of a few slides, will I hope be good enough to furnish me some interesting ones for this occasion to illustrate, if possible the actual photographs of the eclipse!—If any of your most recent, wonderful researches in the polarization of artificial skylight, could be made visible to a large audience, the explanation of polarization, now known to be a common condition of the light reflected from the atmospheres of both the Earth and the Sun, would be a most excellent opportunity to illustrate their physical importance; of which I would be most glad to avail myself, and I would value and esteem the privilege most highly!

I hope that you will kindly excuse this scrawled sheet of paper; which I have, in haste, taken up first by the wrong page. A lecture on Mechanics this evening[27] obliged me to close it, rather quickly, without attempting to rewrite it; as I would very much wish to do.

With the best wishes of the Season, and of a Happy New Year to you, I remain,

Dear Professor Tyndall | Yours very truly | A S Herschel.

RI MS JT/1/H/83
RI MS JT/1/TYP/2/584–6

---

1. *NB:* North Britain.
2. *Paper:* see letter 2905, n. 7.

3. *Professor Roscoe:* Henry Roscoe.
4. *"opalescence" of the sky:* on 1 June 1866, Roscoe gave a Friday Evening Discourse at the RI, 'On the Opalescence of the Atmosphere' (H. Roscoe, 'On the Opalescence of the Atmosphere', *Roy. Inst. Proc.*, 4 (1862–66), pp. 651–9).
5. *Brücke's:* Ernst Brücke (see letter 2926, n. 4).
6. *experiments with solution of gum-mastic in Alcohol:* E. Brücke, 'Ueber die Farben, welche trübe Medien im auffallenden und durchfallenden Lichte zeigen', *Poggendorff's Annal.*, 88 (1853), pp. 363–85.
7. *Manchester Philosophical Society:* the Manchester Literary and Philosophical Society, founded in 1781.
8. *my Brother's:* John Herschel (1837–1921), an English military engineer and astronomer. Herschel spent the years 1864–72 working for the Great Trigonometrical Survey in southern India. Elected FRS in 1871 and a fellow of the Royal Astronomical Society in 1872, Herschel was the son of John Herschel (1792–1871) and grandson of William Herschel (1738–1822) (*ODNB*).
9. *observations on the Photographic intensity of the Sun's light during the eclipse:* J. Herschel, 'Account of the Solar Eclipse of 1868, as Seen at Jamkandi in the Bombay Presidency', *Roy. Soc. Proc.*, 17 (1868–69), pp. 103–25 (read at the RS on 19 November 1868 in Herschel's absence). This paper was followed up with: J. Herschel, 'Spectroscopic Observations of the Sun (Continued)', *Roy. Soc. Proc.*, 17 (1868–69), pp. 506–10 (received at the RS on 10 June 1869).
10. *Bunsen:* Robert Bunsen.
11. *Professor Roscoe's researches:* see n. 4.
12. *Mr. Lockyer:* Norman Lockyer (1836–1920), an English chemist who is most well-known for discovering helium (jointly with Pierre Janssen; see n. 14). After serving in the British War Office, Lockyer turned his attention to astronomical researches, making his observations from his home in Wimbledon, South London. In 1869, the year Lockyer was elected FRS, he founded the journal *Nature* and was its editor until 1920 (*ODNB*).
13. *red-flames may be seen with the spectroscope:* N. Lockyer, 'Sur une méthode pour observer en temps ordinaire le spectre des protubérances signalées dans les éclipses totales de Soleil', *Paris, Comptes Rendus,* 67 (1868), pp. 836–8, and N. Lockyer, 'Sur les protubérances solaires', *Paris, Comptes Rendus,* 67 (1868), pp. 949.
14. *Mr. Janssen's:* Pierre Jules César Janssen (1824–1907), a French astronomer who contributed to stellar spectroscopy and photography. He is credited along with Norman Lockyer (see n. 12) for discovering helium (*CDSB*).
15. *observations:* P. Janssen, 'On the Solar Protuberances', *Roy. Soc. Proc.,* 17 (1868–69), pp. 276–7.
16. *I would like to allude to . . . total eclipse in August last!:* Herschel would give a Friday Evening Discourse at the RI on 22 January 1869, 'On the latest Eclipse of the Sun' (*Roy. Inst. Proc.,* 5 (1866–9), p. 450). While this lecture was not published, Herschel ended the lecture 'by calling attention to the recent solar discoveries of Mr. J. Norman Lockyer and Dr.

Janssen' ('Recent Lectures at the Royal Institution', *The Engineer,* 5 February 1869, p. 108).

17. *Prof. Roscoe showed at the Royal Institution on the 1st of June 1866:* see n. 4.
18. *D^r. Faraday in his experiments on the same remarkable peculiarities of light:* see letter 2920, n. 4.
19. *my lecture, on Friday evening, the 22nd of next month:* see n. 16.
20. *M^r. Ladd:* probably William Ladd (1815–85), an English optician and maker of scientific instruments with a London shop at 11–12 Beak Street (see G. Clifton, *Directory of British Scientific Instrument Makers 1550–1851* (London: Zwemmer, 1995), pp. 125–6, 161).
21. *Dubosq's Polariscope:* Louis Jules Duboscq (1817–86), a French instrument maker known for producing high quality optical instruments (*DBF*).
22. *Nicoll's prism:* see letter 2894, n. 3.
23. *Nüremburg Apparatus:* Nörrenberg apparatus, a type of polariscope invented by Louis Jules Duboscq in 1860.
24. *M^r. Huggin's:* William Huggins.
25. *solitary line spectra of the Nebulae:* see W. Huggins and W. A. William, 'On the Spectra of some of the Nebulae', *Phil. Trans.,* 154 (1864), pp. 437–44.
26. *M^r. Warren DelaRue:* see letter 2576, n. 1.
27. *lecture on Mechanics this evening:* probably the evening class on mechanics that Herschel began in the autumn of 1868. The following is from the minutes of a meeting of the Managers of Anderson's University on 3 November 1868: 'A letter of date 2nd. inst[ant] from Professor Herschell [sic] was read in which he stated that he regretted that he had been unable to form a Day Practical Class on Mechanics at a fee of £2.2/- per session as advertised but he intimated that he was willing to endeavour to form an Evening Practical Class twice a week at a very low fee. The Managers approved of this:—to have a practical evening class, experimentally in the meantime in place of the Day Class, two evenings in the week, on Monday & Wednesday or such other evening as the Hall may be unoccupied at at eight o'c[lock]—the fee to be 5/-.' (Archives and Special Collections, University of Strathclyde, OB/1/1/5: Minutes of Anderson's University 1864–1881, pp. 75–6). This first session of 1868–9 reportedly had 28 students (OB/1/1/5, p. 90).

From Felix Giordano[1] 31 December 1868 2935

Florence le 31 X^bre 68

M^r le Professeur Tyndall

À mon retour de Grèce qui a eu lieu déjà depuis quelques semaines, j'ai retrouvé votre gracieuse lettre du 20 octobre m'accusant réception de la Coupe Géologique du Mt Cervin que je vous avais envoyée.—Depuis cette époque je n'ai plus guère eu le temps de m'occuper de notre ami-sauvage; toutefois j'ai écrit un petit mémoire sur son orographie et géologie que je vous enverrai dès qu'il sera imprimé. En attendant je me permets de vous adresser une autre copie de la Coupe du Cervin un peu retouchée, avec 2 autres planches qui doivent l'accompagner dans le mémoire sur-dit. Je vous prie de les agréer en témoignage de mon estime et de ce que vous avait fait dans le temps nous l'ascension de ce pic.

Quant à moi je dirai que vraiment j'ai trouvé cette fois le pic assez difficile à cause de la grande quantité de neige qui était déjà tombée pendant le mois d'août. J'ai surtout trouvé difficile la traversée de l'arrête qui suit la pointe Tyndall (pointe qui est maintenant marquée dans ma figure *[1 word illeg]*) du côté d'Italie. Quant au versant Suisse je l'ai trouvé moins difficile que je ne croyais parce que la neige y était un peu consolidée par la chaleur des premiers jours de septembre. Moi j'ai monté le 4 du dit mois. En descendant le pic du côté de Zermatt j'ai encouru un véritable danger par les avalanches de pierres qui tombaient sur le pied du pic donnant sur le glacier de Furgen. Un de mes 2 guides a eu le havresac coupé en deux par un bloc et moi-même j'ai été un peu contusionné. C'est un danger qui a été signalé par tous les ascenseurs de cette année.

En outre du Cervin, j'ai fait cette année plusieurs autres courses et j'ai traversé plusieurs glaciers dont quelques uns m'ont impressionné comme confirmant en tout point les principes posés par vous sur leur structure et leurs phénomènes principaux. J'ai cru ainsi avoir reconnu en général que

1e Les <u>crevasses</u> se forment avec leur longueur dans le sens normal à la direction de la plus grande tension.

2e La structure <u>veinée</u> se forme ayant les lignes en veines dans le sens normal à la direction de la plus grande <u>compression</u>.

3e Que les bandes sales (dirth-bands[2]) ont généralement naissance au pied des <u>grandes chutes des séracs</u>, et que leur existence aurait relation non seulement avec les plissements de la surface du glacier, mais aussi à la structure veinée même qui se produit au pied de la chute à cause de la grande pression ayant lieu à cet endroit.

Si vous aviez quelque nouvelle, remarque ou objection à me faire au sujet de ces phénomènes glaciaires, je vous en serais tout à fait obligé.

Permettez M^r^ le Professeur que je vous renouvelle mes sentiments de profonde estime—Mr Felice Giordano | Via Venezia N.2

Florence 31 December 68

Professor Tyndall,

Upon my return from Greece, which happened already a few weeks ago, I found your kind letter of October 20,[3] acknowledging receipt of the Geological Cross-section of Mt Cervin[4] that I had sent you. Since then, I have hardly had the time to take an interest in our <u>wild-friend</u>;[5] however, I wrote a short paper on its orography and geology[6] that I will send you as soon as it is printed. In the meantime, I take the liberty to send you another copy of the cross-section of Cervin, a bit retouched, with 2 other charts attached to it in the paper that I mentioned before. I ask you to accept it as a token of my esteem and for what you did during the ascent of this peak.

As for me, I will honestly say that this time I found the peak rather difficult, because of the large quantity of snow that had already fallen during the month of August. I especially found it difficult to cross the crest that follows the Tyndall peak (a peak that is now marked in my *[1 word illeg]* figure) from Italy's side. As for the Swiss slope, I found it less difficult than I thought because the snow was consolidated by the heat of the first days of September. I climbed on the 4th of that month. While descending the peak from the side of Zermatt, I incurred a real danger from the stone avalanches that fell at the foot of the peak overlooking the glacier Furgen.[7] One of my two guides had his knapsack cut in half by a rock, and myself, I was a bit bruised. It is a danger that was reported by all the climbers of this year.

Aside from Cervin, I made several other hikes this year and I crossed several other glaciers, some of which impressed upon me the affirmation, in all respects, of the theories that you posed on their structure and their main phenomena. In general, I believe that I recognized that:

1st The crevasses occur with their length normally in the direction of the greatest tension.

2nd The veined structure is formed with the lines in the veins normally in the direction of the greatest compression.

3rd That the dirt bands were generally created at the foot of the large ice falls, and that their existence is not only linked to the creases on the surface of the glacier, but also to the veined structure that is also produced at the foot of the fall because of the high pressure in that location.

If you have any new comments or any objections to me regarding these ice phenomena, I would be much obliged if you let me know.

Allow me, Professor, to emphasize again my deepest esteem.

Mr Felice Giordano | Via Venezia N.2

RI MS JT/1/G/6

---

1. *Felix Giordano:* Felix Giordano (see letter 2886, n. 1).
2. *dirth-bands:* Giordano misspelled *dirt-bands.*
3. *your kind letter of October 20:* letter missing; possibly Tyndall's reply to Giordano's letter of 16 October (letter 2886).
4. *Geological Cross-section of Mt Cervin:* see letter 2886, n. 4.
5. *our wild-friend:* the Matterhorn (Mt Cervin is the French name).
6. *short paper on its orography and geology:* see letter 2886, n. 7.
7. *glacier Furgen:* now Furgg, the main glacier under the east face of the Matterhorn, on the Swiss side.

From Edward Frankland [1868][1] 2936

My dear Tyndall

It is not known that there is any action between bisulphide of carbon and aqueous hy-drochloric acid when they are brought together.

Yours sincerely | E. Frankland

RI MS JT/1/TYP/12/3959
LT Typescript Only

---

1. *[1868]:* the year is given as it appears this is related to experiments Tyndall was conducting in late 1868. In letter 2918 to John Herschel from 3 December [1868], he wrote: 'I would give a great deal to show you the splendid effects which I obtained yesterday. I employed in the experiments iodine of Allyl, Benzol and bisulphide of Carbon; allowing a small quantity of the vapour of each to enter my tube, and then filling the tube with air which had been passed through aqueous hydrochloric acid'.

BIOGRAPHICAL REGISTER

This register contains the names and biographical details of people who are mentioned three or more times in the letters included in this volume. Biographical information about people mentioned only once or twice is contained in the notes appended to the appropriate letter(s).

Conventions

The sources for the biographical entries are standard biographical dictionaries, such as the *Complete Dictionary of Scientific Biography*, the *Oxford Dictionary of National Biography*, Poggendorff's *Biographisch-literarisches Handwörterbuch der exacten Wissenschaften*, *Historishes Lexikon der Schweiz*, *Neue Deutsche Biographie*, and the *Dictionary of Nineteenth-Century British Scientists.* These have not been specified in the entries unless necessary. Additionally, the editors have found material from the Tyndall correspondence and from the journals of Tyndall and T. A. Hirst very useful. Where additional sources add important material, they are given in parentheses at the end of the entry.

Acland, Henry Wentworth (1815–1900), first baronet, was an English physician educated at Oxford and St George's Hospital, London. He returned to Oxford as a reader in anatomy in 1845 and was later Regius Professor of Medicine at the Radcliffe Infirmary (1857–94). At Oxford Acland played a central role in supporting science education, including the establishment of the University Museum (1861).

Adair, Harriet Eliza (née Atkinson, 1811–78), was the mother of Mary Adair. In 1828 she married Alexander Adair (1791–1863).

Adair, Henrietta Mary (1843–1921), was the youngest daughter of Alexander Adair (1791–1863) and Harriet Eliza Atkinson (1811–78). She attended Tyndall's lectures at the RI and they conversed about scientific matters, as shown by letters in this volume. Tyndall fell in love with Mary and eventually proposed to her in late 1869, which she declined (*The Ascent of John Tyndall*, 257).

Airy, George Biddell (1801–92), was educated at University of Cambridge, graduating senior wrangler in 1823. In 1826 he was appointed Lucasian Professor of Mathematics, and in 1828 to the Plumian Chair of Astronomy and directorship of the University of Cambridge observatory. In 1835 he was appointed Astronomer Royal, a position he held until 1881, and set about improving the instrumentation and research activities of the Royal Observatory, Greenwich. As reflected by letters in this volume, he pursued research on a wide range of topics.

Allen, Robin (1820–99), was a British poet who was appointed as a clerk in the secretary's office at Trinity House in 1837. From 1867 to 1881 he was the secretary of Trinity House. Some of his correspondence is held at the University of Cambridge Library (MS Add.9225) ("The Trinity House," *Times*, 25 April 1881, 9).

Armstrong, William George (1810–1900), first Baron Armstrong, was an English arms manufacturer who founded W. G. Armstrong & Co. (1847). Armstrong was well known for his work on hydraulics and electrostatics, service to the BAAS, and involvement in a number of engineering institutions. He was knighted in 1859 for giving the patent of the Armstrong gun to the Crown and was raised to the peerage as Baron Armstrong of Cragside in 1887.

Babinet, Jacques (1794–1872), was a French physicist and meteorologist educated at the École Polytechnique. He was a supporter of Fresnel and Young's wave theory of light and investigated the interference of light in the atmosphere and the polarization of light, for which he invented the Babinet compensator. He also achieved notoriety as a popularizer of science and regularly wrote articles for popular journals and gave numerous lectures to lay audiences.

Baring, Louisa Caroline (née Stewart-Mackenzie, 1827–1903), Lady Ashburton, was a philanthropist and art collector. In 1858 she married the extremely wealthy Bingham Baring, 2nd Baron Ashburton (1799–1864). After her husband's death Baring amassed a significant collection of paintings and sketches, which filled her numerous residences, where she entertained Britain's social and cultural elites, including Tyndall and Thomas Carlyle.

Baring, Mary "May" Florence (1860–1902), was the daughter of Louisa Baring, Lady Ashburton and Bingham Baring, 2nd Baron Ashburton. She married William Compton, 5th Marquess of Northampton in 1884, and held the title Marchioness of Northampton from 1897.

Barnard, Jane (1832–1911), was a niece of Michael and Sarah Faraday who lived at the RI and assisted the Faradays. She was the daughter of Sarah's brother, John Barnard, and Michael's younger sister Margaret. Like her aunt and uncle, she was a member of the Sandemanian church. As shown in letters of this volume, Tyndall sought Barnard's assistance in writing his memoir of Faraday, and valued her opinion of how he portrayed her uncle.

Baxendell, Joseph (1815–87), was an English meteorologist and astronomer. Largely self-educated, he was heavily involved with the Royal Astronomical Society and the Manchester Literary and Philosophical Society (he became a member of both in 1858). He was appointed astronomer to the Manchester Corporation in 1859. In 1867 he vehemently opposed the RS's decision to discontinue the storm warning system that had been established in 1860.

Bell, Robert (1800–1867), was an Irish journalist, editor, playwright, and historian educated at Trinity College Dublin. Late in his life he became interested in spiritualism and wrote an account of one of Daniel Dunglas Home's séances for the *Cornhill Magazine* in 1860.

Bennet, Olivia (née Montagu, 1830–1922), Countess Tankerville, was the daughter of Millicent Sparrow and George Montagu, sixth Duke of Manchester. In 1850 she married Charles Bennet (1810–99), sixth Earl of Tankerville. Together they had five children (G. E. Cokayne and G. H. White, eds., *The Complete Peerage*, 2nd ed., 14 vols. [London: St. Catherine Press, 1910–98], vol. 12, pt. 1 [1953], 635–36).

Boxall, William (1800–1879), was an English portrait painter. He initially aspired to be a history painter, but eventually focused on portraits as a more stable source of income. A well-known painter already by the 1830s, he was elected associate of the Royal Academy in 1851 and Royal Academician in 1863. He served as director of the National Gallery between 1866 and 1874.

Brewster, David (1781–1868), was a Scottish mathematician, astronomer, editor, and educator, known for his work in optics and his role in helping found the BAAS. He was a regular contributor to numerous periodicals, by which he mostly earned his living. He was one of three founding editors of the *Edinburgh Magazine* and editor of the *Edinburgh Encyclopedia* (1808–30). He was appointed principal of the United College of Saint Salvator and Saint Leonard at the University of Saint Andrews in 1838, and principal of the University of Edinburgh in 1859.

Brown, William, is not identified. Possibly he was a porter at the RI.

Bunsen, Robert (1811–99), was a German chemist who received his PhD from the University of Göttingen, where he then lectured (1833–36). Bunsen subsequently held positions at the Polytechnikum at Kassel, the University of Marburg (1839–51), the University of Breslau, and, from 1852, the University of Heidelberg. He made many contributions to chemistry, including the Bunsen burner and the discovery of cesium and rubidium, and developed a battery that bears his name. A renowned researcher, he was also an extremely popular teacher, mentoring both Tyndall and Edward Frankland at Marburg. He received the RS's Copley Medal in 1860, and its Davy Medal, with Gustav Robert Kirchhoff, in 1877.

Busk, Ellen (1816–90), was the daughter of Jacob Hans Busk. In 1843 she married her cousin George Busk, with whom she had four children.

Busk, George (1807–86), was a naval surgeon who retired in 1855 and studied zoology and paleontology. He was one of the founders of the Microscopical Society in 1839 and edited the *Quarterly Journal of Microscopical Science* (1853–68). He served as treasurer of the RI (1873–86) and secretary of the zoological section of the Linnean Society (1857–68), and was a founding member of the X Club.

Campbell, George Douglas (1823–1900), Duke of Argyll, was a Scottish politician and scientist. He was eighth duke of Argyll in the peerage of Scotland and first duke of Argyll in the peerage of the United Kingdom. Campbell served over ten years as Lord Privy Seal and from December 1868 to February 1874 he was secretary of state for India. He was a proponent of educational reform and led the Argyll Commission (1865–68) regarding the Scottish school system. As an amateur scientist he wrote on ornithology, geology, and evolution. In his work on evolution he defended the special creation of humans apart from other species, publishing *The Reign of Law* in 1868.

Carlyle, Thomas (1795–1881), was a Scottish essayist, novelist, historian, and one of the most notable intellectual figures in the Victorian era. He was a major literary influence on Tyndall, who avidly read his publications. After reading *Past and Present* (1843), Tyndall commented that Carlyle "must be a true hero. My feelings towards him are those of worship 'transcendental wonder' as he defines it" (Journal, 18 July 1847, RI MS JT/2/13a/231). During the 1860s Tyndall became a close friend of Carlyle's, supporting him in the "Eyre Affair" controversy and accompanying him to Lady Ashburton's home in Mentone, France, in December 1866.

Carrel, Jean-Antoine (1834–91), was an Italian soldier, mountaineer, and alpine guide. Tyndall had hired him as a porter during his failed attempt on the Matterhorn in 1862. Carrel made the second ascent of the Matterhorn on 17 July 1865, just three days after Edward Whymper made the first ascent (*Hours of Exercise in the Alps*, 157–65; C. D. Cunningham and W. W. Abney, *The Pioneers of the Alps*, 2nd ed. [London: Sampson Low, Marston, Searle, and Rivington Ltd., 1888], 126–31).

Chladni, Ernst Florin (1756–1827), was a German physicist whose work was mostly focused on acoustics and vibration. He studied the origin of meteorites and conducted experiments to determine the speed of sound in solids and gases. He is best known for his experiments on sound and vibration in which he drew a violin bow across the edge of a plate covered in sand. The vibrations produced a visible pattern, known as a Chladni figure, consisting of symmetrical nodal lines where there was no vibration on the plate.

Clausius, Adelheid (née Rimpau, 1833–75), was the wife of Rudolf Clausius. They married in Zürich in 1859 and had six children. Both Adelheid and Rudolf Clausius corresponded with Tyndall. They named their son John after him and Tyndall became his godfather.

Clausius, Alfred (1868–1950), was the son of Adelheid and Rudolf Clausius.

Clausius, Hedwig (1866–1946), was the daughter of Adelheid and Rudolf Clausius.

Clausius, Helene (1863–1919), was the daughter of Adelheid and Rudolf Clausius. In 1885 she married the theologian Karl Budde (1850–1935), with whom she had six children.

Clausius, Mathilde (1861–1907), was the daughter of Adelheid and Rudolf Clausius. She married the theologian Friedrich Zimmer (1855–1919).

Clausius, Rudolf John (1864–1915?), was the son of Adelheid and Rudolf Clausius. He was named after Tyndall, who was also his godfather. John Clausius completed a doctorate of law in 1890 at Friedrich Wilhelms University Bonn.

Clausius, Rudolf Julius Emanuel (1822–88), was a German mathematician and physicist and one of the founders of modern thermodynamics. His famous paper on the theory of heat "Ueber die bewegende Kraft der Warme" was published in the *Annalen der Physik* in 1850. He held a professorship in mathematical physics at the Polytechnikum in Zurich from 1855, moved to the University of Würzburg in 1867, and finally to the University of Bonn in 1869. Tyndall translated many of his major works into English for publication in the *Phil. Mag.* and *Scientific Memoirs* in the early 1850s. He was very close to Tyndall, as was his wife, and they named their son John after Tyndall.

Cottrell, John, was a laboratory assistant at the RI between 1866 and 1885.

Croll, James (1821–90), was a Scottish geologist and climatologist. In 1859 he took up a position at the Andersonian Museum in Glasgow. After several publications in the *Phil. Mag.* and making connections with prominent men of science, including Archibald Geikie, William Thomson, and Tyndall, he rose to a position of influence as secretary of the Geological Survey of Scotland. He made a number of compelling insights into the accumulation of subtle variations affecting climate over time.

Darwin, Charles (1809–82), was an English naturalist and geologist, best known for his book *On the Origin of Species* (London: John Murray, 1859). Earlier in Tyndall's career Darwin encouraged his research on glaciers and supported him in his disagreement with Forbes. During the 1860s Darwin and his supporters had vociferously promoted the *Origin*, which was in its fifth edition by 1869 and had been translated into French, German, Italian, Dutch, Swedish, and Russian. In 1864 he was awarded the RS's Copley Medal, after being nominated three years in a row. Letters in this volume suggest that Darwin was struggling again with poor health in fall 1868, which had been a recurrent theme of his life, and particularly following the publication and controversy over the *Origin.*

Davy, Humphry (1778–1829), was a chemist and inventor born in Cornwall, England. From 1801 until 1825 he was director of the RI laboratory, where he was also professor of chemistry (1802–12) and honorary professor (1813–23). He achieved fame as a lecturer and for his chemical research that isolated such elements as sodium, potassium, and chlorine. He is often credited with the invention of the miner's safety lamp. He served as president of the RS from 1820 to 1827.

Debus, Heinrich (1824–1915), was a German chemist who studied at Marburg (1845–48), where he met Tyndall, Edward Frankland, and Thomas Hirst. Debus moved to England in 1851 and taught chemistry at Queenwood College until 1867, when he began teaching at Clifton College, Bristol. Later Debus lectured at Guy's Hospital Medical School (1870–88) and became professor of chemistry at the Royal Naval College, Greenwich (1873–88).

de la Rive, Arthur-Auguste (1801–73), was a Swiss physicist and one of the founders of the electrochemical theory of batteries. He served as professor of natural philosophy at the Academy of Geneva and worked principally on electricity, although he also investigated the specific heats of gases and calculated the temperature of the Earth's crust. His principal book was *Traité d'électricité théoretique et appliquée*, 3 vols. (Paris: J.-B. Baillière et H. Baillière, 1854–58).

Doyle, Abbey/Abby, was a hired caretaker for Tyndall's mother, Sarah, and sister, Emma. Tyndall was often not pleased with Doyle, but he received assurances from Elizabeth Dawson Steuart that she was attentive to his mother and sister. Letters in this volume show that Tyndall's view of Doyle soured as she requested money from him.

Egerton, Edward Christopher (1816–69), was an English politician who served as parliamentary under-secretary of state for foreign affairs (1866–68). He was Member of Parliament for Macclesfield (1852–68), followed by East Cheshire (1868–69). He married Lady Mary Frances Pierrepont in 1845 with whom he had five children.

Egerton, Mary Frances (née Pierrepont, 1819–1905), was the daughter of Charles Pierrepont (1778–1860), the second Earl Manvers. In 1845 she married Edward Christopher Egerton (1816–69), with whom she had five children: Mary Alice, Georgina, Charlotte, Hugh Edward, and Charles Augustus. Egerton was a frequent correspondent of Tyndall's and attended his lectures.

Emerson, Ralph Waldo (1803–82), was an American author, poet, and lecturer, educated at Harvard. He was a leading figure in the transcendentalism movement, which he heavily influenced through such essays as "Self-Reliance" (1841), "Nature" (1836), and "The American Scholar" (1837). During the 1860s he continued a productive output, editing posthumous works of Henry David Thoreau (1817–62), publishing the collection of poems *May-Day and Other Pieces* (1867), and delivering hundreds of lectures across the United States.

Eyre, Edward John (1815–1901), was an English colonial governor, controversially known for his violent repression of the 1865 Morant Bay rebellion while he served as governor of Jamaica (1862–65). In the years following the rebellion, London society was divided over whether Eyre had acted justly or criminally, resulting in the founding of the Jamaica Committee, which sought to prosecute Eyre, and the Eyre Defence and Aid Committee, which sought to defend Eyre's actions, of which Tyndall was a vocal member. In the end, Eyre was unsuccessfully tried for private criminal charges brought by the Jamaica Committee.

Faraday, Michael (1791–1867), was a natural philosopher, Fullerian Professor of Chemistry at the RI, and a member of the Sandemanian church. His extensive research focused on electricity and encompassed electrochemistry, electromagnetism, diamagnetism, and magneto-optics (the Faraday effect). Faraday also held the positions of director of the laboratory (1825–67) and superintendent of the house (1852–67) at the RI. He was Tyndall's patron and helped him obtain the position of professor of natural philosophy at the RI in 1853. Despite significant differences in their views on religion and on the nature of magnetism, they held the highest mutual regard for each other. In 1867 Tyndall succeeded Faraday as superintendent of the house and director of the laboratory at the RI. Tyndall's *Faraday as a Discoverer* was his tribute to his mentor.

Faraday, Sarah (née Barnard, 1800–1879), was the daughter of Mary Boosey and Edward Barnard, a successful London silversmith and Sandemanian. She likely met Michael Faraday at the Sandemanian meeting house in Paul's Alley. After they married in 1821 she moved into his rooms at the RI. In later years she took on many of Faraday's administrative responsibilities at the RI.

Farrar, Frederic William (1831–1903), was an Anglican clergyman, writer, and philologist. From 1855 to 1870 he was assistant master at Harrow, during which time he wrote three novels and a number of books on philology. Farrar was a supporter of Darwin and adapted his ideas to the development of language during the 1860s, and it was Darwin who nominated Farrar as FRS in 1866. Upon Darwin's death Farrar, by then a canon of Westminster, arranged for his burial at Westminster Abbey. From 1895 until his death Farrar was dean of Canterbury.

Francis, William (1817–1904), was a chemist, translator, and editor. He was the illegitimate son of the printer and publisher Richard Taylor. Francis learned the printing trade in Germany in the 1830s before training as a chemist with Justus von Liebig and Gustav Rose in Giessen, where he obtained his doctorate in 1842. He made many translations for his father's *Scientific Memoirs* and the *Phil. Mag.*, and became the editor of the latter in 1851. In 1852 he began a partnership with his father, establishing the Taylor & Francis publishing firm. He first met Tyndall in 1850 and became one of Tyndall's close confidants.

Frankland, Edward (1825–99), was a chemist and one of Tyndall's closest friends. He and Tyndall met at Queenwood College before they studied together at the University of Marburg in 1848. After holding a professorship at Owens College, Manchester, Frankland moved to London and held a number of positions including at St Bartholomew's Hospital and the East India Company's military college. He was professor of chemistry at the RI (1863–68) and afterward at the Royal College of Chemistry (1868–85). In 1864 Frankland joined Tyndall, Thomas Hirst, Joseph Dalton Hooker, Thomas Henry Huxley, and four others as members of the X Club, a monthly dining club in which they promoted science in British society.

Gauthier-Villars, Jean-Albert (1828–98), was a French engineer and publisher. Born to a family of printers in Lons-le-Saunier, he studied at the École Polytechnique (1848–50), became a telegraph engineer, and served in the military. In 1864 he purchased the Mallet-Bachelier printing house, which had published the *Comptes rendus hebdomadaires des séances de l'Académie des sciences.* Under his direction the firm, renamed Gauthier-Villars, specialized in mathematical and scientific books and magazines.

Ginty, Frederic (b. ca. 1854), was the son of Margaret (née Roberts) and William Ginty. During the period covered by this volume he was attending Linton House School, Colchester.

Ginty, George (b. 1851), was the son of Margaret (née Roberts) and William Ginty. During the period covered by this volume he was attending Linton House School, Colchester.

Ginty, Margaret (née Roberts, b. ca. 1826), was the widow of Tyndall's friend William Ginty. She married William in Ireland in 1846, after which they relocated to Manchester. In 1854 the family moved to Rio de Janeiro, Brazil, where William died in 1866. Letters in this volume indicate how Tyndall assisted Margaret in managing her husband's will after his death.

Ginty, William Gilbert (ca. 1820–66), was an Irish civil engineer. He met Tyndall in the 1840s while working on the Irish Ordnance Survey. Ginty later worked as engineer and manager of the Hydro-Carbon Gas Company and served as honorary secretary of the Manchester Athenaeum. In 1854, he moved to Rio de Janeiro, Brazil, where he was engineer for the Imperial Gas Works Company established by Baron de Maua. In Brazil he was also involved in building canals, roads, and railways, and was awarded a knighthood in the Brazilian Order of the Rose (1860). Ginty died suddenly in 1866.

Giraud, Étienne (fl. 1850–60s), was a French publisher and bookseller whose firm was located at 20 rue Saint Sulpice in Paris. He published Abbé Moigno's *Les mondes* (1863), as well as the first French edition of Tyndall's *Heat* (1863) which appeared as *La chaleur* (1864). Letters in this volume suggest that Giraud went bankrupt before paying Tyndall for the sales of *La chaleur*.

Greenwood, Frederick (1830–1909), was an English journalist and editor. He edited the *Cornhill Magazine* (1864) and served as the first editor of the *PMG* (1865–80) and the *St James Gazette* (1880–88). Under his direction the *PMG* forcefully defended Tory policies. Greenwood played an important but unacknowledged role in the British government efforts to purchase part of the Suez Canal in 1875 by passing on intelligence of the French government's plans to buy the canal.

Griffith, George (ca. 1834–1902), was an English science educator. He taught at Oxford between 1857 and 1867, but left after the professorship of experimental philosophy was given to Robert Clifton in 1865. He was afterward the first natural science master at Harrow (1867–93). Griffith also took an active role in the BAAS, serving as its assistant general secretary for thirty years (1862–78, 1881, and 1890–1903) (B. P. L., "George Griffith," *Nature* 66 [1902]: 64).

Grove, William Robert (1811–96), was a Welsh judge and electrochemist. He was called to the bar in 1835 but also took an interest in science. He held a professorship of experimental philosophy at the London Institution (1841–47) and developed a gas battery (the first fuel cell), for which he was awarded the RS's Royal Medal in 1847. In the late 1840s financial constraints turned his attention back to practicing law. He became a Queen's Counsel in 1853, was appointed a judge of the court of common pleas in 1871 and then of the Queen's Bench in 1880. He remained a highly respected member of the scientific community and was elected president of the BAAS in 1866.

Helmholtz, Anna (née Mohl, 1834–99), was a Berlin salonnière and translator of scientific works. She married Herman Helmholtz in 1861. With Clara Wiedemann she translated Tyndall's *Heat* (1863) and *Sound* (1867) which appeared as *Die Wärme, betrachtet als eine Art der Bewegung* (Braunschweig: Friedrich Vieweg, 1867) and *Der Schall: acht Vorlesungen gehalten in der Royal Institution von Grossbritannien* (Braunschweig: Friedrich Vieweg, 1869). Although Clara and Anna translated the work, only their husbands were credited.

Helmholtz, Hermann Ludwig Ferdinand (1821–94), was a German physicist and physician. Educated in military medicine at the Friedrich-Wilhelms Institute in Berlin, he was first a teacher of anatomy at the Berlin Academy of Arts in 1848. He became professor of physiology at the University of Königsberg (1849), and then professor for anatomy and physiology at the University of Bonn (1855), where he turned his attention to physics. From 1858 he was professor of physiology at Heidelberg until he became professor of physics at Humboldt University, Berlin. Helmholtz made many contributions to numerous fields, including thermodynamics, electrodynamics, and the conservation of energy, as well as physiological studies of sound and optics.

Herschel, Alexander Stewart (1836–1907), was an English astronomer and son of John Herschel. He was professor of natural philosophy at Anderson's College, Glasgow (1866–71), and professor of physics and experimental philosophy at the University of Durham College of Science, Newcastle (1871–86). He is known primarily for his work on meteor spectroscopy and for identifying the link between the orbits of comets and the occurrence of meteor showers.

Herschel, John Frederick William (1792–1871), was an English mathematician and astronomer. He was the son of the famous astronomer William Herschel (1738–1822) and himself became one of the most eminent scientists in Victorian Britain. He graduated from University of Cambridge in 1813 as senior wrangler, received two RS Copley Medals (1821 and 1847), and was awarded a knighthood (1831) followed by a baronetcy (1838). His *Preliminary Discourse on the Study of Natural Philosophy* (London: Longmans, Rees, Orme, Brown & Green, 1830) was widely read and greatly admired. In 1849 he published *Outlines of Astronomy* (London: Longman, Brown, Green, and Longmans), a popular presentation of astronomy that went through eleven editions by 1871. He made many contributions to the science of photography and also edited the admiralty's *Manual of Scientific Enquiry* (London: John Murray, 1849), a science field guide for the use of the Royal Navy. From 1850 to 1855 he served as master of the mint, and on his death was buried next to Isaac Newton in Westminster Abbey.

Hinrichs, Gustavus (1836–1923), was a meteorologist and physicist born in Holstein. In 1861 he moved to the United States, settling in Iowa, where he taught languages, and in 1862 he became professor of physics and chemistry at the State University of Iowa. He moved to Washington University in Saint Louis in 1886. He was the head of the Iowa Weather Report Service, which gave weather predictions based on barometric observations. His 1868 book *Atomechanics* proposed a mechanical theory of atoms and developed a helical table of elements in which elements were organized by atomic mass (Loren Horton, "Hinrichs, Gustavus Detlef," in *The Biographical Dictionary of Iowa*, ed. David Hudson et. al. [Iowa City: University of Iowa Press, 2008], 239–40).

Hirst, Thomas Archer (1830–92), was an English mathematician, educator, and Tyndall's closest friend and correspondent. His friendship with Tyndall began in Halifax, Yorkshire, in 1845 when they were both working for the surveyor Richard Carter. Encouraged by Tyndall—who acted as a sort of mentor to him—Hirst studied at the Universities of Marburg, Göttingen, and Berlin between 1849 and 1852. In 1853 he replaced Tyndall as mathematics teacher at Queenwood College. He resigned in 1856 to look after his sick wife, Anna, who died in Paris in 1857. During the 1860s Hirst established himself as an active and respected member of the London scientific community. He taught mathematics at University College London (1860–64), and subsequently held professorships of physics and mathematics (1865, 1867–70). He joined the RS Council (1864), was a founder of the X Club, a monthly dining club in which they promoted science in British society, and served as general secretary of the BAAS (1866–70).

Holland, Henry (1788–1873), first baronet, was an English physician and travel writer educated at the University of Edinburgh. He began a practice in London in 1816 serving members of London's wealthy and fashionable society. He became physician-extraordinary to Queen Victoria (1837), and physician-in-ordinary to Prince Albert (1840). From 1865 until 1873 he served as president of the RI.

Home, Daniel Dunglas (1833–86), was a British medium. Born in Scotland, Home moved with his adopted parents to the United States in 1842. During the 1850s Home claimed that he began experiencing poltergeists, including communication with the dead and the sudden movement of objects around him. Cast out by his adoptive mother, Home moved around as a guest of spiritualists performing séances. He traveled to England in 1855 for his health and began to hold séances for London spiritualists. This was followed by several years traveling throughout Europe. Home established for himself a reputation and seats at his séances became highly sought after. In 1866 he became the secretary of the Spiritual Athenaeum, Sloane Street, London, before losing a much-publicized lawsuit over a £60,000 gift he had received from Jane Lyon, who later accused him of manipulating her. As letters in this volume show, Tyndall became involved in a dispute with Home through the pages of the *PMG* over how Faraday had previously responded to spiritualist enquiries from Home.

Hooker, Frances Harriet (née Henslow, 1825–74), was a daughter of Harriet Jenyns and the botanist John Stevens Henslow. In 1851 she married Joseph Hooker, with whom she had four sons and two daughters. She was fluent in French and German, and in addition to making private translations for her husband she published a translation of *A General System of Botany*, by Emmanuel le Maout (London: Longman, 1873). Although no letters between them survive for the period of this volume, she corresponded with Tyndall and shared literary references, poetry, and novels. She was a skilled editor and occasionally read over manuscripts for him.

Hooker, Joseph Dalton (1817–1911), was a British botanist who specialized in taxonomy and the geographical distribution of plants. He became arguably the most famous British botanist of the Victorian era. He was the younger son of William Jackson Hooker (1785–1865), the director of the Royal Botanic Gardens, Kew. After studying medicine at the University of Glasgow he became a naval surgeon and spent the years 1839–43 on James Clark Ross's geomagnetic survey of the southern oceans, where he collected plants in addition to serving as surgeon. Subsequently he explored and collected plants in India (1847–51) and achieved considerable mountaineering success, before joining his father as deputy director of Kew Gardens in 1855. In 1864 Hooker joined Huxley, Tyndall, and others in founding the X Club, a monthly dining club in which they promoted science in British society. He succeeded his father as director of Kew in 1865. He served as president of the BAAS (1868), and was knighted several times and received many scientific awards, including the RS's Royal Medal (1854) and Copley Medal (1887).

Huggins, William (1824–1910), was an English astronomer known for his work in spectroscopy. After selling his family business in the 1850s he established a private observatory at his home, where he made observations and developed methods and instruments for observing celestial spectra. In 1864 he was the first to observe the spectrum of a planetary nebula, finding that they differed from stars much more than previously thought. This work made him a prominent member of the scientific community. He was elected FRS in 1865, received the RS's Royal Medal in 1866, and with William Allen Miller won the Gold Medal of the Royal Astronomical Society (1867).

Huxley, Ethel Gladys (1866–1941), was the youngest daughter of Henrietta and Thomas Huxley. In 1889 she married her sister Marian's widower, the artist John Collier.

Huxley, Henrietta "Nettie" Anne (née Heathorn, 1825–1915), was born in the West Indies to Henry Heathorn and Sarah Henrietta Richardson (née Harris), and was schooled in Neuwied, Germany, for two years while a youth.

In 1843 she, her mother, and her half sister moved to Australia to join her father. She was keeping house for her sister Oriana and brother-in-law William Fanning in Sydney, Australia, when she met Thomas Huxley in 1847, who was then a surgeon and zoologist on HMS *Rattlesnake*. After an eight-year engagement she arrived in London in 1855 and married Huxley. She assisted her husband in his work by translating German and drawing diagrams for his lectures. She was also a poet and at the age of eighty-six published a volume of her collected poems, *Poems of Henrietta A. Huxley with Three of Thomas Henry Huxley* (London: Duckworth, 1913). As with many of the wives of this small group of scientific friends, Henrietta was very much a part of their community; she corresponded with the Darwins as well as with Tyndall ("Death of Mrs. Huxley," *Times*, 6 April 1915, 12).

Huxley, Henrietta (1863–1940), was the daughter of Henrietta and Thomas Huxley. In 1889 she married the engineer Harold Roller.

Huxley, Henry (1865–1946), was the son of Henrietta and Thomas Huxley. After attending St Bartholomew's Hospital he established a medical practice in London.

Huxley, Jessie Oriana (1858–1926), was the eldest daughter of Henrietta and Thomas Huxley. Emma and Charles Darwin were her godparents. In 1878 she married the architect Frederick William Waller (1846–1933).

Huxley, Leonard (1860–1933), was the son of Henrietta and Thomas Huxley. He was the godson of Emma and Charles Darwin. Educated at University College London, St Andrews, and Oxford, he was an assistant master at Charterhouse School (1884–1901), after which he became an assistant and then, in 1916, sole editor for the *Cornhill Magazine.* He wrote biographies of his father, Thomas Huxley (1900) and Joseph Hooker (1918), and edited the published letters of Jane Welsh Carlyle (1924) and Elizabeth Barrett Browning (1929).

Huxley, Marian (1859–87), was the daughter of Henrietta and Thomas Huxley. She was a talented artist and exhibited her painting *The Sins of the Father* at the Royal Academy in 1879. The same year she married the artist John Collier (1850–1934). She became chronically ill during the 1880s and died in Paris in 1887 while on her way to the hospital of Jean-Martin Charcot.

Huxley, Rachel (1862–1934), was the daughter of Henrietta and Thomas Huxley. In 1884 she married the civil engineer Alfred Eckersley.

Huxley, Thomas Henry (1825–95), was a British zoologist, educator, science popularizer, and one of Tyndall's closest friends. Following medical training at Charing Cross Hospital in London, he was a ship's surgeon aboard HMS *Rattlesnake* on an expedition to Australia (1846–50) under Captain Owen Stanley. While in Australia he met Henrietta (Nettie) Anne Heathorn, whom he married in July 1855 after an eight-year engagement. He became professor of natural history at the Royal School of Mines, London (1854–71), and moved with the school to South Kensington, where he taught in a new laboratory until 1885. He joined Tyndall twice at the RI as Fullerian Professor of Natural History (1855–58, 1866–69). In 1864 Huxley joined Tyndall, Edward Frankland, Thomas Hirst, Joseph Dalton Hooker, and four others as members of the X Club, a monthly dining club in which they promoted science in British society.

James, Bartlett (birth and death dates unknown), was manager, or "Gerente," of the Imperial Gas Works Company in Rio de Janeiro in Brazil, established by Baron de Maua, where he worked with Tyndall's friend William Ginty, who was the engineer, or "engenheiro" (*Almanak Administrativo, Mercantil e Industrial da Corte e Provincia do Rio de Janeiro para o anno de 1865* [Rio de Janeiro: Eduardo & Henrique Laemmert, 1865], 414–15). Ginty died in 1866, and according to a 5 December 1882 letter from Ginty's daughter Fanny Unwin (née Ginty) to Tyndall, James had taken on the role of executor to his will (RI MS JT/1/TYP/11/3682).

Jamin, Jules Célestin (1818–86), was a French physicist, educated at the École Normale Superieur. He was professor of physics at the École Polytechnique (1858–81), and from 1863 he was professor of physics at the Faculté des sciences, Paris. He worked in many branches of physics, including optics, acoustics, electricity, magnetism, and radiation, and in 1858 was awarded the RS's Rumford Medal.

Jones, Henry Bence (1813–73), was a wealthy English physician and medical chemist. He was educated at Harrow School and Trinity College, Cambridge, and studied medicine at St George's Hospital, London, after which he began a successful career as a physician. Attracted by the potential applications of chemistry to clinical medicine, he studied organic chemistry with Justus Liebig at Giessen (1841–43) before assuming a career as a successful physician in private practice and at St George's Hospital. In May 1842 he married a cousin, Lady Millicent Acheson, daughter of the second Earl of Gosford, with whom he had seven children. He published important research on protein chemistry and wrote an abstract in English of du Bois-Reymond's work on electrophysiology (1852) that annoyed Carlo Matteucci, the Italian pioneer of electrophysiology. In 1853 he became a manager of the RI and served as secretary from 1860 to 1873. He frequently entertained Tyndall at both his London home and Folkestone, where he kept a summer house.

Longman, William (1813–77) and Thomas (1804–79), were two brothers who ran the publishing firm Longmans after the sudden death of their father, Thomas Norton Longman (1771–1849). While Thomas, the elder brother, served as chairman of the company, William was reputed to be more skilled in the business of publishing. Longmans was one of the largest publishing firms in nineteenth-century Britain, publishing books on history, theology, travel, sociology, novels, and science. Longmans published Tyndall's *Heat* (1863), *Sound* (1867), *Faraday as a Discoverer* (1868), and *Hours of Exercise in the Alps* (1871).

Lubbock, Ellen Frances (née Hordern, 1834–79), was the daughter of an Anglican minister. She married John Lubbock in 1856 and they had three sons and three daughters.

Lubbock, John (1834–1913), was an English banker, politician, and evolutionary anthropologist. In his youth Lubbock befriended Charles Darwin and John Tyndall. An early supporter of Charles Darwin's *On the Origin of Species* when it was published in 1859, Lubbock became associated with Thomas Huxley, Joseph Dalton Hooker, and other friends and supporters of Darwin. He was also one of the founders of prehistoric archaeology—in 1865 he published the highly influential *Pre-historic Times: As Illustrated by Ancient Remains*, in which he coined the terms *paleolithic* and *neolithic*—and an author of books on insects for the popular reader. In 1856 he married Ellen Frances Horden, and the couple were among Tyndall's closest friends. He was the youngest member of the X Club and joined Tyndall on his climbing adventures in the Alps on a number of occasions. When his father died in 1865, he became Sir John Lubbock, fourth baronet. Starting in 1870, he served numerous times in parliament and was made the first Baron Avebury in 1900.

Magnus, Heinrich Gustav (1802–70), was a German chemist and physicist, and professor of technology and physics at the University of Berlin from 1833 until his death. He established a physics laboratory at his home in Berlin, to which he invited young physicists to work, including Tyndall in 1851. Although his early research was primarily in chemistry, by the 1850s he had begun to investigate a range of physical problems, including the absorption of heat by gases, as his letters to Tyndall demonstrate.

Marcet, Jane (née Haldimand, 1769–1858), was a science and political economy writer. Born in London, she was the daughter of a wealthy Swiss merchant. She married Alexander Marcet (1770–1822), a Swiss physician, with whom she had four children. In 1805 she wrote a science textbook, *Conversations on Natural Philosophy*, that although not published until 1819, served as a basis for her other publications, including *Conversations on Political Economy* (1816) and *Conversations on Chemistry* (1805). Guests at the Marcet home included members of London's elite, and her friends included Michael Faraday, Harriet Martineau, Maria Edgeworth, Thomas Malthus, David Ricardo, and Auguste de la Rive.

Maxwell, James Clerk (1831–79), was a Scottish physicist and mathematician. He graduated in the Cambridge mathematics Tripos (undegraduate examination) as second wrangler in 1854, after which he held professorships at Marischal College, Aberdeen (1856–60), and King's College, London (1860–65), and from 1871 he held the Cavendish Professorship at the University of Cambridge. Maxwell is best known for his contributions to the study of electromagnetic radiation. He developed Faraday's field theory and demonstrated that electricity, magnetism, and light were instances of the same phenomenon. Maxwell's work culminated in his 1873 publication *A Treatise on Electricity and Magnetism.*

Mitchell, Edward P. (1812–80), was the business partner and brother-in-law of Hector Tyndale. Mitchell married Elizabeth Tyndale (ca. 1812–66) and joined Hector as a partner in the china and glassware firm established by Elizabeth and Hector's father Robinson Tyndale (1775–1842).

Moigno, François-Napoléon-Marie (1804–84), was a French Catholic priest, physicist, writer, and editor. Educated at a Jesuit seminary, he studied science alongside theology and was given the Mathematical Chair at Jesuit House in Paris in 1836. After leaving the order in 1843 he became the science editor of *L'epoque*, and later became editor of the scientific journal *Cosmos* (1852–62). He founded the journal *Les mondes* in 1863 and undertook the translation and publication of many of Tyndall's works in French, including *Radiation* (1867), *Le son* (1869), *La chaleur* (1864), and *Faraday inventeur* (1868) ("L'Abbé Moigno," *Nature* 30 [1884]: 291–92).

Murchison, Roderick Impey (1792–1871), was a Scottish geologist and geographer who published over 180 scientific papers as well as several books. He was elected fellow of the Geological Society in 1825, and was twice its president. He was one of the founders of the Royal Geographical Society of London (1830) and served as its president for sixteen years between 1843 and

1871. He used his role in the RGS in particular to promote imperial exploration. He was also one of the founders of the BAAS (1831) and was president of the geographical and geological sections ten times in total.

Murray, John (1808–92), was an English publisher and the eldest child of Anne Elliot and publisher John Murray. Murray was educated at Charterhouse School (1819–26) and for a year (1827) at Edinburgh University. He began his career in the family firm at 50 Albemarle Street, London, in 1828 and became head of the firm after his father's death in 1843. Murray published the *Quarterly Review*, a popular series of *Handbooks for Travellers*, and numerous scientific publications. On 24 November 1859 Murray published both Darwin's *On the Origin of Species* and Samuel Smiles's *Self-Help*. He became a member of the Alpine Club in 1858 and in 1860 published Tyndall's *Glaciers of the Alps*.

Nicol, William (1770–1851), was a Scottish physicist and geologist. He worked as an assistant to the blind science lecturer Henry Moyes (1749–1807) from around 1786 until Moyes's death, when Nicol took over his lecture circuit. Nicol is best known for inventing the Nicol prism in 1828, an optical device made of Iceland spar that is used to detect polarized light.

Peel, Emily (née Hay, 1836–1924), was the daughter of George Hay (1787–1876), eighth Marquess of Tweeddale and Lady Susan Montagu (1797–1870). In 1856 she married Robert Peel (1822–95), third baronet. They had three daughters and one son before she left him to live in Geneva and later Florence. Peel was a regular attendee of Tyndall's lectures.

Plücker, Julius (1801–68), was a German mathematician and experimental physicist. He was professor of physics at the University of Bonn (1847–68). After focusing on analytical geometry, he turned his attention to electricity and magnetism. His ideas about diamagnetism and the behavior of crystals in magnetic fields were challenged by Tyndall, which led to ill feeling and controversy that they appear to have resolved by April 1858.

Pollock, Jonathan Frederick (1783–1870), first baronet, was a British judge, politician, and the father of William Frederick Pollock (1815–88). He was educated at University of Cambridge, graduating in 1806 as senior wrangler. He was Member of Parliament for Huntingdon (1831–44), and served twice as attorney general (1843–45, 1841–44). He was Lord Chief Baron of the Exchequer from 1844 to 1866, and received a baronetcy on his retirement.

Pollock, Juliet (née Creed, 1819–99), was the daughter of the Reverend Henry Creed, vicar of Corse, Gloucestershire, and wife of William Frederick Pollock, lawyer and author. They married in 1844 and had three sons. She and her husband were staunch supporters of the RI and frequently invited Tyndall to dinner. She was the author of several books, including *Julian and His Playfellows* (London: Grant and Griffith, 1852), *New Friends* (London: John W. Parker and Son, 1858), and *Macready As I Knew Him* (London: Remington and Co., 1885).

Pollock, William Frederick (1815–88), second baronet, was a lawyer and author. The eldest son of Jonathan Frederick Pollock (1783–1870), and Frances Rivers (d. 1827), he married Juliet Creed in 1844. He was admitted to Trinity College, Cambridge, in 1832, where he obtained his BA (1836) and MA (1840). Admitted to the Inner Temple in 1833 and called to the bar in 1838, he then became a master of the court of exchequer in 1846 and was appointed queen's remembrancer in 1874. Pollock's two-volume *Personal Remembrances* (London: Macmillan and Co., 1887) illustrates how the family circulated among London's elite and entertained at their Montagu Square home. Both William and Juliet were loyal supporters of the RI and Tyndall was frequently among their guests.

Rayet, Georges Antoine Pons (1839–1906), was a French astronomer educated at the École Normale Supérieure. In 1863 he began working as an observer at the Paris observatory, where he collaborated with Charles Wolf (1827–1918) in developing new spectroscopic methods. After observing the phenomenon in a nova in 1866, they observed rare bright emissions in the spectra of three stars in 1867, now known as Wolf-Rayet Stars. Rayet later became professor of astronomy at the University of Bordeaux (1876), where he helped to found its observatory and became its first director in 1879.

Roscoe, Henry Enfield (1833–1915), was an English chemist who studied at University College, London, before earning his PhD under Robert Bunsen at Heidelberg. After Edward Frankland left for London, in 1857 Roscoe replaced him as professor of chemistry at Owens College, Manchester, where he remained until 1886. Roscoe's research included collaborations with Bunsen on photochemistry between 1855 and 1862 as well as his own work on vanadium, of which he was the first to prepare a pure sample. He wrote a number of popular works on chemistry, including *Lessons in Elementary Chemistry* (1866) and *Spectrum Analysis* (1869).

Sabine, Edward (1788–1883), was an Anglo-Irish astronomer, geophysicist, and military officer who led the British effort to establish magnetic observatories all over the globe. A graduate of the Royal Military Academy at Greenwich, he reached the rank of general in the 1850s. He accompanied J. C. Ross's Arctic expedition (1818) and W. E. Parry's expedition (1819). He received the RS's Copley Medal in 1821 for these magnetic observations in the Arctic. Sabine was closely connected with the administration of the RS and served as its treasurer (1850–61) and president (1861–71).

Spencer, Herbert (1820–1903), was an English biologist, philosopher, journalist, and sociologist. Through family connections, at the age of twenty-eight he secured a post as subeditor at the *Economist*, which put him at the heart of literary London. He published his first book, *Social Statics*, in 1850 and his second book, *Principles of Psychology*, in 1855. Spencer was a respected philosopher, but one of his most lasting contributions was his *Principles of Biology* (1864), in which he coined the phrase "survival of the fittest." Along with Tyndall and seven others, Spencer was a member of the X Club, a monthly dining club in which they promoted science in British society (Spencer was the only member who was not an FRS). In 1865 he joined the Jamaica Committee, which advocated the prosecution of Governor Edward Eyre for his violent repression of the Morant Bay rebellion. Tyndall, like Carlyle and Charles Kingsley, joined others in Eyre's defense.

Spottiswoode, William (1825–83), was an English mathematician and physicist. He was also a partner of Eyre & Spottiswoode, the queen's printer. In 1861 he married Eliza Taylor Arbuthnot, eldest daughter of William Urguhart Arbuthnot, member of the council for India. Parties at the Spottiswoode home were considered the height of the London season, attracting men of science, politics, and society. In 1864 he, along with Tyndall, Huxley, Spencer, and five others, formed the X Club, a monthly dining club in which they promoted science in British society. He was treasurer of the BAAS (1861–74), the RI (1865–73), and the RS (1871–78), and served as president of the BAAS (1878) and the RS (1878–83).

Stanley, Arthur Penrhyn (1815–81), was an Anglican clergyman. He was educated at Rugby School and Oxford University (1837), where he obtained a fellowship. He accepted a canonry at Canterbury (1851) before becoming professor of ecclesiastical history at Oxford (1856). In 1863 he married Lady Augusta, one of Queen Victoria's women of the bedchamber. They were subsequently frequent guests at Windsor Castle. From 1864 to 1881 he was dean of Westminster, during which time he continued to publish lectures and sermons.

Steuart/Stewart, Elizabeth Dawson (née Duckett, 1802–93), was born into a very wealthy family with an estate of twelve thousand acres and a gothic mansion (Duckett's Grove, now a ruin) situated several miles northeast of Carlow. In 1820 she married William Richard Steuart and they lived at Steuart's Lodge, Leighlinbridge. She took a keen interest in Tyndall's career and was probably his most consistent and long-lived patron. Letters in this volume show that Steuart frequently helped Tyndall arrange financial support for his relatives and acquaintances in Ireland.

Stokes, George Gabriel (1819–1903), was an English mathematician and physicist. He graduated from University of Cambridge as senior wrangler (1841) where he then served as Lucasian Professor of Mathematics (1849–1903). He served as secretary of the RS (1854–85) during which time he was responsible for refereeing papers for the *Phil. Trans.*, a task for which he frequently sought Tyndall's assistance. In the 1850s Stokes turned his attention to optics, and in 1852 published on fluorescence, a term he coined, showing that the phenomenon was the result of changes in the wavelength of light. For this work he was awarded the RS's Rumford Medal (1852), and was later awarded the Copley Medal (1893) for his contributions to science.

Strachey, Jane Maria (née Grant, 1840–1928), was a British suffragist, author, and the daughter of John Grant (1807–93), a colonial administrator. In 1859 she married Richard Strachey, who was twenty-three years her senior. She was influenced by John Stuart Mill, and a friend of the suffragist Millicent Fawcett (1847–1929) and George Eliot (1819–80). She was a member of the National Union of Women's Suffrage Societies, and was host to the numerous political and scientific guests who frequented the Stracheys' home.

Strachey, Richard (1817–1908), was a scientist, military engineer, and British colonial administrator in India. Educated at the East India College, he joined the Bombay Engineers (1836) and Bengal Engineers (1839), where he worked on the Jumna Canal irrigation project before serving in the Anglo-Sikh War (1845–46). Upon traveling to the Himalayas to recover his health, he began to study botany and geography. He explored Tibet in 1848, collecting botanical specimens and later publishing much-lauded descriptions of the geography, ethnology, and botany he observed. Strachey held numerous appointments in India, including secretary for public works (1862–65) and inspector-general of irrigation (1866) before returning to England in 1871.

Sylvester, James Joseph (1814–97), was an English mathematician educated at University of Cambridge, where he was second wrangler in the mathematical Tripos (undergraduate examination) in 1837. However, as a Jew he was not permitted to take the degree or hold fellowships. He subsequently held professorships at University College, London (1838–41), the University of Virginia (1841), and earned a BA and MA from Trinity College, Dublin (1841). From 1844 to 1854 he worked as an actuary at a life insurance company in London while at the same time pursuing his mathematical research on invariants. He was professor of mathematics at the Royal Military Academy, Woolwich (1855–70) and later at Johns Hopkins University (1876–80).

Tait, Peter Guthrie (1831–1901), was a Scottish mathematician and physicist and was senior wrangler in the mathematics Tripos (undergraduate examination) at University of Cambridge in 1852. He was professor of mathematics at Queen's College, Belfast (1854–60), and held the chair of natural philosophy at Edinburgh University from 1860 until shortly before his death. During the 1860s Tait forged a friendship with William Thomson and began writing their *Treatise on Natural Philosophy* (1867). During the 1860s Tait and Thomson had been in a dispute with Tyndall over his claims about Julius Mayer's priority over Joule in developing the mechanical equivalent of heat. This heated dispute spilled over into popular journals before they resolved their differences at the Dundee meeting of the BAAS in 1867, where Tyndall wrote they "seemed like brothers together." Letters in this volume reflect Tyndall's desire to properly acknowledge Tait, Thomson, and other "North British" physicists while writing *Faraday as a Discoverer* (1868).

Tennent, James Emerson (1804–69), first baronet, was an Irish politician and travel writer educated at Trinity College, Dublin. He was member of parliament for Belfast (1832–45) and Lisburn (1852). He served as colonial secretary of Ceylon (1845–50) and secretary to the Board of Trade (1852–67). He published accounts of his travels in Ceylon and Greece, and was made a baronet on his retirement in 1867.

Tennyson, Alfred (1809–92), first Baron Tennyson from 1883, was poet laureate from 1850 in succession to William Wordsworth. Born in Lincolnshire, he was educated at Trinity College, Cambridge. In 1850 he published *In Memoriam* (London: Edward Moxon). In 1853 he moved to Farringford on the Isle of Wight, where he published *Maud* (London: Edward Moxon, 1855). Tyndall greatly admired Tennyson's poetry and met him in 1858. He subsequently developed a close friendship with the Tennysons and stayed with them at their home in Farringford. Tennyson declined honors or titles from Queen Victoria four times between 1865 and 1880, but accepted a barony in 1883.

Tennyson, Emily Sarah (née Sellwood, 1813–96), Lady Tennyson from 1883, was the wife, manager, and secretary of Alfred Tennyson. She married Tennyson in 1850 and together they had two sons, Hallam and Lionel, in 1852 and 1854. Emily has been recognized by her contemporaries and by historians for the important role she played in Tennyson's work. She answered his correspondence, edited and set his poetry to music, and read French, German, and Italian.

Tennyson, Hallam (1852–1928), second Baron Tennyson, was the eldest son of Alfred and Emily Tennyson. He was educated at home by his mother (as was his brother) before attending Marlborough College (1866–72), followed by Trinity College, Cambridge. Following the death of his father Hallam inherited the barony in 1892, and subsequently wrote his father's biography, *Tennyson: A Memoir* (London: Macmillan and Co., 1897). He served as governor of South Australia (1899–1902) and governor-general of Australia (1903–4).

Tennyson, Lionel (1854–86), was the second son of Alfred and Emily Tennyson. He was educated at home by his mother (as was his brother) before attending Eton College, followed by Trinity College, Cambridge. In 1878 he married Eleanor Locker, with whom he had two children, Alfred (b. 1878) and Charles (b. 1879). He worked for the India Office. While traveling in India in 1885 he developed a fever and died the following year.

Thomson, William (1824–1907), was a distinguished Irish-Scottish mathematician, physicist, and inventor. He was educated at the University of Glasgow and Peterhouse College, Cambridge, where he was second wrangler and Smith's Prizeman in 1841. He was professor of natural philosophy in Glasgow from 1846, paying frequent visits to London. He first met Tyndall at the Edinburgh meeting of the BAAS in 1850, and from the start the two physicists had a strained relationship. For much of his career Tyndall regarded Thomson as a scientific rival; however, the two physicists resolved their differences by the Dundee meeting of the BAAS in 1867, where Tyndall wrote that they "seemed like brothers together."

Thornton, William Thomas (1813–80), was an English economist and civil servant. He worked for the East India Company at its London headquarters (1836–38), and after rising in the company's ranks became the first secretary of the India Office (1858) when the British Government took over administration of the East India Company's territories following the Indian Rebellion of 1857. He wrote numerous works on economics and was a close friend and supporter of John Stuart Mill despite significant and public disagreements.

Trayer, James John (ca. 1816–77), was a doctor in Bagenalstown, County Carlow, Ireland. He was educated at Trinity College, Dublin where he earned a BA (1837) and MB (1839). In 1855 he married Jane Carey.

Tyndale, Hector (1821–80), was an American businessman, Union General, and cousin of John Tyndall. He was the son of Tyndall's father's brother, Robinson Tyndall (1775–1842), who had emigrated to Philadelphia from Leighlinbridge around 1804 and founded a glass and ceramics import business. Hector inherited the business and first met his physicist cousin in 1854 on one of his frequent business trips to Europe. They immediately struck up an intellectual rapport and began an occasional correspondence. During the American Civil War (1861–65), Tyndale rose to the rank of brigadier general, and after his resignation in 1864 due to health, he was promoted to brevet major general in acknowledgment of his service (1865).

Tyndale, Julia (née Nowlen, 1823–97), was the wife of Hector Tyndale. They married in 1842.

Tyndall, Caleb (ca. 1798–1879), was an Irish farmer and uncle of John Tyndall. In 1841 he achieved notoriety by shooting Mary McAssey while facing a pro-Repeal mob. At his trial he was cleared of malicious wounding. He subsequently served as high constable for the Barony of Idrone West and overseer of the poor.

Tyndall, Emma (1817–1904), was John Tyndall's sister. She visited her brother in Preston in August and September 1843. She does not appear to have married and later in life suffered from a mental illness that manifested as an extreme form of religiosity.

Tyndall, Frances "Fanny" (née Stone, ca. 1839–89), was the wife of William Tyndall. They married in Kilkenny in 1860.

Tyndall, John (ca. 1815–75), was a relative of Tyndall's in Gorey, County Wexford, Ireland. A doctor and member of the Royal Historical and Archaeological Society of Ireland, he died in 1875 of acute bronchitis ("Obituaries," *Lancet*, 1 May 1875, 633).

Tyndall, Sarah (1844–1917), was John Tyndall's cousin and the daughter of Caleb Tyndall and Dorothea Shirley. In 1869 she married David Miller. Letters in this volume indicate that Tyndall had committed to providing a "little marriage portion" for Sarah but had concerns over her suitor's expectation of the money.

Tyndall, Sarah (née Macassey [Macasey/McAssey], 1793–1867), was John Tyndall's mother. She was descended from a prosperous farming family with lands in Ballybrommell, County Carlow, Ireland, that included in earlier generations Quakers by the name Malone. In 1813 she married John Tyndall senior (1792–1847). After her husband's death she remained in the family cottage at Leighlinbridge, Ireland, supported financially by Tyndall. Sarah died on 30 April 1867, and Tyndall traveled to Ireland to attend her funeral.

Tyndall, William (ca. 1826–1902), was John Tyndall's cousin and the son of Caleb Tyndall and Dorothea Shirley. He was a civil bill officer and married Frances Stone in 1860.

Vieweg, Heinrich Rudolf (1826–90), was a German publisher. He was the son of Eduard Vieweg (1797–1869), and grandson of Friedrich Vieweg (1761–1835), who had founded the family publishing business in Berlin (1786) and later Brunswick (1799). Heinrich Vieweg worked and studied natural sciences in Heidelberg before he joined his father as a partner in the family firm in 1853 and took over the firm in 1866 when his father had a stroke. On taking over the firm Vieweg continued his father's emphasis on scientific works and published German translations of a number of Tyndall's books (Paul Forman, "The Archive of Friedr. Vieweg & Sohn, Braunschweig: Correspondence of Nineteenth-Century Physical Scientists, Especially German Chemists," *Isis* 60 [1969]: 384–86).

Wallace, Alfred Russel (1823–1913), was a British naturalist known for independently formulating a theory of evolution by natural selection. Wallace traveled in the Amazon (1848–52) and in the Malay Archipelago (1854–62) collecting specimens and studying the flora and fauna. It was during the second expedition that Wallace arrived at his theory of natural selection, which he forwarded to Darwin, prompting Darwin to publish his own work on evolution—*On the Origin of Species* (1859)—which he had begun formulating twenty years earlier. On his return to England, Wallace organized some of the 126,000 specimens he collected, which included over a thousand new species. He also set about writing an account of his travels—*The Malay Archipelago* (1869)—and was awarded the RS's Royal Medal in 1868 for his contributions to zoology. The two letters in this volume for Wallace, one to and one from, relate to his interests in spiritualist phenomena, especially séances.

Weber, Wilhelm Eduard (1804–91), was a German physicist educated at the University of Halle. He and his brothers Ernst Heinrich Weber (1795–1878) and Eduard Weber (1806–71) were all accomplished scientists. In 1831 Weber was appointed professor of physics at Göttingen and in 1833 he constructed the first electromagnetic telegraph with Carl Friedrich Gauss (1777–1855). After being dismissed from Göttingen for his criticism of the government he eventually found a position at Leipzig before returning again to Göttingen in 1848. Weber's work was wide-ranging and included physiology and acoustics; however, for the majority of his career he focused on electricity and magnetism, for which he won the RS's Copley Medal in 1859.

Wheatstone, Charles (1802–75), was an English experimental physicist, inventor, and businessman. He had been a musical instrument maker before becoming professor of experimental philosophy at King's College, London (1834–75) where he was a reluctant lecturer. There he developed the electric telegraph with William Fothergill Cooke as well as a number of optical instruments, including the stereoscope which led him into controversy with David Brewster. He married Emma West in 1847 and together they had five children by 1855. The family settled in Hammersmith, where Wheatstone frequently entertained Tyndall, whom he first met in 1852.

Wiedemann, Clara Laura (née Mitscherlich, 1827–1914), the daughter of the chemist Eilhard Mitscherlich (1794–1863), was a German translator who married Gustav Wiedemann in 1851. With Anna Helmholtz, she translated Tyndall's *Heat* (1863) and *Sound* (1867), which appeared as *Die Wärme, betrachtet als eine Art der Bewegung* (Braunschweig: Friedrich Vieweg, 1867) and *Der Schall: acht Vorlesungen gehalten in der Royal Institution von Grossbritannien* (Braunschweig: Friedrich Vieweg, 1869), respectively. While Clara and Anna translated the work, only their husbands were credited (*HLS*; B. Meijer et. al., eds., *Nordisk familjebok*, 2nd ed. [Stockholm: Nordisk familjeboks förlags, 1904–26]).

Wiedemann, Gustav Heinrich (1826–99), was a German physicist and editor. He was trained by Heinrich Gustav Magnus at the University of Berlin, where he came to know Hermann Helmholtz. He held professorships of physics at Basel (1854–63), Brunswick (1854–66), and Karlsruhe (1866–71) before taking the chair of physical chemistry at Leipzig in 1871. After the death of Johann Poggendorff in 1877, Wiedemann took over as editor of the *Annalen der Physik und Chemie*, which subsequently took on the name *Wiedemann's Annalen*.

Young, Emma (née Tyndall, b. ca. 1825), was the cousin of John Tyndall and daughter of his uncle Caleb Tyndall. She married George Young in 1841, and William Hargrave in 1865 after her first husband's death.

INDEX

The index refers to the letters using their serial numbers. The letter 'n' denotes a footnote reference and appears if an item is not otherwise mentioned in the letter to which the footnote pertains. A number in bold signifies a letter to or from a correspondent.